Digital Electronics Guidebook

Digital Electronics Guidebook

Myke Predko

McGraw-Hill
New York Chicago San Francisco Lisbon London Madrid
Mexico City Milan New Delhi San Juan Seoul
Singapore Sydney Toronto

Cataloging-in-Publication Data is on file with the Library of Congress.

2 3 4 5 6 7 8 9 AGM/AGM 0 7 6 5 4 3 2

P/N 0-07-137769
PART OF
ISBN 0-07-137781-6

The sponsoring editor for this book was Scott Grillo, the editing supervisor was Carol Levine, and the production supervisor was Pamela Pelton. It was set in Century Schoolbook by McGraw-Hill Professional's composition unit, Hightstown, N.J.

Printed and bound by Quebecor/Martinsburg.

This book is printed on recycled, acid-free paper containing a minimum of 50% recycled de-inked fiber.

Contents

Preface

I know this will date me, but when I was a teenager, I was always fascinated by Steve Wozniak and the Apple computer. If you ever get the chance to look inside an original Apple, study it and look at the rows upon rows of simple transistor-to-transistor logic (TTL) chips built into it. In fact, when you look at the board, you will see that there are only *three* large chips; they are the 6502 processor for the computer, the character generator ROM for the video output, and the BIOS and BASIC ROM for the computer.

The myriad of TTL chips provided all the other functions required by the Apple computer including:

- Processor reset
- Processor/system clocking
- Memory subsystem control with video memory output multiplexing
- Bus interface and control
- Cassette hardware interface

The computer seems to be the ultimate realization of digital electronics. Over the years, I have watched computers change from collections of devices that take up a large air-conditioned room to small 8-pin DIPs or massive chips that can outperform "supercomputers." Computers are now the building blocks of our modern society. I am amazed at how computer processors have become the means of controlling basic mechanical and electrical functions that were once controlled by mechanical devices or simple digital logic.

The result of this change to computer-based controls is a loss of focus on basic digital electronic design skills. Today, it is not unusual for engineers to think they have a basic understanding of how digital electronics work, only to get themselves into situations where the circuitry doesn't work, or it is marginal. In going forward, much of the basic information that is required for properly designing digital circuits has been lost.

This book is a push back to explaining TTL and CMOS logic, both from a digital electronics as well as analog electronics perspective. I consider the information contained within this book to be a necessary prerequisite to understanding how to create electronic interfaces to computer systems as well as designing "glue" logic to build your own applications.

The *logic* label applied to the TTL chip's name indicates that these chips' output values are a result of combining the inputs they are given in different ways. In addition to providing specific outputs based on the current input conditions, some of the more complex chips keep track of their internal states and use these as inputs as well. These chips and the circuits that are built from them take a bit more effort to get working correctly, but you will find that after a while you will quickly understand how they work and how to take advantage of them.

In a modern PC (or even an Apple(e), which came out several years after the original Apple), you will find that all the functions that were originally implemented in TTL digital logic chips are combined into a single chip known as an *application specific integrated circuit* (ASIC). ASICs are remarkable devices, with the ability to perform the same functions as devices with up to ten *million* logic functions.

Despite the remarkable capabilities of ASICs, simple TTL chips with less than 100 (most have less than 10) standard logic functions are still available and a lot of fun to use. These chips offer you the ability to develop your own digital logic applications at very modest costs and are very easy to work with [ASICs can have nonrecurring expenses (NREs) on the order of $100,000 and require expensive development and simulation tools.]

I will discuss each circuit's interface electrical characteristics in detail. In the projects, I will show the impact of using different technology devices in different applications. For many applications, the device choice will not affect the behavior or performance of the circuit; in others, choosing the device with the proper electrical parameters will be critical.

Along with using TTL logic chips to make basic decisions, I discuss how they can be interfaced into "real world" circuits. It is very easy to create a TTL chip circuit that flashes lights in an attractive order, but unless you can connect the chips to other devices and circuits, they are not very useful. This information can be extrapolated to include intelligent digital devices (such as microcontrollers or even your PC). The theory provided with each device will be directly applicable to interfacing them with intelligent devices.

If you have read any of my other books, you will notice one important difference between this one and the others—I have not provided any software on CD-ROM so that I can demonstrate the complex functions that are possible with straight digital logic.

When choosing the parts to be used in projects, I have relied on my copy of the 1981 *TTL Data Book for Design Engineers,* published by Texas Instruments. I bought my first copy of this book in school and it has remained the best reference I have found for TTL logic. The second criterion that I used to choose parts was whether or not they could be bought from Digi-Key. All the parts referenced in this book meet these two restrictions except in one or two cases that I will discuss.

I am pleased to be able to provide a PCB with this book. The PCB will help you resolve one of the biggest problems you will have when you first start working with TTL logic: finding a suitable power supply. The +5-V power supply that is included with this book will allow you to inexpensively, safely, and easily power the breadboard circuits given in this book and can also be used to power your own digital logic prototype circuits. Procuring the parts and building the breadboard power supply will be one of the first tasks you will have to perform to work through this book.

The second PCB, which is "snapped away" from the +5-V power supply will provide you with a simple means of interfacing to digital electronics as well as provide you with some oscillators that can be used for driving more complex circuitry. Using the two PCBs that come with the book, you will be able to create your own applications in very short order; this will allow you to focus on the circuitry you are working on, rather than on the mundane peripheral issues.

If you have looked through the contents, you will have discovered that I have included the design for a simple (8-bit) computer system. This is probably the most complex application in this book, but it should be the most helpful in understanding the concepts that have been presented. It will provide you with an interesting tool for going on with your education in digital electronics. The computer is extremely crude, but it can carry out all the functions required of a small microprocessor or microcontroller.

I hope that as you work through the projects presented in this book, you will discover how much fun they are to create. While most of the applications people work with today are "intelligent" and are controlled by microcontrollers or large computer processors, there are still a lot of interesting and creative things that can be done with chips that were designed more than 30 years ago.

Myke Predko

Acknowledgments

I would like to thank McGraw-Hill and specifically my editor, Scott Grillo, for the opportunity to write this book. This book might be considered a "throwback" because it deals with chips and concepts that have been available for 30 years or more. Scott, I appreciate the time you spent on the phone discussing the concept as well as working through the idea of including a printed circuit board (PCB) to allow the reader to build easily many of the projects. I hope you are pleased with the result.

Peter Crowcroft of DIY Electronics spent a number of hours helping me to understand what kind of parts are easily (and inexpensively) available and also offered advice on designing PCBs that would make it easier for people new to electronics to build. The end result was the +5-V Power Supply and Interface PCB that comes with this book.

To my family, as always, for giving me the time to work on this book and being as enthusiastic about it as if it were my first. Marya, Joe, and Elliot, you're all great kids.

The biggest "thank you" goes to my wife, Patience. Somehow this book got written while there were major renovations, holidays, school problems, social events, doctor's appointments, various meetings, and business trips. Somehow you always manage to keep things going while I locked myself away to work. You are the best.

Chapter

1

Introduction to Electronics

Electricity is often referred to as "flowing" in a circuit, which probably doesn't seem to be that good an analogy to us in our physical world. When given the word *flowing,* we tend to think of a stream in which water flows to a lower level. This analogy is only 50 percent accurate in describing how circuits work.

To really get an appropriate analogy, we have to go back to grade school and look at the water cycle in Fig. 1.1. In the water cycle, oceans are warmed by the sun (the power source) and water evaporates from them to become clouds. These clouds then move over land where they rain and form the flowing streams mentioned above. The water itself flows from a higher point to the lowest possible point (the oceans) and starts the process all over again.

The path water is taking is actually a repeating "circuit" and, as surprising as it seems, this process is remarkably analogous to how electronic circuits work. A basic electronic circuit is shown in Fig. 1.2. In the electronic circuit, electrons are given energy from the power supply in order for them to move through the circuit (this is the same as the sun evaporating the ocean's water

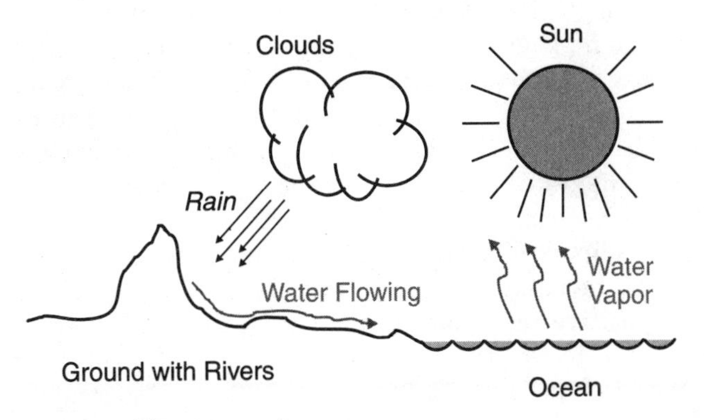

Figure 1.1 The water cycle.

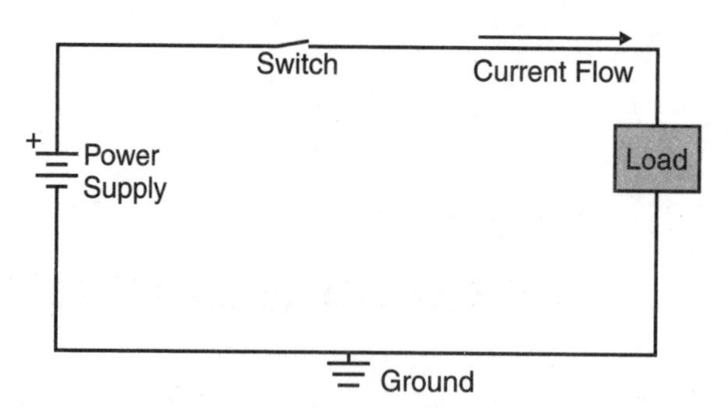

Figure 1.2 Basic electronic circuit.

so it can form clouds, rain, and eventually flow back into the ocean). The power supply gives the electrons a *potential* for moving within the circuit. The electrons that are given this potential will then flow through the circuit to the point of lowest potential (*ground* in Fig. 1.2). This process is repeated as long as power is applied to the circuit.

This process was worked out by Benjamin Franklin, and the notion that electrical current flows from positive (+) to negative (−) has become the basic convention used for all circuits. Unfortunately, it is wrong.

It is wrong because Franklin assumed that it was positively charged particles that make up electrical current flow in the circuit. In actuality, these particles (called *electrons*) that flow in the circuit are negatively charged and actually flow in the opposite direction. We keep with this convention simply because it has been used for over 2 centuries now, and we are comfortable with it.

If you take microelectronics, you will be introduced to the concept of *holes,* spaces within atoms that can accept electrons. As electrons flow from negative to positive, the holes must flow from positive to negative to provide a place for the electrons to flow into. While holes actually exist (as empty electron positions in an atom's electron "cloud"), I find that they are really more of a crutch for people that can't accept that current actually flows in the opposite direction.

As you work through the circuits in this book (and through other electronic circuits), I recommend that you just accept the convention that current flows from positive to negative, with the positive terminal of the power supply being at the highest electrical energy potential and the negative terminal the lowest.

Looking back at Fig. 1.2, I want to discuss a few of the conventions I used when making the diagram, as I will continue to use these conventions for the rest of the book.

The first is the idea of electrical potential. As I discussed above, electrical current flows from the point of highest potential to the lowest. With Franklin's convention, the positive terminal (marked with the + sign) is at the highest potential and is therefore at the top of the diagram. After the current has gone through the load, it is now at ground potential. This is normally referred to as just *ground* because it has nowhere to flow down to (it is at the bottom) and because in many circuits, this ground is also the actual electrical potential of the planet. In some countries, such as Britain, the term *ground* is replaced with the term *earth* but both terms mean the same thing.

For safety's sake, in many circuits ground is connected to the earth to provide a safe path for current to flow if there is ever any problem. Going back to Ben Franklin, remember that one of his most useful inventions was the lightning rod, which is a connection directly from a high point into the ground to provide a safe path for the current from a lightning bolt (instead of through a house or a barn). In most homes, electrical ground is provided by a connection to the cold water inlet pipe.

In many electrical circuit diagrams, you will see a circuit like Fig. 1.3 in which there is no continuous path between the negative terminal of the power supply and the load's negative terminal. Instead, it is assumed that there is one common path between the ground (or Gnd) symbols that is not shown, to keep the diagram from being cluttered with many lines. This convention is also followed for the power, with an upward arrow or a closed circle being used to describe a connection to the positive power terminal.

Current is able to flow only if the circuit is "closed." This means that there is a continuous path for electrical current to flow from the power supply positive terminal to its negative terminal. This is done by closing the switch that establishes the continuous current path.

In Fig. 1.3, the *load* is the circuit that is being powered. This may seem like an ingenuous way of referring to the digital electronic circuits that are going to be discussed in this book, but it is an accurate representation from the power supply's perspective.

As will be shown in the next section, the current through the load is calculated from its electrical *resistance* to the applied electrical force (which is known as *voltage*). An important aspect of designing any electrical circuit is knowing how much current is going to be passed through the circuit. The amount of current depends on the electrical resistance of the load (the greater the resistance, the less current that can pass through it). Like water in a stream, electrical current also follows the path of least resistance, which means if you have two loads wired side by side (that is, in *parallel*), more current will flow through the load with the lesser resistance.

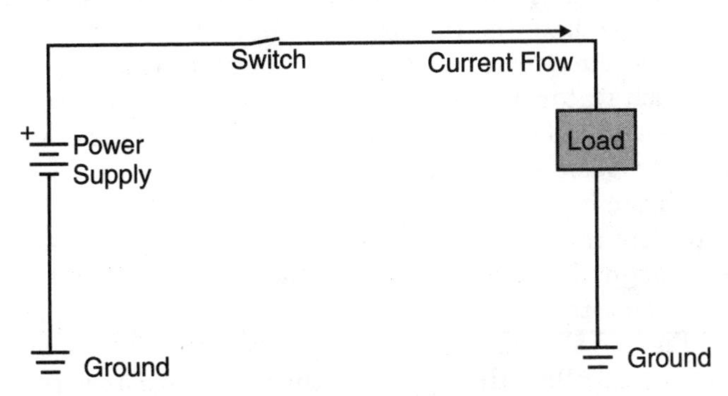

Figure 1.3 Basic electronic circuit with common ground.

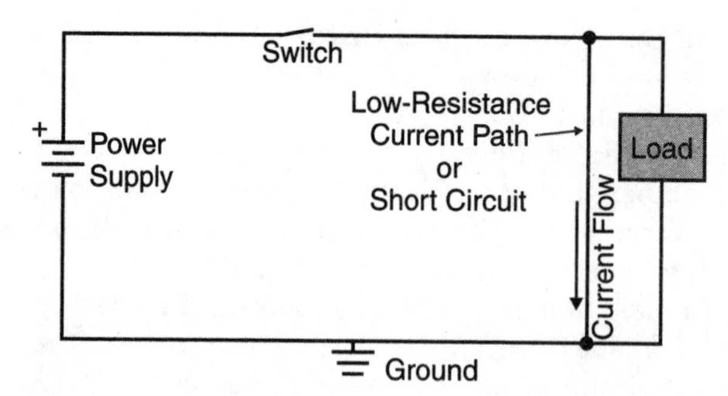

Figure 1.4 Short circuit.

These are important concepts to understand because there is one condition that you will encounter when building your circuits that can be potentially damaging to your power supply and components as well as yourself. If an error is made in the circuit wiring and a path is allowed to exist directly between the positive terminal of the power supply and the negative terminal (Fig. 1.4), a great deal of current will flow through this path. This low-resistance path (known as a *short circuit*) could be caused by a wire fragment that has fallen on your circuit, a defective component, or an error in the assembly of the circuit and can cause damage to the power supply or the application itself. You must always be careful around short circuits, because large amounts of current can pass through them, resulting in very high temperatures that could burn you or start a fire.

For this reason, power supplies should always be protected with a *fuse* or the ability to "crowbar" (turn itself off when the current draw is too great). For added protection, I recommend placing an indicator [such as a light-emitting diode (LED)] into a circuit that will allow you to see if there is any current available for the circuit or if it is being lost to a short circuit.

Ohm's and Other Basic DC Laws

In the previous section, I pointed out that *current*—the movement of electrons in a wire—is analogous to the flow of water in a stream or a pipe. Like electron current, water current is also subject to similar laws that determine how fast and how much water flows. In this section, I will introduce you to "*Ohm's law,*" which is the most basic voltage/current law. I will also introduce you to some of the other basic electricity laws and how they can be visualized as water flowing through pipes.

For water to flow through a pipe, some force must be exerted on it. As the water flows through the pipe, it will experience resistances that impede its flow and lessen the total volume that flows through the pipe over a given period of time. As Fig. 1.5 shows, if the resistance to the water is small, then large amounts of water can flow through it. If there is a greater restriction (as in

Fig. 1.6), then a smaller volume of water will flow through the pipe over a given period of time.

This situation is analogous to the simple circuit in Fig. 1.7. Electrons are supplied by the power supply with a force known as *voltage* (which has the units volts and the label V). *R* is a resistor, or load, that restricts the flow of electrons in a circuit. All electrical devices have some resistance built into them, but for most circuits, only resistors are assumed to have electrical resistance—all the others are assumed to be "ideal" and not have any resistance at all.

For most circuits, this is a reasonable assumption. While other devices have *parasitic resistance* (i.e., the wires that connect them and are used inside them to resist current flow), for the purpose of this book (and most circuits), this is assumed to be zero. *Active devices* require power to operate, and this is calculated from the amount of current they consume. This concept will become clearer as I discuss power supplies later in this book.

In the water pipe examples (in Figs. 1.5 and 1.6), if more force (pressure) is applied to the water entering the pipe and restriction, common sense dictates that the volume of water going through the pipe in a given period of time increases as well. This is also true for the simple circuit in Fig. 1.7—as the force applied to the electrons (voltage) is increased, the electron current through the circuit will also increase. In Fig. 1.7, I wrote this relationship as:

$$I = V/R$$

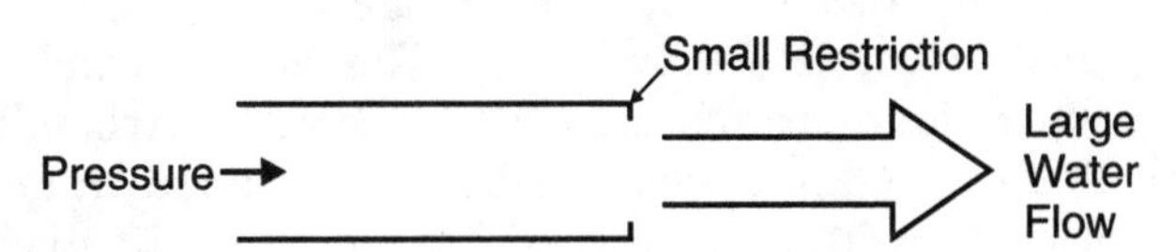

Figure 1.5 Water analogy with a small restriction.

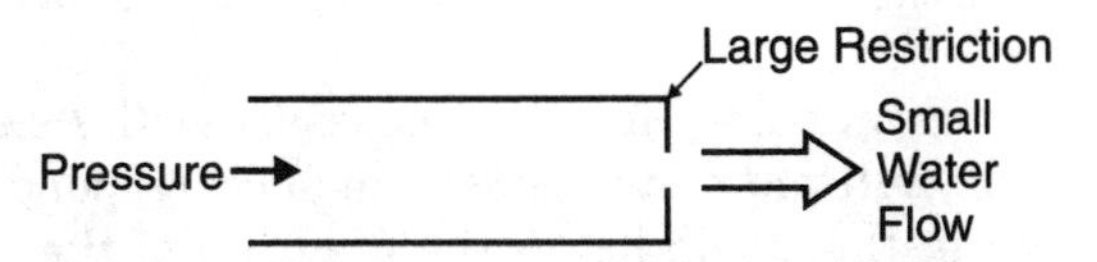

Figure 1.6 Water analogy with a large restriction.

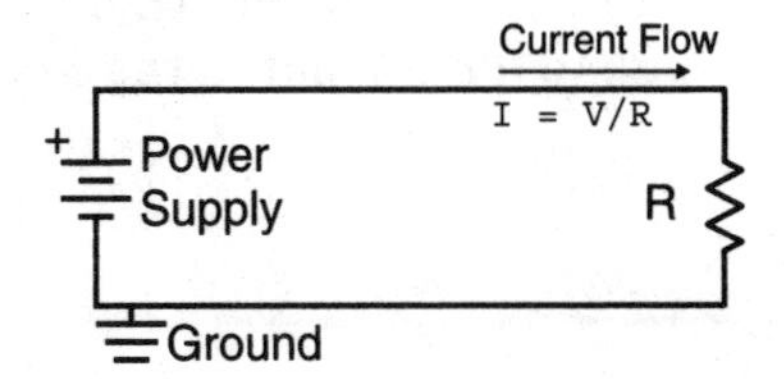

Figure 1.7 Basic resistance circuit operation.

where current (I) (which is measured in *amperes* (and has the short form *amps,* or uses the label A) is the volume of electrons flowing through the wire leading to the resistor (R) in a given period of time. Resistance is measured in ohms, and often is referred to by the Greek letter omega, which is the character Ω. As I indicated above, V is the force applied to the electrons by the power supply that is known as *voltage* and is measured in volts.

The formula above can be used to calculate the current expected to flow through the circuit. For example, if 3.5 V is applied by the power supply and the resistance is 2 Ω, the formula can be used to find the current flowing through the circuit:

$$I = V/R$$

$$= 3.5\ V/2\ \Omega$$

$$= 1.75 \text{ amps}$$

This is known as Ohm's law and is normally expressed as

$$V = I \times R$$

This formula is something you will have to commit to memory because it is basic to the operation of all electronics. Once you understand it and can apply it along with the concepts I will explain in the rest of this section, you will understand almost all of the electronics presented in this book.

Ohm's law can be rearranged to find the voltage, current, or resistance of a circuit if any two of the parameters are known. The formula can be rearranged by using algebraic techniques, or you can use the Ohm's law triangle in Fig. 1.8.

I first learned about the Ohm's law triangle when I was in grade 10, before I had learned any formal algebra. The triangle is a tool that will allow you to easily figure out what is the formula for finding a given parameter by placing your finger over the parameter you want to calculate. As is shown in Fig. 1.9, when you place your finger over the I, V over R is left, which is the formula used to calculate I.

The water analogy can also be used for other cases as well. As in Fig. 1.10, a pipe can be drawn with multiple restrictions, which correspond to multiple pressure drops as shown in the graph below the diagram of the pipe. This is analogous to the situation where there are multiple series resistances in a circuit as in Fig. 1.11.

The resistances are said to be in *series* because the electron current has to flow through the series of them. To calculate the current in the circuit, the cur-

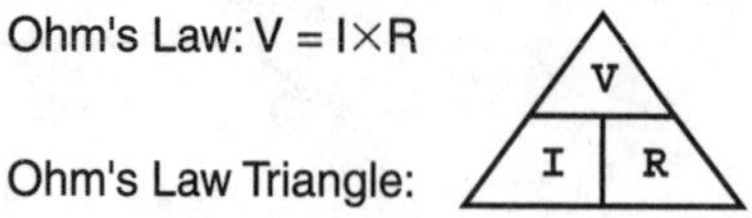

Ohm's Law: V = I×R

Ohm's Law Triangle: V / I | R

Figure 1.8 Ohm's law triangle.

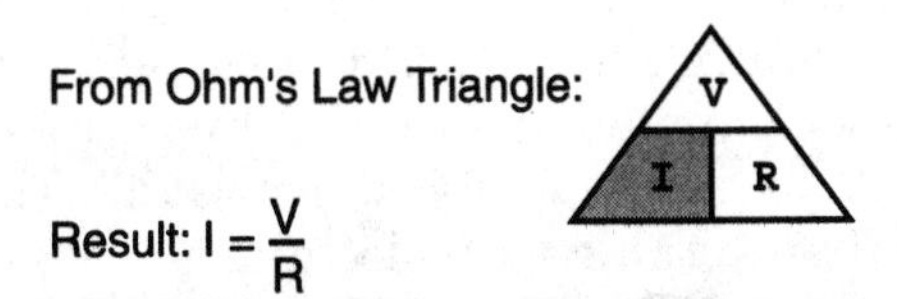

Figure 1.9 Ohm's law triangle example.

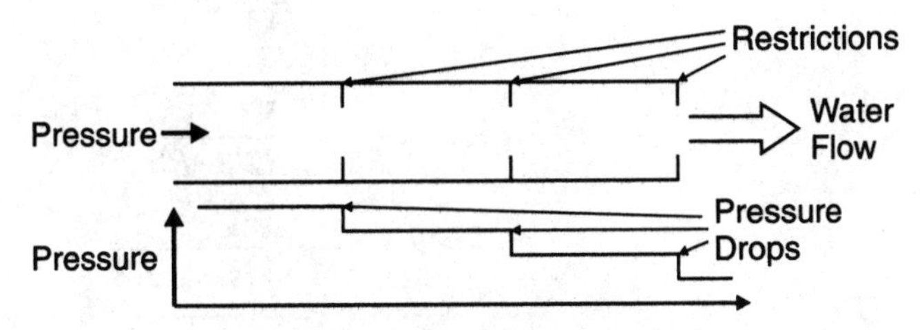

Figure 1.10 Water analogy with multiple restrictions.

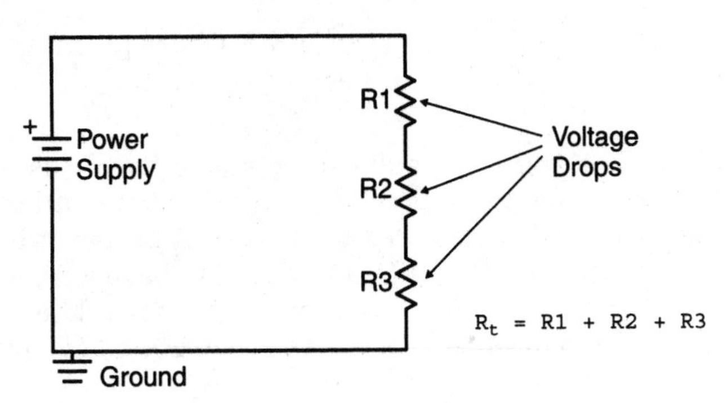

Figure 1.11 Series resistance circuit.

rent through the *equivalent resistance* is calculated. The equivalent resistance is the sum of the resistances. So, for the circuit in Fig. 1.11, the equivalent resistance to the three resistances is

$$R_t = R1 + R2 + R3$$

Each resistor has a voltage drop across it, just as the resistances in the pipe have a corresponding pressure drop across them. This drop can be calculated by just finding the current through the equivalent resistance (which is V/R_t) and then multiplying it by the individual resistor values. So, for the voltage drop across $R2$, the formula

$$V2 = \frac{R2 \times V}{R1 + R2 + R3}$$

is used.

The voltage across a resistor is measured at the current input and output points, and the ground level is not used unless that is one of the two points of the resistor. In the example circuit in Fig. 1.11, if the voltage at the current input of $R2$ is measured along with ground, the voltage drop across $R2$ *and* $R3$ would be returned (which would be higher than just across $R2$). This is a common mistake made by people first working with electronics.

When you total the voltages across $R1$, $R2$, and $R3$, you will find that this value equals the voltage applied to the circuit, and this can be confirmed very easily by summing the voltage drops across each resistor:

$$\begin{aligned} V_{\text{resistors}} &= V1 + V2 + V3 \\ &= R1 \times \frac{V}{R1 + R2 + R3} \\ &\quad + R2 \times \frac{V}{R1 + R2 + R3} \\ &\quad + R3 \times \frac{V}{R1 + R2 + R3} \\ &= (R1 + R2 + R3) \times \frac{V}{R1 + R2 + R3} \\ &= V \end{aligned}$$

This is known as *Kirchhoff's law* and can be stated simply as "The voltage applied to a circuit is equal to the sum of the voltage drops within it."

Kirchhoff's law can be demonstrated with an LED that is lit by a 5-V supply. As is shown in Fig. 1.12, the LED is assumed to have a 2.0-V drop and a maximum current through it of 5 mA. The problem is to find the correct resistor value for the LED to have 20 V applied to it and 5 mA of current flowing through it.

From Kirchhoff's law, the voltage across the resistor and LED must equal the applied voltage. Since we know the applied voltage and the voltage across the LED, we can find the voltage across the resistor as

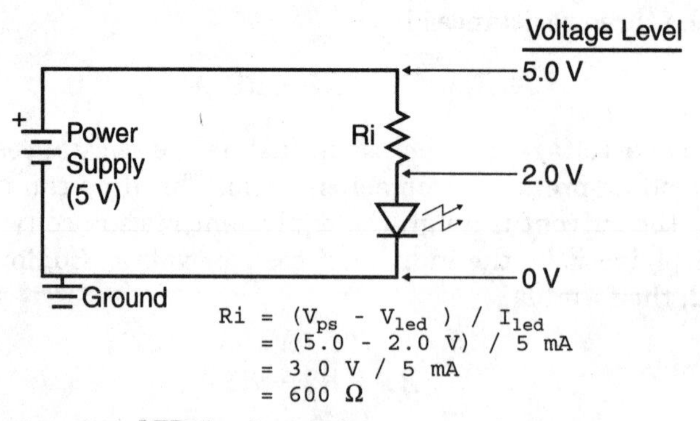

Figure 1.12 LED circuit operation.

$$V_{\text{applied}} = V_r + V_l$$

$$5.0\ V = V_r + 2.0\ V$$

$$V_r = 5.0\ V - 2.0\ V$$

$$= 3.0\ V$$

Now, to find the appropriate resistance that will limit the current in the circuit (and through the LED) to 5 mA, Ohm's law is used:

$$R = V/I$$

$$= 3.0\ V/5\ mA$$

$$= 600\ \Omega$$

The resistor used to hold down the current passing through the circuit is called, not surprisingly, a *current-limiting resistor* and ideally should be 600 Ω. In this type of application, I use a 220- or 470-Ω resistor, although other values can be used, depending on the actual voltage drop and current requirements of the specific LED. Either value is a good convention to use for LEDs as it will allow most LEDs to light properly without any possibility that there will be less than the required current available to the LED.

To lower the resistance in a pipe, either it can be made smoother (actually lessen the resistance) or the cross-sectional area (the area the water flows through) could be increased. In Fig. 1.13, I show a pipe situation where the resistance is lessened by splitting the pipe into two *parallel* paths for the water.

Like series resistance, parallel resistances work exactly the same as the water analog. In Fig. 1.14, two resistances are wired in parallel as the load, with the equivalent resistance defined as

$$R_p = \frac{R1 \times R2}{R1 + R2}$$

It is important to note and remember that the equivalent parallel resistance will always be less than either one of the two resistances. This is a good check

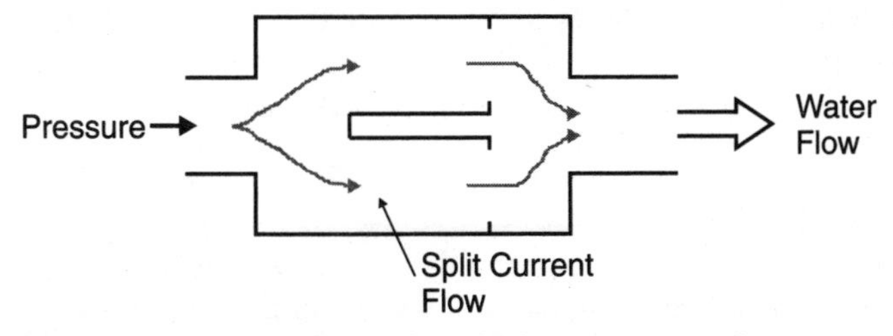

Figure 1.13 Water analogy with multiple restriction paths.

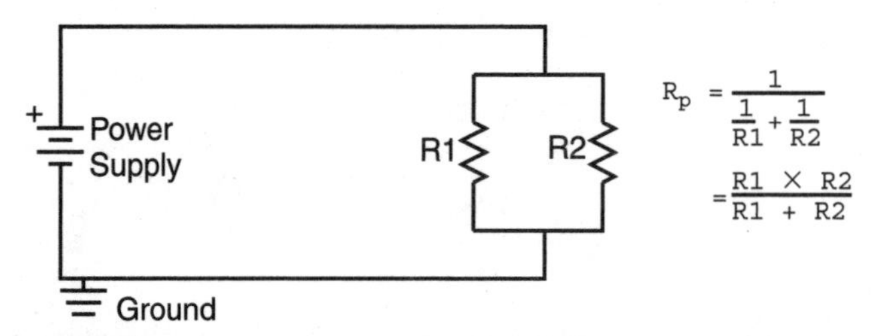

Figure 1.14 Parallel resistance circuit.

when you are calculating the equivalent resistance and you aren't sure if you have the right answer. For example, if you had 2-Ω and 5-Ω resistors in parallel, their equivalent resistance would be

$$\begin{aligned} R_p &= \frac{R1 \times R2}{R1 + R2} \\ &= \frac{2 \times 5}{2 + 5}\ \Omega \\ &= \frac{10}{7}\ \Omega \\ &= 1.43\ \Omega \end{aligned}$$

Multiple resistors put in parallel are defined by the formula

$$R_p = \frac{1}{1/R1 + 1/R2 + 1/R3 + \cdots}$$

Always remember that the equivalent parallel resistance is always less than the lowest resistor.

The last concept and law that I would like to introduce to you is Thevenin's law, which states that any series of loads that are ultimately connected to a power supply by two terminals can be represented as a single two-terminal load. This is illustrated in Fig. 1.15. The figure shows how $R1$, $R2$, and $R3$ can be combined into an equivalent resistance (Re) by using the series and parallel resistance formulas presented above.

With Thevenin's law, the equivalent resistance can be calculated and used to find the current supplied by the power supply, and from this current, the voltages across $R3$ and the currents through $R1$ and $R2$ can be calculated.

These laws should seem quite simple and easy to apply—they are, and once you understand them you will be able to work through 80 percent or more of the theory of the circuits that you are presented with.

Measuring Voltage and Current

Measuring voltage and current probably strikes you as something that is very easy to do, but there are a few things you should be aware of to help you to understand what your digital multimeter is telling you.

The basic meter consists of a coil and a piece of iron that is drawn into the coil by the magnetic force of current flowing through it. The basic operation is shown in Fig. 1.16. The higher the current through the coil, the more force that is applied to the piece of iron. As more force is applied to the iron, it pulls against a spring, which moves the needle proportionately to the current going through the coil. This is a basic ammeter.

In an ideal situation, the ammeter's coil has a resistance of zero Ω. Obviously, this is not possible, so there will be some voltage drop across the coil. A good rule of thumb is: The lower the voltage drop, the better the quality (and the higher the cost) of the meter.

The ammeter is inserted in series with the circuit whose current is to be measured, as I show in Fig. 1.17.

A voltmeter works similarly, but places a high-value resistor in series with the coil to allow a small amount of current to deflect the needle. The voltmeter is shown in Fig. 1.18. The in-line resistance (R in Fig. 1.18) is typically 100 kΩ or greater. The voltmeter is connected in parallel to the device being checked. Ideally, the resistance should be infinite because the act of putting the voltmeter across a device will affect voltage and current through the device—but in the real world, the current drawn by the voltmeter is very small. The connection for measuring the voltage across a device in a circuit is shown in Fig. 1.19.

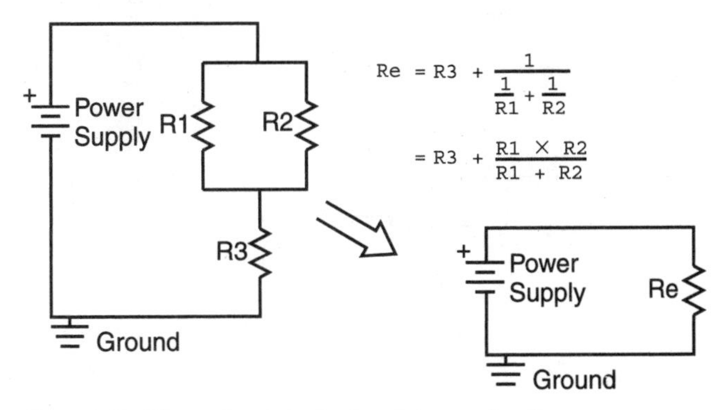

Figure 1.15 Thevenin equivalent resistance circuit.

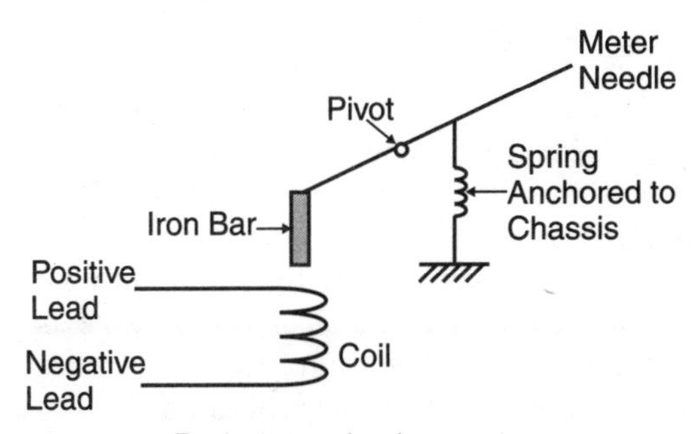

Figure 1.16 Basic meter circuit.

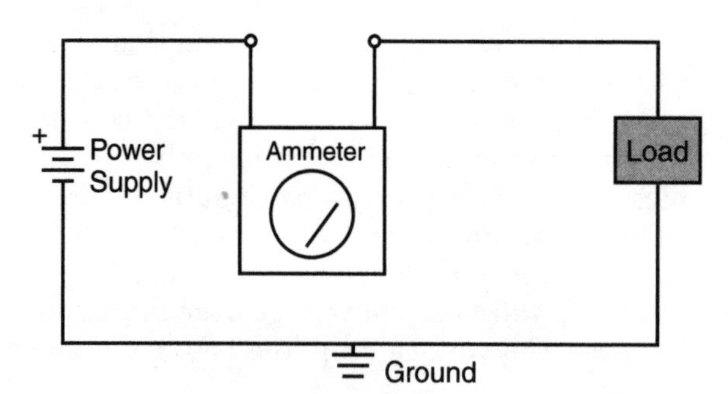

Figure 1.17 Basic electronic meter circuit.

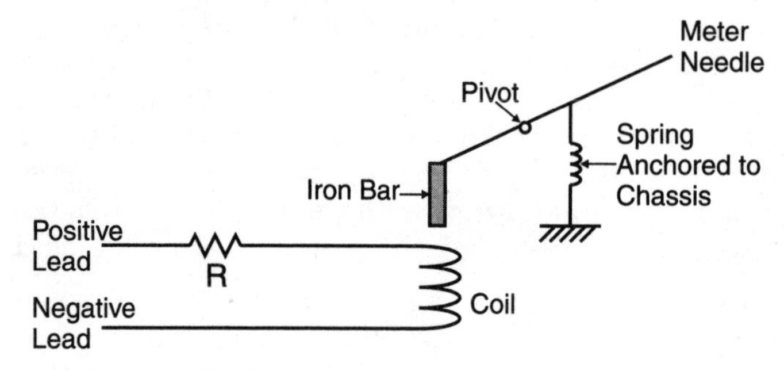

Figure 1.18 Basic voltage meter circuit.

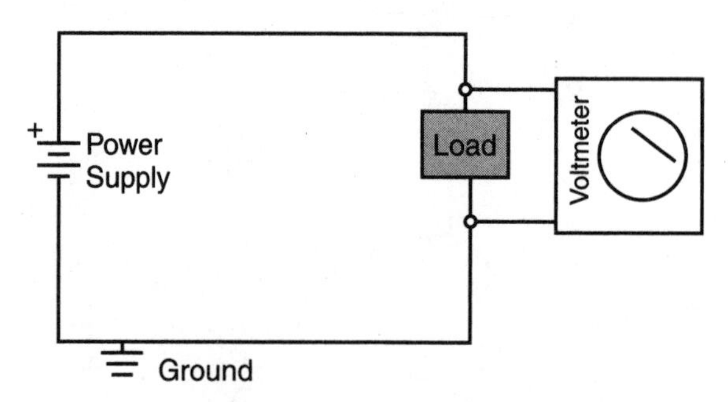

Figure 1.19 Basic Electronic voltage meter circuit.

Both an ammeter and a voltmeter can be used to measure the resistance of a device by combining the two previous circuits into the one shown in Fig. 1.20. In this circuit, the resistance of the load can be calculated from the values determined by the ammeter and the voltmeter by using Ohm's law.

The reason why I'm showing how these meter types work is that modern digital multimeters (DMMs), which have voltage, current, and resistance measurement capabilities built in, emulate these functions. Usually in a DMM,

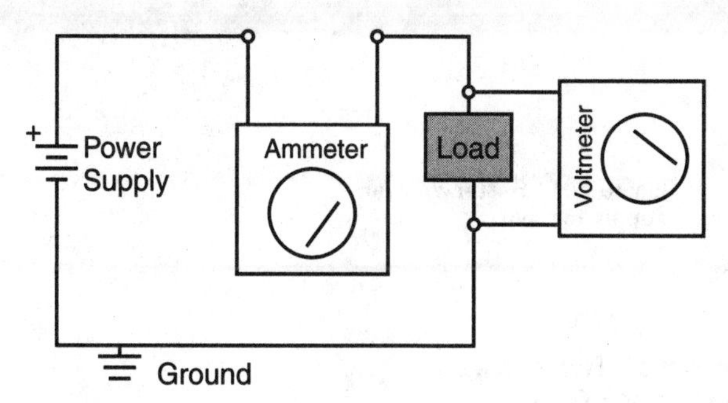

Figure 1.20 Voltage current meter circuit.

the current measurement circuitry has an internal resistance of 0.1 Ω or less. For the voltmeter function, the internal resistance is 1 MΩ or greater (10 MΩ or more is not unheard of).

Basic Electronic Devices and Their Symbols

Before working through all the different circuits, I want to go through some of the basic electronic devices, describe their characteristics, and mention things to watch out with them. I am going to focus on the practical aspects of the devices rather than the theory behind them and how they are manufactured. With this information, I will expand on various devices later in the chapter.

The basic two devices in any circuit are the power supply and the load. Earlier in the chapter, these devices were presented as necessary for calculating current power requirements. The power supply provides an electron source, giving the electronics a specific voltage potential for a maximum rated current. Voltage can either be unchanging (*direct current* or DC) or changing according to a sine wave above and below zero volts. This is known as *alternating current* (AC).

Digital electronics projects will largely focus on DC power supplies with 5-V output. The symbol used for power will be the *battery* symbol shown in Fig. 1.21, with the positive terminal connected to V_{cc} (or V_{dd} for CMOS devices). The negative terminal is connected to *ground* (*Gnd*) (or V_{ss} for CMOS). The battery symbol is meant to represent the multiple plates in a physical battery; I always remember positive because the symbol looks like an actual cylindrical radio battery with the small ground (which is positive in the physical device) connected backward.

A *resistor* can characterize a *load* that is a device that impedes current flow through a circuit. Its symbol is shown in Fig. 1.22. Resistors are not polarized, and you should have a pretty good idea of how they work from the previous sections. The only point I would like to add about them concerns their power rating.

Occasionally when you place a resistor in a circuit, you will find that it gets very hot. This is due to more power being dissipated than is specified for the

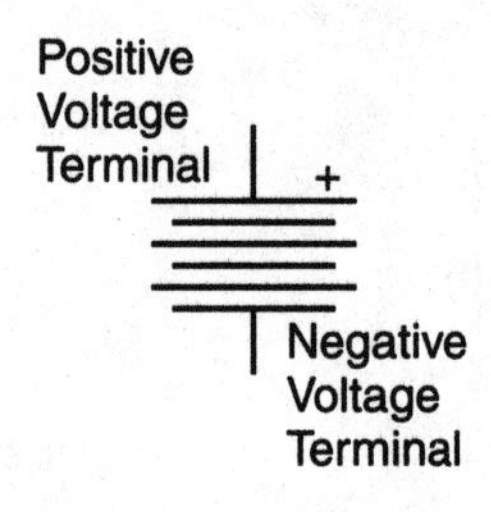

Figure 1.21 Battery/power supply symbol.

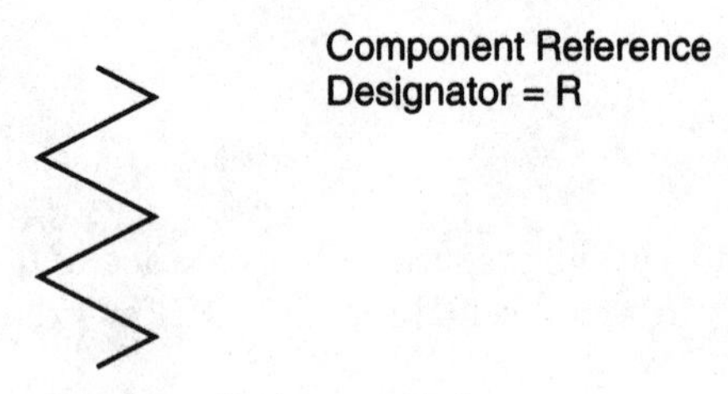

Figure 1.22 Resistor symbol.

resistor. Resistors are normally rated at $^1/_8$, $^1/_4$, $^1/_2$, 1 W and larger power dissipation ratings, with the rating being dependent on the power through it according to the power formula:

$$P = I \times V$$

Or, with Ohm's law substituted,

$$P = \frac{V^2}{R}$$
$$= I^2 \times R$$

For most applications, a $^1/_4$-W resistor will be more than adequate, but occasionally, you will have situations that you should be aware of. To determine what the power rating of the resistor you use is, you should apply the formula above and always use the next larger value of resistance.

For example, in a circuit that has a 220-Ω resistor as a current-limiting device when dropping a power supply from 13.4 V to 5.1 V, the maximum current through the resistor is 40 mA. From the power formula:

$$P = I^2 \times R$$
$$= 0.040 \times 0.040 \times 220$$
$$= 0.35 \text{ W}$$

The power dissipated by the resistor is greater than $^1/_4$ W. For this reason, in this circuit a half-watt-rated resistor should be used instead of a normal $^1/_4$ W.

Resistors normally have a series of four or five color bands printed on them. These bands are used to specify the resistance of the circuit according to the formula:

$$\text{Resistance} = [(\text{band 1} \times 100) + (\text{band 2} \times 10) + (\text{band 3} \times 1)] \times 10^{\text{band 4}}$$

The following table lists the values for each band and what they mean in the resistor specification formula:

Number	Color	Band 1	Band 2	Band 3	Band 4	Optional band 5
0	Black	N/A	0	0	10^0	N/A
1	Brown	1	1	1	10^1	1% tolerance
2	Red	2	2	2	10^2	2% tolerance
3	Orange	3	3	3	10^3	N/A
4	Yellow	4	4	4	10^4	N/A
5	Green	5	5	5	10^5	0.5% tolerance
6	Blue	6	6	6	10^6	0.25% tolerance
7	Violet	7	7	7	10^7	0.1% tolerance
8	Gray	8	8	8	10^8	0.05% tolerance
9	White	9	9	9	10^9	N/A
N/A	Gold	N/A	N/A	N/A	10^{-1}	5% tolerance
N/A	Silver	N/A	N/A	N/A	10^{-2}	10% tolerance

So, for a 220-Ω resistor, the bands would have the colors: red, red, black, and black.

Remembering what the color codes is usually accomplished either by noting that the colors are in the order of the colors of the rainbow or by an obscene mnemonic. For all practical purposes, nobody today reads the bands and tries to figure out what the resistance is; instead the Ohmmeter function of a digital multimeter is used to read the resistance value. This process is usually faster and definitely more accurate in terms of what the actual resistance value is.

Variable resistors, or *potentiometers,* are used to change the resistance of a circuit or provide a user-selected voltage. The symbol is shown in Fig. 1.23.

I usually use high-value potentiometers (10 kΩ or more), so the issues with regards to power in potentiometers in my circuits are not as significant as they are in simple resistors. Most printed circuit board (PCB) mount potentiometers can dissipate up to $1/4$ W or so of power. Potentiometers are available that can handle much larger amounts of current, but these devices tend to be expensive and difficult to work with. If I have a circuit where I want to control a DC device with varying power levels, I typically use a *pulse-width-modulated* control rather than a potentiometer.

In this book, I will be concerned with two types of switches. Push buttons are normally referred to as *momentary on* or *momentary off* switches. Their

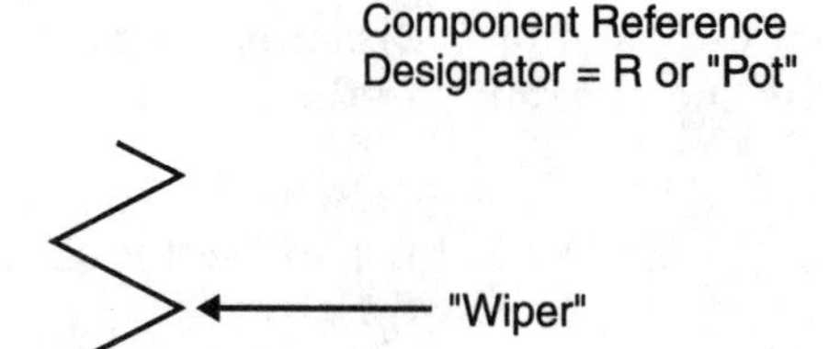

Figure 1.23 Potentiometer variable resistor symbol.

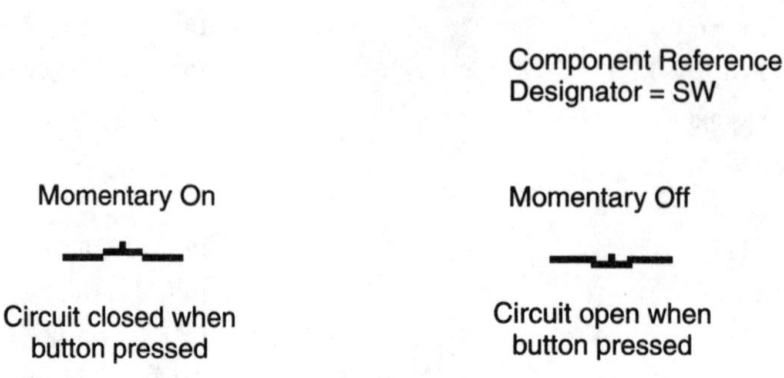

Figure 1.24 Momentary switch symbols.

symbols (i.e., operation), are shown in Fig. 1.24. Momentary-on switches close the circuit when they are pressed and open the circuit when they are released. Momentary-off switches work in the opposite manner, with the circuit opened when the button is pushed.

Momentary switches are used to satisfy a variety of different digital application input requirements, which are described elsewhere in the book.

Slide switches are used to make a specific contact based on the position the switch is left in. The most basic switch has one or two "throws," or positions, that result in a circuit closed. In Fig. 1.25, single- and double-throw switches are shown.

There can be more than one throw in the switch, and multiple switches can be ganged together as multipole switches. There are many different combinations that can be built out these basic types of switches.

In this book, I will be mostly concerned with single-pole, single-throw (SPST) switches for turning on and off power or controlling basic signals. The symbol for the SPST is shown in Fig. 1.26.

When you look for slide switches, you will find that most of them are single-pole/double-throw (SPDT), in which case, just the common connection and one switched connection are used.

You will have to wire multiple boards together, and this is accomplished by the use of *connectors*. There seems to be an infinite variety of different connectors available. In many cases, application developers select the connectors they feel most comfortable with. For this reason, except for some very specific

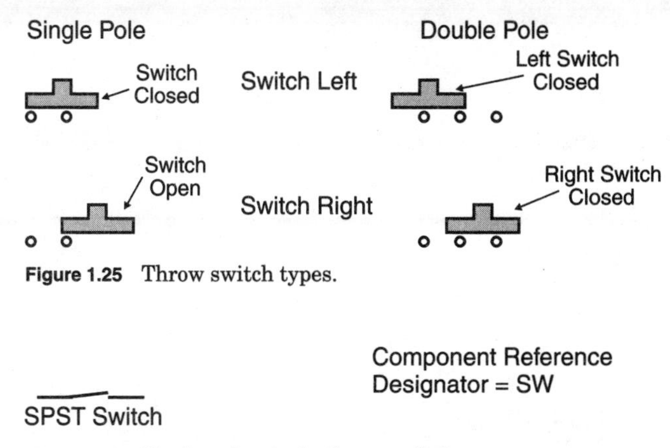

Figure 1.25 Throw switch types.

Figure 1.26 Single-pole, single-throw switch.

instances where high current or convenience is required, I have not specified the type of connector to be used.

The connector symbol label used in most circuits is *J*.

Fuses can be used within your application to limit current draw and protect circuits, wiring, and power supplies. Like home fuses, electronic ones are rated in amps and are in a variety of different board-mountable or in-line packages and *cartridges*. The symbol is shown in Fig. 1.27.

Fuses are *not* rated in volts, but for a joke you can always ask somebody to go out and buy you a "six-volt, half-amp" fuse and tell them not to come back until they have it. If somebody asks *you* to go out for a six-volt, half-amp fuse, take their money and buy yourself a good dinner with it and return hours later saying all the stores you went to were sold out.

In my own applications, I prefer to avoid specifying fuses except when absolutely necessary. Instead I prefer to use voltage regulators that shut down, rather than burn out, when too much current is drawn through them. This does not eliminate the need for fuses in situations where they are indicated for safety or regulatory reasons, but it does mean they are not required in experimental applications.

I could probably write an entire chapter on capacitors, their operation, dos and don'ts, and hints. It has actually been a lot of work for me to come up with just a simple explanation of the types of capacitors and their characteristics.

Capacitors are designed for storing electrical charge. They usually consist of two metal plates separated by a nonconductive material called a *dielectric*. The capacitor symbol is shown in Fig. 1.28.

The charge placed in a capacitor is measured in farads (symbol F) in the units of coulombs/volt. A *coulomb* is a charge of 1.6×10^{19} electrons, making 1 farad a very large number. This very large number is why most capacitors are measured in millionths (μF) or trillionths (pF) of a farad.

The dielectric that separates the plates in Fig. 1.27 is a material that doesn't conduct electricity. The type of dielectric used in a capacitor usually gives it its name. The most common types of capacitors are:

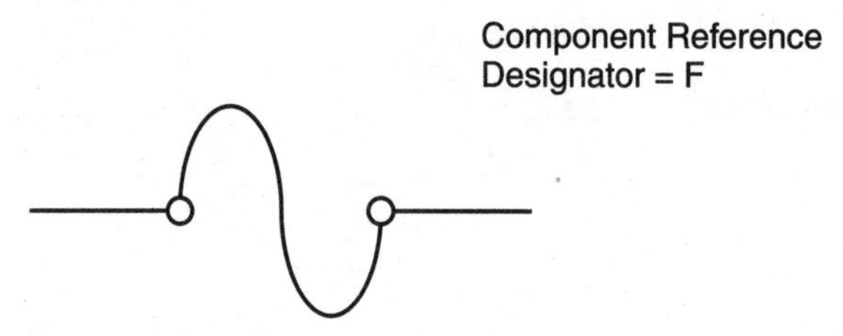

Figure 1.27 Fuse symbol.

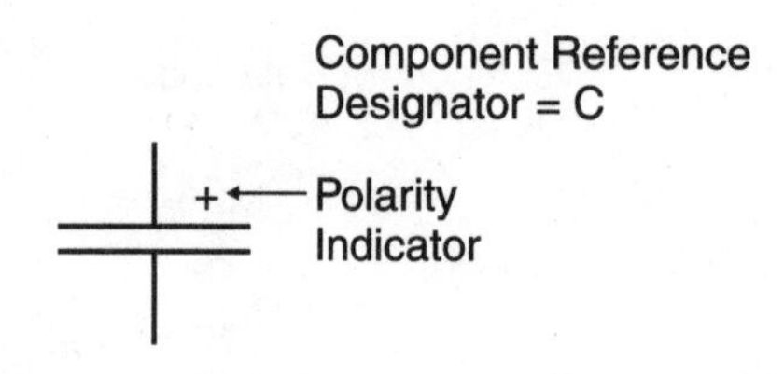

In some references, the symbol is:

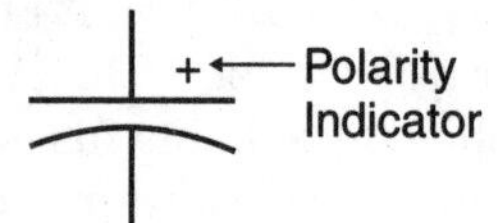

Figure 1.28 Capacitor symbol.

1. Ceramic disk
2. Polyester
3. Tantalum
4. Electrolytic

Ceramic disk and polyester are the most "benign" type of capacitors you can work with. They are typically used for applications requiring small capacitance values at reasonably high frequencies. The reason for this is that these dielectrics do not enhance the capacitance of the two plates as significantly as the other two, and large capacitances (such as microfarads) would be measured in inches across and would probably weigh a pound or more. These types also tend to be the least expensive of the capacitors you can buy.

Tantalum capacitors are a relatively new invention and offer excellent capacitive densities at high frequencies. This is due to their very low *equivalent series resistance* (or ESR), which minimizes the equivalent *RC* effects of the capacitor in the circuit. Tantalum capacitors are well suited for use as decoupling capacitors, and I keep an electronics drawer full of 0.1-μF, 16-V tantalum capacitors at all times.

Tantalum capacitors do not have a good reputation among many people because of the tendency of early devices to explode. "Pinhole" shorts between the plates in the capacitor caused this. These shorts would heat up, burning away the dielectric, grow larger in surface area, burn away more dielectric and so on until there was a catastrophic failure, which is *very* spectacular if you happen to see it. This problem is accelerated by reverse biasing the capacitor; always make sure the side marked + is connected to V_{cc}.

New tantalum capacitors do not have this problem if they are properly *derated*. All tantalum (and electrolytic) capacitors are given an operating voltage for which they are designed. By derating the input voltage specification, you are actually selecting devices with thicker dielectrics, which makes the likelihood of *burn-through* much more remote.

A good rule of thumb is to derate tantalum capacitors by 60 percent. This means that a tantalum cap used as a decoupling cap in a 5-V system should be rated at 12.5 V or more (which is why I normally use 16-V-rated parts).

Electrolytic capacitors rely on a chemical liquid for their dielectric and are in sealed metal packages. The plates and the chemical are designed to be biased in one way to reach the desired capacitance. Reverse biasing an electrolytic capacitor will cause the liquid to break down, which, while not as spectacular as a tantalum capacitors explosion, could hit you with debris or spray you with hot electrolytic. At the very least, you will have to clean the residue from your circuit board.

Electrolytic capacitors' big advantage is their dense capacitance in a small, inexpensive package. They do not have very fast responses and are not well suited to be used for decoupling and other high-speed filtering applications.

Some capacitors have been given a set of colored bands similar to the markings used on resistors. Making the issue confusing is that there are at least two sets of standards for marking capacitors. I recommend that you either buy capacitors with their values printed on them or segregate the parts in different marked bags or bins until you need them. Many modern digital multimeters can read capacitor values down to 30 pF or less, which can be useful in figuring out what that old bag of parts that you have gathering dust has in it.

While capacitors store energy in the form of a charge, *inductors* (or *coils*) store energy in the form of a magnetic field. Inductors (Fig. 1.29) are not often used in digital electronic circuits, except for *switch-mode power supplies,* which increase or decrease incoming power by using the characteristics of coils to resist changes in current flow. Like the capacitor's farad unit of measurement, the inductance unit, the henry (H), normally has values in millionths for the devices to be effective in a circuit.

Inductors can be difficult to work with if you are new to electronics. I would recommend that they not be used except when a chip to be used in a specific application specifies them in its datasheet.

Component Reference
Designator = L

Figure 1.29 Inductor symbol.

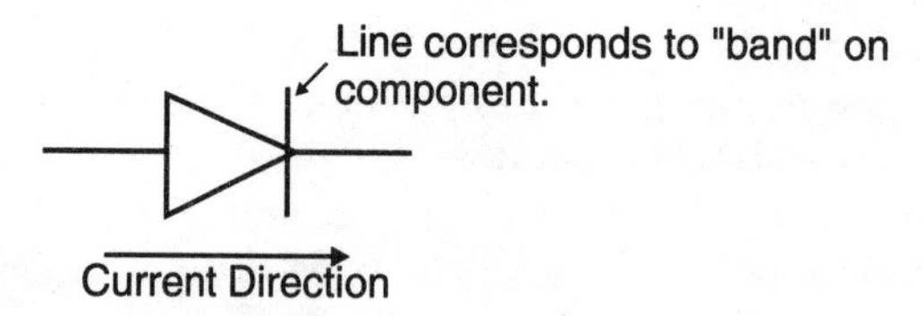

Figure 1.30 Diode symbol.

In modern applications, switching is accomplished by *semiconductors,* so named because their ability to conduct current can be controlled in different situations. The most basic of these circuits is the *diode* (shown in Fig. 1.30), which will only pass current in one direction.

Diodes are also known as *rectifiers* because they can be used to convert or "rectify" *alternating current* (AC) voltages into *direct current* (DC). A diode is made out of a piece of silicon that has been chemically impregnated ("doped") with two different materials. These materials create an interface in which the energy of electrons is much higher in one than the other. This interface is known as the *PN junction* because of how the silicon is doped.

To cross the boundary, electrons can flow only from the high-potential (n-doped) side to the low-potential (p-doped) side. If electrons enter the device from the p side, they cannot climb the "wall" to the n-doped side, thus current is limited to flowing only in one direction. This is shown in Fig. 1.31.

The electron potential drop results in some voltage drop in the applied current. For silicon diodes, this drop is 0.7 V, and I use this function of PN junctions (and diodes in general) to create simple voltage references for power supplies.

When electrons fall down from the high-potential side to the low-potential side, they release energy. This energy release, according to Professor Einstein, is a set amount and is released from the diode as a *quantum* of energy and takes the form of a *photon.*

In Fig. 1.31, the release of photons is shown as the electrons go from a higher energy state to a lower one. Standard silicon diodes release photons in the far-infrared spectrum of light, but by changing the diode materials, the photon released can be in a visible frequency of light. These diodes are known as *light-emitting diodes* (LEDs) and come in a variety of colors. Red was the first color of LED available and is the least expensive; green, yellow, orange, blue, and white are also available, although the higher the output frequency of the light, the more expensive the device.

LEDs will have varying voltage drops across them—always more than the 0.7 V of a silicon diode.

Transistors

When I was in university, the primary type of transistor was the *bipolar* device. This type of transistor is quite difficult to understand—both how it

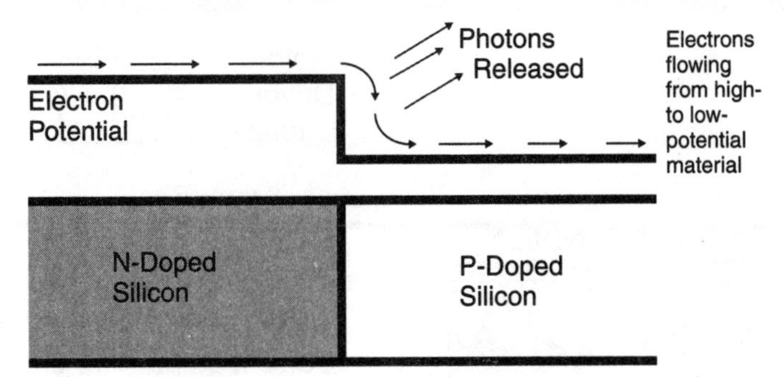

Figure 1.31 Diode operation.

works and how it is manufactured. Despite this, the bipolar transistor became the primary device used in the 1960s and 1970s. In the 1980s, *field-effect transistors* (FETs) became the most popular type of transistor, and this trend continues today.

When I was in school, roughly 60 percent of all transistors built into integrated circuits were bipolar devices and 40 percent were FETs. Today, it is something like 99 percent are FETs and 1 percent is bipolars. In the cases where bipolar transistors are used, they are used for very specific reasons that have to do with their electrical characteristics. FETs are easier to manufacture and can be placed in much smaller amounts of silicon chip "real estate," which allows for the amazing circuit densities of modern very large scale integration (VLSI) chips. I remember my third-year microelectronics professor stating that FETs were just a passing fad—this was probably the most colossally incorrect statement I received from any of my instructors.

In this section, I will introduce you to the two different classes of transistors and their characteristics. In addition, I will discuss how integrated circuits are made with transistors and what you would expect to see if you could magnify a modern silicon chip and look at what it consists of. I will concentrate on how transistors work in computer systems, but I will discuss some of the concepts that are necessary to understand analog applications.

The most basic transistor is the NPN, which consists of two pieces of n-doped silicon sandwiching a piece of p-doped silicon. N-doped silicon has a material added to it that causes electrons to be available in the crystal lattice. P-doped silicon has a material added that causes an affinity for accepting electrons in the crystal lattice.

As I showed in the previous section, these materials can be combined to form a rectifier, or diode, which allows current to pass in only one direction. In the NPN transistor, this doped silicon prevents current from passing through unless current is injected at the p-doped region (which is known as the *base*). The NPN transistor is shown in Fig. 1.32.

As I indicated above, transistors are switches. Bipolar transistors are actually variable switches in which the amount of current passing through the *collector* and *emitter* can be controlled by controlling the amount of current passed

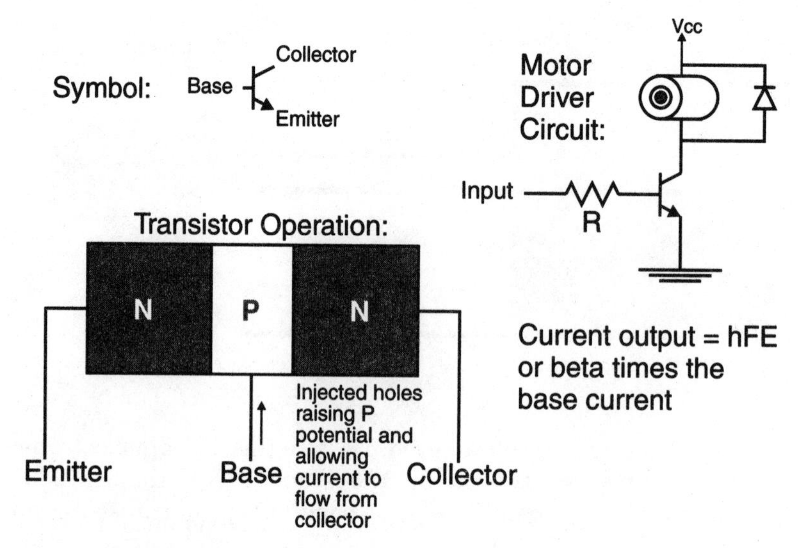

Figure 1.32 NPN transistor operation.

through the *base.* In Fig. 1.32, I show a simple circuit for a DC motor driver. When current is passed through the base to the emitter (which is connected to ground), the collector will be able to accept current from the V_{cc} source.

This current is dependent on the formula

$$I_{ce} = I_{be} \times h_{FE}$$

where I_{ce} is the current through the collector and emitter and I_{be} is the current through the base and emitter. h_{FE}, which is also known as *beta,* is the multiplication factor for the current flowing through the base and emitter to the collector and emitter.

For a common transistor like the 2N3904, h_{FE} is in the range of 150 to 200. This means that a significant amount of current reduction through the base and emitter is required so as not to burn out the transistor (or cause other problems with the PICmicro, which are discussed elsewhere in the book). If a 2N3904 were used to control a motor in the circuit of Fig. 1.32, then the value of *R,* the base-emitter current-limiting resistor would have to be calculated for the current passing through the motor.

For this example, assume that the 2N3904 has an h_{FE} of 150 and the motor requires 100 mA when it is running. To calculate the value of *R,* the current flowing through the base and emitter has to be calculated by using the formula above:

$$I_{ce} = I_{be} \times h_{FE}$$

$$I_{be} = I_{ce}/h_{FE}$$

$$= 100 \text{ mA}/150$$

$$= 667\ \mu\text{A}$$

Now, assuming that the voltage applied to the base is V_{cc} (+5 V for most applications), the resistance can be calculated from Ohm's law (although it should first be recognized that there is a 0.7-V drop across the base-emitter junction to ground). The resistance is:

$$
\begin{aligned}
R &= V/I \\
&= (5\text{ V} - 0.7\text{ V})/667\ \mu\text{A} \\
&= 4.3\text{ V}/667\ \mu\text{A} \\
&= 6447\ \Omega
\end{aligned}
$$

So, a 6.5k resistor would be used in this application to allow a 2N3904 to drive an electric motor requiring 100 mA with 5 Volts applied to it.

When the current is limited (but not shut off) by a transistor, then there will be a definite voltage drop across it with a specific current passing through it. These values can be translated into a specific amount of power dissipated by the transistor. This is the reason why the relatively small base-emitter current-limiting resistor is used. With 13 mA driving into the transistor, any current that is available at the collector will be passed through to the emitter with a minimum of resistance. When this is done, the transistor is said to be in *saturation mode* and can pass the current applied to the collector with a minimum of resistance.

For most of the applications that use bipolar transistors in this book, I specify a 330-Ω base-emitter current-limiting resistor. From calculations above, this will work out to an I_{be} of 13 mA with a maximum I_{ce} of 1.96 amps! For most applications that use transistors, any external devices (that is, resistors) and the transistor itself will limit the I_{ce} current.

Besides the NPN bipolar transistor, which turns on when current is *applied* to it, there is the PNP bipolar transistor, which passes current between its collector and emitter when current is *drawn* from it. This device, shown in Fig. 1.33, behaves like the NPN transistor, with the amount of current passing through the collector and emitter being proportional to the current being drawn from it.

A PNP transistor will *source* current and the NPN transistor will *sink* current. Note that for both devices, the emitter should be connected to the power supply "rails." For the NPN transistor, the emitter should be connected to ground, and for the PNP transistor, the emitter should be connected to positive power, or V_{cc}.

Bipolar devices have the advantage of being quite fast, but unfortunately take up a lot of space on a silicon integrated circuit. This is due to the need for each transistor to be built in a "tub" in the silicon to make sure it is isolated from all other transistors built on the chip. In addition, for proper operation, the transistors also require current-limiting resistors connected to their bases, which are very hard to build on silicon chips.

The field-effect transistors do not have these limitations. The most popular of these devices are the *metal oxide semiconductor field-effect transistors*

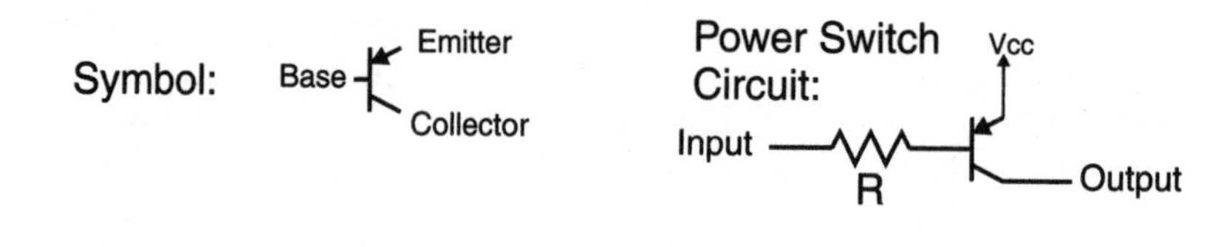

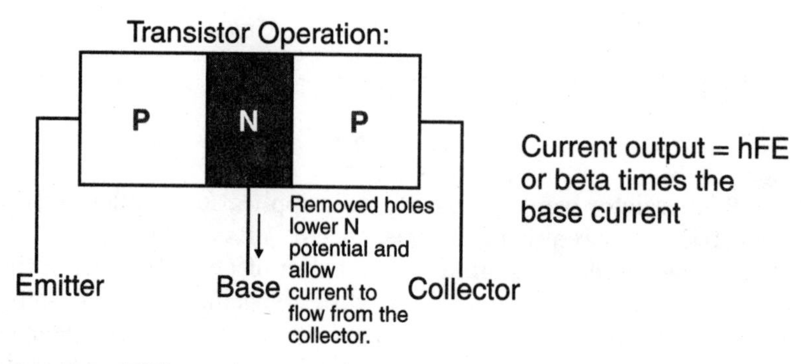

Figure 1.33 PNP transistor operation.

(MOSFETs). These components are much smaller and use a fraction of the power of the bipolar transistors. Their only real drawback is the amount of time they require for switching; bipolar transistors can "natively" switch must faster, and for this reason, bipolar transistors were the primary transistors used in supercomputers up until about 10 years ago. MOSFET devices have replaced bipolar ones as they have become much faster, and newer computer architectures with parallel circuits have allowed MOSFETS to become the building block of choice for these systems.

The basic type of MOSFET is the *N-channel* device, shown in Fig. 1.34. When a positive voltage is applied to the *gate,* an electrical field is set up in the p-doped silicon substrate. This field will cause the electrical characteristics in the substrate below the gate (the *Conducting Region* in Fig. 1.34) to mimic those of n-doped silicon.

N-doped silicon is a conductor and allows current to pass between the *source* and the *drain.* No current passes from the gate to either the source or the drain, so there is no bipolar I_{be} current analog in the MOSFET.

The gate consists of a metal plate that is separated by a layer of silicon oxide (also known as *glass*) from the substrate. FETs were originally patented in the 1870s, but it was the difficulty in growing the silicon oxide on the substrate that delayed demonstration of working FETs until the 1960s (almost 100 years later). The silicon oxide layer must be as thin as possible to maximize the performance of the MOSFET.

The size of the conducting region can be controlled by the amount of voltage applied to the gate. For digital applications, a set amount of voltage is constantly applied, making the N-channel behave like an on/off switch in digital applications.

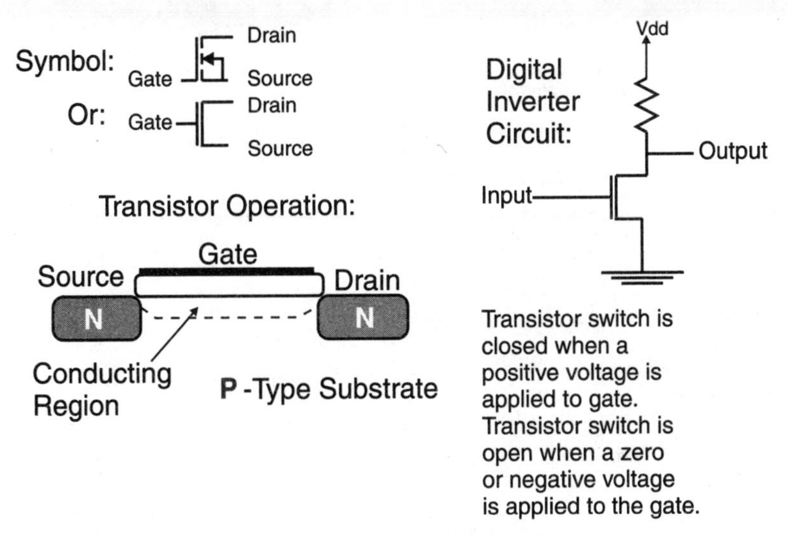

Figure 1.34 N-channel MOSFET transistor operation.

MOSFETs are usually defined by the resistance between the drain and the source when the transistor is on, or conducting. For an N-channel device, this resistance is normally measured in fractions of an ohm.

In Fig. 1.34, I have shown an N-channel transistor being used as a simple switch that "inverts" the logic state of the input signal to the output. This is an example of what was known as NMOS logic.

The N-channel MOSFET is very easy to build on a p-doped silicon substrate and became the first type of MOSFET logic used. If you look at many older integrated circuit data sheets, you will see that this type of logic was used for them. While N-channel MOSFET devices are very easy to place on the substrate, there was still the need for placing resistors on the silicon; for NMOS logic this was accomplished by doping regions of the substrate with n-type doping materials. The characteristics of the resulting resistor could vary by a few tens of percent, and devices were therefore unsuitable for analog applications.

N-channel MOSFETs also have a complementary device, the P-channel MOSFET. These transistors normally conduct when a zero voltage is applied to them because a conducting tub of n-doped silicon has been placed under the gate, as I've shown in Fig. 1.35. When a positive voltage is applied to the gate, the n-doped silicon changes its electrical characteristics to those of p-doped silicon and stops conducting. When the depletion region grows to the point where the entire n-doped silicon behaves like p-doped silicon underneath the gate, the transistor is no longer conducting and is turned off.

Varying the amount of voltage applied to the gate can control this *pinching off* of the N-channel tub. The P-channel MOSFET has an on resistance of several ohms or more, which is much higher than the on resistance of the N-channel MOSFET, and it can be difficult to match the two devices for analog applications.

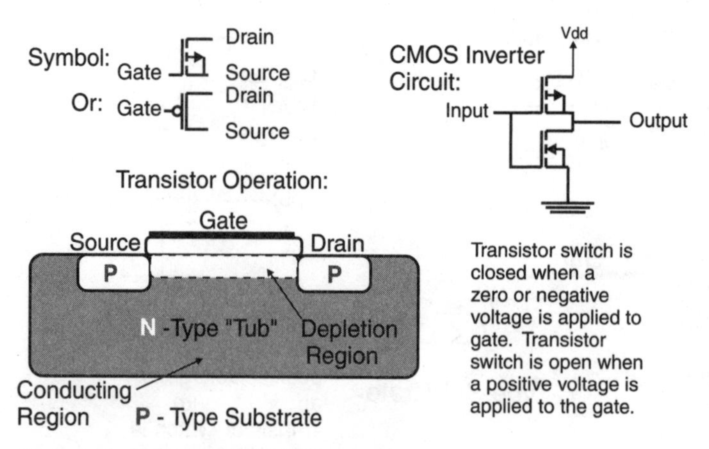

Figure 1.35 P-channel MOSFET operation.

Where P-channel MOSFETs have found a niche is in working with N-channel MOSFETs in complementary metal oxide semiconductor (CMOS) logic. As I've shown in Fig. 1.35, a P-channel MOSFET can be combined with an N-channel MOSFET to produce an inverter and not require the current-limiting resistor of the NMOS inverter. This has been an important breakthrough—by providing P-channel and N-channel MOSFETs in the same circuit, the need for resistors is eliminated. When the N-channel transistor is conducting (and pulling the output to ground), there is current being passed through a current-limiting resistor to ground.

When you see the symbols for MOSFETs, it is easy to forget which is which. Always remember that the N-channel device has the arrow going *iN*. This is not true for NPN transistors, where the arrow indicates current output.

In CMOS circuits, the only real opportunity for current flow is when the gates are switching and stored charges are passed through the transistors. This accounts for the phenomenally low current (and power) requirements of CMOS circuits and why the current requirement goes up when the clock frequencies go up. As the number of switch transitions per second increases, the amount of charge moved within the chips goes up proportionally. This charge movement averages out to a current flow.

Integrated Circuits

Over 40 years ago, the idea of combining transistors and other parts was put forward with the idea of simplifying the task of developing circuits. The idea was to use a common silicon substrate for the transistors and link their connections together in the form commonly required circuits. The silicon chips these devices were built upon became known as *integrated circuits.*

The first integrated circuits combined a few transistors together to form basic *logic gates.* Logic gates change the electrical state of one or more inputs into another one. For bipolar integrated circuits, a logical NAND, which outputs a high voltage only if both inputs are low, is the basic circuit from which other circuits are built.

There are two problems with using bipolar transistors in an integrated circuit. The first is that they have to be built in tubs, which isolate them from the other devices on the silicon chip. This requires extra manufacturing steps and also takes up more space than just what the single transistor requires. The second problem with using bipolar transistors in integrated circuits is that it is difficult to build resistors on the silicon.

The solution to these problems is to build integrated circuits using MOSFETs. Currently, most integrated circuit technologies use CMOS circuits as discussed in the previous section. To produce a NAND gate using N-channel and P-channel MOSFETs, the circuit becomes the one shown in Fig. 1.36.

When integrated circuits are built, the transistors are linked together by aluminum traces put down onto the silicon substrate. Some new electronic technologies use copper instead of aluminum for this function. The aluminum is deposited onto the chip and then etched away by a process similar to that for building a printed circuit board (PCB), which is described in a later chapter. This is known as the *metallization* layer of the integrated circuit.

When silicon chips are manufactured, large silicon "wafers" have a large number of the single chip "dies" imprinted on them. During the manufacturing process, circuits are laid out on the dies by a photographic process. Once the circuits are laid out, chemicals are applied and will convert the pictures on the dies into circuits. When the manufacturing process has finished, the dies are cut apart and put into packages for what is normally known as *chips*. This is shown in Fig. 1.37.

The vast majority of the time the wafers and dies spend in the manufacturing process is devoted to the chemical processes that build the circuits. These chemical processes do not change with the number of circuits on a wafer. This means that the manufacturing process cost per die goes down the more dies that can be put through the process (or the more dies that can be put on a single wafer).

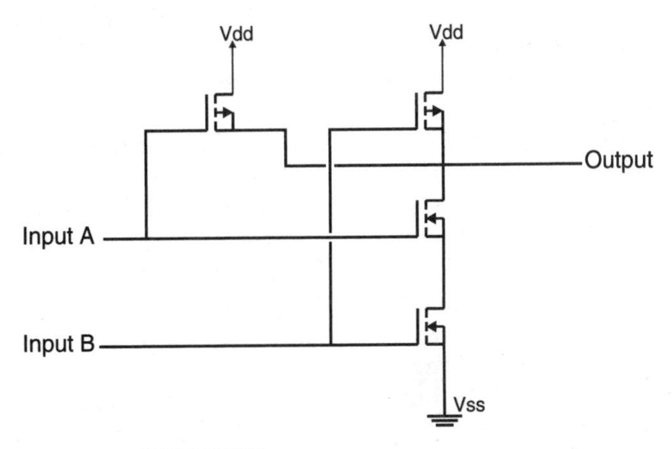

Figure 1.36 CMOS NAND gate.

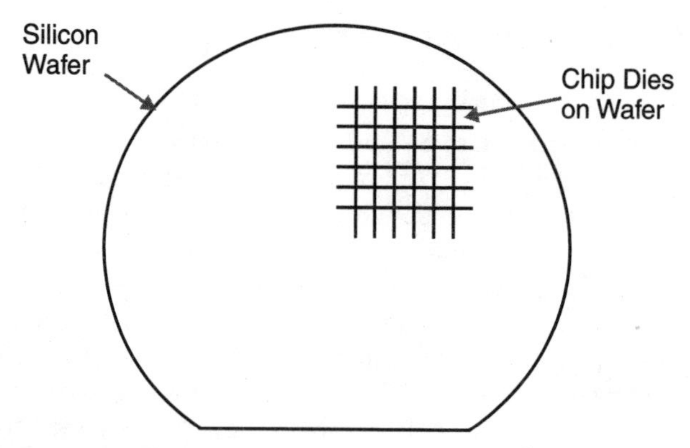

Figure 1.37 Wafer and die relationship.

For example, if a reduction of 25 percent can be achieved on a die axis, an area reduction of almost 50 percent per die can be achieved. This means that by reducing the size of a chip by a quarter on each side, almost twice as many chips can be put on a wafer, and the chip cost will be a little more than one half of what it was originally to the manufacturer.

The silicon integrated circuits in dies are then *encapsulated* in a plastic (epoxy) or ceramic package. Plastic encapsulants are the most prevalent and use an epoxy potting compound that is injected around a chip after it has been wired to a lead frame.

The lead frame becomes the pins on the package and is wired to the chip via very thin aluminum wires ultrasonically bonded to both the chip and the lead frame. Some chips are attached to the lead frame using *C4* technology, which is described later.

Once the encapsulant has hardened, the chip is protected from light, moisture, and physical damage from the outside world. Most logic and memory chips are built in plastic chips, as this is the cheapest method for building the part. Electrically programmable read-only memory (EPROM) microcontrollers and EPROM memories that are encapsulated in a plastic package are generally referred to as one-time programmable (OTP) devices (shown in Fig. 1.38). Once the EPROM memory has been programmed, the device cannot be used for anything else.

The primary purpose of putting a chip into a ceramic package is to allow a quartz window that is built into it to be used to allow erasing EPROM memory, as I show in Fig. 1.39. When a ceramic package is used, the chip is glued to the bottom half and is wired to the lead frame. Ceramic packaging is normally only available as a pin through-hole (PTH) device, while plastic packages can be in a very wide range of different card attachment technologies.

Ceramic packaging can drive up the cost of a single chip dramatically (as much as 10 times more than the price of a plastic-packaged OTP device). This makes this type of packaging suitable only for such uses as application debugging, where the advantage of the window for erasing outweighs the extra cost of the package.

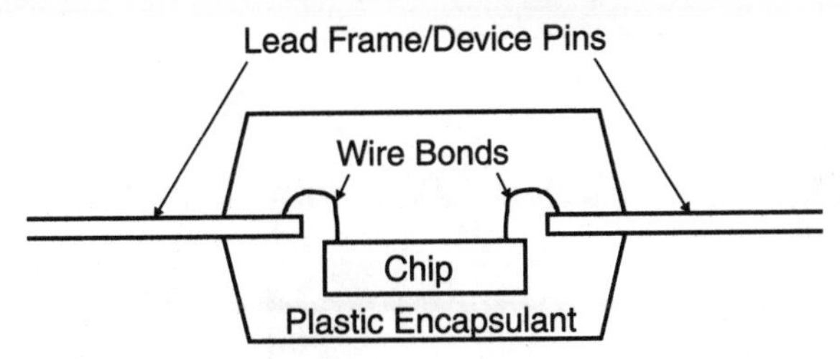

Figure 1.38 T P plastic package.

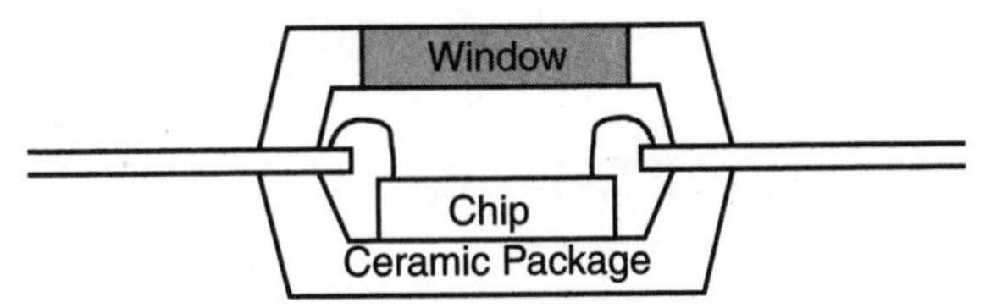

Figure 1.39 Windowed ceramic package.

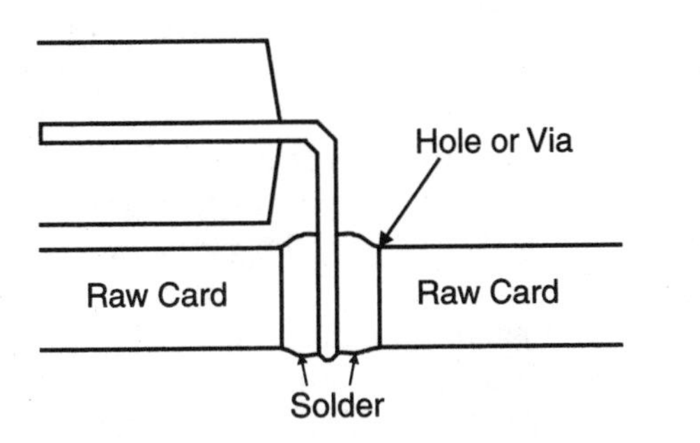

Figure 1.40 Pin through-hole connection.

The technology used to attach the chip to the board has changed dramatically over the past 20 years. In the '70s, most devices were only available in *pin through-hole* technology, shown in Fig. 1.40, in which the lead frame pins are soldered into holes in the raw card.

In this book, all the chips that I will be working with are packaged exclusively in plastic encapsulant (OTP) *dual inline packages* (DIPs), as shown in Fig. 1.41. This type of package is very commonly used for electronic prototyping and training. It is easy to work with and well-suited to the "breadboard" prototyping systems that will be used to build the projects shown in this book.

All the chips are placed in a circuit according to the location of pin 1. Pin 1 on a DIP chip is a corner pin of the device and its position is marked either by the semicircular indentation in the top of the package shown in Fig. 1.41 or by an indented (or printed) dot. While in Fig. 1.41 I show both these pin 1 indicators, in the wiring diagrams presented in this book, just the semicircular indentation will be shown.

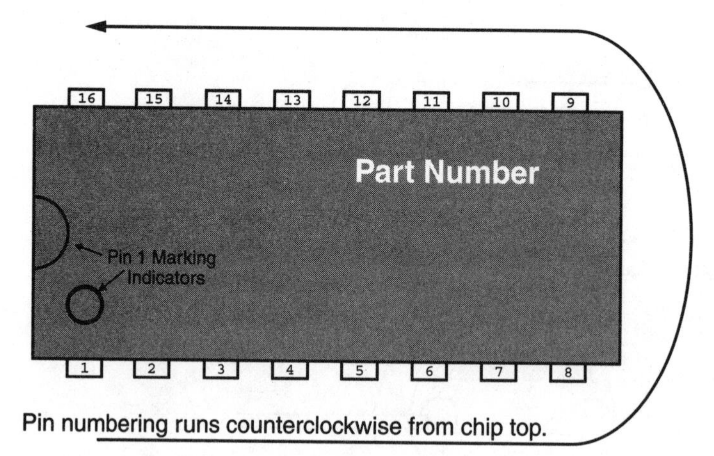

Figure 1.41 Features of a DIP chip package.

When you are building circuits from schematic diagrams, you will have to look up the part's pinout (including pin 1) from datasheets. In Appendix B, I have included the pinouts for many of the TTL chips used to build circuits in this book.

PTH attachment technology has the advantage that it is very easy to work with (very little specialized knowledge or equipment is required to manufacture or rework boards built with PTH chips). The primary disadvantage of PTH is the amount of space required to put the hole in the card. In addition, the requirements for space around each hole makes the spacing between lead centers quite large in comparison to *surface mount technology* (SMT), in which the pins are soldered to the surface of the card. See Fig. 1.42.

Pin through-hole packages are normally built with pins 0.100 in (100 thousandths of an inch) between pin centers. For some *pin grid array* parts (in which the pins are put on a two-dimensional matrix), lead centers can be as low as 0.071 in (71 thousandths of an inch) between pin centers. The measurement *between lead centers* is a critical one for electronics because it is directly related to how densely a board can be populated with electronic components.

There are two primary types of SMT leads used: the *gull wing* and the *J lead,* shown in Fig. 1.43. Each type of package offers advantages in certain situations. The gull wing package allows for hand assembly of parts and easier inspection of the solder joints. The J lead reduces the size of the part's overall footprint. Right now, gull wing parts are significantly more popular because this style of pin allows easier manufacturing and rework of very small leads.

The smaller size and lead centers of the SMT devices has resulted in significantly higher board densities (measured in chips per square inch) than PTH. As noted above, typical PTH lead centers are 0.100 in apart while SMT starts at 0.050 in and can go as low as 0.16 in. The SMT parts with small lead centers are known as *fine-pitch* parts.

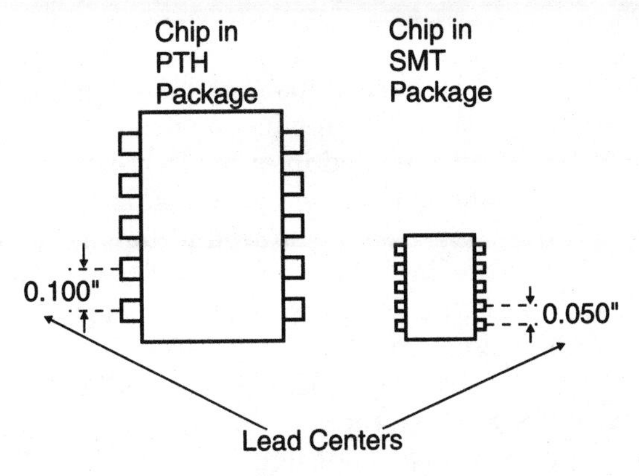

Figure 1.42 PTH package versus SMT.

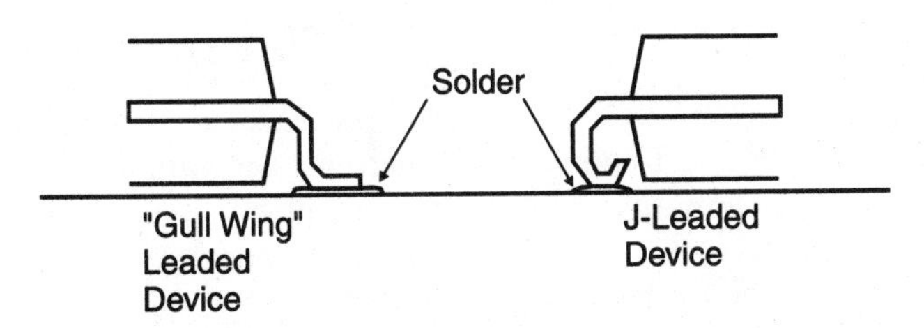

Figure 1.43 Surface mount technology packages.

To give an idea of the advantages of SMT in terms of board density, consider a PTH package that has pins at 0.100-in lead centers and an SMT package with pins at 0.050-in lead centers. With the smaller lead sizes, the SMT package can be about half the size of the PTH part in each dimension (that means that four SMT parts can be placed in approximately the same space as one PTH part). Moreover, without holes through the card, components can be put on both sides of the card, meaning that in the raw card space required for one PTH part, up to eight SMT parts can be placed on the card.

Assembly and rework of SMT parts is actually easier in a manufacturing setting than PTH. Raw cards have a solder/flux mixture called *solder paste* "screened" onto the SMT pads of the boards. This screening process uses a metal stencil with holes cut into in the locations where the solder paste is to be put. A squeegee-like device spreads the paste over the stencil, and the paste is deposited on the card where there are holes.

Once the paste has been deposited, the parts are placed onto the paste and then run through an oven to melt the solder paste, and the parts are soldered to the board. To rework a component, hot air (or nitrogen gas) is flowed over the solder joints to melt the solder, allowing the part to be pulled off. While SMT is easier to work with in a manufacturing setting, it is a lot more difficult for the hobbyist or developers to work with (especially if parts have to be pulled off a board to be reprogrammed).

Programmable Logic Devices

Programmable logic devices (PLDs) are chips that have logic gates and flip-flops built in, but not interconnected. The application designer will specify how the gates and flip-flops are interconnected in order to create a portion of the application's circuit. Most people feel that programmable logic devices are a relatively new invention, but they have been around for many years. It has only been quite recently (in the last 10 years or so) that PLDs based on reusable chip technology (that is, EPROM and flash memory) have been available at prices hobbyists and small companies could afford.

The simplest type of PLD is an array of logic gates and devices, known as a PAL or GAL (I generically refer to them as PALs). The chips themselves are quite simple and relatively easy to design into circuits. These circuits are normally arranged as a "sum of products" in which signals on the chip can be easily interconnected to form more complex logic functions. The chips are normally blocked out as a series of inputs and outputs, as shown in Fig. 1.44.

The vertical lines, or buses in Fig. 1.44 are referenced to the gates and I/O pins they are connected to. To form logic functions, the sum of products is used. In Fig. 1.44, a simple four-I/O, 12-gate PAL is shown. Every output is driven on a bus in both positive and negative format. Connections are made between the gates and the buses to create logic functions.

For example, the XOR gate, which is characterized by

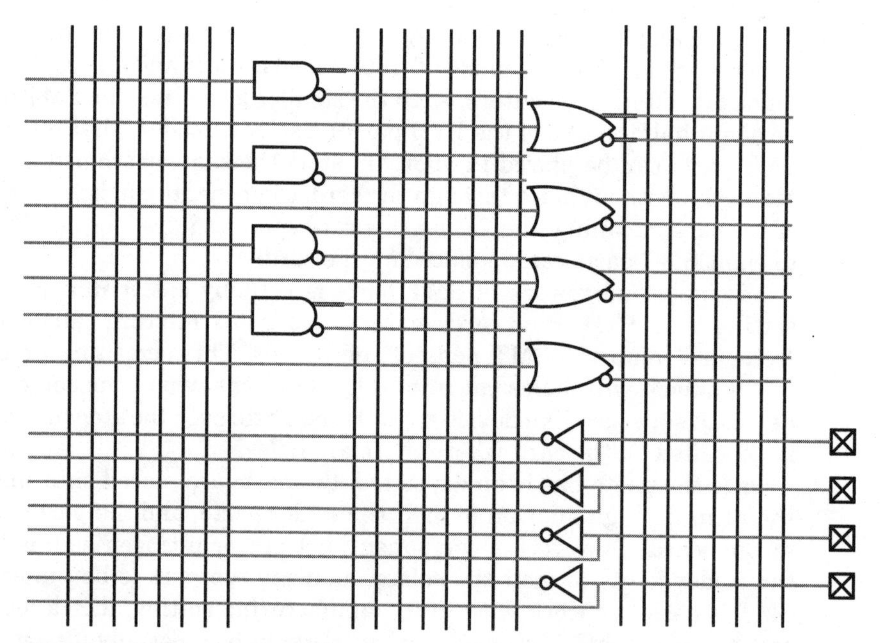

Figure 1.44 PLD matrix connections.

$$\begin{aligned} \text{A XOR B} &= \text{A} \wedge \text{B} \\ &= \text{_A AND B OR A AND _B} \\ &= (\text{_A} * \text{B}) + (\text{A} * \text{_B}) \end{aligned}$$

is not often available in standard logic. Taking the chip in Fig. 1.44 and connecting the buses to the different I/O pins and gates within the PAL, I can implement the XOR gate as shown in Fig. 1.45.

Note in Fig. 1.45 that an I/O pin can be changed from an input to an output by simply connecting it directly to a gate output. This feature allows the pins to be used to provide feedback to the circuit.

Options for PALs include varying numbers of inputs to the internal AND and OR gates. For the PLD in Fig. 1.43, I have left open the option that any of the pins can be used for any purpose. This is a bit unusual, and normally in PALs the number of inputs to a gate is restricted. Another option is to include built in flip-flops to store states and turn the PAL from a combinatorial circuit into a sequential one.

PALs may seem simple, but they can result in large decreases in the chip count for an application. In some cases, PALs may be more expensive than the chips they replace, but they reduce application power and board space chip requirements. These savings could result in an overall product savings. It is not unusual for 10 TTL chips to be replaced by a single PAL, resulting in huge PCB and power supply cost savings.

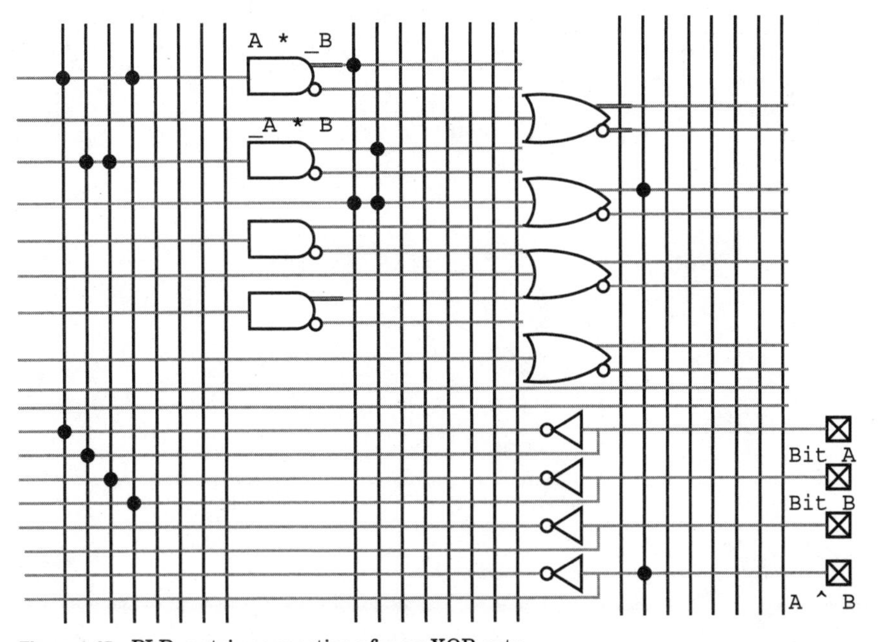

Figure 1.45 PLD matrix connections for an XOR gate.

At the high end of the programmable logic device family range, some devices are virtually application-specific integrated circuits (ASICs), and use the same programming language (VHDL) and development tools as ASICS. These complex parts generally have their functions broken up into "macros." An ASIC/PLD macro can be an AND, or XOR not, logic gate, a flip-flop, or a collection of functions (such as multiplexers and arithmetic logic units). Macros simplify the task of circuit development and eliminate the need for wiring individual gates into basic functions.

The high-end programmable logic devices' programming information is often directly transferable to ASIC technology. This allows initial production to use programmable logic devices that require little cost to program, and, when the design is qualified, ASICs can be built at a chip foundry for reduced per-unit costs.

Programmable logic devices have the advantage of being able to implement fast (less than 10 nsec) logic switching, but they do not have the ability to store more than a few bits of data.

Often programmable logic devices are used in proprietary circuits because their functions cannot be easily traced and decoded.

Programmable logic device and ASIC development tools are generally function-text-based applications, as opposed to graphically based applications (such as a schematic drawing). This means that a text format, like the XOR definition above, must be used to define the functions. Most compilers for these statements are intelligent enough to pick the best gates within the device to work with and pick the best paths without your intervention. They are typically much more sophisticated (and expensive) than the compilers used for converting high-level program statements into instructions for a processor.

Chapter

2

Engineering Mathematics

If you subscribe to Internet list servers or news groups, you will periodically see an email that has the message:

> I am interested in learning about electronics, but I do not want to have to learn math because I'm not very good at it. Can anybody point me in the right direction?

Depending on your expertise in electronics, you will either be sympathetic with the sender or sigh in frustration.

Having a good background in mathematics is critical for being able to design your own electronic circuits of any type. It is also critical for understanding how to program a computer. Actually having good basic math skills is critical for many aspects of your personal life, from trying to fill out your tax return to understanding at the grocery store what package will give you the most product for the least amount of money.

Before you put this book back on the shelf, let me note one thing about the mathematics that I am going to present in this book. If you are comfortable with adding, subtracting, multiplying, and dividing then you will not have any problems with this book or most digital electronics that you will work with. Most mathematics associated with electronics, as you will see in the following chapters, is quite basic with not a lot of difficult operations or concepts. I would expect anyone who is at the high-school junior (grade 11) level of mathematics to understand what is being presented in this book and apply it in circuit calculations.

For many situations in this book, I will present you with just formulas, not how they were derived. If you want to learn more about analog electronics (where most of the complex formulas are), you will either have to look at some of the texts referenced at the end of this book or enroll yourself in a college engineering or technology program.

I should note that the data and mathematical formulas used in electronics might be presented here in a different format than you are used to. This is not to say that it will be difficult for you to learn—just that you will have to learn a new set of conventions. In this chapter, I will introduce you to many of the basic concepts of electronic mathematics, while in a later chapter, I will introduce you to boolean logic, which is used by digital electronic circuits to process information.

Numbering Systems

When working with digital electronics (as well as computer programming), you will be inundated with different numbering systems in different formats. This can make understanding what is happening in your application difficult or the data you are seeing confusing. Having a good understanding of what the different numbering systems are and how data is converted between them is critical to being able to develop a working application and debugging one that isn't.

All our lives, we have been taught to work in the base 10 system. *Base 10* means that numbers are represented by the 10 digits 0, 1, 2, 3, 4, 5, 6, 7, 8, and 9. If a number greater than nine needs to be written down, then multiple digits are used, with each digit representing the power of 10 of the digit. For example, 123 is one hundreds (10 to the power 2), two tens (10 to the power 1) and three ones (10 to the power 0). It could be written out as:

$$123 = (1 \times 100) + (2 \times 10) + (3 \times 1)$$

$$= [1 \times (10^2)] + [2 \times (10^1)] + [3 \times (10^0)]$$

Each digit in a number is given a position starting from the right-most digit (working left and numbered starting with 0). The digit position is actually the exponent the base is multiplied by to get the value for the digit. When you learned base 10 mathematics, each digit was referred to as ones, tens, hundreds, and so on instead of using the exponent to base 10 (which these values actually are).

When you work with digital electronics and want to change specific digits, you will find it easier to work with numbers in terms of the base's exponent rather than the actual value. I am making this point because for most of this book, numbers will be represented as binary values with each digit 2 times (in base) greater than the digit to its right.

In a computer, numbers are represented by a series of bits, where a bit (or *digit*) can have the value 0 or 1. Binary values are represented as a series of ones or zeros with each digit representing a power of 2. Using this system, binary 101100 can be converted to decimal by using the binary exponent of each digit.

$$\text{Binary } 101100 = [1 \times (2^5)] + [0 \times (2^4)] + [1 \times (2^3)] + [1 \times (2^2)] + [0 \times (2^1)] + [0 \times (2^1)] \quad \text{(Decimal)}$$

$$= 32 + 8 + 4 \quad \text{(Decimal)}$$

$$= 44 \quad \text{(Decimal)}$$

For convenience, binary numbers are normally written as *hexadecimal* digits instead of individual bits. Hexadecimal digits combine 4 bits into a single value. The numbers 0 to 9 and A to F are used to represent each hex digit. For multidigit hex numbers, each digit is multiplied by 16 to the appropriate power:

$$\text{Hex } 123 = [1 \times (16^2)] + [2 \times (16^1)] + [3 \times (16^0)] \quad \text{(Decimal)}$$

$$= 256 + 32 + 3 \quad \text{(Decimal)}$$

$$= 291 \quad \text{(Decimal)}$$

When working with the different numbering systems in this book, I will use three different formats to let you know which numbering system is being used. When the number is enclosed in single quotes (' ') and prefixed by a B or by the string 0b0, then the number is a "Binary" value. When 0x0 is in front of the number, then the number is in hexadecimal format. If the number isn't modified in any way, then it is decimal.

Number format	Number type
0b0...	Binary
B'...'	Binary
0x0...	Hexadecimal
No prefix	Decimal

As an aside, as you may know, *hex* is the prefix for 6, not 16. The correct prefix for 16 is *sex,* and in the early days of programming, base 16 numbers were known as *sexadecimal.* When IBM released the System/360, all the documentation referred to base 16 numbers as *hexadecimal.* This change was made because IBM felt that the System/360 would be used by a large group of different users, some of whom would object to the word *sex* as the basic numbering system of the computer. To be fair, I'm sure many of the early programmers (some programmers can be pretty immature) snickered at the term *sexadecimal,* creating a feeling that a reputable computer company should not sell a machine that was programmed in "sex."

Along with decimal, binary, and hex numbers, there is also a base 8 numbering system known as *octal.* Octal numbers were made popular by DEC in its VAX line of minicomputers, as well as the C language. Octal numbers are awkward to work with because each byte character requires 3 digits. Most software development tools support octal, but it is not popular for digital electronics.

To help documenting binary numbers with specific bits set easier, I suggest using the arithmetic expression

$$1 << \text{bit number}$$

rather than trying to figure out which bit is which value (for the different radixes). The << symbol is the C language symbol for shifting the left parameter to the left by the right parameter number of bits.

I find it easier to shift one up by the bit digit number. Thus to create a byte constant where bit 4 is set, I could use the expression

$$1 << 4$$

This is a very useful trick for working with digital electronics, as well as for assembly language programming where individual bits are set and reset for hardware control. Constants with specific bit settings can be created much more quickly and accurately by using this trick than by creating a binary number from the bits you want to set and converting to hexadecimal or decimal.

Despite little tricks like this, you will still have to convert numbers from one base to another as you are programming. To aid in this operation, I recommend that you buy a calculator that has base conversion [as well as bitwise operation (AND, OR, XOR, and NOT)] capabilities. I have owned two HP-16C calculators for over 15 years (one for home and one for work) and they are two of the best investments I have ever made.

Some modern calculators with numbering system conversion (and bitwise operations) capabilities that are built in are:

Hewlett Packard. HP-20S

Texas Instruments. TI-85, TI-86

Sharp. EL-546L, EL-506L

When writing mathematical expressions and formulas for digital electronic devices, you will have to be familiar with how mathematical operations work. Addition and multiplication, for integer operations, are quite simple to understand. In subtracting numbers, if a digit is greater than the one above it, the number base is borrowed from the digit to the left of it. For example, to subtract 15 from 22, in subtracting the ones, 5 is greater than 2, and 10 is borrowed from the tens:

$$\begin{array}{r} 22 \\ -15 \end{array} \quad = \quad \begin{array}{r} 10 + 12 \\ \underline{-(10 + 5)} \\ 0 + 7 \end{array}$$

If the magnitude of the number subtracted is greater than the value it is subtracted from, then a negative number is the result:

$$\begin{array}{r} 22 \\ \underline{-25} \\ -3 \end{array}$$

In digital electronics, there is no such thing as a negative number; instead, when a value is subtracted from a smaller value, the borrowed value results in a "two's complement" negative number. This can be shown in the previous example, by converting the two decimal numbers into binary:

$$\begin{array}{r} 22 \\ -25 \\ \hline \end{array} \quad = \quad \begin{array}{r} 00010110 \\ -00011001 \\ \hline \end{array}$$

For the right-most digit (bit 0), 1 cannot be taken away from 0, so the digit to the left (bit 1) is borrowed from. In this case, 2 is borrowed from bit 1 (subtracted from bit 1), and the result of 1 subtracted from 2 is placed as the result.

For bit 3, the 1 in the number being subtracted has to borrow from bit 4, leaving bit 4 without anything to be subtracted from it. Because the higher bits are all zero, this is not possible, so the bits are checked to the end of the byte (bit 7) until a bit is found to borrow from. In most computer processors, if no other bits are available to be borrowed from, the hardware behaves as if the first bit that is larger than the most significant bit of the byte is set. So, in actuality, the subtraction is:

$$\begin{array}{r} 1\ 00010110 \\ -0\ 00011001 \\ \hline \end{array} \quad = \quad \begin{array}{r} 1\ 00010110 \\ -0\ 00011001 \\ \hline 11111101 \end{array}$$

The value B'11111101' is the two's complement representation of −3. This value, when added to a positive value, or another two's complement negative number, will result in a valid value.

For example, when −3 is added to 5, the result is 2. With the two's complement representation of −3 above, this can be shown with binary numbers:

$$\begin{array}{r} -3 \\ +5 \\ \hline 2 \end{array} \quad = \quad \begin{array}{r} 11111101 \\ +\ 00000101 \\ \hline 1\ 00000010 \end{array}$$

Like the virtual ninth bit made available for the borrow, the ninth bit produced is ignored (or used as the "carry" in the circuit). With the ninth bit ignored, the result is B'00000010' or 2 (decimal).

Two's complement negative numbers can be generated from positive numbers by using the formula:

$$\begin{aligned} \text{Two's complement negative} &= \text{NOT (positive)} + 1 \\ &= \text{!positive} + 1 \\ &= (\text{positive} \wedge \text{0x0FF}) + 1 \end{aligned}$$

The NOT or ! character will cause any "set" (1) bits to change to "reset" (0) and vice versa. The ^ character is used to represent the XOR operation. Positive ^ 0x0FF will invert, or complement, each bit in the positive variable. XOR is a boolean logic operator and will be described in more detail in the next chapter.

For −3, this formula can be used to generate the bit pattern

$$\begin{aligned} -3 &= \text{NOT (3)} + 1 \\ &= \text{NOT (B`00000011')} + 1 \\ &= \text{B`11111100'} + 1 \\ &= \text{B`11111101'} \end{aligned}$$

which is the same as the value calculated above.

With two's complement negative numbers, the most significant bit (bit 7 in a byte) is often referred to as the *sign* bit. If the sign bit is set, the number is negative. If it is reset, the number is positive. The remaining seven bits are the number's *magnitude.* Two's complement bytes can be in the range of −128 to +127.

Now that I've shown that computer processors can handle positive and negative two's complement numbers, I have to point out that most processors *cannot* recognize negative numbers *natively* (without any special instructions or software). Two's complement negative numbers are a representation of a negative number—they are not actually "negative numbers" to the processor. The difference is subtle, but important; the computer processor only works with bit patterns.

There is one other numbering system that is popular with many people, and that is known as the *binary-coded decimal* (BCD) format. In this numbering system, each base 10 digit of a number is represented as a nybble. Using BCD, 1 byte can represent the decimal numbers 00 to 99. BCD is quite popular for people working with mainframe processors, where there is essentially unlimited memory available and processor support is built in for handling the numbers natively. This format is very rarely used in digital electronics because of the difficulty of creating circuits that work with this data format.

Representing a number in BCD is straightforward. For example, the number 23 decimal is simply stored in a byte as 0x023. This means that 100 numbers can be represented in a byte. Compared to the possible 256 numbers using all 8 bits of a byte without restriction, BCD cannot represent 60 percent of the values that binary numbers can. For larger numbers, this inefficiency becomes even more acute. For example, 2 bytes (which have 4 nybbles) can represent the 10,000 BCD values 0000 to 9999, while the 2 bytes could represent 65,536 values if the data is stored as straight binary. In the 2-byte case, the efficiency of BCD in representing all possible values has dropped to just over 15 percent compared to binary.

Handling BCD numbers can be very difficult as well. While addition will seem simple, subtraction is more complex (especially with negative numbers). Additional complexity comes into place if more than 2 digits per number are used or if multiplication or division operations are required on the BCD numbers.

Number Bases

In this section, I will take the number system concepts of the previous section and expand on how they are converted between *bases* (or *radixes*). Along with this, I will also discuss how different values are represented by engineers, which is an issue because most values are either very large or very small, and to avoid problems with confusing orders of magnitude, the SI (metric system) conventions are used. These topics will be discussed in this section to help you to understand what is happening when I am working with the different values.

The usual number base (or radix) that we humans use is 10—probably because we have 10 fingers, and this made it easy to count and keep track of things. Other bases can be used, and are often used in electronics and computer science.

For example, if we had eight fingers, the *base* we would use would be 8. The actual numbers in each digit would be 0 through 7 and *not* eight. Eight would not exist in a base 8 numbering system (just as 10 doesn't exist in our base 10 numbering system); it would actually be written as 10. This can be somewhat confusing and to avoid problems, the base is usually identified when a number is presented in this (and most other) books.

In base 8 case, the "ones" would be the digits 0 through 7, the next digit (the "eights") would be 0 through 7 multiplied by 8, the second digit (the "sixty-fours") would be 0 through 7 multiplied by 8 times 8 (8 to the power of 2), and so on.

To convert between numbering systems, the value would be divided into individual digits until the actual result was found. Going back to the base 10 (decimal) number, it can be converted to base 8 (octal) by working through each of the digits.

To start off, the decimal number is divided by powers of 8 until they are larger than the number, then the power is backed off and the next power down is divided into the number, and this continues until there is an equivalent *order of magnitude.*

For example, if 123 decimal was to be converted into octal, the first operation would be to find the most significant digit of the octal number of the same order of magnitude. Starting with an exponent of 0 (the ones), the conversion number would be compared until the base and exponent are greater than the number to be converted. For decimal 123, this operation is:

$$\text{Exponent 0: } 8^0 = 1 < 123$$

$$\text{Exponent 1: } 8^1 = 8 < 123$$

$$\text{Exponent 2: } 8^2 = 64 < 123$$

$$\text{Exponent 3: } 8^3 = 512 > 123$$

In this case, the fourth digit (exponent 3) is greater than the conversion value. This means that the octal equivalent will be 3 digits in length and the order of magnitude would be 64.

Now, the value of each digit is divided into the number until the number is less than the value of the digit. This is shown starting with the third digit (exponent 2):

Exponent	Test	Remainder	Current octal
2	$123 - 8^2 = 123 - 64 = 59$	59	100
2	$59 - 8^2 = 59 - 64 = -5$	Can't be done	100
1	$59 - 8^1 = 59 - 8 = 51$	51	110
1	$51 - 8^1 = 51 - 8 = 43$	43	120
1	$43 - 8^1 = 43 - 8 = 35$	35	130
1	$35 - 8^1 = 35 - 8 = 27$	27	140
1	$27 - 8^1 = 27 - 8 = 19$	19	150
1	$19 - 8^1 = 19 - 8 = 11$	11	160
1	$11 - 8^1 = 11 - 8 = 3$	3	170
1	$3 - 8^1 = 3 - 8 = -5$	Can't be done	170
0	$3 - 8^0 = 3 - 1 = 2$	2	171
0	$2 - 8^0 = 2 - 1 = 1$	1	172
0	$1 - 8^0 = 1 - 1 = 0$	0	173
0	$0 - 8^0 = 0 - 1 = -1$	Can't be done	173

So, 173 octal is the equivalent to 123 decimal. Notice that the octal number is greater than the equivalent decimal number; this is true for any conversion from a higher base to a lower base.

While octal is a possible base for working with computer systems, I don't recommend it. Octal was popular in the 1970s and early 1980s because DEC VAX minicomputer systems used it, but I find it to be a very confusing numeric base to work with.

The most popular base for working with computer systems is base 2 (binary) and base 16 (hexadecimal) because of the binary operation of digital systems. Conversions between decimal, binary, and hexadecimal numbers can be carried out by many common hand-held calculators or, once you are familiar with working with them, done in your head.

Digital logic circuits can only operate in one of two states, set (or on) and reset (or off), which are usually represented as 0 and 1, respectively. These state circuits (known as *bits*) are usually grouped together into sets of four (nybbles) or eight (bytes). Grouping 4 bits together is the basis for providing a base 16 (hexadecimal) number, and this is the normal way of expressing binary values.

Decimal to binary conversions are carried out exactly the same way as converting a decimal number to an octal value. First the exponent greater than the value is found, and the conversion value is generated from the lower values. This can be shown with decimal 123, which will only require eight bits to be represented. The eighth bit (2 to the power 7, or 128) is greater than the value we are converting. The conversion can use a table in the same way the decimal to octal conversion was carried out:

Exponent	Test	Remainder	Current binary
6	$123 - 2^6 = 123 - 64 = 59$	59	1000000
6	$59 - 2^6 = 59 - 64 = -5$	Can't be done	1000000
5	$59 - 2^5 = 59 - 32 = 27$	27	1100000
5	$27 - 2^5 = 27 - 32 = -5$	Can't be done	1100000
4	$27 - 2^4 = 27 - 16 = 11$	11	1110000
4	$11 - 2^4 = 11 - 16 = -5$	Can't be done	1110000
3	$11 - 2^3 = 11 - 8 = 3$	3	1111000
3	$3 - 2^3 = 3 - 8 = -5$	Can't be done	1111000
2	$3 - 2^2 = 3 - 4 = -1$	Can't be done	1111000
1	$3 - 2^1 = 3 - 2 = 1$	1	1111010
1	$1 - 2^1 = 1 - 2 = -1$	Can't be done	1111010
0	$1 - 2^0 = 1 - 1 = 0$	0	1111011
0	$0 - 2^0 = 0 - 1 = -1$	Can't be done	1111011

Normally, binary numbers are written in groups of 8 bits, so the binary equivalent of 123 decimal is 0b001111011.

These 8 bits are usually converted into two hexadecimal (base 16) digits. The hexadecimal digits are 0 through 9, followed by A, B, C, D, E, and F for the remaining 6 digits. So, the hexadecimal number 0x07B can refer to the binary value 0b001111011.

The hexadecimal digits, A through F, are usually referred to by their phonetic names, which are listed in the table below.

Letter	Decimal value	Phonetic name
A	10	Able
B	11	Baker
C	12	Charlie
D	13	Dog
E	14	Easy
F	15	Fox

I can go through the same conversion operation on decimal 123 to hexadecimal to show that 4 bits grouped together work out to the same value. The second digit of a hexadecimal number is groups of 16 and the third digit is 256. The hex conversion of decimal value 123 is going to only be two digits in size. The conversion operation is shown below.

Exponent	Test	Remainder	Current hexadecimal
1	$123 - 16^1 = 123 - 16 = 107$	107	10
1	$107 - 16^1 = 107 - 16 = 91$	91	20
1	$91 - 16^1 = 91 - 16 = 75$	75	30
1	$75 - 16^1 = 75 - 16 = 59$	59	40
1	$59 - 16^1 = 59 - 16 = 43$	43	50
1	$43 - 16^1 = 43 - 16 = 27$	27	60
1	$27 - 16^1 = 27 - 16 = 11$	11	70
1	$11 - 16^1 = 11 - 16 = -5$	Can't be done	70
0	$11 - 16^0 = 11 - 1 = 10$	10	71
0	$10 - 16^0 = 10 - 1 = 9$	9	72
0	$9 - 16^0 = 9 - 1 = 8$	8	73
0	$8 - 16^0 = 8 - 1 = 7$	7	74
0	$7 - 16^0 = 7 - 1 = 6$	6	75
0	$6 - 16^0 = 6 - 1 = 5$	5	76
0	$5 - 16^0 = 5 - 1 = 4$	4	77
0	$4 - 16^0 = 3 - 1 = 3$	3	78
0	$3 - 16^0 = 3 - 1 = 2$	2	79
0	$2 - 16^0 = 2 - 1 = 1$	1	7A
0	$1 - 16^0 = 1 - 1 = 0$	0	7B
0	$0 - 16^0 = 0 - 1 = -1$	Can't be done	7B

In this case, going from a small base to a larger one, note that the number of digits in the number is decreased (which is expected).

When working with binary (or hexadecimal) values, note that the first number that is valid is zero. This is different from what we are usually taught in grade school, where you start counting at one. In electronics, numbers are equivalent to states, and the first one is zero (or 0b000000000 in binary). This is important to remember when you start working with electronics.

Counting can be confused by the concept of Gray codes. *Gray codes* are a way of encoding data in such a way that increasing a number changes only one bit at a time. In the table below, I show a 16-number (4-bit) progression in traditional decimal and binary counting along with the Gray codes associated with them:

Decimal	Binary	Gray code
0	0000	0000
1	0001	0001
2	0010	0011
3	0011	0010

(*Continued*)

Decimal	Binary	Gray code
4	0100	0110
5	0101	0100
6	0110	0101
7	0111	0111
8	1000	1111
9	1001	1011
10	1010	1010
11	1011	1110
12	1100	1100
13	1101	1101
14	1110	1001
15	1111	1000

The Gray codes presented above are arbitrary; the important thing to notice about them is that only 1 bit changes at a time, and each of the 16 different states is represented, just not at the same time. Going from state 15 back to state 0 changes only 1 bit as well.

Gray codes are used in equipment like optical rotary encoders, where both the position and direction can be decoded by comparing two sequential codes. For example, if 0b00111 was received followed by 0b00101, it can be determined that the optical encoder's last position is 6 and it is turning in a downward (from high to low number) direction.

So far, I have discussed integer values, which do not have values less than 1. Fractions can also be expressed in binary as well as decimal values.

This is accomplished by using negative exponents to the base numbers. For example, the decimal value 123.4 can be represented in binary numbers by recognizing that 2^{-1} is 0.5 (decimal), 2^{-2} is 0.25, and so on. The table used above can continue for the negative numbers:

Exponent	Test	Remainder	Current binary
6	$123.4 - 2^6 = 123.4 - 64 = 59.4$	59.4	1000000.000
6	$59.4 - 2^6 = 59.4 - 64 = -4.6$	Invalid	1000000.000
5	$59.4 - 2^5 = 59.4 - 32 = 27.4$	27.4	1100000.000
5	$27.4 - 2^5 = 27.4 - 32 = -4.6$	Invalid	1100000.000
4	$27.4 - 2^4 = 27.4 - 16 = 11.4$	11.4	1110000.000
4	$11.4 - 2^4 = 11.4 - 16 = -4.6$	Invalid	1110000.000
3	$11.4 - 2^3 = 11.4 - 8 = 3.4$	3	1111000.000
3	$3.4 - 2^3 = 3.4 - 8 = -4.6$	Invalid	1111000.000

(*Continued*) Exponent	Test	Remainder	Current binary
2	$3.4 - 2^2 = 3.4 - 4 = -0.6$	Invalid	1111000.000
1	$3.4 - 2^1 = 3.4 - 2 = 1.4$	1	1111010.000
1	$1.4 - 2^1 = 1.4 - 2 = -0.6$	Invalid	1111010.000
0	$1.4 - 2^0 = 1.4 - 1 = 0.4$	0.4	1111011.000
0	$0.4 - 2^0 = 0.4 - 1 = -0.4$	Invalid	1111011.000
−1	$0.4 - 2^{-1} = 0.4 - 0.5 = -0.1$	Invalid	1111011.000
−2	$0.4 - 2^{-2} = 0.4 - 0.25 = 0.15$	0.15	1111011.010
−2	$0.15 - 2^{-2} = 0.15 - 0.25 = -0.15$	Invalid	1111011.010
−3	$0.15 - 2^{-3} = 0.15 - 0.125 = 0.025$	0.025	1111011.011

As the negative binary exponents are worked through, note that the result is not an even value. As you work through the fractions in base 2, you will find that fractional values (like 0.4), which are very easily expressed in base 10, are not easily represented in other numbering systems.

For very large and very small numbers, I will express them in *scientific notation* with a base ten exponent. This format consists of a single whole number (with a two digit fraction) multiplied by an exponent of 10. For example, 123 can be expressed in scientific notation as

$$1.23 \times (10^2)$$

The advantage of scientific notation is that its order of magnitude can be very easily converted to a logarithm and then be multiplied or divided with another number. I will use scientific notation where appropriate throughout the book with 2 or 3 *significant digits* (the fraction displayed in the number).

Of more use are the standard unit exponent multipliers. In the SI (metric) measurement system, a change of 3 orders of magnitude is indicated by a prefix to a unit. The prefixes are listed below.

Prefix	Exponent multiplier
p (pico)	10^{-12}
n (nano)	10^{-9}
μ (micro)	10^{-6}
m (milli)	10^{-3}
k (kilo)	10^{3}
M (mega)	10^{6}
G (giga)	10^{9}
T (tera)	10^{12}

With this system, very large units can be expressed in more convenient terms. For example, 10,000 ohms (Ω) is usually expressed as 10 kΩ or just 10k. In another case, 0.0000001 farads is expressed as 0.1 microfarads or 0.1 μF.

The only other engineering mathematics that you should be aware of is how logarithms work. Logarithms are the complementary operation to exponents. Passing a value to a logarithmic function will return the exponent (to the specific base) along with a logarithm value. In the text above, I refer to the exponent as the *order of magnitude*; the operation of finding a logarithm also finds the order of magnitude of a number.

In this book, and for most electronics, you have to be concerned with the returning of only the exponent from a logarithmic operation. For example, you should be aware that 10,000 logarithm base 10 is 4.

Along with knowing the base 10 exponent logarithms, you should also be aware of the base 2 exponent logarithms. These values for the first 17 exponents of 2 are shown in the table below.

Decimal value	Base 2 logarithm/ base 2 exponent
1	0
2	1
4	2
8	3
16	4
32	5
64	6
128	7
256	8
512	9
1,024	10
2,048	11
4,096	12
8,192	13
16,384	14
32,768	15
65,536	16

Chapter

3

Setting up Your Own Digital Electronics Lab

If the vision that a "digital electronics lab" conjures up is one of white-lab-coated scientists staring at expensive equipment—don't worry; the lab I am going to help you set up will be quite modest. In fact, you can probably start working through the projects in this book with a cash outlay of $20 (US) or less. I use the term *lab* or *laboratory* because you will be running experiments on the equipment and it will provide you with the capability of developing your own circuits.

For all of the projects presented in this book, you will require a +5-V power supply and a prototyping system that you can use for testing out the circuits. The +5-V power supply is used to power the TTL logic that will be used for most of the projects and will be created from the PCB that comes with this book. The power supply works with a wall-mounted AC/DC adapter (often known as a "wall wart"); you probably have a few lying around the house. Even if you don't find a suitable AC/DC adapter handy, you can buy it and the parts for the power supply for less than $10.

The prototyping system that I will recommend that you use is known as a "breadboard" and consists of a matrix of spring-loaded holes that chip pins and wires can be pressed into to make a circuit. Building circuits this way isn't extremely fast, but the tool does not damage chips or wires and can be reused many times. A small breadboard can be purchased from an electronics store for less than $5.

Once you have power supply and breadboard, you will have to get some parts. A basic set of parts requires a couple of chips, a few LEDs, a few resistors, a switch or two, and a couple of capacitors. These parts can be bought at electronic retailers such as Radio Shack, but you could probably save some money by going to an electronics surplus store. All of these should be available for purchase at $5 or less.

With this $20 investment, you should have enough equipment to create your own digital electronics lab. Of course there is more equipment that you can (and probably should) buy, but this modest investment will get you through the first five or so chapters of the book. I suggest that you read through this chapter and before continuing, purchase the necessary parts, build the power supply, and get ready to start experimenting with digital electronics!

Electronic Components

Before going on to start building some circuits, which I call *projects* in this book, including assembling the printed circuit boards (PCBs) that are included in the book, I want to first introduce you to the components that you are working with. In the previous chapter, I introduced you to how these components work. Here I want to introduce you to their physical characteristics before you start to work with them in circuits.

The silicon "chips" that are used in this book all come in black plastic (epoxy actually) packages that have two rows of metal pins coming out of them. These packages are given the name DIP for *dual in-line package,* and are normally 0.300 or 0.600 in wide. Figure 3.1 shows what these components actually look like, while Fig. 3.2 will give you an idea of what features you should look for on the chips.

When you are wiring a silicon chip into a circuit, the two most important things you will have to look for are the part number and the Pin 1 indicator. The chip part number (usually starting in 74LS for the projects in this book) references a TTL chip with a specific digital logic function. This part number *must* be matched with the one that is given in the text, or the circuit will not work properly or not work at all. Normally the chip's part number is printed on the top of the chip, but sometimes it is burned in with a laser and is very hard to read. You may want to have a magnifying glass available to help you read the part number imprinted on the chip.

Figure 3.1 74LS. TTL NAND logic chips.

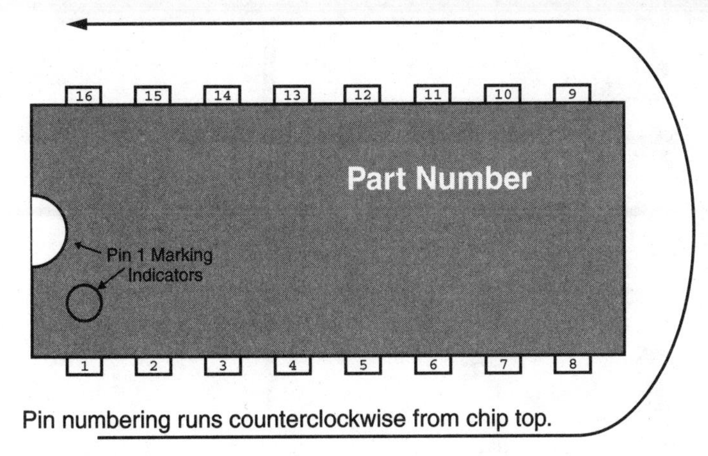

Figure 3.2 Different features of a DIP chip package.

The different functions of the chip are available only if power is supplied to it correctly and it is wired into the circuit correctly. To do this, the different pins are numbered as I have shown in Fig. 3.2, and must follow *exactly* the connections given in the schematic and wiring diagrams in the book.

The counterclockwise pin wiring convention is a carryover from vacuum-tube pin numbering, where the pins were arranged circularly around a common axis. DIP and SMT parts follow this convention, and you will become very comfortable with it after working with electronic components for a short while.

The most basic type of semiconductor component is the diode, and it looks like the component shown in Fig. 3.3. The purpose of a diode is to allow current to pass in only one direction (this process is also known as *rectifying*). The anode band around the device corresponds to the direction current flows through the device.

The most basic passive (or non-power-requiring component) that you will have to work with is the resistor, and I have shown a diagram of it in Fig. 3.4. When I specify resistors in the project circuits in this book, I will indicate what their resistance in ohms is. The bands printed on the side of the resistor indicate what the resistance of the component is.

The first three bands indicate the value of the resistor. The colors of the first two bands indicate the value, while the color of the third indicates the multiplier the value of the first two bands is raised to. The fourth band on the resistor indicates the tolerance the resistor has been manufactured to. The chart below lists the different colors and their meanings for the four bands of the resistor.

Color	Band 1 (1st digit)	Band 2 (2nd digit)	Band 3 (multiplier)	Band 4 (tolerance)
Black	N/A	0	1	
Brown	1	1	10	1%
Red	2	2	100	2%

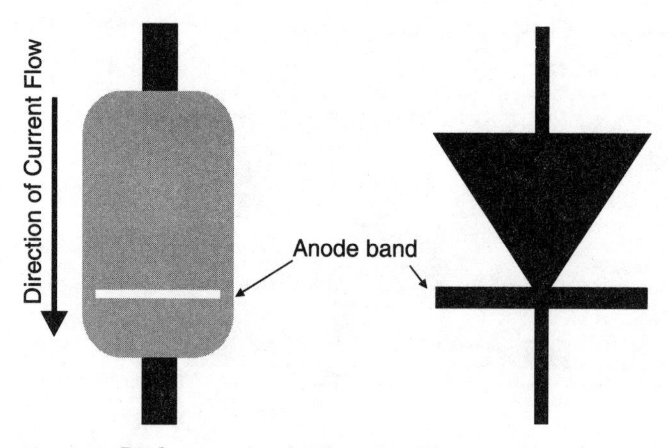

Figure 3.3 Diode appearance and markings.

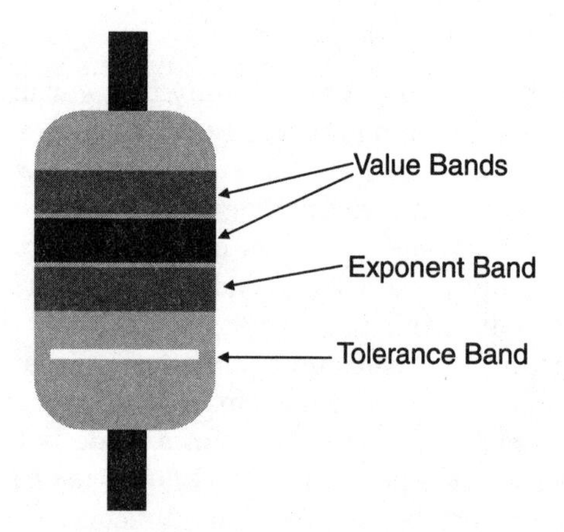

Figure 3.4 Resistor appearance and marking.

Color	Band 1 (1st digit)	Band 2 (2nd digit)	Band 3 (multiplier)	Band 4 (tolerance)
Orange	3	3	1,000	
Yellow	4	4	10,000	
Green	5	5	100,000	
Blue	6	6	1,000,000	
Violet	7	7	10,000,000	
Grey	8	8	100,000,000	
White	9	9	1,000,000,000	
Gold				5%
Silver				10%

For example, if you had a resistor with the bands colored yellow, violet, red, and gold, you would have a 4700-Ω (4.7k) resistor that has been built to 5 percent tolerance. You will find some other resistors that have five bands on them, or the tolerance band seems to use colors for tolerance values that are not listed above. In these cases, you will have to consult with the resistor manufacturer to find out exactly what values they are.

I really don't know of very many technicians or electrical engineers who "read" the color bands. With the advent of digital multimeters that can tell the resistance of a given component within seconds and do not require the chart above (or an obscene mnemonic to remember it by), reading resistors has become a lost skill. I have really included the chart above only for your edification and reference.

One type of resistor that you should be aware of is the single in-line package (SIP), which combines a number of resistors together in one package (shown in Fig. 3.5). This device can really simplify the wiring of an application in a number of different situations and you will see quite a few of the projects presented in this book that use SIPs to eliminate the need for wiring multiple resistors in an application.

SIPs are available in a number of different configurations (the component shown in Fig. 3.5 has a common resistor terminal) and number of pins. When you are buying SIP devices, I recommend that you confirm the internal wiring of the device because you may find that it is incompatible with the circuit that I have given you.

The single common resistor terminal is the only type of SIP that is specified in the book. Personally, I like to use a 10-pin SIP, which has nine resistors built into it as I have shown in Fig. 3.5. This is the largest size of SIP that can be bought easily. Larger sizes may be available, but they will probably be designed for specialized applications. Smaller SIPs are available, but I find that they are not as useful as the 10-pin part.

In the various applications, you will be using a number of different types of capacitors. In Fig. 3.6, I have shown the three different types that I will specify in the different circuits in this book.

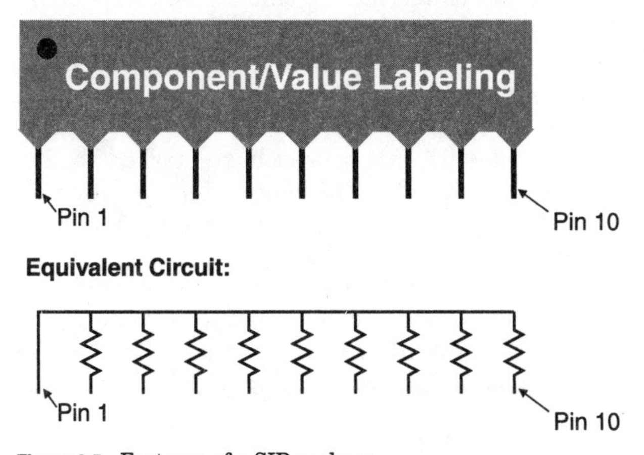

Figure 3.5 Features of a SIP package.

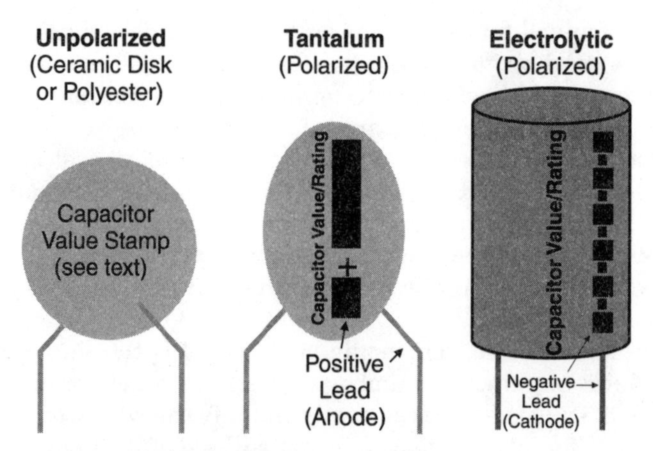

Figure 3.6 Capacitor appearance and markings.

While being very simple, capacitors can be the least standardized parts that you will have to work with when it comes to markings and ratings. The biggest culprits of this problem are the simple, unpolarized *ceramic disk* and *polyester* capacitors; these devices are often just simple two-leaded devices with either numbers stamped on them or colored bands like resistors. These markings are not standardized in any way and for some manufacturers, the printed values are based on picofarads (pF), while other manufacturers base the values on nanofarads (nF). To avoid confusion with these parts, I generally buy and keep them separately until I get home, and when I do get home, I place them in labeled plastic tubs, to avoid confusing them with different parts. I should point out that, depending on the technology used to manufacture a capacitor, smaller-value capacitors could actually be larger in size than larger-value capacitors. This means that you cannot depend on the size of a capacitor as an indicator of its value.

Polarized capacitors generally have their value (along with their maximum voltage rating) and units printed on the side of the component, which eliminates the confusion of the nonpolarized parts. There is one thing that you should watch for, and that is that tantalum capacitors have the positive (anode) lead marked while electrolytic capacitors have the negative (cathode) lead marked.

On the PCBs that come with this book, I have marked the anode of all the polarized capacitors that are used on the boards. I am mentioning this because the marking of the electrolytic capacitor used in the +5-V power supply will be facing *away* from the + symbol on the PCB.

Along with the components that I have presented, I work with LEDs, connectors, transistors, crystals, and ceramic resonators. Each one of these components is built slightly differently or can have many different permutations (a great example of this is connectors). Trying to identify the physical characteristics of each one is close to impossible and could take up a major fraction of this book.

Instead of trying to show you what every possible component looks like, I will just rely on this simple introduction and ask you to check the markings on the PCBs that come with the book; the actual orientations of the components are *silk-screened* on them. Secondly, check the text for any special instructions on using the components and note that in the wiring diagrams, I will show how the components are inserted and oriented in the breadboard. You can find more detailed information on different components in manufacturers' datasheets that are available on the Internet, or you can get information in distributors' catalogs. You can also ask an expert (but please remember that it may take me a day or so to get back to you via e-mail, so look for somebody closer that you can rely on).

Breadboard Prototyping System

Probably the easiest way to create test applications is to use a breadboard. This product consists of spring-loaded holes, which will grip a component and provide connections to other holes so that circuits can be built out of connected components (Fig. 3.7). In this book, I have designed the experiments and many of the products to use breadboards for simple build-up and tear-down

Figure 3.7 Breadboard without any components.

of digital logic circuits. The breadboard consists of two sets of connected pins for DIP chip packages. Each row of pins, along with two columns across the top and bottom, is interconnected as is shown in Fig. 3.8.

To make power distribution easier, most breadboards are shipped with *bus bars* across the top and bottom. These rails connect by plastic tabs and receptacles and lock the parts together. Multiple rails and breadboards can be connected together to create different-size prototype assembly areas.

The rails are available with one or two common strips. I normally use the ones with two rows and place V_{CC} and Gnd on both sides of the breadboard. These rails may be marked with red and black (or blue) stripes to indicate that the pins are common across them. Figure 3.9 shows how this is done.

If you look at the PCB for the +5-V power supply circuit that is included in the book as I describe how to build in the next section, you'll see that I've

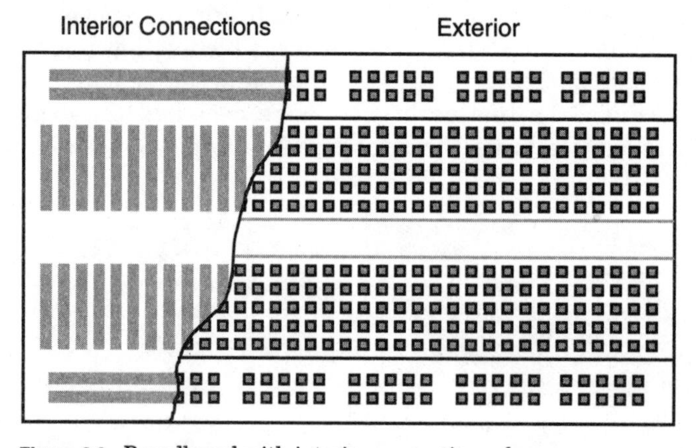

Figure 3.8 Breadboard with interior connections shown.

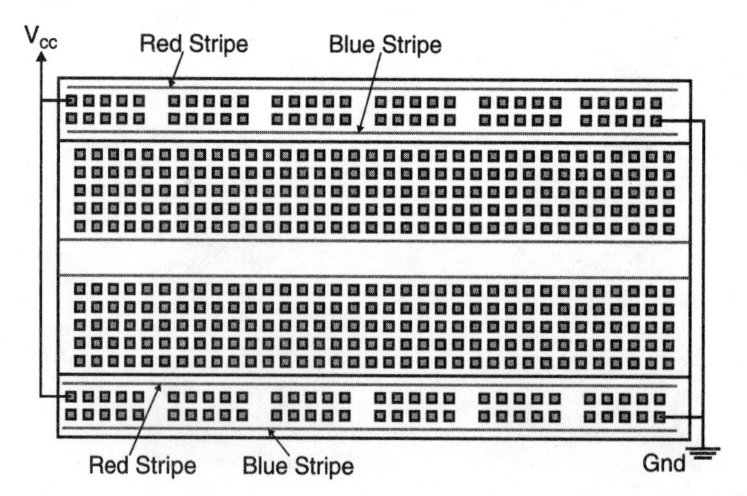

Figure 3.9 Breadboard power connection.

included two rows of 0.100-in connectors. This allows the power supply to be connected to the breadboard common strips without any wiring to provide power to the breadboard and application.

Wiring circuits is accomplished by inserting component pins into different positions in the breadboard and then using short strips of wire to interconnect the parts. Making circuits by this technique is fast, but not as reliable as other forms of "prototyping." Breadboard circuits should never be used permanently; instead the circuit should be re-created on another form of prototyping circuit or a PCB.

Care should also be taken in selecting where different components are placed to ensure that there is enough space on the breadboard for the circuit and the wiring doesn't get so complex that you cannot follow a wire to try and understand what a problem is. For most of the projects, I will give you both a schematic diagram and a suggested breadboard wiring diagram for the circuit.

This messiness can be alleviated somewhat by planning the layout of your application to keep the number of wires in an area to a minimum. By choosing different-color wires, signals can be traced more easily as well.

Of the different project prototyping methods I present here, breadboarding is the easiest to create circuits on. It is also the method that can have the greatest problems with noise, parasitic capacitances, and *crosstalk.* This is because of the large connections in close proximity to each other. For this reason, I have kept the circuit speeds for the various projects built on a breadboard to a minimum (never more than 2 MHz). I have run circuits on breadboards at higher speeds, but 2 to 4 MHz is a good clock speed in which the signals will be just about perfect.

Project 1—Breadboard Power Supply

One of the most useful circuits you can build for yourself if you want to work with digital electronic circuits is a simple 5-V power supply. To make things very easy for you, I have provided the PCB for the power supply in this book. I designed the first version of this circuit and PCB about a year ago, and since then it has undergone two revisions. The version that comes with the book (Fig. 3.10) is the final design and one that I believe will be very useful for you beyond using it for the projects presented in this book.

The power supply can be powered either from a "wall wart" power supply or a 9-V alkaline radio battery and provides up to 1 amp at +5 V. The circuit itself is very simple and is shown in Fig. 3.11.

The bill of materials (parts list) for this circuit is:

Part	Description
7805	7805 5-V regulator
1N4001	1N4001 silicon diode
LED	0.1-in, 5-mm red LED
10 μF	10 μF, 35-V electrolytic capacitor
0.1 μF	0.1 μF, 16-V tantalum capacitor

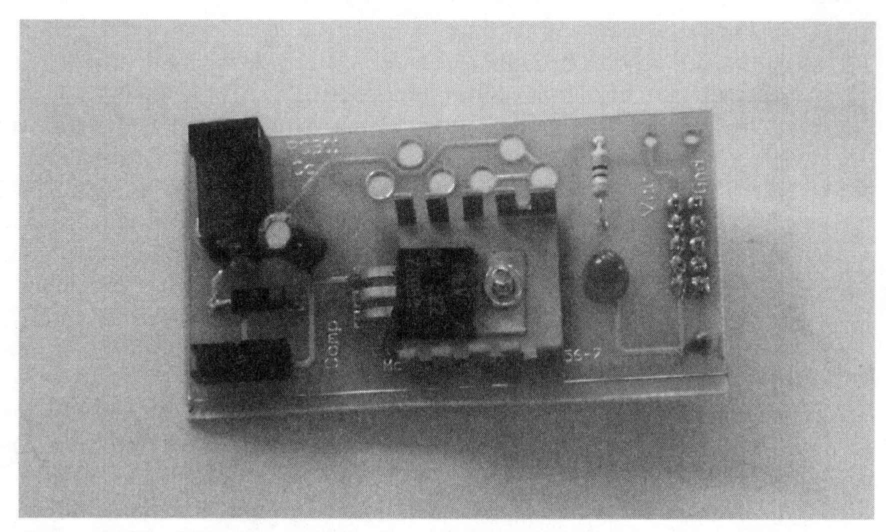

Figure 3.10 The 5-V power supply for the projects in this book.

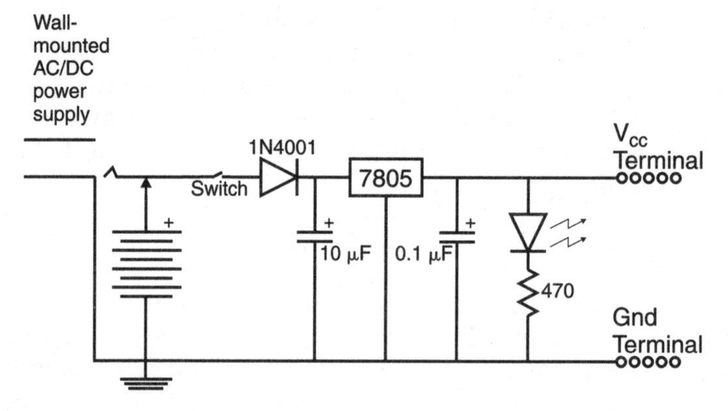

Figure 3.11 Breadboard power adapter.

Part	Description
470	470 Ω, ¼-W resistor
Power input	2.5-mm power plug
Battery	9-V battery PCB mount connectors, Digi-Key part numbers BSPCF-ND and BSPCM-ND
Terminals	Two-wire screw terminals and 2 × 5 IDC male connectors
Miscellaneous	PCB (supplied with book), 8+ V, 300+ mA AC/DC wall-mounted power supply with 2.5-mm power plug, 7805 heat sink with screw and nut
Tools	Needle-nose pliers, wire clippers, flathead screwdriver, soldering iron, solder, digital multimeter (optional)

In this circuit, I use a center pole power connector, which has the battery ground interrupted when a power plug is inserted into the connector. The power plug is used to prevent shorting and arcing when the plug is inserted or withdrawn. Disconnecting the ground (negative) connection of the battery may seem unnatural (especially when you see the theory behind its use later in the book), but it serves the same purpose as disconnecting the battery's positive connection.

The unregulated power sources can be a home electronics 8- to 15-V wall-mounted transformer that can be bought new for just a few dollars, or, chances are, you have more than a few lying around your home. These power supplies are used for personal music machines, calculators, PDAs, external modems, and so on.

If just this power supply circuit is going to use the wall transformer adapter, as little as 8-V output is required. Note that a DC output is required with up to 1 amp of current if the full capabilities of the 7805 voltage regulator are to be realized.

The higher the input voltage, the greater the voltage drop across the 7805 regulator and the more power (heat) that will have to be dissipated. While I have added a heat sink on the bottom of the PCB design, you may want to put an external heatsink on the 7805 to reduce its temperature and the opportunity for thermal shutdown.

You may be considering building your own transformer/rectifier circuit instead of buying a wall wart, and I want to discourage you from doing that. I am personally amazed at how cheaply you can buy a commercial wall transformer. At a local (Toronto) surplus store, I can buy a new 0.3-amp, 9-V power supply for $3 Canadian ($2 US). The "Cadillac" AC/DC converter with multiple voltages and output plug options is $10. When I can buy the parts this cheaply, it doesn't make sense for me to build one myself. Using a safety certified commercial unit will save you time and money and avoid any shock hazard that a circuit you build yourself might present.

Building this circuit on the PCB is very simple and will take you as little as 15 minutes once you have all the parts together. Soldering the parts onto the board is quite simple, with only two "tricks" to watch out for.

The first trick is that the 9-V battery connectors are polarized with the wider (bell-shaped) connector on the side away from the 2.5-mm power connector. If you reverse these two connectors, the battery will be providing reverse-bias current to the circuit and the diode (CR41) will block it.

The second trick is to remember that the 5 × 2 IDC connector is mounted on the bottom (solder) side of the board. Depending on how you kit your parts and work through the assembly, it is easy to put this connector on the same side as the rest of the components.

The PCB layouts are shown in Fig. 3.12.

To assemble the breadboard power supply, I recommend that you work your way through the following procedures that will help you to "debug" your power supply as you build it. If you are unfamiliar with soldering, I recommend that you jump to App. C and review my comments on soldering and also find someone you can consult with for soldering the board.

Note: In the instructions, as you build the power supply, I am going to ask you to plug the partially completed PCB into the AC/DC power converter to

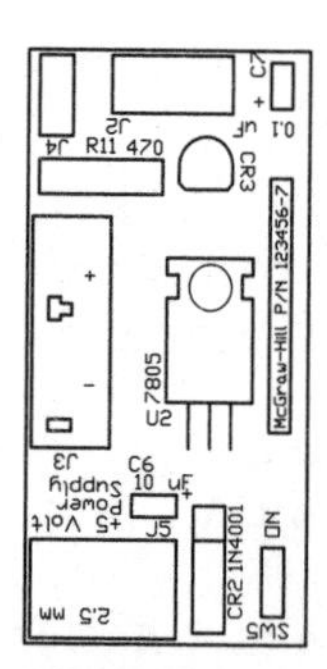

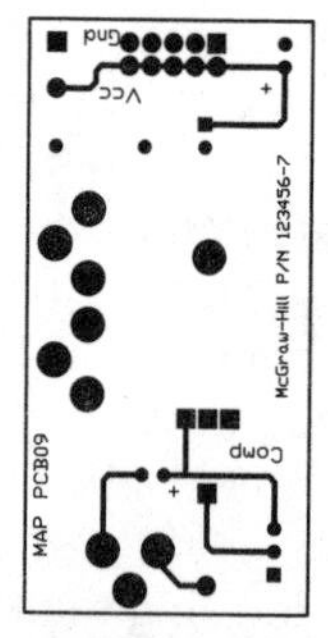

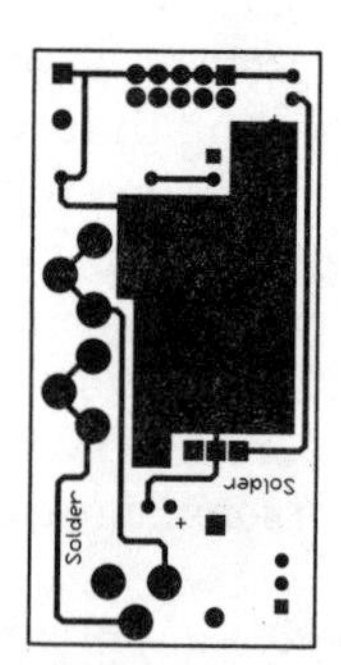

POWER Overlay POWER Top Layer POWER Bottom Layer

Figure 3.12 PCB design for 5-volt power supply.

check its operation. After each step, make sure that the AC/DC power converter is unplugged from the PCB before soldering in any new components. Making sure no power is being applied to the circuit is a safety precaution to make sure there is no path between the main power source and you.

1. Plan on building this board on a weekend afternoon when you have nothing else planned. This is obviously not a hard board to build, but any time you are going to be soldering, make sure you are well rested and you have planned out what you are going to do without feeling rushed. If something comes up that requires your attention, stop work on the power supply and return to it when the problem is resolved—mistakes are made when you feel like you have to get something done in just a few seconds.
2. Get all the parts listed in the bill of material above and place them on your workbench along with your soldering iron and solder.
3. Read through these instructions and make sure you understand each step. Do not proceed if you have any questions.
4. Solder in the 2.5-mm power socket (J3) and 1N4001 diode (CR41). Note that the band on the diode faces the center of the card. Plug in the AC/DC power converter and plug it into J3. Using the digital multimeter set to a 20-V scale, with the red lead touching the lead closest to the band on CR41 and the black lead touching the center pin of the 7805 (U41), you should read 8 V or more. If you don't read 8 V or more, check the AC/DC power converter to ensure that it is plugged in and that it is driving voltage out. Next, check the polarity of the 1N4001 diode (CR41).
5. Next, solder in the two 9-V battery connectors to the J4 connectors. As noted above, make sure that the wide, or bell, connector is placed away from the J3 2.5-mm power connector. When you have finished, connect a 9-V alkaline battery to the connectors and measure the voltage at the same points as in the previous step. This step is optional and can be done later, if you don't feel that you will be using 9-V battery power with your power supply.

6. Solder in the SPDT power switch (SW1) and the 10-μF electrolytic capacitor (C5). Note that the side marked with the stripe with the – character in it should point away from the + character on the PCB. The stripe with the – character should face the side of the PCB where the AC/DC power plug is to be inserted. Check the operation of the switch by connecting the AC/DC power supply to the PCB and touching the red lead of the DMM (which should still be set to 20 V) to the inside lead of the 7805 (U41) and the black lead to the center lead of 7805 (U41) while opening and closing the switch. As you close and open the switch, you should see the power level change from zero on the digital multimeter to 9 V or more.
7. The 7805 (U41) is to be installed next. If you are going to use a metal heat sink with the 7805 voltage regulator, place this underneath the 7805 and screw both down before soldering the pins. After the 7805 is soldered in, connect the AC/DC power converter and check the voltage at the 0.1 μF capacitor position (C6) using the digital multimeter with the red lead touching the hole marked +. Anywhere from 4.9 to 5.1 V should appear on the digital multimeter. You should also observe the voltage changing when the switch (SW1) is opened and closed.
8. Next, solder in the 470-Ω resistor (R7), the LED (LED1), and the 0.1-μF capacitor (C6). When soldering in the LED, note that the flat side of the device must match the flat side of the symbol on the PCB. The pin marked + on the 0.1-μF capacitor must be put into the hole marked + on the PCB. To test these parts, connect the AC/DC power converter and open and close the switch. The LED should turn on when the switch is closed.
9. Solder in the two-screw terminal connector (J5). This connector should be mounted with the wire openings to the outside of the PCB. Soldering this component onto the PCB is optional and can be done at any time.
10. As the final step, solder in the 2 × 5 IDC connector, but with the pins pushed into the bottom (solder) side of the PCB and soldered from the top (Comp or component side).

That's it! If you have followed the 10 instructions above, you will have a working breadboard power supply that you will be able to use for the projects that follow in this book.

When the power supply is operational, you will find the 7805 (J1) will get quite hot. This is especially true when you add the switches and LEDs in Project 2. Your will understand the reason for the high temperature when I explain more about Ohm's law and how power is dissipated in a circuit, but the bottom line is, the 7805 will get very warm and could get so warm (hot) that you can be burned if you touch it. This production of heat is why I specify that a heat sink must be used with the 7805.

I do not have extra power supply PCBs ready to send out to people who have ruined theirs. Please make sure that you follow the instructions given above and are careful as you work through the building of the circuit.

Project 2—Interface PCB

Along with the +5-V power supply PCB included with this book, I have included a simple interface PCB that is designed to be used with the initial projects presented in this book. The circuit (Fig. 3.13) probably appears to be quite complex, but I think you will find that you can comfortably assemble it (solder parts onto it) in no more than an hour—even if you are a complete beginner.

The interface board provides the following functions that can be used to learn about digital electronic circuits:

- Eight switch-controlled LED input/output (I/O) pins that are designed to work with 74LS logic
- Two button-controlled LED I/O pins that are designed to work with 74LS logic
- 2-MHz ceramic resonator–based oscillator
- 1-kHz relaxation oscillator
- One delayed I/O that can be used for application reset/debounced button input

The actual layout of the PCB is shown in Fig. 3.14. I will be explaining the operation of the circuits on this board and the theory behind their operation.

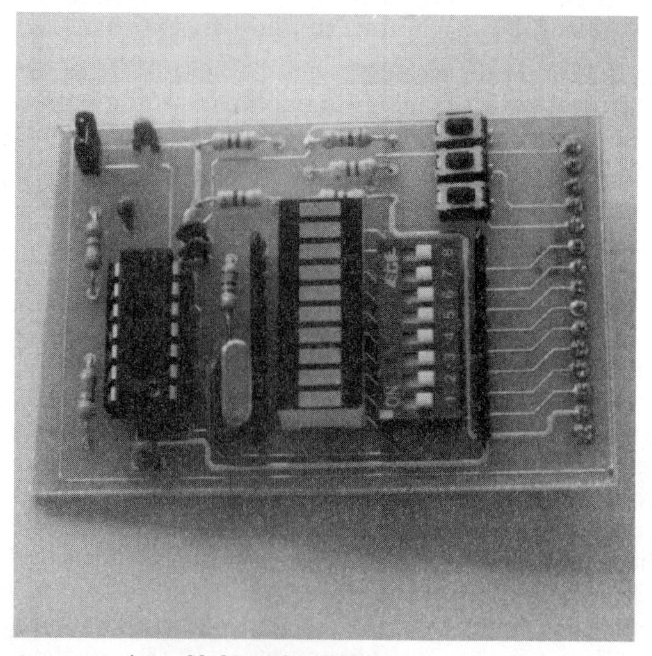

Figure 3.13 Assembled interface PCB.

In this section, I will explain how the PCB should be soldered together step by step and include simple tests that you can make to ensure that there aren't any problems with the work that you have done.

The bill of materials for this PCB is:

Part	Description
U1	74HC04 CMOS hex inverter
CR1	10-LED bargraph display
Y1	2-MHz ceramic resonator with internal capacitors (Digi-Key part number PX200MC-ND; Panasonic part number EFO-MC2004A4)
R1–R2	470-Ω, 8/9 resistor SIP
R3–R6	470-Ω, ¼-W resistors
R7	1 MΩ, ¼-W resistor
R8	47k, ¼-W resistor
R9, R12	4.7k, ¼-W resistors
R10	10k, ¼-W resistor
C3–C5	0.1-µF, 16-V tantalum capacitors
J1	15 × 1 pin strip
SW1	8-pin DIP switch
SW2–SW4	PCB mount momentary-on push button switches

Note that the component designators that I have used for this PCB don't start at 1 (such as C1) on this circuit. To simplify the book's manufacture by having one PCB placed in it rather than two PCBs, I have put them together, just to be snapped apart when you are ready to build the cards. When they are placed together, they are assumed to be one card, and component designators cannot be shared.

1. The first step in assembly is to get all the components together. I recommend this for all the projects presented in this book and that is why I have

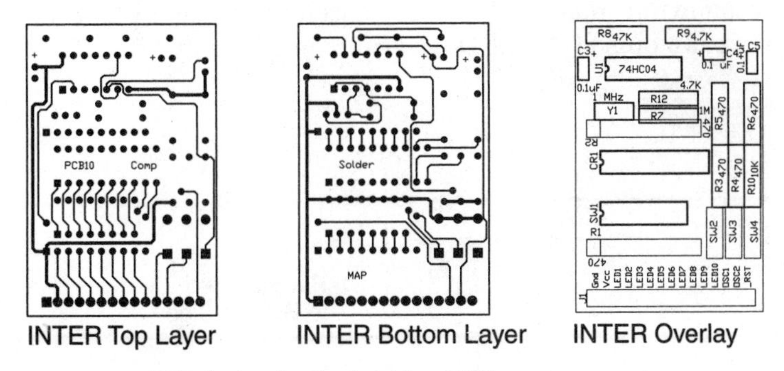

Figure 3.14 PCB design for the interface PCB.

created the next section, with one large bill of materials for all the parts needed for the first 15 projects shown in this book.

2. Next, solder in the 15-pin strip. This strip and how it is assembled on the PCB are shown in Fig. 3.15. The 15 pins on this strip will be used to plug the board into a breadboard for prototyping. Whereas for the +5-V supply required a 5 × 2 strip, for this PCB you will require a 15 × 1 strip, which can be cut down from a 36-pin (or post) 3M 2.54-mm "straight, single-row male" strip (Digi-Key part number 929834-02-36-ND). You can also buy this part from other suppliers.

   Make sure, when you buy this strip, that the pins are 0.100 in (2.54 mm) between centers. There are similar products, but with different pin spacing, which cannot be used with this PCB or the breadboard.

   The Berg strip should be soldered to the back (Solder) side of the interface PCB, with soldering taking place on the top (Comp) side as shown in Fig. 3.15.

   Make sure that the Berg strip is perpendicular to the interface PCB. The actual holes in the PCB were intentionally made quite large to ensure that you could easily fit in the posts. To make sure the pins are perpendicular, I recommend just soldering one pin first and then, by remelting the solder, get the pin properly lined up. When this has been done, you can solder in the remaining pins. You might want to go back to the original pin and touch it up—by moving the part while the solder is hardening, you might end up with a "cold solder joint".

3. Solder in the two 470-Ω SIP resistor strips (R_x and R_x), the 10-LED bargraph display (CR1) and the eight-switch DIP (SW2). Note that the common pin of the resistors are marked and should be put into the square pads of the interface PCB. Make sure that the notch cut out of the corner of the LED bargraph display matches the notch on the PCB's silkscreen (the white graphics and lettering on the PCB).

   Note that the SIP resistors can either be 9 or 10 pins long. The PCB is drilled for a 10-pin SIP, but the last position is not wired, so either part can be used interchangeably without any changes elsewhere in how the board is assembled.

   I recommend that when you are assembling this PCB that you solder in the first two pins of the LED bargraph display only after the R1 and R2 are

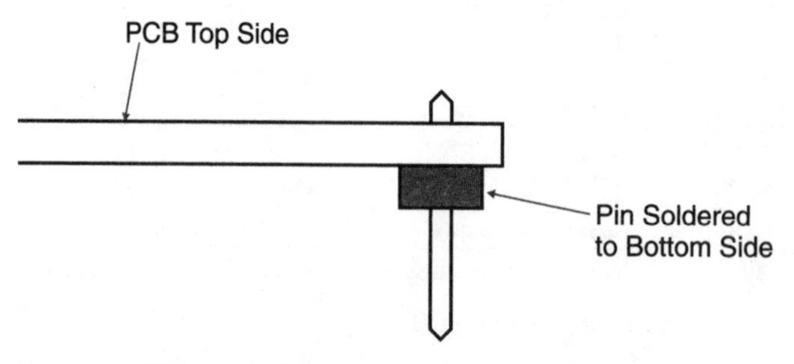

Figure 3.15 Fifteen-pin by one-strip soldered to PCB.

soldered in. When this is done, plug the board into a breadboard as I've shown in (Fig. 3.16) and turn on the +5-V power supply. If the first LED does not light, then check the orientation of the LED bargraph display or the two SIP resistors.

Note that the first two pins of the interface PCB are Gnd and V_{cc}. Please make sure these pins are connected correctly to the Gnd and V_{cc} "rails" on the breadboard. In Fig. 3.16, I have shown Gnd and V_{cc} being passed from the +5-V power supply on the top rails of the breadboard to the bottom rails and then the interface board being powered from the bottom rails.

When the LED lights without any problems, solder in the remaining pins and the eight switch DIP.

Once these components are soldered in, you can plug the board into the PCB as I have shown in Fig. 3.16 and play around with the switches controlling the LEDs.

4. Install the four 470-Ω, ¼-W resistors into R3, R4, R5 and R6) along with the two push button switches (SW3 and SW4). When you have done this, you can test out their operation by installing the interface PCB into a breadboard as was done in the previous step.

   These two push buttons are used as additional digital input/outputs. Like the other eight, these pins can be connected to 74LS inputs and outputs. If these circuits are connected to a digital input, then the switch or button (with the associated LED) will turn send a high (1) or low (0) to the input. If the switch is left open (not on) and the circuit is connected to an output, the 74LS logic will drive the LED on or off.

   The two buttons are *momentary on,* which means they are normally open, and only closed when they have been pushed.

   These 10 circuits will be used by most of the first 10 applications as input and output pins. I will explain exactly how they work in more detail later in the book.

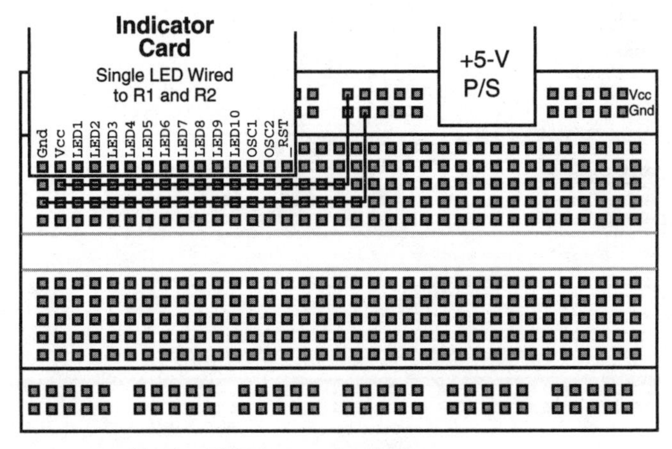

Figure 3.16 Testing LED bargraph polarity.

5. Install the third push button (SW5) along with its associated *RC* network as well as the 74HC04 and its 0.1-μF decoupling capacitor. This circuit provides two functions, a time delayed negative active reset for circuits built onto a breadboard as well as a "debounced" button input. Later in the book, I will be explaining these functions.

   To test the push button, run a wire, as shown in Fig. 3.17 from pin 15 of the Interface board to pin 3. When the push button is pressed, the LED should turn off and turn back on when it is released.

6. Solder in R8 (4.7k) and R9 (47k) along with the 0.1-μF capacitor (C_x) to the PCB. These three components, along with the 74HC04, form a 1-kHz relaxation oscillator and are known as OSC2 on the interface PCB. The simple way to test these components is to run a wire from pin 14 of the interface PCB to pin 3 as shown in Fig. 3.17.

   When pin 14 is connected to pin 3 and power is applied, you should see the first LED become somewhat dimmer than the other on LEDs on the interface PCB. You may want to turn off the adjacent LEDs to better see the change when the oscillator is running.

7. The final step is to solder in the 2-MHz ceramic resonator and the 1 MΩ and 4.7K, ¼-W resistors on the PCB. Once this is done, you can test the operation of this oscillator (OSC1 on the PCB) by connecting pin 13 of the Interface card to pin 3. As in the previous step, you will have a dimmed (but not off) LED.

With the interface PCB tested out, you are now ready to wire it onto a breadboard along with the +5-V power supply to be ready for the following projects, which demonstrate how TTL works. To do this, you will place the +5-V power supply onto a breadboard at the power supply rails. This will connect the +5-V power supply to the top V_{cc}/ground rails of the breadboard. When this has been done, put in two wires for the interface PCB V_{cc} and Gnd to the two left-most breadboard "columns" as shown in Fig. 3.18.

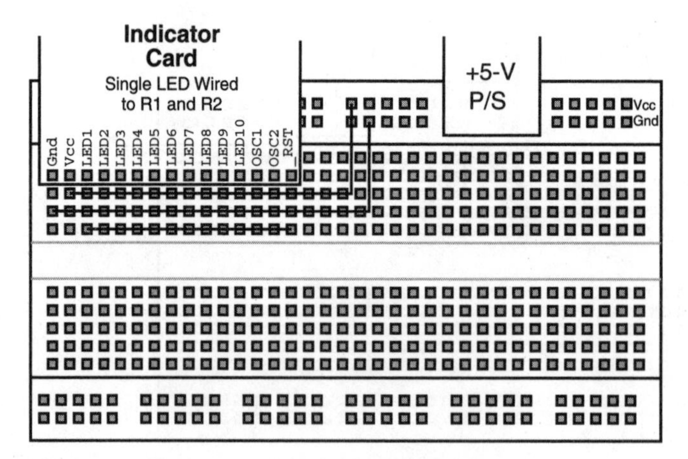

Figure 3.17 Testing _RST button operation.

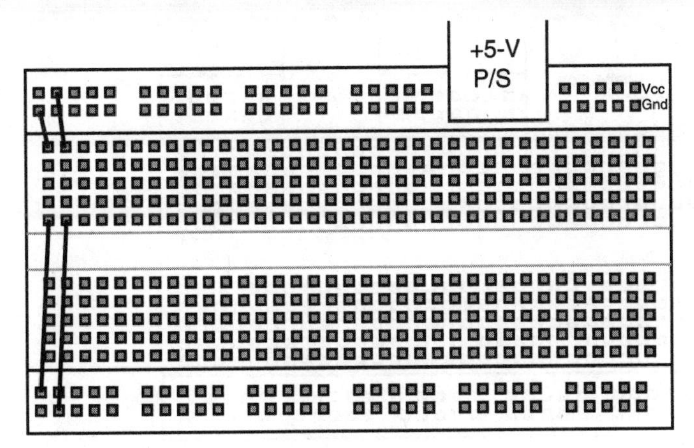

Figure 3.18 Connecting +5-V power supply.

Next, place the interface PCB with its left-most pin beside the Gnd connection made in the previous step. Once this is done, pass the V_{cc} and Gnd connections from the interface PCB connection to the bottom set of rails as shown in Fig. 3.19.

You are now ready to start wiring in the applications that I will work through, starting in the next chapter.

You will notice that more of the breadboard will be covered and obscured by the +5-V power supply PCB and interface PCB than what I show in the wiring diagrams for the different applications. This is deliberate, as there will be cases where I will show wiring to the V_{cc}/Gnd rails that will actually be under the PCBs. This is not to say that the holes in the card will be shared between the +5-V power supply PCB, just that the holes that you are to wire the chips to will be obscured by the PCB itself.

Before going on, I want to explain the theory behind the switch operation, because I'm sure that it will be confusing to you. I will discuss the oscillator circuitry on the board later in the book.

The switch input is modeled after the simple pull-up switch shown in Fig. 3.20. In this circuit, as long as the switch is open, the voltage at the input pin is at V_{cc} (there will be negligible current flow, if any, through the pull-up resistor). If the switch is closed, the voltage at the input pin will be at ground. With the input pin driven to V_{cc}, the digital electronic device will read a 1, and it will read a 0 when the input pin is pulled to ground.

This circuit works quite well, but it does not give you any feedback as to the state of the pin. Ideally, there should be an LED connected to the pin to make sure that you are aware of the logic level of the pin. One way of adding an LED is to wire the circuit shown in Fig. 3.20 to a TTL "buffer" and let it drive the LED. The problem with this method is the extra real estate required for the TTL buffer, along with the extra wiring it involves, and the added cost and power requirements.

The solution that I have found works best for me is to apply the basic electronics knowledge that was discussed in the previous chapter and use the

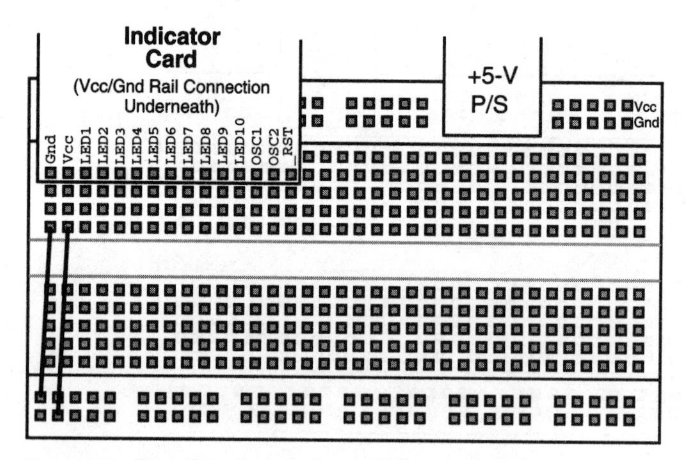

Figure 3.19 Breadboard ready for project wiring.

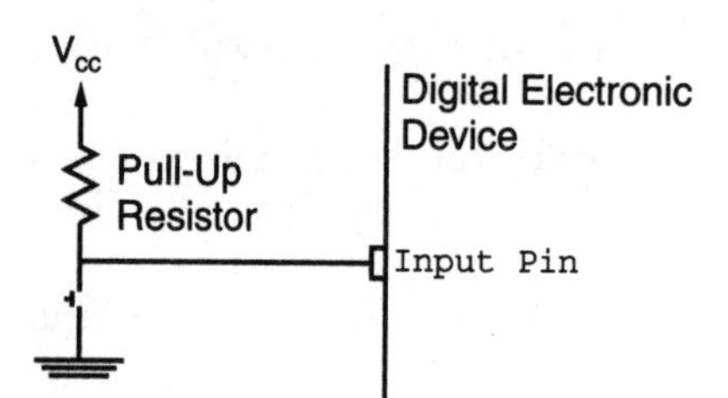

Figure 3.20 Pull-up switch.

switch to control the state of an LED without an additional buffer. The circuit that I use is shown in Fig. 3.21.

This circuit uses the same pull-up switch circuit as in Fig. 3.20, but it is used not only to control the voltage level being passed to the TTL input pin, but also to control the passage of current through the LED. When the switch is open, the current supplied by the power supply at V_{cc} can pass only through the LED (CR1). Along with the diode, a resistor is placed in series to limit the current passing through the LED.

This circuit works on the assumption that current flows through the path of least resistance. When the switch is open, the low resistance path for the current is through the LED. When the switch is closed, the switch provides a zero-ohm path to ground from the R1 current-limiting resistor—when the switch is closed, there is no current available for the LED (CR1), and it doesn't light.

Looking at this circuit, you will probably be wondering about the voltage level's output to the TTL input pins. Going back to the previous chapter, I can apply Kirchhoff's law to calculate the voltages when the switch is open (1 being driven) and when it is closed (0 being driven).

In the first case, the switch open, current flows through the two resistors and the LED. As I will discuss later in the book, LEDs are a special type of diode—they are not manufactured with P- and N-type silicon like a regular diode.

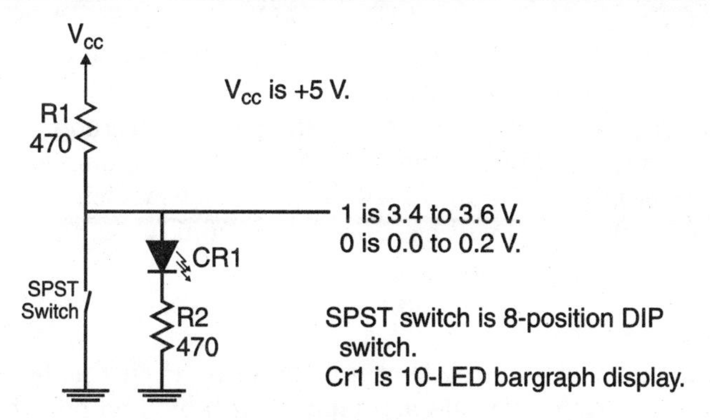

Figure 3.21 Switch input circuit with LED.

Instead, different materials are used so the photons released by the diode are in the visible light spectrum. This means that instead of the 0.7 V of a standard silicon diode, there is about 2.0 V (rule of thumb) across the LED when current is passing through it.

Since I know that I am applying 5 V to the two resistors and the LED, I can use Kirchhoff's law to divvy up the voltages across these three devices:

$$V_{cc} = V_{R1} + V_{R2} + V_{CR1}$$

Since we know V_{cc} (5.0 V) and V_{CR1} (2.0 V), we can state the voltage across the two resistors as:

$$5\text{ V} = V_{R1} + V_{R2} + 2\text{ V}$$

$$V_{R1} + V_{R2} = 3\text{ V}$$

Because R1 and R2 are the same value, the same voltage drop will be across each one of them:

$$2V_R = V_{R1} + V_{R2}$$

$$= 3\text{ V}$$

$$V_R = 1.5\text{ V}$$

So, in this circuit, the voltage drop across R1 and R2 is 1.5 V.

When the switch is open, the voltage applied to the input pin is the voltage across R2 and CR1. Since we know these, we can express the switch open voltage as:

$$V_{open} = V_{CR1} + V_{R2}$$

$$= 2.0\text{ V} + 1.5\text{ V}$$

$$= 3.5\text{ V}$$

3.5 V is greater than the nominal “threshold” voltage of a TTL input (it is usually around 1.4 V) so the 1 output by this circuit will be sensed as a valid voltage.

When the switch is closed, the current will bypass CR1 and R2 and will flow through the zero-ohm resistor at the SPST Switch. A zero-ohm load has a zero-volt drop across it and the input pin will be pulled low, which is below the TTL input threshold voltage and will be sensed as a valid 0.

Project Kit Parts List

Before going ahead and starting through the projects in this book, I suggest that you make a photocopy of the accompanying table and buy the parts listed here. The parts list will provide you with all the parts needed for the next 15 projects and will avoid the need for you to run out and buy just one more part. When you have this kit together, I think you will find that it will be an invaluable kit for creating your own applications.

When you look at the projects, you'll wonder if the 74LS193 4-bit counter is my favorite part. The 74LS193 is a very handy part for different applications because of its flexibility. The same goes for the 74LS74, 74LS138, 74LS139, 74LS174, 74LS244, 74LS245, and 74LS374—all these parts offer some significant capabilities for working with digital electronics that are not immediately obvious. Hopefully, you will see some of the things that can be done with these parts in this book that will seem to be a result of “out-of-the-box” thinking.

I have not included the parts for the last two projects (Hexadecimal Bus Interface and TTL Chip Computer) because their complexity puts them in a different league from the other projects presented here. These two projects will need much larger breadboards (if, indeed they are used at all instead of wire wrapping or other prototype circuits) and will take several hours to build.

The initial projects (whose parts are listed here) are designed to be implemented in just a few minutes on a standard “single short” (35-row) or “single-long” (65 row) breadboard with the +5-V power supply and interface PCB boards that come with the book.

Quantity	Part	Comments	Have/Notes
1	+5-V power supply PCB	Included with book (Project 1)	
1	Interface PCB	Included with book (Project 2)	
1	Short, 35-row breadboard		
1	Long, 65-row breadboard		
1	Large, 3-set minimum breadboard	For experimentation and follow-on projects/ your own applications	

Quantity	Part	Comments	Have/Notes
3	Breadboard wire kits	See text for information	
1	500-ft (150-m) Spool of 22-gauge solid-core wire		
1	4060B	CMOS oscillator/counter	
1	74HC04	CMOS hex inverter	
2	74LS00	Quad two-input NAND gate	
1	74LS02	Quad two-input NOR gate	
1	74LS04	Hex inverter	
1	74LS08	Quad two-input AND gate	
1	74LS32	Quad two-input OR gate	
1	74LS74	Dual *D* flip-flop	
3	74LS85	4-bit comparator	
1	74LS86	Quad two-input XOR gate	
1	7LS125	Quad Tristate buffers	
1	74LS139	3-to-8 decoder	
1	74LS139	Dual 2-to-4 decoder	
1	74LS174	Hex *D* flip-flop	
7	74LS193	TTL 4-bit binary counter	
1	74LS283	4-bit adder with fast carry	
2	74LS374	Octal D flip-flop with tristate outputs	
1	NE555	Bipolar (or CMOS) oscillator chip	
4	2N3904 NPN transistors		
2	1N914 silicon diodes		

Quantity	Part	Comments	Have/Notes
3	10-LED bargraph display	Same part used in interface PCB	
1	32.768-kHz crystal	Watch Crystal	
1	120-Ω, ¼-W resistor		
20	470-Ω, ¼-W resistor		
1	1k, ¼-W resistor		
1	1.5k, ¼-W resistor		
1	3.9k, ¼-W resistor		
2	10k, ¼-W resistor		
1	100k, ¼-W resistor		
1	470k, ¼-W resistor		
2	4.7MΩ, ¼-W resistors		
1	10MΩ, ¼-W resistor		
3	470 Ω, 8/9 resistor SIP		
1	10k, single-turn potentiometer	PCB board mount recommended to be wired onto breadboard	
50	0.1-µF, 16-V tantalum capacitors		
1	1.0-µF, 16-V electrolytic capacitor		
1	100-pF capacitor (any type)		
1	15-pF capacitor (any type)		
1	Momentary-on SPST push button	Solder wire to it so it can be used with breadboard	

In the table above, notice that I have included a space for you to place a check mark when you have the part or to write a note to yourself about where to find it.

All these parts are available at most electronics stores. In selecting the parts for the various projects, I have tried to be as conservative as possible and have chosen parts that are as widely available as possible. If you have any problems finding any of these parts, please contact the part supply houses listed in Appendix D. If none of the suppliers can furnish a specific part, please contact me or look at my web page to see if I have any information about them.

For the resistors, I recommend buying a bulk pack containing resistors of various values. Most supply houses have these for just a few dollars, and it's nice to have a lot of different-value resistors; you never know when you will need a specific part. This is not as true for capacitors, where I have tried to keep the number of different types to a minimum. The reason for minimizing the number of capacitors is their lack of consistent labeling. By just working with four or five different values, you will simplify your life immensely.

It is probably surprising that I specified that you buy three breadboard wire kits. These kits consist of a number of precut, prestripped wires that will be used to wire your circuits. When you first look at the kits, you may feel that one is sufficient, but as you work through the projects, you will find that wires will break, you will have to twist them in order to form the circuit, or you will deform them to attach probes to them. In addition, there are a few projects that will require more wires than are available in one kit. That is why I suggest that you have additional kits available as well as a spool of wire and some wire strippers for adding your own wires.

The number of 0.1-μF tantalum capacitors seems large, but you will go through them, lose them, and end up using them for many different purposes. In the different projects, I have tried to use just 0.1-μF devices anywhere capacitors are required to minimize the number of different parts that you have to keep on hand.

When you have all these parts, I suggest you store them in some kind of labeled container. The bipolar technology TTL digital logic parts that you will (mostly) be using is quite insensitive to electrostatic discharge (ESD) so you can save these parts in a large multidrawered bin you can buy at a hardware store. This is shown in Fig. 3.22.

For absolute safety, I recommend using the "ESD-safe" chip tubes that the chips originally came in. Most electronic stores end up throwing away these tubes, so you should be able to get some simply by asking for them. I try to keep one part number to a tube, and I have them clearly labeled on the outside so I can find parts quickly as can be seen in Fig. 3.22.

Please note that if you live in the Toronto, Canada, area, I get first dibs on all spare chip tubes. Sorry, but that's the way it is.

Store these parts neatly as soon as you get them; then, with the +5-V power supply and interface PCBs, you will have a pretty good TTL development kit for your own use.

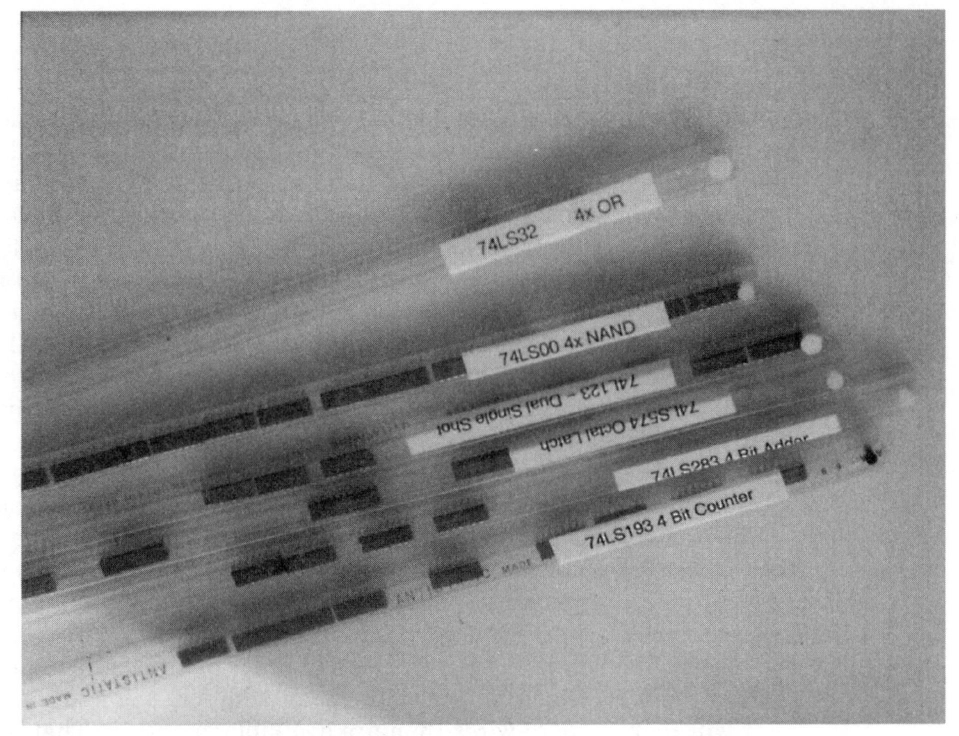

Figure 3.22 Chips stored in tubes for easy access and locating.

In terms of hand tools, to carry out these projects, you should have:

- Wire clippers
- Wire strippers (adjustable from 14 gauge to 30 gauge)
- Small needle-nose pliers
- Small flat-bladed screwdriver
- Tweezers
- Magnifying glass

The flat-bladed screwdriver is invaluable for prying up small components. The tweezers and magnifying glass are for inspecting and retrieving parts as well as making small repairs (lead bending).

Along with these hand tools, you should also have:

- Digital multimeter (DMM) with AC/DC voltmeter, milliammeter, and ohmmeter at a minimum
- Logic probe

You will not need an oscilloscope or a logic analyzer, as I will present oscilloscope pictures and timing diagrams that should help you with understanding what is happening in the circuit.

Datasheets

When you begin to design your own applications using TTL logic chips, you will want to find documentation about the parts explaining how they work and how they can be used. If you are new to electronics, you will find that datasheets are very confusing and seem to have more information than you could ever want—or need. This may seem like a paradox, but the chip datasheets provide information that you will probably not really understand or find relevant.

For example, looking at the datasheet for the 7400 quad two-input NAND gate chip, in my 1981 vintage Texas Instruments *The TTL Data Book for Design Engineers* book, I can look up the following information about the part:

- Pinout
- Packaging information with dimensions
- Supply voltage ranges
- Chip supply current requirements
- Operating temperature ranges
- Output current levels
- Input voltage levels
- Input clamp voltages
- Input currents
- Clamping voltages
- Switching (low-high and high-low) characteristics (time for a test output load)
- Gate schematic

For other, more complex circuits, this list will be expanded to include:

- Truth tables
- Logic diagrams
- Operating waveform diagrams
- Input/output equivalent circuits

If you look at the different pages that encompass these different parameters and include the permutations of the information based on different families of parts (i.e., straight TTL versus S, LS, C, HC, HCT, AS, ALS and so on), you will be very quickly overwhelmed. For many chips, you will have to look at multiple datasheets or even in multiple books.

To be honest, the Internet has done a lot to simplify the information search with "pdf" files containing all the information you require for a chip in one location. The information is still confusing and quite overwhelming, but at least it is all in one place.

Once you have read through this book and built the applications presented in it, you should be comfortable with reading through datasheets and understand what kind of information they are giving you. Throughout the book, I will be explaining the different aspects of digital electronic chips and how they affect application design. When you have finished with the book, you should be able to read through a datasheet and find the information that is important for your application's design.

I will be referencing the pertinent datasheet information for a given application or description if the information is important to why a component is selected or how it is used. I will not be expecting you to look up the information in the chip's datasheet to understand why I designed or used something in a specific way. Instead I will discuss what are the important characteristics and why I chose the chip that I did.

As you work through the book, your understanding of the different chips will go from black boxes to building blocks that you will feel comfortable with using in your own applications. In understanding how the chips work, you should be comfortable with being able to look through their datasheets and see how to design circuits in such a way as to avoid many of the problems that new developers encounter.

One final word about datasheets—many unscrupulous distributors take datasheet books that they are given free of charge from chip manufacturers and place them for sale. I can't tell you how angry this practice makes me. Before buying a datasheet from a distributor, remember that the information can be downloaded for free over the Internet or, by contacting the manufacturer, you may be able to get the same datasheet book free of charge.

Chapter

4

Boolean Logic and Digital Electronics

In the previous chapters, I introduced different numbering systems and the mathematics that are used to process data in order to solve a real-world problem. In this chapter, I am going to expand on these concepts and show how *binary data* is manipulated in digital systems.

Before going on, I should caution you that you are going to have to think of data in much more fundamental ways than you have ever really been expected to before. While on the surface this will seem easy, as I get in to more complex operations, you will find that you have to return to this chapter as well as the information in the appendices to better plan how your applications are going to work.

Most aspects of engineering design and development have been automated by the use of computer applications that simulate the behavior of the system or of the code. *Simulators* (as these applications are known) are available as digital logic as well as computer processors and other devices. There are quite a few simulators that can be downloaded free of charge from the Internet, and many manufacturers have developed simulators for sale that are often integrated into chip and PCB development systems.

I have not included a CD-ROM with a logic simulator in this book because I would like you to understand intimately how digital logic works and be able to understand and debug circuits on your own. I find that simulators are often used by new designers in a way that doesn't optimize the gates, and problems are addressed without totally understanding what is happening. Later in the book, I will show how simulators work and what features you should look for when developing your own applications.

Boolean Logic

Boolean logic, or *boolean arithmetic,* is a branch of mathematics where an expression is evaluated to a single-digit result. Unlike traditional mathematics, a boolean logic statement expresses parameters and result as a two-value single digit known as a *bit.* A *bit* can be a 0 or 1, true or false, or voltage levels. In this book, I will characterize a bit in terms of all of these different representations, depending on what is appropriate. In this section, I will introduce you to the simple bit and the boolean logic functions that are used to manipulate data and create meaningful circuits that can be converted into complex functions.

Bits can be hard to conceptualize because you are probably most comfortable with *numbers,* which can be of quite a large range. For example, the basic mathematical equation:

$$2 + 2 = 4$$

cannot be represented by a single boolean logic expression. The bit value of 0 or 1 is 2 orders of magnitude less than the value 4. Despite this apparent limitation, boolean logic can manipulate multiple bits of output combined to implement a meaningful function or result, such as the add function (known as an *adder*) shown above. Later in this chapter, I will be showing how circuits can be placed in parallel to provide multiple-bit outputs.

With a value of 0 or 1, you might wonder what can be done on a single bit. On its own, obviously not a lot—the only function that is available to a bit is *inversion* (which is also known as *complement*). This is the operation of transforming a bit from one value to another and is given the name NOT and the symbol !. In some texts, you may see the complement operation being represented as a bar over the term that it is inverting. In circuit designs, you will see an underscore character (_) to indicate that the bit is *negative active* (which means it is active when the signal is 0, reset, or low).

In this book, when I am writing down inverted boolean logic statements, I will use the ! character with optional parentheses to indicate the scope of the inversion.

To show how the boolean logic NOT operation works, I can write it in the mathematical form known as the *output equation:*

$$B = !A$$

which is used to indicate that the bit B is given the complemented contents of A. This means that if A is equal to 1, then B will be assigned the value 0 and vice versa. When boolean logic expressions are written out, they either use the output equation form shown above, or they are written out in *truth tables.*

Truth tables show the outputs for a given logic function along with the output for each different set of inputs. For the equation above (which is a simple *inverter*), the truth table is:

A	B
0	1
1	0

Besides the complement, there are two other basic boolean logic operations that can be performed on bits. These three operations form the basic operations provided in all digital electronic systems. The two additional operations are known as AND and OR and take two bit inputs and output a new value based on these inputs.

The AND operation outputs only a 1 when both of its inputs are a 1. When you see boolean logic statements, the *dot* (.) or *splat* (asterisk: *) are used to represent the AND function. In this book, I will use the . (dot) symbol to represent the AND in boolean logic equations and statements:

$$A = B \,.\, C$$

AND can be thought of as a multiply operation because the only time the result will be nonzero is when both terms are nonzero. Multiplying anything by zero will result in zero.

The truth table for the AND operation is:

B C	A
0 0	0
0 1	0
1 0	0
1 1	1

Note that in this truth table, I have treated B and C like a two-bit number and simply run through the four possible values. This is one convention that can be used for truth tables, but I will modify this format slightly to use *Gray codes* instead of incrementing values.

The Gray code version of the AND truth table is:

B C	A
0 0	0
0 1	0
1 1	1
1 0	0

The OR operation returns a 1 if either input is equal to 1. The symbol used to represent OR is the plus sign (+) because the result of an OR is nonzero if any of the inputs are nonzero:

$$A = B + C$$

The truth table for the OR operation is:

B C	*A*
0 0	0
0 1	1
1 1	1
1 0	1

It is hard to believe, but the three basic boolean logic operations presented so far in this section are the basis for all digital logic. As I present new functions in this section, and in the rest of the chapter as well as the book, I will try to always to go back and show how these functions are based on these three basic boolean logic operations.

In some texts on digital electronics and virtually all texts on programming, you will see AND represented by the ampersand (&) and OR represented by the vertical bar (|) characters. These two character representations come from high-level programming languages where they are used to represent the AND and OR functions and are used with logic to avoid having to establish two different conventions for what is considered essentially the same thing. Personally, I do not consider the AND, OR, and NOT functions in programming to be equivalent to boolean logic AND, OR, and NOT functions, so I differentiate them by using the different characters.

The AND and OR functions are available as chips with more than two inputs. In this book, I will concentrate on AND and OR being two input devices only. This will allow the use of standard TTL chips in the projects. When you look in datasheet pages, you'll notice that AND and OR gates that have more than two inputs are used within the chips to provide basic functions without having to resort to displaying the functions as multiple two-input devices.

As I've noted in the previous paragraph, multiple gate functions can be combined to form more complex or different boolean logic functions. Along with the AND, OR, and NOT boolean logic operations, there are three other operations that are considered to be part of the basic boolean logic operations. These three additional boolean logic operations use a combination of AND, OR, and NOT to provide the new functions.

The primary boolean logic operation that you will be working with is the NAND, which inverts the output of a simple AND gate. When I write out the NAND operation, I use an AND and NOT together:

$$A = !\,(B\,.\,C)$$

The truth table for NAND is:

B C	*A*
0 0	1
0 1	1
1 1	0
1 0	1

You were probably surprised by the comment "the primary boolean logic operation that you will be working with is the NAND." The reason for saying this is that all bipolar TTL logic is based on the NAND operation, which is the most inexpensive function, or *gate,* to build on a silicon chip.

In actuality, virtually all the other operations built on a silicon chip are based on the NAND. Later in this chapter I will discuss how complex functions are built out of this simple gate. For now, I just want to point out that the functions in commercial bipolar transistor TTL chips are usually built with just NAND circuits (known as *gates*) instead of the AND, OR, NOT, NOR, or XOR that are described in this section.

Along with the NAND operator, there is also the NOR. This operation is analogous to the OR operator the same way NAND is related to AND. In the NOR operation's case, the output of an OR operator is inverted. As with NAND, I use a complemented OR operator to indicate that the operation is a NOR:

$$A = !\,(B + C)$$

The truth table for the NOR operation is:

B C	A
0 0	1
0 1	0
1 1	0
1 0	0

The NOR gate is the basic function for some types of CMOS logic.

The last type of basic gate is the XOR, which outputs a 1 any time one (but not the other) input is equal to 1. In many texts and references, the symbol you will see is a plus sign (+) with a circle around it. In this book, I will use the symbol ^ to indicate the XOR function. The XOR operation:

$$A = B \wedge C$$

has the truth table:

B C	A
0 0	0
0 1	1
1 1	0
1 0	1

XOR is not a basic operation—in fact, it is normally written out as a compound operation in the *sum of products* form:

$$\text{XOR}\,(A, B) = (!\,(A)\,.\,B) + (A\,.\,!\,(B)\,)$$

The sum of products is a standard form for writing out combinatorial circuits' functions. The most basic way of developing a combinatorial function is to take the truth table and OR (add) together the product (AND) of each term that produces a 1 for the output. From the XOR table, I can see that NOT A AND B produces a 1 for the output as does A AND NOT B. These two terms are ORed together to create the XOR function.

Product of sums is also possible but not used simply because it can be quite difficult to decipher and the actual function of the boolean logic can be lost. For example, by using De Morgan's theorem (one of the techniques that I will present later in the chapter), the XOR boolean logic function could be represented in product of sum form as:

$$\text{XOR}(A, B) = !\,(\,(A + !\,B)\,.\,(!\,A + B)\,)$$

In actual TTL implementations, De Morgan's theorem is applied again to characterize the XOR as a NAND-based output equation. As I will discuss later in the book, an *Inverter* (NOT gate) is a NAND gate with all its inputs tied together.

$$\text{XOR}(A, B) = !\,(!\,(!\,A\,.\,B)\,)\,.\,!\,(A\,.\,!\,B)\,)$$

Except where I have noted, the sum of products form will be used throughout this book in explaining complex boolean logic functions.

With this, you should have a basic understanding of how boolean logic works from a mathematical/theoretical perspective. Using this background, I will explain how boolean logic provides complex functions in digital electronics. As I work through my description of boolean logic, I will provide you with optimizations that will make the development of the necessary circuits simpler.

The Basic TTL Gates

In this book, I will primarily be presenting you with circuits designed from *transistor-to-transistor logic* (TTL). Along with TTL, there are a number of other forms of digital logic, including CMOS and RTL. TTL logic is by far the most popular method of implementing simple digital logic circuits. In this chapter, I would like to introduce you to the basic TTL gates and the different technologies that are available within them. In later chapters, I will expand on this basic understanding, demonstrating how multiple chips can be combined to implement complex functions.

The six basic gates are listed in the following table, along with the TTL and CMOS chips that I am most likely to use with them. Also included is a simple *truth table,* showing how the gates work.

<table>
<tr><th>Operation</th><th>Electronic symbol</th><th>Description and state table</th></tr>
<tr><td>NOT (! A)</td><td>A Output</td><td>Invert the input signal.
TTL chip: 7404
CMOS chip: 4049
<table>
<tr><th>Input</th><th>Output</th></tr>
<tr><td>0</td><td>1</td></tr>
<tr><td>1</td><td>0</td></tr>
</table></td></tr>
<tr><td>AND (A . B)</td><td>A
B Output</td><td>Return true if both inputs are true.
TTL chip: 7408
CMOS chip: 4081
<table>
<tr><th>A B</th><th>Output</th></tr>
<tr><td>0 0</td><td>0</td></tr>
<tr><td>0 1</td><td>0</td></tr>
<tr><td>1 1</td><td>1</td></tr>
<tr><td>1 0</td><td>0</td></tr>
</table></td></tr>
<tr><td>OR (A + B)</td><td>A
B Output</td><td>Return true if either input is true.
TTL chip: 7432
CMOS chip: 4071
<table>
<tr><th>A B</th><th>Output</th></tr>
<tr><td>0 0</td><td>0</td></tr>
<tr><td>0 1</td><td>1</td></tr>
<tr><td>1 1</td><td>1</td></tr>
<tr><td>1 0</td><td>1</td></tr>
</table></td></tr>
<tr><td>NAND (! (A . B))</td><td>A
B Output</td><td>Return false if both inputs are true.
TTL chip: 7400
CMOS chip: 4011
<table>
<tr><th>A B</th><th>Output</th></tr>
<tr><td>0 0</td><td>1</td></tr>
<tr><td>0 1</td><td>1</td></tr>
<tr><td>1 1</td><td>0</td></tr>
<tr><td>1 0</td><td>1</td></tr>
</table></td></tr>
</table>

Operation	Electronic symbol	Description and state table
		(Continued)
NOR (! (A + B))	A, B — Output	Return false if either input is true. TTL chip: 7402 CMOS chip: 4001
XOR (A ^ B)	A, B — Output	Return true if only one input is true. TTL chip: 7486 CMOS chip: 4030

NOR state table:

A B	Output
0 0	1
0 1	0
1 1	0
1 0	0

XOR state table:

A B	Output
0 0	0
0 1	1
1 1	0
1 0	1

These functions are available as discrete functions in a chip or combined to provide complex functions in a chip. Later in the book, you will see how they can be combined to form more complex functions. Here I want to introduce you to the different logic functions as well as discuss some of the interfacing issues of digital circuits.

When you look at a catalog of TTL chips, you'll find that there is a bewildering array of different chip "families" available. Each chip is identified by a code starting with 74, followed by a letter and a number. The letter denotes the family the chip belongs to. The letter (which may not be present, or is S, LS, HC, HCT, AL, or ALS) indicates the type of circuit and driver. Each of these families work slightly differently and at different speeds (and requires different amounts of current to operate).

Standard CMOS logic starts with a 40 for the part number. CMOS logic does not have letter modifiers like TTL.

The term *threshold* is used to specify the voltage at which an electrical signal is determined to be a 1 or 0. For positive logic (which TTL is), a 1 is a high voltage, while a 0 is low voltage. The table below lists some different types of logic and their threshold and output voltages. Later in the book, I will go into more detail about how TTL and CMOS logic works. For now, the voltage thresholds listed below should be considered the point at which an input transitions from a 0 to a 1 and vice versa.

Type	Threshold, volts	0 output, volts	1 output, volts
TTL	1.4	0.3	3.3
HC	2.4	0.1	4.9
HCT	1.4	0.1	4.9
CMOS	2.5	0.1	4.9

The circuits presented in this book are (mostly) designed to work with LS logic, which stands for *low-power Schottkey,* a reasonably low power logic family that works fast enough with enough current drive for most application functions. If another type of logic family is required, I will make note of it and why. The table below lists the gate transition times (in terms of nanoseconds), which is the speed measurement of the different families, as well as their current sink capability, which is the amount of current each gate can pull to ground. I consider the current sink capability to be the most important consideration of each logic family because the circuits driving an LED have to have sufficient capability to light it.

In the table below, I have given general values for each logic family. Note that the transition time is really per gate—multiple-stage chips will have much slower transition times. As I will explain later in the book, this is an important factor in deciding how an application is to be designed.

Family	Transition time, nsec	Maximum current sink, mA
Straight TTL (no letter)	8	12
L TTL	15	5
LS TTL	10	8
S TTL	5	40
AS TTL	2	20
ALS TTL	4	8
F TTL	3.5	20
C CMOS	50	1.3
HC/HCT CMOS	9	8
4000 CMOS	30	0.5

For the rest of the chapter, I will be presenting information and projects about how basic TTL circuits operate. Later in the book, I will discuss the actual operation of TTL and CMOS gates. In Appendix B, I have provided the pin-outs for the most common TTL devices, along with a list of the CMOS logic chips that provide equivalent functions.

Project 3—Testing Gates

So far in this book, I have loaded you up with theory. Now it is time to look at how TTL logic gates actually work and compare their operation to what I have

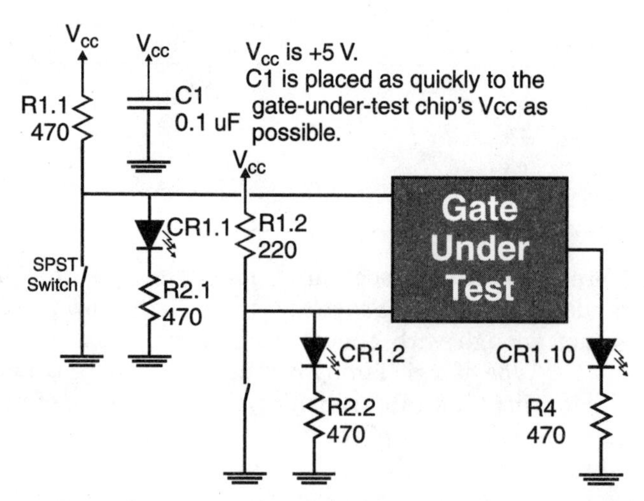

Figure 4.1 Circuit to test gate operation.

discussed up to this point. The circuits presented in this project will probably seem to be almost unreasonably trivial, but I recommend that you go through them because they will give you some practical experience in wiring chips into the breadboard and help you to see the different gates in actual operation.

The basic circuit for this project is shown in Fig. 4.1 Two switch inputs are passed to a TTL chip that consists of just the six basic logic gates that I have presented so far. The output of this gate is used to drive an LED to indicate the actual state output. I found that I could wire each of the six first circuits in less than a minute and the last circuit (which is an example of combining gates) took less than 5 minutes.

In Fig. 4.1, I have shown R1, R2, R4, and CR1 as separate devices. You will be wiring this circuit with the interface PCB that has these components built into it.

The bill of materials (parts list) for the seven circuits is:

Part	Description
C1–C2	0.1-μF tantalum capacitors
7400	74LS00 quad NAND (wired in as gate under test)
7402	74LS02 quad NOR (wired in as gate under test)
7404	74LS04 hex inverter (wired in as gate under test)
7408	74LS08 quad AND (wired in as gate under test)
7432	74LS32 quad OR (wired in as gate under test)
7486	74LS86 quad XOR (wired in as gate under test)
Miscellaneous	Breadboard, +5-V power supply, interface PCB, wire

The first gate to work with is the inverter, the wiring of which is shown in Fig. 4.2 After the circuit is wired into the breadboard and power is applied, the opposite of the state of the left-most switch will be displayed on the right-most LED. Note that only the left-most switch is used for input in this circuit because an inverter has only one input pin. When the left-most switch is on (its LED is lit), the right-most LED will be off and vice-versa. The function provided by the simple inverter is not that useful, but as I will show later in the book, it does become very useful when combined with other logic gates.

The second gate to test out is the AND, which outputs a 1 when both of its inputs are 1. The wiring for the circuit that I used for testing out the function of the 7408 TTL quad (or four times) two-input AND gate is shown in Fig. 4.3 The two left-most switches are used for input, and you will find that when their respective LEDs are on, it is the only time the right-most LED is on.

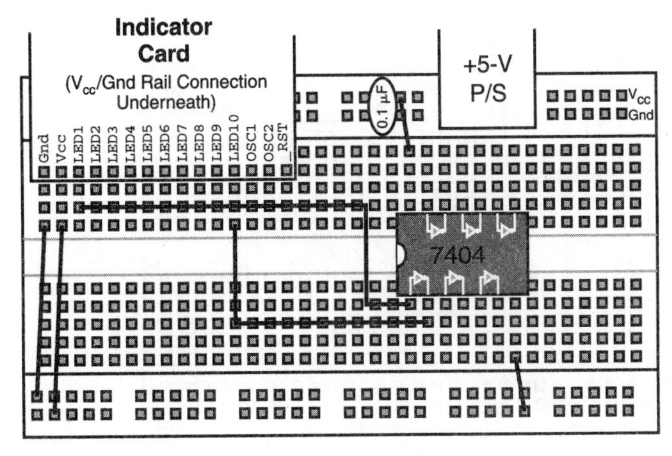

Figure 4.2 Project 3—inverter test wiring.

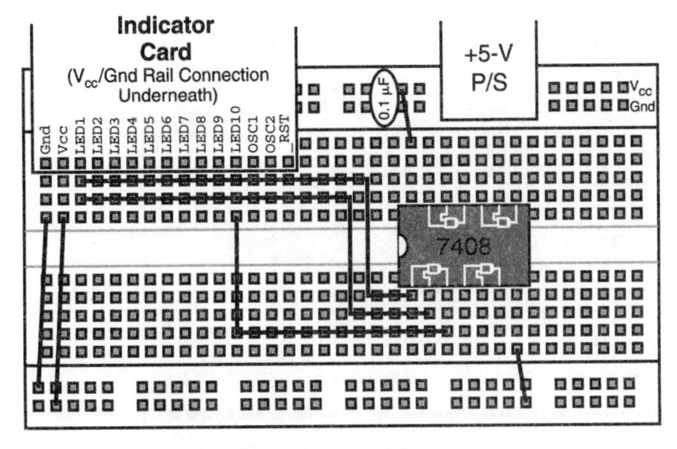

Figure 4.3 Project 3—AND gate test wiring.

The OR gate outputs a 1 anytime any of its inputs are a 1. This can be seen when a 7432 is wired into the test circuit, as I've shown in Fig. 4.4

I have mentioned that the NAND gate is the basic gate used by bipolar TTL logic—all other gate types are derived from this. The NAND outputs a 1 as long as both inputs are not equal to 1. Later in the book, I will show how this is done and what it means to the operation of the TTL logic and the circuits that you design. Fig. 4.5 is the circuit used to demonstrate the operation of the 74LS00 NAND gate.

While the NAND gate is the basis for bipolar TTL logic, the NOR gate is often the basis for CMOS logic. Both these gates will seem like they are quite limiting, but as I work through the methods for combining gates into new functions, you will find that all the different gate functions can be implemented

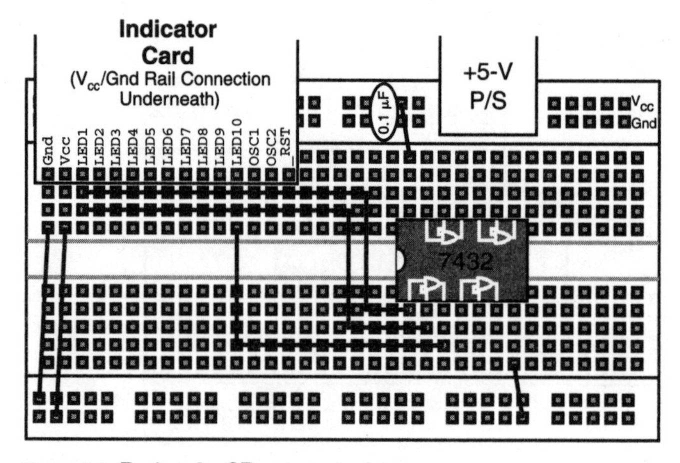

Figure 4.4 Project 3—OR gate test wiring.

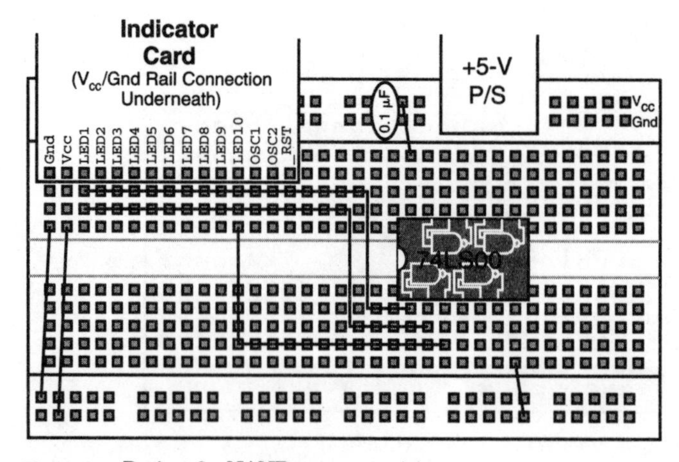

Figure 4.5 Project 3—NAND gate test wiring.

with just one type of gate. The NOR gate outputs a 1 when both inputs are at 0. Fig. 4.6 shows the wiring to demonstrate the operation of the 74LS32 NOR gate.

The last basic gate is the XOR, which outputs a 1 if the two inputs are different. Fig. 4.7 shows the gate test circuit with a 74LS86 quad two input XOR gate chip.

The wiring for the last circuit to test out is shown in Fig. 4.8, This circuit is a combinatorial circuit in which the outputs from an AND and a NOR gate are processed by another NOR gate. The output equation for this circuit is:

$$\text{Output} = \text{NOR}\,(\text{AND}\,(A, B)\,,\ \text{NOR}\,(A, B)\,)$$

$$= !\,(\,(A\,.\,B) + !\,(A + B)\,)$$

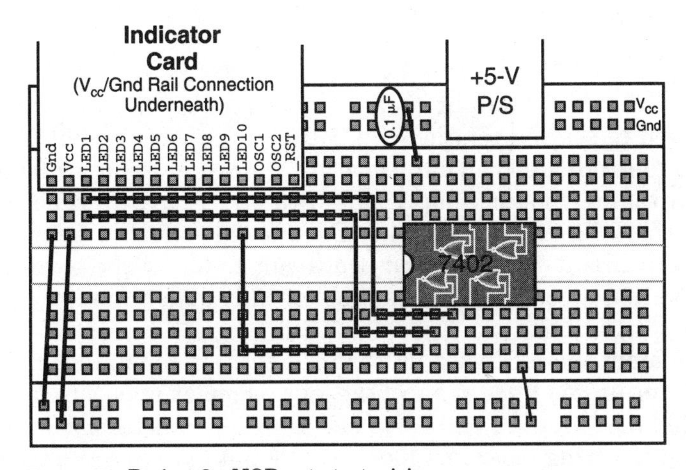

Figure 4.6 Project 3—NOR gate test wiring.

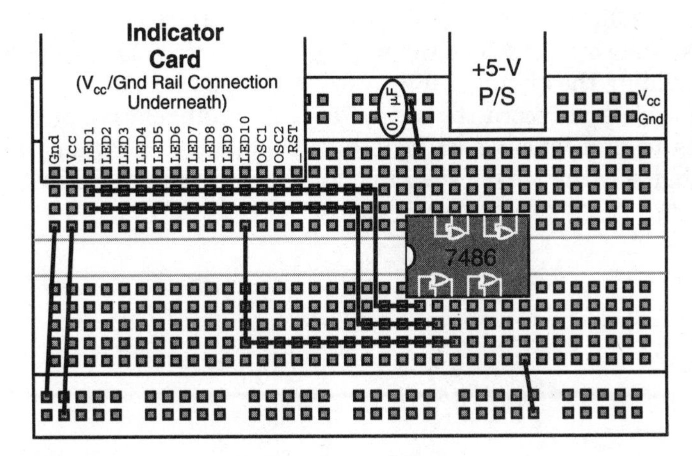

Figure 4.7 Project 3—XOR gate test wiring.

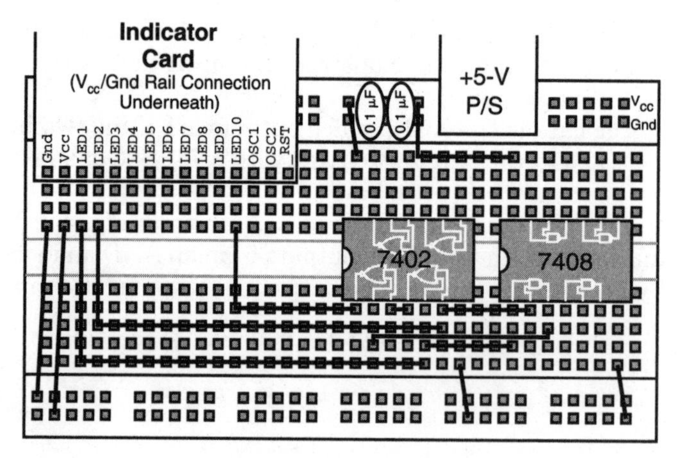

Figure 4.8 Project 3—mystery circuit test wiring.

When you compare the output equation above to Fig. 4.8. you will probably be hard pressed to see the relationship between the wiring and the equation. I have tried to make it somewhat easier to follow the wiring by laying it out in such a way that no lines overlap. Along with this, I have overlaid the functions built into the TTL chips in Fig. 4.8. Despite this effort to improve the readability of the circuit, you will find it difficult to follow the wires and see the output equation listed above. This problem only gets worse as you work with more complex circuits.

Unfortunately, there is not a lot that can be done to improve the readability of the circuits. As you build some of the more complex applications in this book, you will find that they are real "rat's nests." Where it is feasible, I will continue to present the wiring diagrams like Fig. 4.8, but once more than three 14-pin chips are used in the application, I will not be able to recommend how the wiring should be done. In later projects I will work with you to understand how to wire the applications in such a way that the opportunities for mistakes are minimized and you can follow the wiring to find errors as easily as possible.

After building the circuit in Fig. 4.8, I recommend that you write down the following table and record the output of the circuit relative to the four possible inputs to find what basic logic gate functions it provides. In the table, assume that *A* is the left-most switch and *B* is the one beside it:

A B	Output
0 0	
0 1	
1 1	
1 0	

If you have wired the circuit correctly, you will have one of the common logic functions that were examined earlier in the chapter. The method used for producing this function will probably seem convoluted, but it will make sense as you work through the material later in the next section.

Combining and Optimizing Boolean Logic Operations

So far in this chapter, I have shown how the basic six boolean logic operations work and can be combined together to form different functions. When multiple boolean logic operations are combined together, you will find that they are not always done in the most efficient manner or do not take advantage of left-over boolean logic gates on the circuit. In this section, I would like to introduce you to some of the techniques used for optimizing combined digital circuits into specific functions and show how they can be designed in the most efficient manner.

The reasons for optimizing circuits are probably pretty easy to guess. The primary reason for optimizing digital logic circuits is to minimize the number of gates required in the circuit to reduce the total cost of the circuit. The fewer the gates, the fewer the number of chips needed and the lower the wiring complexity of the board, and leftover gates can be taken advantage of. All these factors will result in an application that is cheaper and easier to prototype and wire as well as cheaper to manufacture. Fewer chips also mean that the total power dissipated by the application is reduced, resulting in the need for a smaller power supply.

It is a situation that actually feeds on itself. You may find that eliminating 1 percent of the chips on a board will result in a 5 to 10 percent overall reduction in product cost because of the savings resulting from the lower chip count.

In this section, I will go through a sample combinatorial circuit and explain the different tools that are available for reducing the overall circuit. The sample circuit that I would like to optimize is shown in Fig. 4.9. In this circuit, I

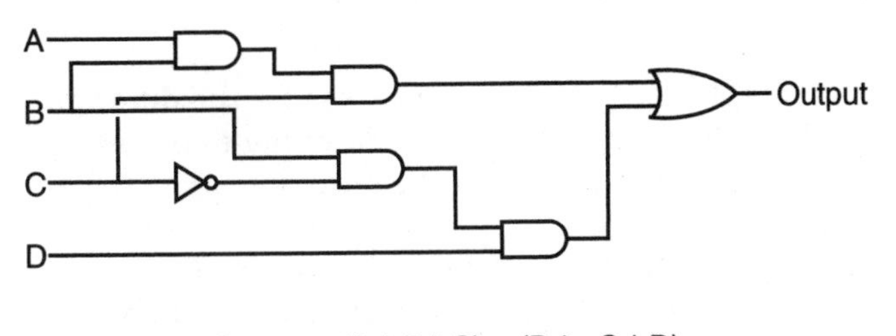

Output = (A * B * C) + (B * _C * D)

Output has 3 or 4 gate delays.

Figure 4.9 ANDs, NOT, and OR combined as logic.

combine four inputs together using an inverter, three AND gates, and an OR gate to implement the function (given in sum of products format):

$$\text{Output} = (A \,.\, B \,.\, C \,.\, !\,(D)\,) + (B \,.\, !\,(C) \,.\, D)$$

which could be used for such things as selecting a value or decoding input data. Different logic circuits can be implemented for an infinite number of requested functions.

To understand how the digital logic function presented above works, the first tool that you should use is the truth table. This table takes all the possible inputs and lists the intermediate values as well as the final output. For the output function above, I can create the standard truth table:

D C B A	(*A* . *B* . *C* . ! (*D*))	(*B* . ! (*C*) . *D*)	Output
0 0 0 0	0	0	0
0 0 0 1	0	0	0
0 0 1 0	0	0	0
0 0 1 1	0	0	0
0 1 0 0	0	0	0
0 1 0 1	0	0	0
0 1 1 0	0	0	0
0 1 1 1	1	0	1
1 0 0 0	0	0	0
1 0 0 1	0	0	0
1 0 1 0	0	1	1
1 0 1 1	0	1	1
1 1 0 0	0	0	0
1 1 0 1	0	0	0
1 1 1 0	0	0	0
1 1 1 1	0	0	0

In the truth table, I usually put in the values for the different AND filters of the sum of products logic function. This allows me to better see what the circuits are doing before being passed to the output.

Most logic circuits are designed from a black box, like the one shown in Fig. 4.10. The black box takes a series of inputs (*A, B, C,* and *D* in Fig. 4.10 and performs the logic comparisons that were discussed above to get a specific "Output." The best way to understand these outputs is to place them into a *Karnaugh map,* as I show in Fig. 4.10.

A Karnaugh map is a tool used to find relationships in inputs and outputs. For the example circuit shown in Fig. 4.10, the truth table is:

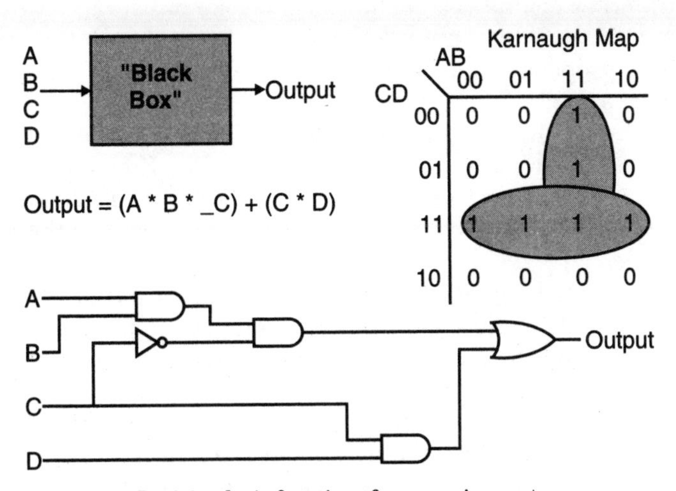

Figure 4.10 Deriving logic functions from requirements.

D C B A	Output
0 0 0 0	0
0 0 0 1	0
0 0 1 0	0
0 0 1 1	1
0 1 0 0	0
0 1 0 1	0
0 1 1 0	0
0 1 1 1	0
1 0 0 0	0
1 0 0 1	0
1 0 1 0	0
1 0 1 1	1
1 1 0 0	1
1 1 0 1	1
1 1 1 0	1
1 1 1 1	1

This truth table isn't very helpful in showing what the relationships are between the circuits. Looking at the truth table, you might come up with an output equation that looked like:

$$\text{Output} = (A \,.\, B \,.\, !\,(C) \,.\, !\,D) + (A \,.\, B \,.\, !\,(C) \,.\, D) + (C \,.\, D)$$

which will require 7 two-input AND gates, 2 two-input OR gates, and 2 inverters. The propagation delay through the function will range from two to four gate delays.

By making up the Karnaugh map shown in Fig. 4.10 from the truth table, I can look for relationships in the output settings. The Karnaugh map is drawn with the bits of each axis changing by only one value at any time (basically using Gray codes). This way relationships with bits staying the same can be more easily seen.

This is done by circling all the cases where the output is 1 with the cases that are side by side or up and down. Diagonal relationships are not circled. The result of circling the cases is to make relationships become more readily apparent.

For example, I can see that when A and B are both 1, the output will be 1 except when the C input is 1. Going further with this, I can ignore the case when A and B are set and D and C are set because it is already encompassed in the AND function $C * D$.

The Karnaugh map allows me to reduce the output equation to:

$$\text{Output} = (A \,.\, B \,.\, !\, C) + (C \,.\, D)$$

which, as can be seen in Fig. 4.10, requires only three AND gates, one inverter, and one OR gate. The propagation delay is improved to two or three gate delays as well.

Karnaugh maps probably seem pretty easy to work with in this way but do take some practice. When you first start working with them, you will discover that you may have to redraw them multiple times to best see what the most efficient results are. Over time, you will probably find that Karnaugh maps are less and less useful because you will learn to see relationships and visualize the desired functions in your head.

One of the ways of optimizing circuits is to look through their output equations and try to find relationships that you can take advantage of. To do this, you should be aware of the boolean *identities*. These are simple formulas that you should remember or keep the list of them in the appendix handy when you are working with digital logic. I have listed the important boolean logic identities in Appendix C.

Going back to the example output equation that I came up with before optimizing with the Karnaugh map,

$$\text{Output} = (A \,.\, B \,.\, !\,(C) \,.\, !\,(D)\,) + (A \,.\, B \,.\, !\,(C) \,.\, D) + (C \,.\, D)$$

I could not help but notice that the first two terms are almost identical. The only difference is that one uses the positive value of D and the other uses the inverted value of D. Going to the appendix, I can look up the *associative law*, which states that

$$A \,.\, B \,.\, C = (A \,.\, B) \,.\, C$$

and use it to make the first two terms even more similar:

$$\text{Output} = ((A \,.\, B \,.\, !\,C) \,.\, !\,(D)) + ((A \,.\, B \,.\, !\,C) \,.\, D) + (C \,.\, D)$$

Next, I can use the *distributive law:*

$$((A \,.\, B) \,.\, C) + ((A \,.\, B) \,.\, D) = (A \,.\, B) + (C \,.\, D)$$

to merge the first three values of the first two terms together:

$$\text{Output} = (A \,.\, B \,.\, !\,C) + (\,!\,(D) \,.\, D) + (C \,.\, D)$$

Finally, using the *complementary law,* I know that:

$$A \text{ AND } !\,A = 0$$

I can reduce the output equation to:

$$\text{Output} = (A \,.\, B \,.\, !\,C) + (C \,.\, D)$$

which is identical to what I came up with using the Karnaugh map.

Using the identities in Appendix C is actually quite straightforward, and there should only be one surprising law in there—De Morgan's theorem. The theorem is used to state the relationship between negated AND and OR gates:

$$!\,(A \,.\, B) = !\,A + !\,B$$

and

$$!\,(A + B) = !\,(A) \,.\, !\,(B)$$

This theorem probably does not seem that useful, but as you become more familiar with boolean logic, the uses for it will become more obvious.

For example, in the previous section, I asked you to test out a circuit that seemed arbitrary (indeed it required one AND gate and two NOR gates) if you built the circuit and tested it, how did the circuit perform?

The circuit should have performed like an XOR gate, but at the start of the chapter, I defined an XOR gate as having the output equation:

$$\text{Output} = !\,(A) \,.\, B + A \,.\, !\,(B)$$

It does not seem likely that the XOR output equation above matches the one I gave you:

$$\text{Output} = !\,((A \,.\, B) + !\,(A + B))$$

But, using De Morgan's theorem as well as the other identities in Appendix B, I can go through the following manipulations to prove they are all equal:

$$
\begin{aligned}
\text{Output} &= !\,(\,(A \,.\, B) + !\,(A + B)\,) \\
&= !\,(A \,.\, B) \,.\, !\,(!\,(A + B)\,) && \text{Using } !\,(A + B) = !\,(A) \,.\, !\,(B) \\
&= !\,(A \,.\, B) \,.\, (A + B) && \text{Using double negation law} \\
&= (!\,(A) + !\,(B)\,) \,.\, (A + B) && \text{Using: } !\,(A \,.\, B) = !\,(A) + !\,(B) \\
&= !\,(A) \,.\, A + !\,(A) \,.\, B + A \,.\, !\,(B) + !\,(B) \,.\, B && \text{Using distributive law} \\
&= !\,(A) \,.\, B + A \,.\, !\,(B) && \text{Using complementary law}
\end{aligned}
$$

The final result is identical to the output equation for the XOR gate.

These boolean arithmetic manipulations take a bit of experience to learn and work with efficiently. When you are trying to simplify an expression, always remember that there are 10 laws and theorems that you can take advantage of.

The last tool that I would use is the one that I have already shown you, and that is to create the truth table by using Gray codes. Once this is done, the truth table can be reduced for each case where the output doesn't change for changing bits.

With Gray codes instead of an incrementing counter, the function's truth table looks like:

D C B A	Output
0 0 0 0	0
0 0 0 1	0
0 0 1 1	1
0 0 1 0	0
0 1 1 0	0
0 1 0 0	0
0 1 0 1	0
0 1 1 1	0
1 1 1 1	1
1 1 0 1	1
1 1 0 0	1
1 1 1 0	1
1 0 1 0	0
1 0 1 1	1
1 0 0 1	0
1 0 0 0	0

From this table, I can see that when D and C (and B and A are any value) the output is 1. Along with this, I can see that there are two 1s that are not encompassed by this rule, but are very similar. I can rearrange the truth table so that these two terms are side by side (while keeping the $C \,.\, D$ outputs together):

D C B A	Output
0 0 0 0	0
0 0 0 1	0
0 0 1 1	1
1 0 1 1	1
1 0 1 0	0
1 0 0 0	0
1 0 0 1	0
1 1 0 1	1
1 1 0 0	1
1 1 1 0	1
1 1 1 1	1
0 1 1 1	0
0 1 0 1	0
0 1 0 0	0
0 1 1 0	0
0 0 1 0	0

Note that as I rearranged the truth table, I kept the feature of the Gray codes intact—each position is only one bit changed from the previous. The point of this is to bring all the different inputs where a 1 is output together so the relationships can be observed.

In the first two 1s output, 0b00011 and 0b01011, bits *A, B,* and *C* are in common with bit *D* being both 0 and 1. Knowing this, I can simplify the truth table to:

D C B A	Output
0 0 0 0	0
0 0 0 1	0
x 0 1 1	1
1 0 1 0	0
1 0 0 0	0
1 0 0 1	0
1 1 0 1	1
1 1 0 0	1
1 1 1 0	1
1 1 1 1	1
0 1 1 1	0
0 1 0 1	0
0 1 0 0	0
0 1 1 0	0
0 0 1 0	0

The new truth table shows that D can be any value, and I can state that a 1 is output by using the boolean logic expression:

$$A \,.\, B \,.\, !C$$

Looking at the second group of 1s, I can see that there are four cases with the only commonality being that both bits C and D are equal to 1. By simplifying the truth table, I can eliminate A and B to make the boolean logic expression that produces this result much more obvious:

D C B A	Output
0 0 0 0	0
0 0 0 1	0
x 0 1 1	1
1 0 1 0	0
1 0 0 0	0
1 0 0 1	0
1 1 x x	1
0 1 1 1	0
0 1 0 1	0
0 1 0 0	0
0 1 1 0	0
0 0 1 0	0

Now knowing that bits A and B can be any value, I can state that a 1 is output using the boolean logic expression:

$$C \,.\, D$$

Going back to the truth table, there are no more situations where there are 1's output, so I have found the products in the required boolean logic expression—I just have to sum them for the final equation:

$$\text{Output} = (A \,.\, B \,.\, !\,C) + (C \,.\, D)$$

This method does produce the same results as the previous two, but it can be somewhat tedious and error prone, especially if you are new to boolean logic. It is not a bad method of checking your work, however.

When I am looking to simplify a combinatorial circuit, the method I usually use is to first develop a basic sum of products for the output equation, using a Karnaugh map. Next, I use the identities and laws in Appendix C to manipulate the equation as many different ways as I can and look at the gate requirements.

Going back to the XOR gate output equation manipulation I showed above, I can take these manipulations and plot alongside them the number and types of gates required, along with the number of gate levels the function has to work through.

While I have presented the concept of gate delays for different types of logic technologies, in most logic families, the time required for data to pass through a gate is approximately constant, regardless of the gate type. With this constant gate delay, the deeper the circuit is in terms of gate levels, the longer the delay will be for the function. Each gate the signal has to pass through is characterized as being delayed by one gate delay. For circuits that are to be used in very fast applications, it is critical that the number of gate delays the data experiences on its way through to the output is minimized. To do this, carrying out an analysis like the one in the table below will help you to understand what is the best method for implementing the function.

For the example of the XOR gate:

	NOT	AND	OR	NAND	NOR	Levels
Output $= !(A) . B + A . !(B)$	2	1	1	0	0	3
$= !(!(!(A) . B) . !(A . !(B)))$	2	0	0	3	0	3
$= !((A + !(B)) . !(!(A) + B)$	2	0	1	1	1	3
$= !((A . B) + (!(A) . !(B)))$	2	2	0	0	1	3
$= !((A . B) + !(A + B))$	0	1	0	0	2	2

From this analysis, what is the best solution to the problem of implementing the XOR gate?

The correct answer would be that it depends on the application it is going into. If speed were a critical factor, then only the last output equation (and the circuit you built in Project 3) would be acceptable. If low cost were an issue, you might choose the second method because it uses the least costly parts of all the methods. The circuit may have a couple of NOTs, ANDs, and a NOR available for the XOR's gate use, so it would be best. Creating a table like this gives you options to help tailor your circuit to maximize the efficiency of your application.

Performing this type of analysis can also be quite a bit of fun, but it also requires some experience to know what to look for. As you gain more experience with digital electronics, using the identities and laws in Appendix B will become second nature, as will understanding the improvements that can be made to the circuit.

Project 4—Sum of Products Decoder

In the previous project, I showed how basic logic circuits could be used to emulate each other, and in the previous section, I discussed how boolean arithmetic can be manipulated in such a way as to show how specific functions can be re-created in simpler or more convenient forms. In this project, I would like to use this knowledge to show how a decoder circuit can be optimized from a

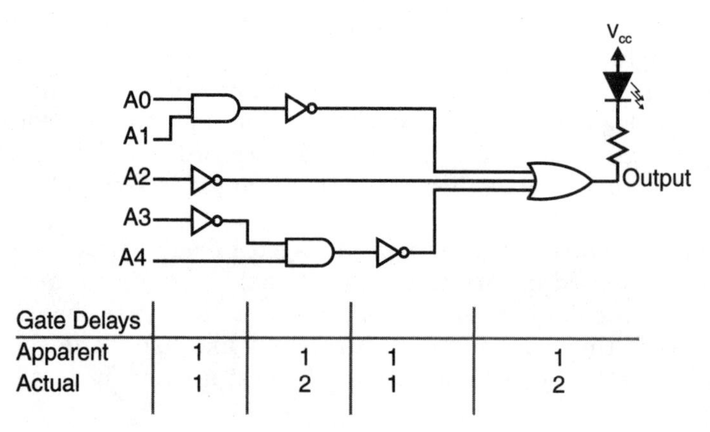

Figure 4.11 Proposed decoder circuit.

circuit that is quite wasteful of circuit gates into one that performs the same function, but uses fewer gates and provides the function more quickly.

The circuit that I would like to use is the sum of products logic circuit shown in Fig. 4.11. This circuit combines five digital inputs using the basic boolean logic functions presented above into a single output that is "negative active" when specific input conditions are met.

Negative active means that the output is valid when it is driving out a low voltage. As you understand TTL and other types of logic circuits better, you will find that this is often a preferable method of indicating that an input or output is active. When the output of the decoder circuit shown in Fig. 4.11 is low, the LED will turn on, indicating that the decoded condition is active.

The purpose of a decoder is to take a number of digital inputs and indicate whether or not they meet a predetermined value. Decoders are often used in computer systems for decoding addresses in peripheral devices to decide whether to respond if they are being addressed by the processor. The peripheral devices are often "keyed" by a decoder's negative output. I will discuss (and use) decoder circuits quite a bit later in the book.

The output function of the decoder will be 0 or negative active when the correct input conditions are met. The output equation that is specified for this circuit is:

$$\text{Output} = !\,(A0\,.\,A1) + !\,A2 + !\,(!\,A3\,.\,A4)$$

In this circuit, I have specified a three-input OR gate. Up to now, I have been concerned only with two input gates. Multiple input gates do exist, and I will show how they are implemented later in the book. In this case, you just have to accept that the output of the three-input OR gate will be 1 any time *any* of the three inputs are a 1.

To understand exactly when the output is active (low), you could create the truth table I show below and solve it for the different equations.

A0 A1 A2 A3 A4	! (A0 . A1)	! A2	! (! A3 . A4)	! (A0 . A1) + ! A2 +! (! A3 . A4)
0 0 0 0 0	1	1	1	1
0 0 0 0 1	1	1	0	1
0 0 0 1 1	1	1	1	1
0 0 0 1 0	1	1	1	1
0 0 1 1 0	1	0	1	1
0 0 1 1 1	1	0	1	1
0 0 1 0 1	1	0	0	1
0 0 1 0 0	1	0	1	1
0 1 1 0 0	1	0	1	1
0 1 1 0 1	1	0	0	1
0 1 1 1 1	1	0	1	1
0 1 1 1 0	1	0	1	1
0 1 0 1 0	1	1	1	1
0 1 0 1 1	1	1	1	1
0 1 0 0 1	1	1	0	1
0 1 0 0 0	1	1	1	1
1 1 0 0 0	0	1	1	1
1 1 0 1 0	0	1	1	1
1 1 0 1 1	0	1	1	1
1 1 0 0 1	0	1	0	1
1 1 1 0 1	0	0	0	0
1 1 1 1 1	0	0	1	1
1 1 1 1 0	0	0	1	1
1 1 1 0 0	0	0	1	1
1 0 1 0 0	1	0	1	1
1 0 1 0 1	1	0	0	1
1 0 1 1 1	1	0	1	1
1 0 1 1 0	1	0	1	1
1 0 0 1 0	1	1	1	1
1 0 0 1 1	1	1	1	1
1 0 0 0 1	1	1	0	1
1 0 0 0 0	1	1	1	1

It was a lot of work to create this truth table, and the circuit is actually quite modest.

Instead of going through each possible combination, I can look at the products that are summed by the OR gate. In the truth table above, note that I broke out each of the three products and then, to find the final output, I simply ORed them together.

This could be done to find the active output condition. The first product is !($A0$. $A1$). This is essentially a NAND gate, and from the NAND gate truth tables presented above, you can look up the output characteristics to find that the output is equal to 0 when both inputs are equal to 1. This means, that for the output to be 0, $A0$ and $A1$ must both be 1. Looking at the second summed term, ! $A2$, we know that when $A2$ is equal to 1, the inverted value for it is equal to 0. Lastly, ! (! $A3$. $A4$) outputs 0 when $A3$ is equal to 0 (the input to the NAND gate is 1) and $A4$ is equal to 0.

Looking back at the truth table for the function above, you can see that when this input condition is active, the output is 0.

Putting these three products together, we know that $A4$, $A2$, $A1$, and $A0$ being set to 1 and $A3$ set to 0 will cause a 0 to be output from the circuit. This method of evaluating the products of a sum of products output expression can be a lot faster than working through the truth table and chances are it will be more accurate. The truth table could be used to confirm that the input states you have found are correct.

The decoder circuit that I have presented in Fig. 4.11 can be built with two-thirds of a 7404, one-half of a 7408, and one-third of a 7427 (triple three-input OR Gates). There are two problems with building this circuit with these parts. The first problem is, it is not very efficient in terms of the number of chips that are used. To provide this function, parts of three chips are required. Depending on the circuit, these chips may be available, or they may have to be added to the application.

The second problem with this circuit is that the actual speed of the circuit is not readily apparent. In Fig. 4.11, I have included a small table underneath the circuit indicating the apparent number of gate delays (time taken to drive the signal through a gate) along with the actual.

The actual number of gate delays is probably surprising, but goes back to the comment I made earlier in the book that TTL chips are made up of multiple NAND gates, rather than the six different gates. As is shown in Fig. 4.12, two 2-input NAND gates are used to implement the NAND function. To implement an OR gate, three 2-input NAND gates are used as indicated in Fig. 4.13. For both these cases, two gate delays are actually provided in each of the logic functions. This makes the total number of gate delays for the circuit six and not four as you would have expected looking at the original circuit.

To see if I could come up with a better method of implementing this function, I started working through my boolean equivalents to see how the circuit could

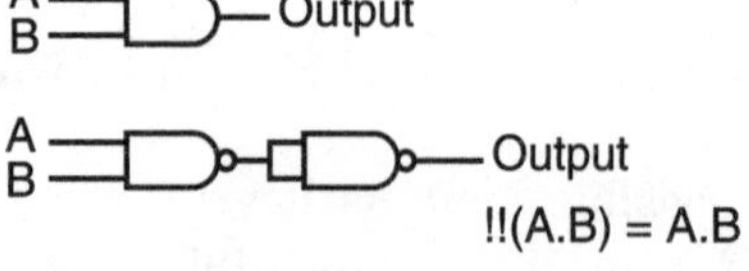

Figure 4.12 AND gate analog using NAND gates.

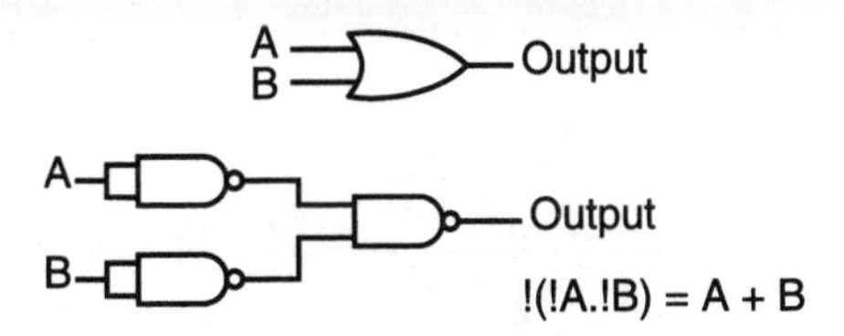

Figure 4.13 OR gate analog using NAND gates.

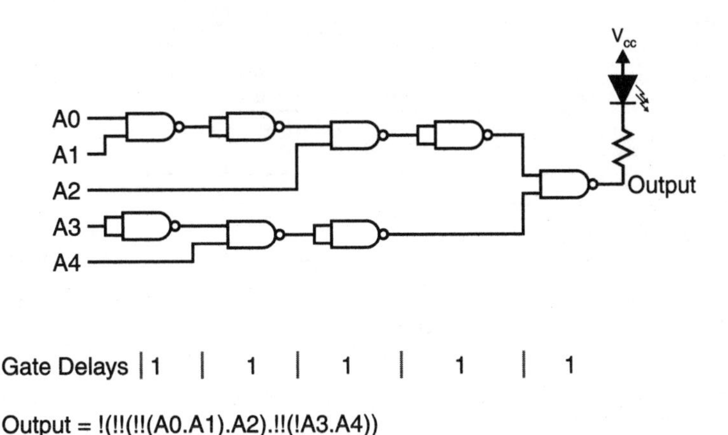

Figure 4.14 NAND-optimized decoder circuit.

be converted into straight NAND and NOT gates instead of the combination of ANDs, ORs, and NOTs. I should not have to point out that a NAND gate with both its inputs tied together behaves as if it were a NOT gate.

$$\begin{aligned}\text{Output} &= !\,(A0\,.\,A1) + !\,A2 + !\,(!\,A3\,.\,A4) \\ &= !\,(!\,!\,(A0\,.\,A1)\,.\,!\,!\,A2) + !\,(!\,A3\,.\,A4) && \text{De Morgan's theorem} \\ &= !\,(!\,!\,(A0\,.\,A1)\,.\,A2) + !\,(!\,A3\,.\,A4) && \text{Double negation law} \\ &= !\,(!\,(!\,!\,(A0\,.\,A1)\,.\,A2)\,.\,!\,!\,(!\,A3\,.\,A4)\,) && \text{De Morgan's theorem}\end{aligned}$$

The final result looks quite unwieldy, but it can be converted into a logic circuit quite easily as shown in Fig. 4.14. This circuit uses eight 2-input NAND gates, which means that it can be implemented with two 7400 chips. Besides using fewer chips, the chip incurs only five gate delays.

This circuit can be built with the breadboard tools, as shown in Fig. 4.15. A separate LED is wired into the circuit and turns on when the decoder is "active." The separate LED was used instead of the LEDs on the interface PCB because the interface PCB LEDs turn on when the logic level is "high."

The bill of materials for this project is:

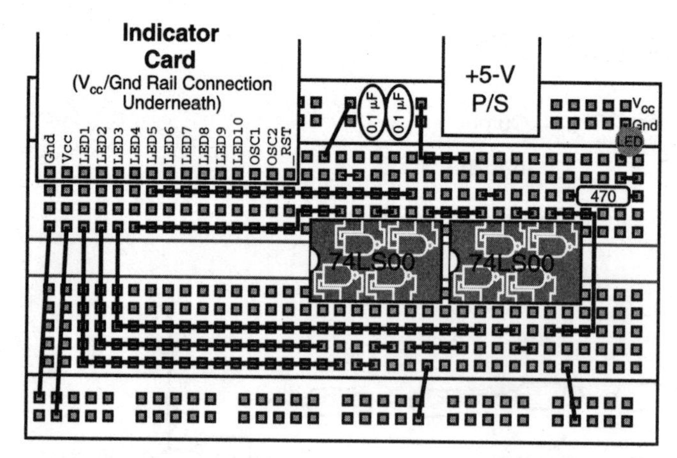

Figure 4.15 Project 4—NAND decoder wiring.

Part	Description
C1–C2	0.1-µF tantalum capacitors
CR1	5-mm LED
R1	470 Ω, ¼-W resistor
7400	Two 74LS00 quad NAND gates
74138	74LS138 three-to-eight decoder
Miscellaneous	Breadboard, interface PCB, +5-V power supply, wire

When you have correctly wired this circuit, you will find that when switches 1, 2, 3, and 5 are on with switch 4 off, the decoded output will light the LED. No other combination of switches will cause the LED to light. This demonstrates the function of the decoder—it will only become active for a specific signal.

While this circuit is an improvement to the original, you can do better by looking for TTL chips that provide you with the functions that you require. In Fig. 4.16, I show how a 74LS138 can be used instead of the three different chips or two 7400s to provide the same function.

The 74LS138 has six input pins. Three of the inputs are used to "enable" the chip (pins 4 and 5 must be held low and pin 6 must be held high) while three other pins are used to select one of eight outputs to be (negative) active. This chip was originally designed to be used in computer systems and decode I/O, but you will find that it can be very useful in a number of different applications to specify events. For the more complex circuits shown in this book, I will be using the 74LS138 quite a bit.

Using the 74LS138, instead of requiring two chips with a five-gate delay, I can get a single chip that provides the same function and has only a four-gate delay from input change to output. Along with providing the basic output, there are seven other output pins that can be used for decoding other conditions.

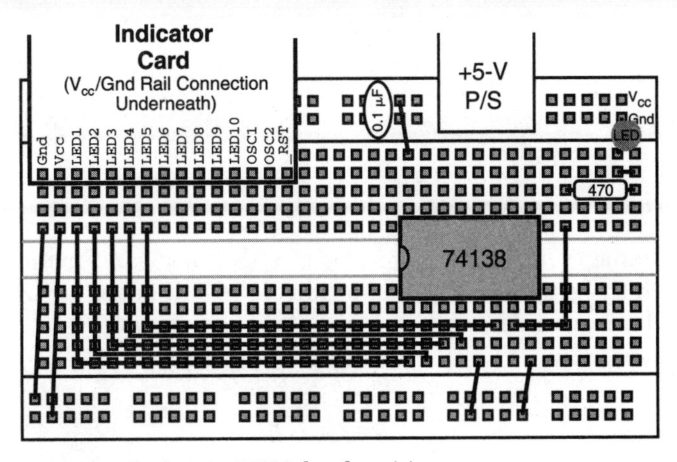

Figure 4.16 Project 4—74138 decoder wiring.

Project 5—Bit Adders

To demonstrate how gates can be combined to form useful circuits, I wanted to use the bit adders. *Adders* are combinatorial digital circuits that add two values together. Adder circuits can be enhanced to subtract, multiply and divide, as well as the other functions that are required to implement the *arithmetic/logic unit* (ALU) functions of a computer processor. In this section, I will start with demonstrating how an adder works and how the basic gates are combined to form the functions.

The most basic adder is the *half adder.* This circuit, shown in Fig. 4.17, takes two inputs and processes them to provide two outputs. The first output is the sum (or S in Fig. 4.17) and is output according to the truth table:

A B	S
0 0	0
0 1	1
1 1	0
1 0	1

where A and B are the inputs to the circuit. This is the XOR boolean function.

If both A and B are set, then the result will not be a single bit, but instead 2 bits. The function that defines whether or not the second bit is set is defined as:

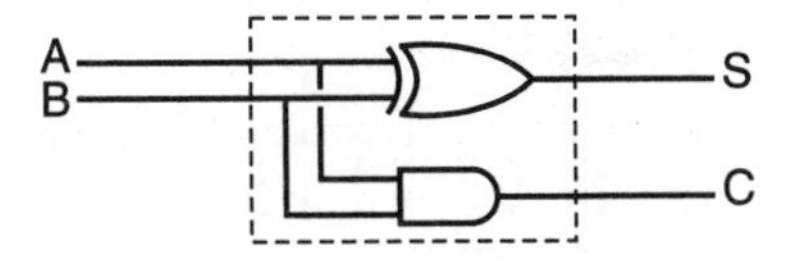

Figure 4.17 Half-adder circuit.

A B	*C*
0 0	0
0 1	0
1 1	1
1 0	0

This is the same truth table as that for the AND boolean function.

Along with the graphical representation of the half adder shown in Fig. 4.17, the half adder can be described by the output equations:

$$C = A \,.\, B$$

$$S = A \wedge B$$

The half adder can be demonstrated by creating the circuit shown in Fig. 4.18.

In this section, I will describe three circuits that make up Project 5; the bill of materials for the three circuits is listed below.

Part	Description
C1–C2	0.1-μF tantalum capacitors
CR1	Two 10-LED bargraph displays (see text)
R5	470-Ω by 8 SIP resistor (see text)
7408	74LS08 quad AND
7486	74LS86 quad XOR
74283	74LS283 4-bit adder with fast carry
220	220-Ω, ¼-W resistor
Miscellaneous	Breadboard, +5-V power supply, interface PCB, wire

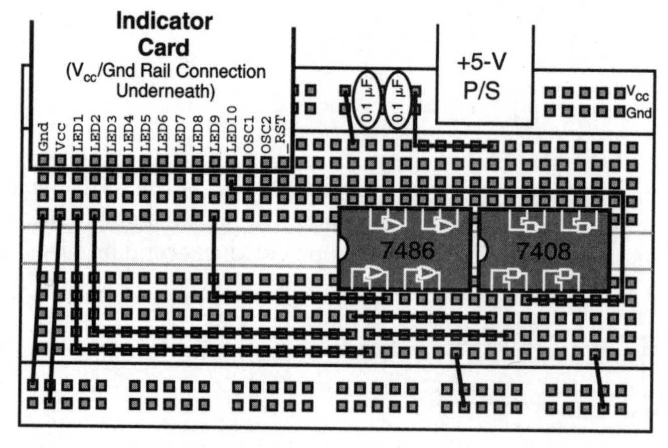

Figure 4.18 Project 5—half-adder wiring.

In Fig. 4.18, I have used the right-most two LEDs of the 10-LED bargraph display as the S and C bits.

In playing around with this circuit, you should find that it works quite well—the problem comes in when you want to add together two values that are larger than 1 bit in size. In this case, you will have to take the carry values from lower bits and add them to the current bit. By combining two half adders with an XOR gate, as shown in Fig. 4.19, you can create a full adder, which adds together three values, A, B, and a carry in (C_{in}).

The outputs for the full adder are the same two bits of the half adder and are defined with the output equations:

$$S = (A \wedge B) \wedge C_{in}$$

$$C = (A \,.\, B) \wedge (C_{in} \,.\, (A \wedge B))$$

The full adder circuit can be built from the half adder circuit that you built originally for this project and is shown in Fig. 4.20.

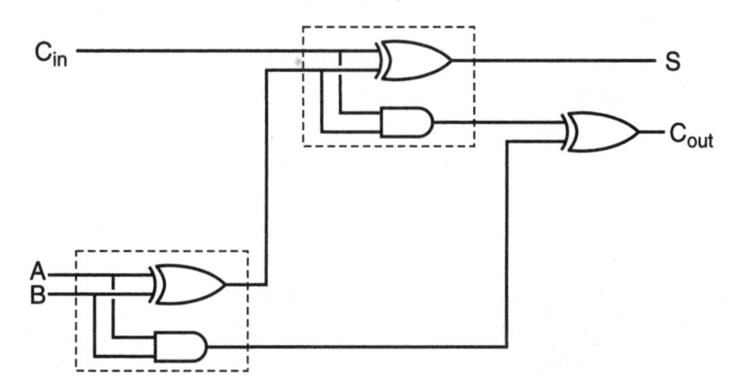

Figure 4.19 Full adder circuit.

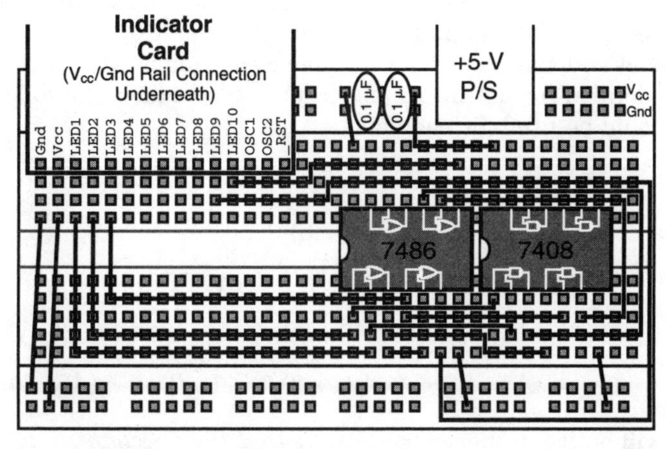

Figure 4.20 Project 5—full adder wiring.

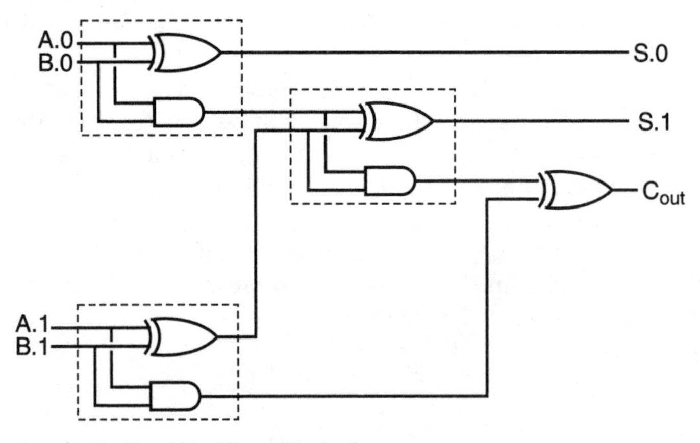

Figure 4.21 Two-bit adder with ripple carry.

After playing around with this circuit, you may want to use the first 4 bits of the switch/LED bargraph combination for a half adder's carry (*C*) bit feeding an input to a full adder. In this case, the second bit and the final carry value will be dependent on the second 2 bits as well as the first 2. In Fig. 4.21, I show how a half adder can be combined with a full adder to add two 2-bit values together.

When I use multiple bit values, I tend to label them as a collection of bits. To differentiate the bits, I will add a (zero-based) bit number after the value label and a "." (dot). This convention will be used throughout the book. By this convention, *A*.0 is the least significant bit of the collection of bits known as *A*. *B*.7 will be the eighth least significant bit of *B* or the most significant bit of the byte labeled *B*. This is a very common convention in digital electronics.

The only time I will dispense with this convention is in cases like the one below where I am ANDing two bits together. In this case, I dispense with the dot and just put the bit number at the end of the label. Switching between the two conventions is not as complex as I have probably made it seem, and if there is anything that seems overly complex, I will note it in the text.

This type of adder is known as a *ripple carry* adder, and, if the output bits were to be defined for a 4-bit (plus carry) output adder, the following output equations could be used for each bit:

$$S0 = A0 \wedge B0$$

$$S1 = (A1 \wedge B1) \wedge (A0 \,.\, B0)$$

$$S2 = (A2 \wedge B2) \wedge (\,(A1 \,.\, B1) \wedge (A0 \,.\, B0)\,)$$

$$S3 = (A3 \wedge B3) \wedge (\,(A2 \,.\, B2) \wedge (A1 \,.\, B1)\,) \wedge (A0 \,.\, B0)\,)$$

$$C_{out} = (A3 \,.\, B3) \wedge (\,(A2 \,.\, B2) \wedge (A1 \,.\, B1)\,) \wedge (A0 \,.\, B0)\,)$$

What you will notice in these equations is that they become more complex, the more significant the bit. In the 4-bit example above, the least significant bit has only one gate delay to calculate the sum, but this goes up to four for the most significant bit and the carry output.

If a gate delay in a chip is 10 nsec, this means that the total result output is not valid for 40 nsec, even though for the least significant 2 bits, the delay is 20 nsec or less. For a generic full adder, the delay will be three gate delays, or 30 nsec. For 4 bits this isn't a problem, but for a large computer processor (say a 64-bit Itanium) this delay becomes intolerable.

The solution to this problem is to come up with a *fast carry* circuit for use in full adders. The fast carry adder uses all the inputs for not only the current bit but also all the less significant bits to calculate a carry value for the current bit. By doing this, a constant number of gate delays through each bit of the adder is achieved.

There are two aspects of this circuit that could be considered ironic. The first is that, even though the fast carry adder is designed to make the circuit implementation simpler, it is much more complex that a ripple carry adder—to the point where looking at the circuit diagram or reading the output equations will not help you. The second aspect of a ripple carry adder is that it typically has a much higher total gate delay than the simple cases shown here. For example, the 74283 that I will be demonstrating below has a constant four-gate delay for each of the 4 bits that are being added together.

This probably seems like I have gone backward in terms of improving the circuit. The reasons for going with a fast carry adder is devices like the Itanium processor, where a 64-bit fast carry adder with a four-gate delay is a tremendous improvement over what would happen with a ripple carry adder.

In the last circuit of this project, I want to show you how to implement a 4-bit adder with carry. This circuit will require all 8 bits of input from the DIP switch along with one additional bit (controlled by a push button), as shown in Fig. 4.22. These 9 bits will be the inputs for the 74LS283 used in the circuit. The five outputs ($S0$ through $S3$ and C_0) are passed to a second 10-LED bargraph display, which is wired according to Fig. 4.22.

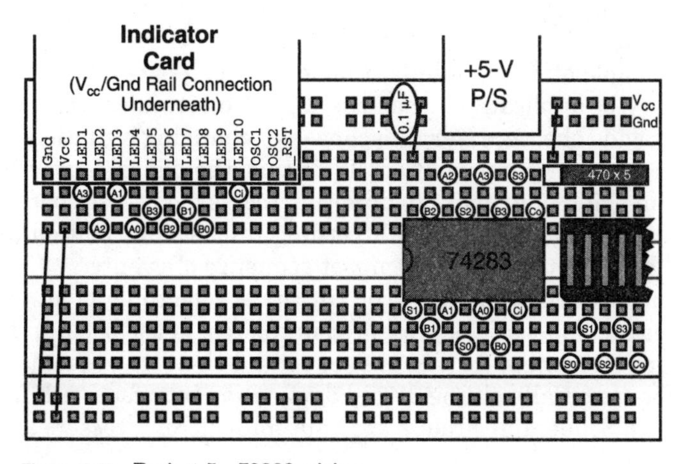

Figure 4.22 Project 5—73283 wiring.

In Fig. 4.22, notice that I did not bother to draw the wiring between the LEDs and the 74LS283. The reason for this was to try and keep the drawing as simple as possible. Instead, the circled letter/number combinations should be matched up between the LEDs and the 74LS283.

The DIP switch/LED combination consists of two inputs; the least significant 4 bits are *A* and the next 4 bits are *B*. The push button bit is considered to be the carry in bit. The five LEDs of the second LED bargraph display are the four sum bits along with the high-order carry bit.

When you try out this circuit, you will probably discover that it doesn't work as you expect. For example, if you set both *A* and *B* to 2 (0b00010), you will discover that the result displayed on the second LED bargraph display is 5 (0b00101) and not 4 (0b000100) as you would expect. The reason for this discrepancy is the use of the momentary-on push button instead of an SPST switch for the carry in. If you were to disconnect the C_i bit between the 74LS283 and the push button, you would notice that the chip behaves as if the C_i bit were still connected to the push button. For example, the result of 2 added to 2 would still be 5.

TTL input pins are "pulled up" internally; they behave as if there is a resistor connected to V_{cc}, and if there is no connection to the pin, then it will behave as if a 1 was input into this line. So far in the book, I have been pretty cavalier about not explicitly wiring unused pins to V_{cc} and ground. This is not good practice and one that you should avoid. Later in the book I will discuss how the input/output pins of TTL chips are designed and how they should be wired when they are not in use.

Sequential and Memory Circuits

The boolean logic circuits that I have so far presented are known as *combinatorial* and provide an output based on a specific set of inputs. This output does not change over time unless the inputs change. Developing combinatorial circuits is quite easy to do and can be a lot of fun playing around with Karnaugh maps and other tools for reducing the functions.

For some applications, combinatorial logic circuits are adequate—but for the vast majority of applications that use digital electronics, combinatorial logic does not provide the necessary responses to changing situations with respect to previous states and information. To do this, the digital logic circuits have to be given some kind of memory in order to process along a set path. Digital logic circuits that are able to respond according to past information are known as *sequential* because they follow a preset sequence of events.

Developing sequential digital logic circuits is something of an art and will be the focus for the applications presented in the rest of the book. In fact, all of the following projects except for one are sequential circuits. Sequential circuits are the basis for essentially all the digital devices you interface with in your day-to-day activities. Sequential circuits are built into microprocessors and other large computer chips.

The basic sequential circuit has input to some logic, which also has input from *state* memory as is shown in Fig. 4.23. The output state is dependent on both the input state as well as the current value of the circuit state memory. With this type of circuit, highly complex operations can be implemented.

For example, a traffic light controller could be implemented by using a modification of this circuit. For this example, I would use a 5-bit counter (which counts from 0 to 31 and starts over at zero again) that is driven by a 1-Hz (1 cycle per second) clock. The circuit logic for this application (shown in Fig. 4.24) would turn on one of the three traffic lights according to the value of the counter.

The *light state counter* consists of a sequential digital logic circuit that increments each time the clock cycles. Later in this chapter, I will show how a counter can be built with the flip-flop circuit that will be presented below along with the full bit adder. The outputs of the light state counter are labeled *C*0, *C*1, *C*2, *C*3, and *C*4 for this example. *C*4 is the most significant bit.

The 5 bits output from the counter can be represented as a 5-bit number, in the format 0b0#####, where # is one of the binary digits. In this data format, the counter output of "5" is written as 0b000101.

Previously in this chapter, I have shown how binary circuits could be used to implement a key that has an active output when specific input conditions are met. This circuit will be used to select which traffic light will be on according to the counter output that I have given above.

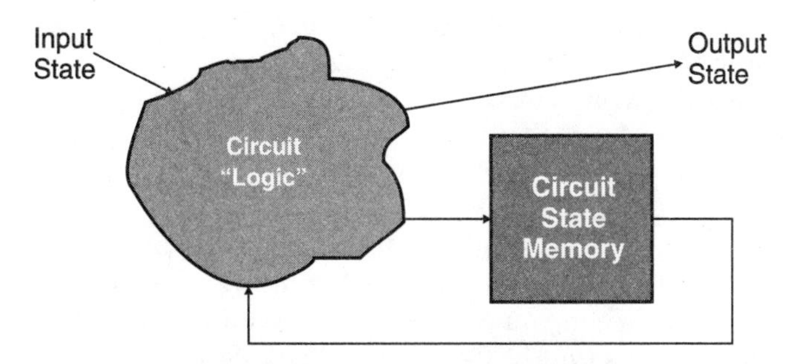

Figure 4.23 Basic sequential circuit block diagram.

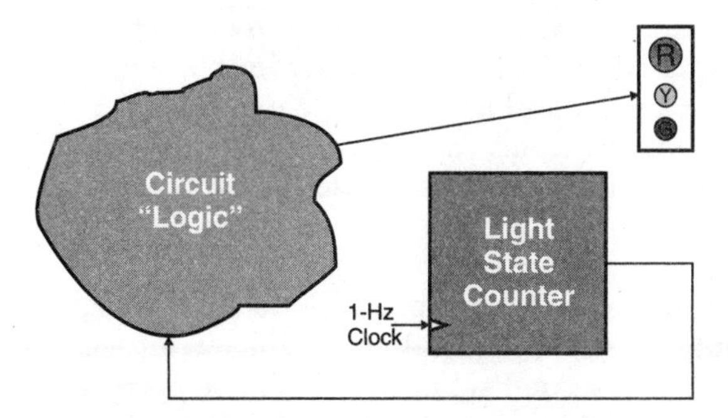

Figure 4.24 Traffic lights as a sequential circuit.

For this example, I am going to assume that the green light is on for 16 of the 32 possible seconds; next, the yellow light is on for 4 seconds and is followed by the red for 12 seconds. When the counter "overflows," or changes from 31 to 0 (0b011111 to 0b000000), the traffic lights will switch from red to green.

Using the data in the previous paragraph, I can write out the key conditions for each light to be on. Green is on when the counter outputs 0b000000 to 0b001111. Yellow is on when the counter outputs 0b010000 to 0b010011, and red is on when the counter outputs 0b010100 to 0b011111. These values can be expressed as sum of products:

$$\text{Green} = !\,C4$$

$$\text{Yellow} = C4\,.\,!\,C3\,.\,!\,C2$$

$$= !\,(!\,C4 + C3 + C2)$$

$$\text{Red} = C4\,.\,C3 + C4\,.\,C2$$

$$= C4\,.\,(C3 + C2)$$

$$= !\,(!\,(C4\,.\,C3)\,.\,!\,(C4\,.\,C2)\,)$$

Looking at the values I have selected, you should be able to see that anytime the most significant bit of the counter ($C4$) is equal to 0 (or reset), then the green light is on. When $C4$ is equal to 1 (or set), then $C3$ and $C2$ must be tested to determine whether the yellow or red light is on.

Note that when I wrote out the equations, I also included equivalent expressions based on De Morgan's theorem as well as the associative theorem for boolean logic. As you begin working with more complex logic functions, this is a useful exercise to carry out because you don't know how many gates and which type are going to be available to you. Looking at the equation for red, the function can be implemented with either an OR and two AND gates or just three NAND gates. When you are creating a complex function out of standard TTL chips, having this type of flexibility will allow you to avoid adding unnecessary chips and make the most of the ones you already have.

This example may seem artificial on two points. The first is that in Fig. 4.24, I do not show any data from the current logic being used to control the counter circuit. The counter circuit (as I will show later in this chapter) is a sequential circuit all on its own and is used as a subsystem in the total sequential circuit system. You have to understand that Fig. 4.23 is really a general-case diagram. It is not unreasonable to have portions of the diagram missing or to have the application sequential circuit built out of smaller "sub" sequential circuits, as is the case with this one.

The second point where this example probably seems artificial is with regard to the simplicity of the application. If this circuit were used in a real-life set of traffic lights, a second set of lights (for the perpendicular direction) would also be required. By using the same clock (and many of the same boolean functions), the second set of lights could be created by assuming green was active from 0b010000 to 0b011111, yellow was active from 0b000011 to 0b000011, and red was on from 0b000100 to 0b001111. A new set of equations for the perpendicular lights could be written out as

$$\text{Green}_p = C4$$

$$\text{Yellow}_p = !\,C4\,.\,!\,C3\,.\,!\,C2$$

$$= !\,(C4 + C3 + C2)$$

$$\text{Red}_p = !\,C4\,.\,C3 + !\,C4\,.\,C2$$

$$= !\,C4\,.\,(C3 + C2)$$

which allows for the total traffic light circuit shown in Fig. 4.25.

While the entire circuit requires only a 5-bit counter, 1-Hz clock, and three each NOT, AND, and OR gates, it will perform the duties of most traffic lights installed in cities up into the 1970s. Not only will it perform the duties of most traffic lights; this is the amount of intelligence that was built into them. It has only been in the last 20 years or so that microprocessors have been built into this type of application to allow response to more complex situations (such as pedestrian buttons or emergency vehicle overrides). Before, circuits like this, which could be implemented in just a few digital electronic gates, were widely used for basic control functions.

The memory portion of the sequential circuit is based on the flip-flop, which is a circuit that provides feedback from the outputs into the inputs. When you first look at a flip-flop circuit, its operation will seem unusual at best and it probably seems like some law is violated in any case. This isn't true, but as you can see in Fig. 4.26 the circuit appears to be somewhat strange.

The output of each of the two NAND (AND gates with an inverter on the output) gates is used as an input of the other. When I first saw this, I imagined a signal racing around the two NAND gates. But this is the wrong way to

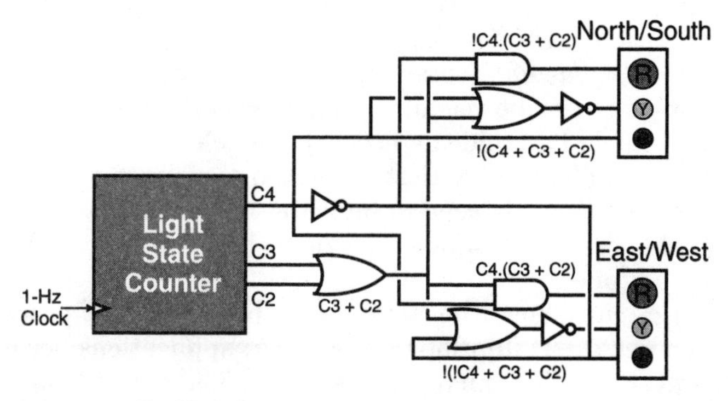

Figure 4.25 Traffic lights circuit.

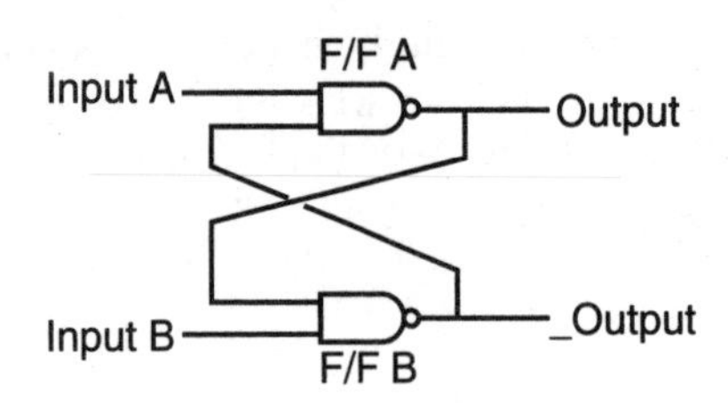

Figure 4.26 Simple flip-flop.

visualize the circuit. Signals do not "race" between the gates, but state information is stored when the two input lines go high.

When input *A* and input *B* are high, then the previous state of the NAND gates is stored in the flip-flop. For example, if input *A* was initially set (equal to 1 or at a high voltage) and input *B* was reset (equal to 0 or at a low voltage) and then both inputs were set, the flip-flop would output a high voltage on output *A* and it would output a low voltage on output *B*.

Only one of the two inputs can be pulled low at any one time to load in a bit to save in the flip-flop. Once the state is set, then both lines are pulled high and the state is stored. The truth table for the flip-flop is:

Inputs	Outputs	
A B	*A B*	Comments
1 1	A_o B_o	Saved
0 1	1 0	Set *A* high
1 0	0 1	Set *B* high
0 0	N/A N/A	Invalid

Both inputs 0 are invalid because they will result in both the outputs being high. When the two input lines are brought high, the outputs will initially go low and will enter an unstable state in which the flip-flop will jump to a data state based on any imbalance. This imbalance can be caused by one line rising faster than the other line or even an errant electron charge that is not balanced. The condition of the flip-flop at this unstable state before it goes to one state is known as *metastable* because it can go equally easily to either state. Metastable flip-flops can be useful in some specific applications, but for the most part they should be avoided wherever possible.

In the truth table above, note that the saved state outputs have the value *o* beside their letters. This *o* is used to indicate that the values being output are the same as the previous values—previous being whatever was the state before the current one.

The flip-flop circuit shown here, while usable, isn't that practical in most circuits. I often use the D flip-flop (Fig. 4.27) in applications because it can be more easily interfaced to a microprocessor bus. The data in state is stored in the input when the clock line is brought from low to high. This is usually described as catching in data on the rising edge of the clock.

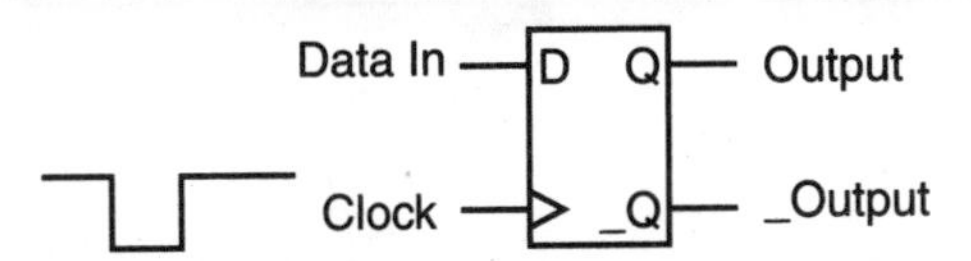

Figure 4.27 D flip-flop.

The next project will show how the D flip-flop is implemented and how it works. The D flip-flop is often used in counters as well as other sequential circuit chips because it is very easy to work with and changes state on the rising edge of the input clock or latching signal.

The toggling flip-flop, or T flip-flop (Fig. 4.28), will change the output state each time the clock input is a rising edge. T flip-flops are often used to divide a clock by 2. This is because, if a clock is input, the outputs will change states at one half the rate of the clock. As can be seen in Fig. 4.28, this results in outputs that have 2 times the period of the clock.

The set/reset flip-flop (Fig. 4.29) is often referred to as an *RS flip-flop.* The output (*Q*) is set on reset depending on whether or not an input is pulled low. The truth table for the RS flip-flop is:

Inputs	Outputs	
R S	*Q _Q*	Comments
1 1	Invalid	
0 0	Q_o $_Q_o$	Saved
0 1	0 1	
1 0	1 0	

The last flip-flop is the JK (Fig. 4.30), which provides a similar function to the RS flip-flop except when both lines are high, the outputs are toggled. The JK truth table is:

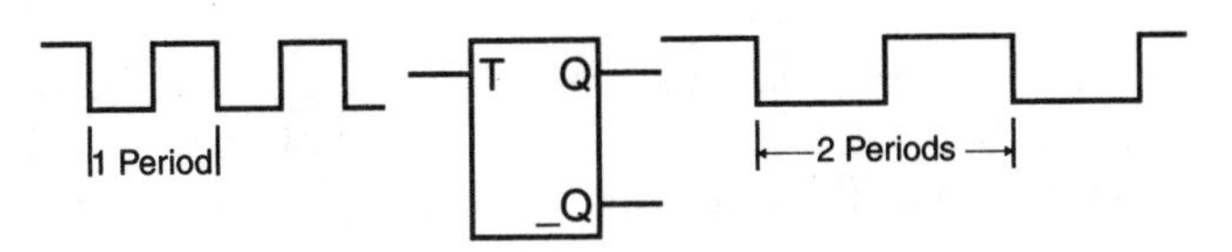

Figure 4.28 T flip-flop.

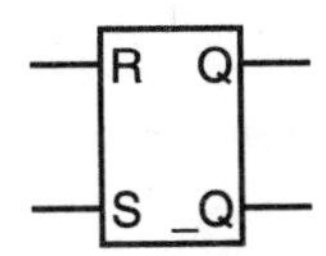

Figure 4.29 RS flip-flip.

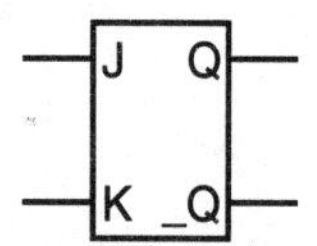

Figure 4.30 JK flip-flop.

Inputs	Outputs	
R S	*Q _Q*	Comments
0 0	Q_o $_Q_o$	Saved
1 1	$_Q_o$ Q_o	Toggled
0 1	0 1	
1 0	1 0	

Later in the book I will discuss the topics of clocks and oscillators and how they can be used to help enhance sequential circuits. Before leaving the topic of sequential circuits, I want to make you aware of a situation that can arise that will cause problems with your application and should be avoided at all costs.

A very simple digital electronic oscillator, known as a *ring oscillator,* can be created from nothing more than a pair of inverters as I have shown in Fig. 4.31. This oscillator will run at the switching speed of the inverter.

There are times when ring oscillators are required, but for the most part they should be avoided at all costs. Looking at Fig. 4.31, you probably feel that this should be pretty easy to avoid because there won't be very many cases where you will intentionally feed back the output of an inverter to itself.

The problem comes in when the situation like the one shown in Fig. 4.32 is encountered and there is an inadvertent feedback path created with the input being inverted to the output that is fed back. These situations can be quite insidious because the circuit will oscillate only when the inputs to the two NAND gates are set (high or 1 levels). In this case, you may find oscillations at specific instances of the circuit's operation, which can be very difficult to find.

I am pointing this out to you because once you find that you have an oscillating circuit, you can look back through your schematics to find a situation where an output feeds back to one of the inputs in the circuit that generated it. To fix the

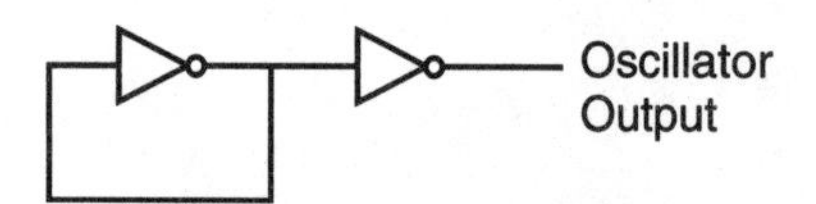

Figure 4.31 Ring oscillator circuit.

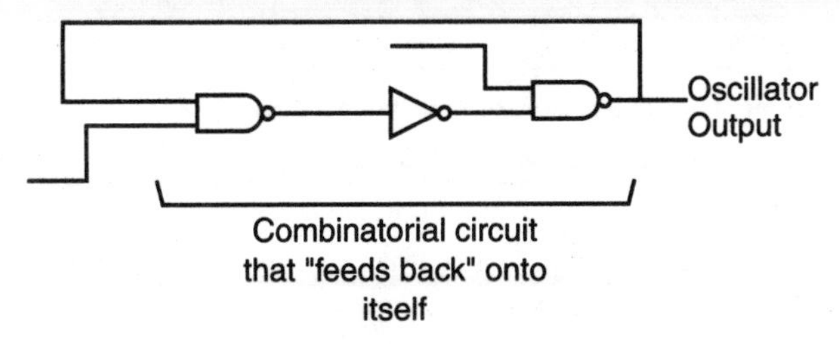

Figure 4.32 Inadvertent ring oscillator.

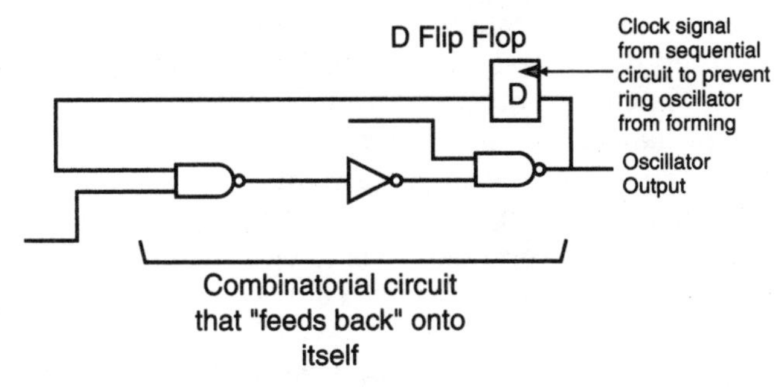

Figure 4.33 Fixed inadvertent ring oscillator.

problem, you may have to put in a flip-flop, as in Fig. 4.33, to prevent the ring oscillator from being able to start up.

I recommend that, as a matter of course, you put in the flip-flop as part of your sequential circuit if there is any combinatorial output that feeds back to one of the inputs that generates its signal.

Project 6—D Flip-Flop

When I develop my own applications, I usually turn to the D flip-flop first because of its ease of working use as well as the flexibility that it has when added into a circuit. The D flip-flop can be used as a straight storage device in a digital logic circuit or it can be used in an intelligent system as the basis of a peripheral. In this project, I will show how the D flip-flop is implemented and show how a common integrated D flip-flop chip works.

In this book, I will be using several chips with D flip-flops built in. Each chip is configured differently, but they all contain the negative-edge-latch D flip-flop. At the end of this section, I will discuss what type of situations each of these chips is best suited for. The table on the next page lists the different D flip-flop chips that are used in this book and their characteristics:

TTL number	Characteristics
7474	Dual D flip-flop chip with presets and clears
74174	Hex D flip-flop chip with common clock and clear
74373	Octal D flip-flop with transparent latches, common clock, and tristate outputs
74374	Octal D flip-flop with common rising-edge clock and tristate outputs
74573	Same operation as 74373, but with different pin-out
74574	Same operation as 74374, but with different pin-out

The D flip-flop is a basic memory device that stores the data driven onto its D input when the clock signal is falling (transitioning from a 1 to a 0). Figure 4.34 shows how the D flip-flop is designed with *Q* and *!Q* being the stored outputs.

The storage (or "latching") of the bit value input at *D* on the clock's falling edge is actually an advantage. In the traditional flip-flop, the two input pins must be controlled in order to store a logic value. In the D flip-flop, the state of *D* can change, but it will not affect the flip-flop's output values (*Q* and *!Q*) until a falling clock edge is received. This allows the D flip-flop to be used as a bus device, in which information is saved only when a specific device address appears on the bus.

The D flip-flop can be created very simply with 7400 two-input NAND gates. As I have noted above, the NAND gate is the basic TTL technology gate, so the circuit shown in Fig. 4.34 is actually at the lowest gate level possible. This means that the circuit will pass the signals with a minimum number of hidden gate delays. The D flip-flop circuit can be built into the application shown in Fig. 4.35.

The circuit I created for this application is a bit "squished," but you should be able to manage to build it quite easily. The breadboard wiring diagram, Fig. 4.36, follows the pin numbering in the schematic shown in Fig. 4.35.

The bill of materials for this project is:

Part	Description
C1–C2	0.1-μF tantalum capacitors
7400	Two 74LS00 quad NAND gates (labeled U1 and U2 in Fig. 4.35)
7474	74LS74 dual D flip-flop
Miscellaneous	Breadboard, +5-V power supply, wire

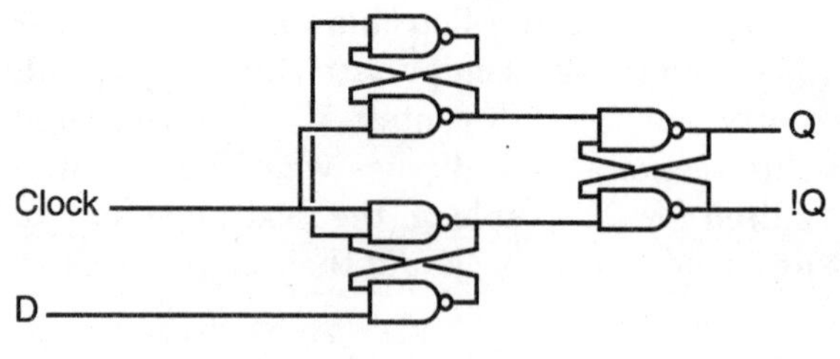

Figure 4.34 D flip-flop circuit.

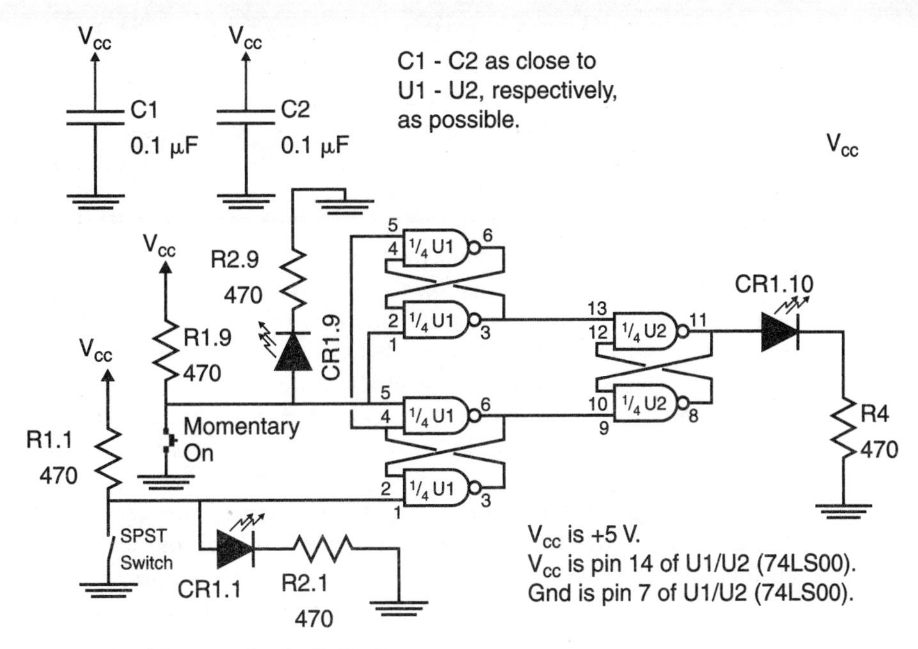

Figure 4.35 Schematic for the D flip-flop.

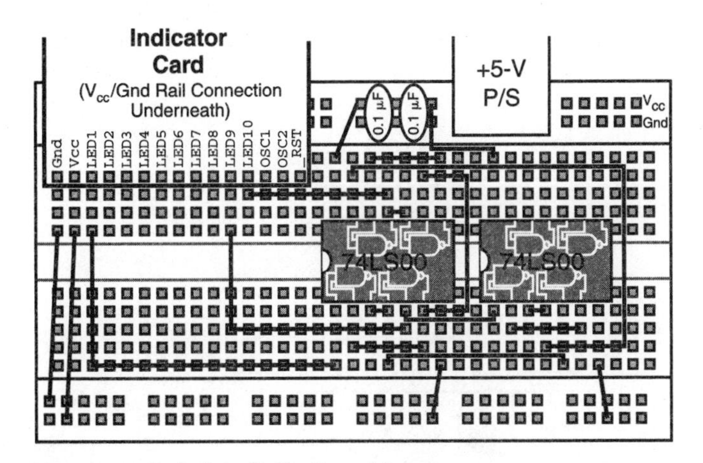

Figure 4.36 Project 6—D flip-flop wiring.

When this circuit is in operation, you will find that it will power up according to the state of the left-most switch. When the state of this switch is changed, you will see the corresponding LED on the bargraph display change, but not the right-most LED. To change the right-most LED, you will have to push the momentary-on switch. When you do this, you will notice that, when the push button's LED turns off, the new data is saved in the flip-flop. The bit value saved in the flip-flop will remain unchanged until either the D input is changed and the clock line goes low or power is removed from the circuit.

As I pointed out at the start of this section, along with building the D flip-flop yourself out of basic TTL gates, you can also buy TTL chips that have these functions built into them. The 7474 has two D flip-flops built in and can have its *Q* output preset or put into a clear (or 0) state. To demonstrate how this chip works, I created the circuit shown in Fig. 4.37 and wired it as I have shown in Fig. 4.38. This circuit works exactly the same as the previous circuit in this project—when the momentary-on button is pushed, the output *Q* is set at the state of the *D* (left-most switch) input.

You could put push buttons on the _PR (preset—set *Q*) or _CLR (clear—reset *Q*) to demonstrate how they work, but I found that simply toggling the switches indicates what their function is. _PR will set the *Q* output, or make *Q* output a 1 or high voltage. _CLR will reset the *Q* output. These two pins give the 7474 similar capabilities to the RS flip-flop, but with the ability to latch data in. This is why I consider the D flip-flop one of the most useful memory devices that you can work with.

At the start of this project, I gave you a list of the different chips that are based on the D flip-flop. I want to go through each part number and give you an idea of what they are good for and some applications they are used in.

The 7474 can be used in a variety of different general-purpose applications, including state storage or button debouncing (this will be shown later in the book). I find the 74174 six bit (or "hex") flip-flop most useful as a small "shift in, parallel out register. If I need a full 8 bits for a computer system bus, then I tend to use the 74374 (or the 74574, which is exactly the same in operation,

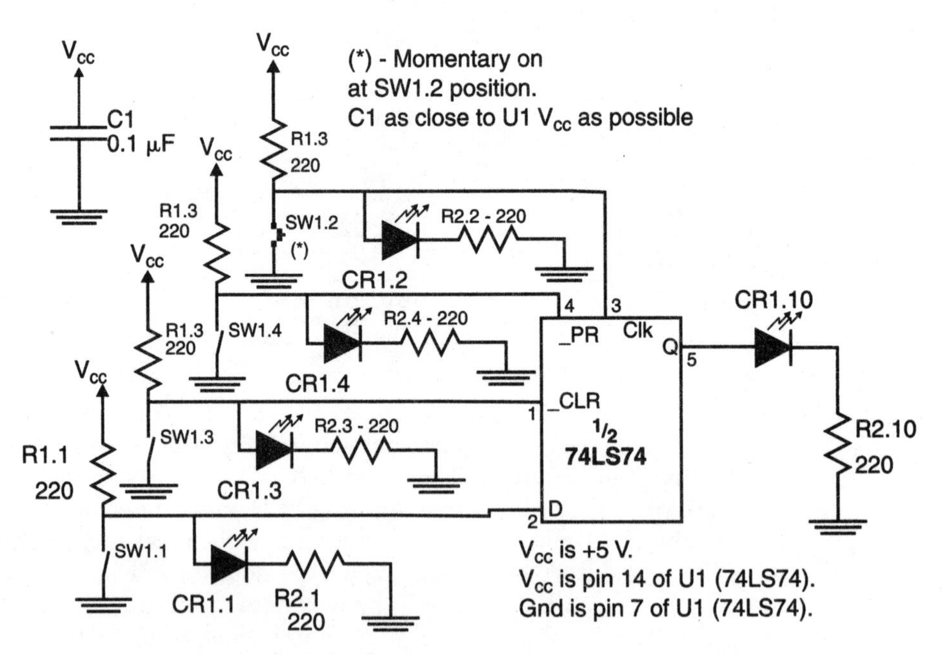

Figure 4.37 Schematic for the D flip-flop using 7474.

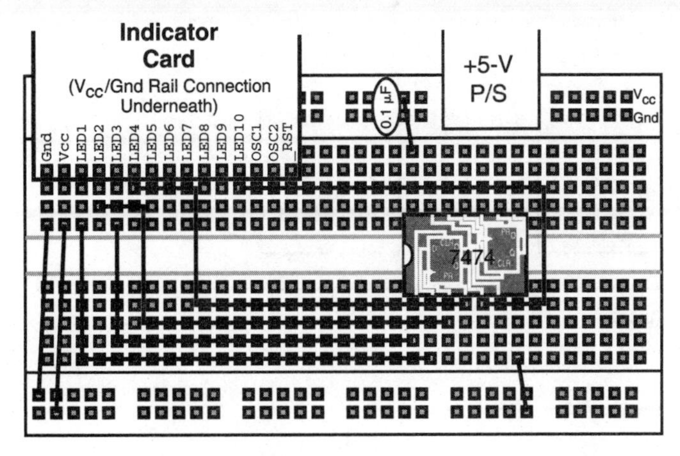

Figure 4.38 Project 6—7474 test layout.

but has a different pin-out). The 74374 can also be used as an 8-bit shift register, and, with its tristate outputs, it can be used for saving or loading data from a processor's bus.

The 74373 (and its pin-incompatible cousin, the 74573) do have D flip-flops, but they are considered to be "transparent" latches, which is to say that when the clock is low (0), the data that is being driven on *D* is passed to *Q*. When the clock goes high (1), the data driven out on *Q* is saved and will not change until the clock goes low again.

The 74373 is used in applications where a computer processor has a multiplexed address and data bus, with the address being driven first on the bus. By using a transparent latch, the 74373 passes the address to the system's address decoders and memory circuits, giving them some extra time to set up before the read/write is requested. In describing what the 74373 is best suited for, I have discussed some concepts that I have not yet explained.

Project 7—Binary Counter

One of the most useful functions that you will use when you develop digital electronic circuits is the counter. The counter will be a basis for many of the following circuits and will be used in a variety of different situations to time applications and help them run in an orderly fashion. I will also be using the counter a lot in the state machines that I will build into the projects presented later in this book.

The basic counter circuit consists of a set of flip-flops that drive into and are driven from an adder. This counter circuit is shown in Fig. 4.39.

The *counter clock* will be discussed at length later in the book. This circuit is a repeating digital signal that will cause the value in the D flip-flops to be added to 1 and saved into the flip-flops as the next count value. In the circuit, I hold the second set of inputs to 0 and add the 1 as the carry.

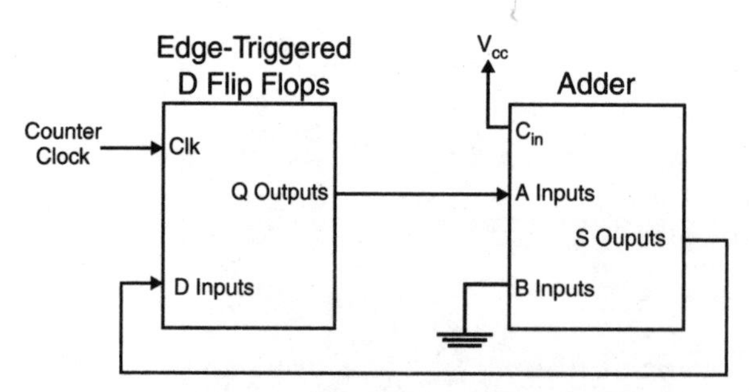

Figure 4.39 Basic counter circuit.

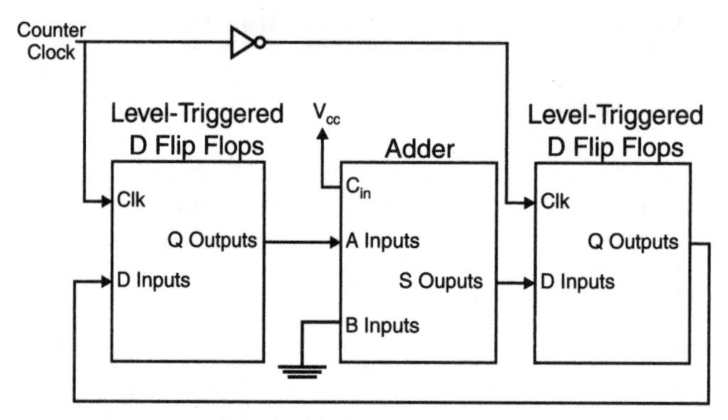

Figure 4.40 Needlessly complex counter circuit.

The use of edge-triggered flip-flops is a very important aspect of the circuit shown in Fig. 4.39 and one that you should keep in mind. When the counter clock changes state, the output value of the adder (which is the D flip-flop value plus 1) is saved in the D flip-flop as the new saved value.

When you first look at the circuit, you probably wonder why, when the new value is saved in the flip-flop, it isn't incremented in the adder, saved in the flip-flop, and so on. This is the reason why I have specified an edge-triggered flip-flop in Fig. 4.39.

If an edge-triggered flip-flop were not used in the circuit, then you would have to use something like Fig. 4.40. In this circuit, I have put in two-level triggered flip-flops, each one out of phase with each other. This is to say that when the clock is high, one flip-flop is storing the data while the other is changing to the new value. When the clock goes low, the first flip-flop changes to the value in the second flip-flop and the second stores its current data.

I feel that implementing a counter in this way is unnecessarily complex and potentially very slow—the extra set of flip-flops will slow down the performance of the counter and limit its maximum speed.

The basic counter circuit can be built with the 74LS174 hex D flip-flop and the 74LS283 that were used in earlier projects. The circuit shown in Fig. 4.41 will demonstrate how the counter works.

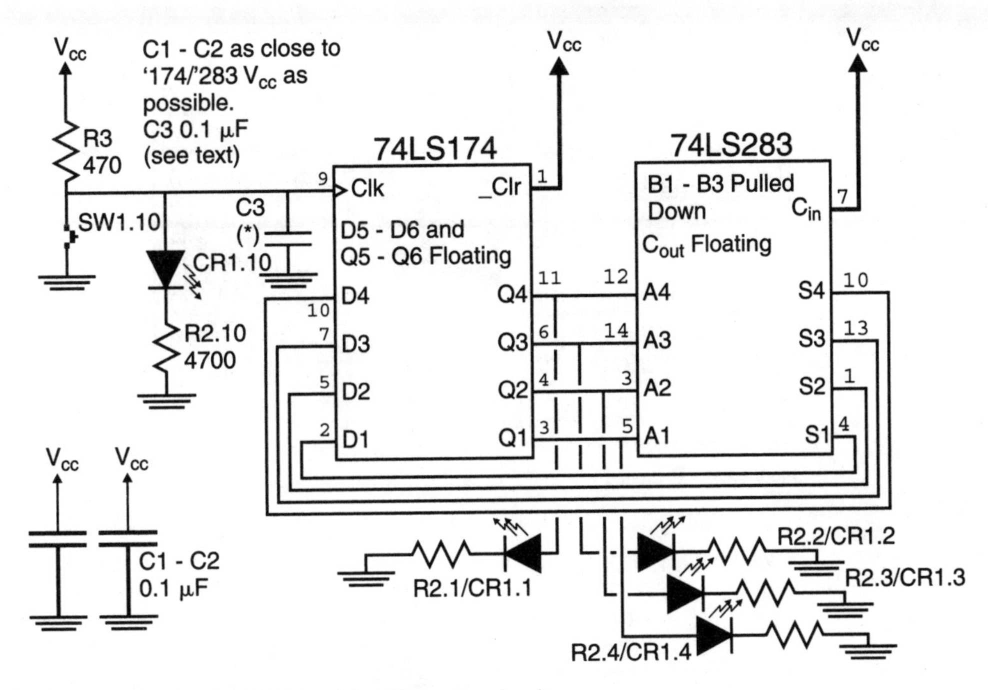

Figure 4.41 Project 7—74174 and 74283 counter circuit.

The bill of materials for Project 7 is:

Part	Description
C1–C3	0.1-µF tantalum capacitors
74174	74LS174 hex D flip-flop
74283	74LS283 4-bit adder
74193	74LS193 four-bit counter
Miscellaneous	Breadboard, interface PCB, +5-V power supply, wire

The circuit itself is fairly easy to wire, but with a few twists that can be seen in the breadboard wiring diagram Fig. 4.42. In this circuit, I have used the push button and discrete resistor as the clock (this is similar to its use in Project 6). The difference is the optional capacitor that I have added to the circuit. I will discuss the need for this capacitor below.

When you try out this circuit, the first thing that you will probably notice is that when you press the button, the LEDs will not increment by 1, but by 2, 3, or even 4. The reason for this is *switch bounce,* which is caused by the switch not making immediate contact with the internal contactors and instead bouncing off them several times. Later in the book I will discuss different strategies for eliminating switch bounce, but for now, if you wire a 0.1-µF tantalum capacitor between the clock pin (LED10) and ground, you should minimize this problem (although you will probably not eliminate it).

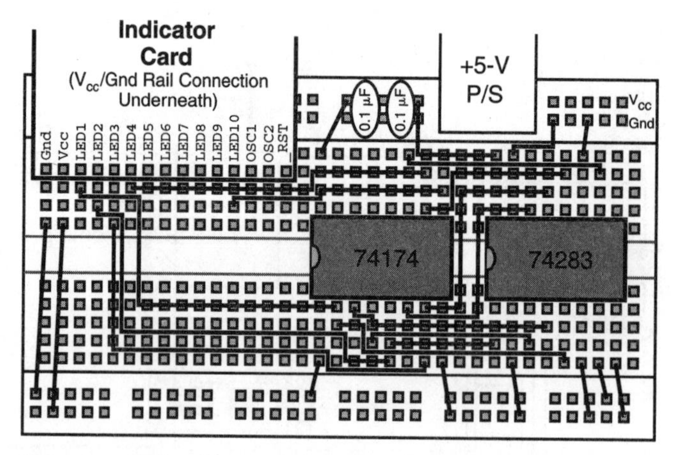

Figure 4.42 Project 7—two-chip counter circuit.

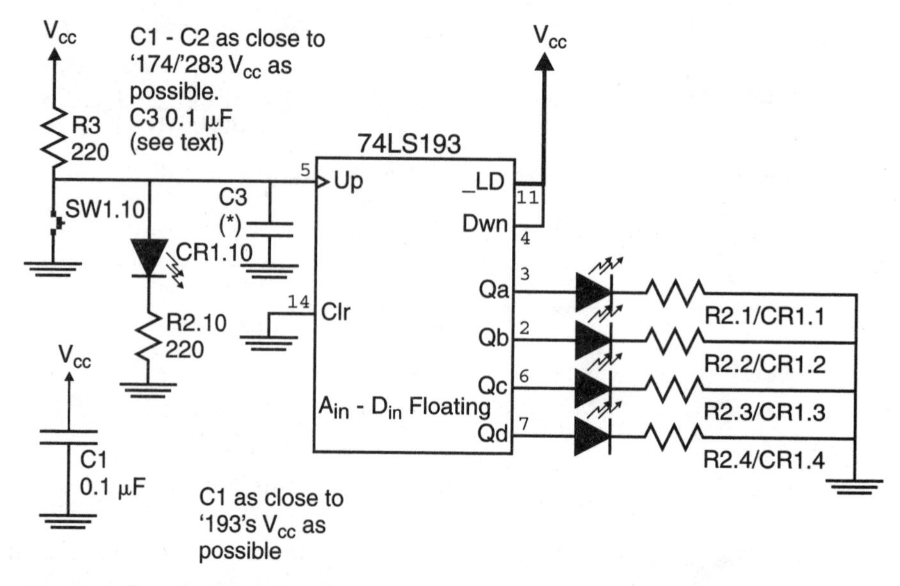

Figure 4.43 Project 7—74193 counter circuit.

The counter circuit should work well for you. As with the previous projects, a single chip can be used where multiple chips would otherwise be required. The counter chip that I usually work with is the 74LS193, which combines the D flip-flop and the adder along with the ability to decrement the result. Later in the book, I will show how this chip can be used with others to *cascade* from a 4-bit counter to an 8- and 16-bit counter.

With the 74LS193, the 4-bit counter project becomes the circuit shown in Fig. 4.43. This circuit is quite a bit simpler than the previous circuit for this project and the wiring diagram, Fig. 4.44, reflects this.

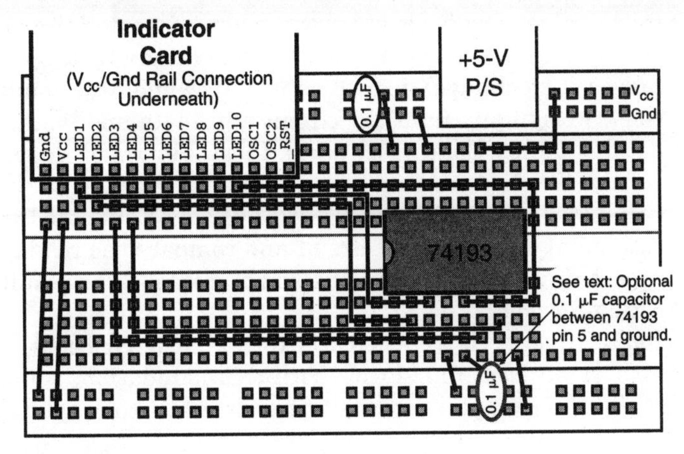

Figure 4.44 Project 7—74193 test wiring.

When I built this circuit the first time, I had a number of different problems centered on the part I chose to use. The first time I built the circuit, I used a 74HC193 instead of a 74LS193 because I had the CMOS (HC) part on hand. I admit I was in a hurry and I thought the CMOS part would substitute reasonably well for the TTL part. This bit of arrogance (I neglected to check a datasheet) turned out to be a big mistake and cost me about two hours in trying to get what I thought would be a simple circuit working.

As I have discussed in this chapter, I have left unused inputs disconnected (or *floating*) because I relied on the internal pull-ups of the TTL parts. The CMOS (HC) parts do not have built-in pull-ups on the input pins. Instead, they float. This caused problems when the switch was pressed, and the down counter of the chip seemed to toggle, causing the value after the switch press to be less than the value before. I could also see the problem by tapping on the chip with my finger and watching the LEDs change state for no apparent reason. Tying down (pin 4) to V_{cc} eliminated this problem.

But this wasn't the end of my problems. Another problem was that the part could not provide enough current to drive the LEDs safely. These parts are limited to about 5 to 8 mA of sink capability, which was not enough for the LEDs, and the chips seemed to behave and reset unexpectedly when the counter was working. When I disconnected the chips from the LEDs and used a logic probe to monitor what they were doing, I found that the chip was working much better—all the outputs could toggle.

I replaced the 74HC193 with a 74LS193 and I found that the problems I had encountered before were pretty much eliminated. The chip could work with down left floating, and it drove LEDs without any problems. The lesson in this exercise for me was that I have to remember that different families of 74 logic parts work differently in their input and output characteristics.

Of course I *knew* this before I started the project. You will find that you make mistakes like this when you are in a hurry and don't have the correct parts on hand. The net result for me was 2 hours of wasted time trying to figure out why the counter was behaving strangely (and why tapping on it with my finger caused it to change) as well as why it would only count to 3 before the LEDs reset. I originally built the circuit quite late and instead of waiting until the next day to buy the correct parts, I lost 2 hours of sleep.

When you have finished reading through this section and working with the counter circuit, please don't disassemble it. In the next chapter I will use some basic test equipment to look at the circuit and compare the readings to what I expect. I suggest that you carry out the same process—the results should be quite interesting for you.

In this chapter, I have introduced you to the basic concepts, circuits, and chips behind boolean electronic logic. With this foundation, you are ready to start looking at more complex circuits and some practical applications.

Chapter

5

Power Supplies

One of the most overlooked areas of digital electronic application development is the specification and design of power supplies. While many applications are very insensitive to the power supply design used, a properly specified power supply can make an application safer, much more inexpensive, reliable in hostile environments, and better able to handle changing power loads.

For DC circuits, the power required by a circuit is defined by the equation

$$P = V \times I$$

Where P is the power consumed (in watts), V is the voltage applied to the circuit, and I is the current (in amperes, or amps) drawn from the power supply. For example, in a +5-V digital circuit, if 0.15 amps is drawn from the power supply, then the circuit is *dissipating* 0.75 W or 750 milliwatts (mW). Since most digital electronic circuits require +5 V, the primary focus of power supply specification and design is how much current is required by the application and the parts used within it.

As you may be aware, many new high-speed CMOS logic families require 1.8 or 3.3 V. For the purposes of this discussion, I will just concentrate on +5 V, as it will be the logic supply you will most likely work with. Other voltages may be required, but the currents they supply are often measured in milliamperes; the +5-V supply (which I normally refer to as V_{cc}) is often measured in amps. This can be shown in the example of a 250-W PC power supply. You will find that the −12- and −5-V supplies can output 100 mA or less (for a total of 1.7 W out of the total) and the +12-V supply provides up to 1 amp (for 12 W supplied by the power supply). This leaves over 236 W of power devoted to the primary logic—which, for the formula above and a primary logic voltage of 5 V, is over 47 amps. If the PC's processor runs on 3.3 V, the total available current is more than 71 amps! To give you an idea of how much current this is, note that most houses have what is known as a 100-amp service. This means they

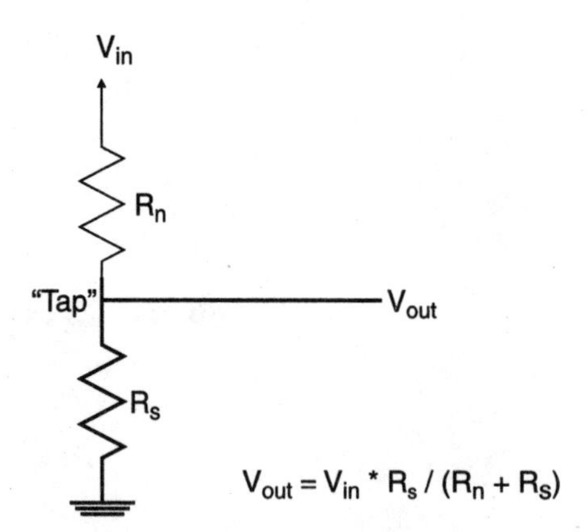

Figure 5.1 Voltage divider circuit.

have circuits that are capable of supplying current at 100 amps, only about twice what a PC running with a +5-V, 250-W power supply requires.

Earlier in the book, I introduced the *voltage divider* (see Fig. 5.1), which is used to convert an input voltage to a smaller output voltage. This circuit could be used with a *rectifier*, which converts alternating current into direct current (and will be discussed in this chapter). If the input voltage is 110 V, from the formula for the voltage divider:

$$V_{\text{out}} = V_{\text{in}} \times \frac{R_s}{R_s + R_n}$$

and

$$V_{\text{out}} = 5 \text{ V}$$

$$V_{\text{in}} = 110 \text{ V}$$

To find R_s and R_n, I could go back to Ohm's law (assuming I wanted 50 amps at +5 V):

$$R = V/I$$

where $V = 5$ V
$I = 50$ amps
$R = 0.1\ \Omega$

This R is actually R_s in the voltage divider formula and can be used to calculate R1:

$$V_{\text{out}} = \text{V}_{\text{in}} \times \frac{R_s}{R_s + R_n}$$

$$5 \text{ V} = 110 \text{ V} \times \frac{0.1\ \Omega}{R_n + 0.1\ \Omega}$$

$$R_n + 0.1\ \Omega = \frac{110\ \text{V}}{5\ \text{V}} \times 0.1\ \Omega$$

$$R_n = (22 \times 0.1\ \Omega) - 0.1\ \Omega$$

$$= 2.1\ \Omega$$

Depending on your familiarity with electronics, these values for R_n and R_s may not seem unreasonable. In actuality, there are five problems with powering the PC with this circuit. From the formula at the start of this chapter, the power dissipated in the R_n resistor can be defined as:

$$P = V \times I$$

$$= V \times V/R$$

$$= V^2/R$$

$$= (105^2)/2.1$$

$$= 5250\ \text{W}$$

If a common household hair dryer provides 1500 W of heat, this means R_n will be dissipating the same amount of heat as more than three hair dryers. Or, if a horsepower is 741 W, more than seven horsepower of heat will have to be dissipated by this circuit. This is a huge amount of heat to dissipate and will be a significant problem to deal with.

The second problem is the efficiency of the circuit. If 250 W is used out of the total 5250 W available, the efficiency of this power supply is somewhere around 5 percent. Coupled with the heat that must be dissipated by the Voltage Divider of the power supply, this makes this type of power supply very expensive to run.

The third problem with this circuit is that it will be very hard to build in the real world. In the voltage divider circuit, the PC could be approximated to the 0.1-Ω load, but this draw changes over time as different circuits in the PC are used. It is not unusual to see a PC drawing 10 percent less current in some circumstances.

In this case, if the current required by the PC is reduced by 10 percent (5 amps), the voltage divider changes. This changes the voltage across the R_n resistor to:

$$V = R \times I$$

$$= 2.1 \times 45\text{V}$$

$$= 94.5\ \text{V}$$

or, if 110 V is input, 15.5 V is applied to the PC logic. Going in the other direction (10 percent more current), the voltage applied to the PC's logic drops dramatically (to almost zero volts). Thus a decrease in current requirements may not be a problem, but increasing current requirements results in the PC not being able to run properly (if at all).

Another aspect of this problem is that most household wiring will not handle 50 amps of current. You will find that the resistance of your household wiring will become a major factor in the circuit (if it doesn't get so hot as to set fire to your home first).

The fourth problem is the variability of household power supplies. It is not unusual in North America, where the household voltage is nominally 110 V, for it to range between 100 and 120 V. This gives almost ±10 percent variability to the PC logic power, which will definitely cause problems with the PC's operation. This problem is exacerbated by situations that have only 220 V available (as in Europe) and how a dual-source power supply would be implemented.

The fifth issue with this power supply is safety. Obviously, the heat generated by the 2.1-Ω resistor is going to be a problem to deal with, and the current passing through the home wiring will cause the household wiring to get very hot. A significant issue is that 110 V at 50 amps will be available and is more than enough to burn or kill somebody unlucky enough to accidentally touch the power going into the circuit while connecting the PC while the power is on. Ideally, the power supply should be able to detect a problem and "crowbar," or shut off, when there is a problem.

The solution to these issues is the application of a specialized circuit, known as a *power supply.* In this chapter, I want to introduce you to some simple power supply circuits that have these characteristics:

1. They are safe for their users and designers.
2. They are relatively efficient in terms of the amount of power that is lost converting household current to logic level currents.
3. They provide very accurate voltage levels, independent of the voltage input or the current required by the application.
4. They are inexpensive.
5. Their design is optimized for the application that they are providing power for.

A few important points about what is presented in this book: I will be focusing on creating power supplies for applications that require 1 amp or less of current. While I used a 250-W supply as an example in this introduction, the methodologies and circuits required for producing this much power are quite a bit different than what is required for the simple power supplies presented here. Advanced degrees are normally required for properly designing high-current power supplies that work at high efficiencies.

In this chapter, I will focus on converting household AC current to DC for use in a digital application. Alkaline radio batteries may be considered an optimal method of powering a digital circuit, but they have a limited life and are much more expensive to use than power from the household mains. Batteries should be considered only for portable applications or ones that require a backup, in case the main power is lost.

Note that the information provided here is also not appropriate for use in audio electronic circuits. While the information provided here may seem appropriate for low-power audio applications, I would recommend that they not be used for these applications simply because they may pass "noise" from the original power source to the circuit. Chances are if you used one of these circuits in an audio application, you would get a significant amount of 60-Hz "hum."

The information presented in this chapter will be enough for you to design your own power supplies for your own simple digital logic circuits. The information will give you enough of an understanding so that when you are building somebody else's circuit, you will understand whether or not the power supply is appropriate.

Mains Voltage Conversion

By far, the most popular way of providing power to an application is by simply plugging the circuit into a wall socket. For the most part, this method is cheap and reliable, but working with mains power does have some risks associated with it. In this section, I want to introduce you to the issues of working with 110 (or 220) V alternating current (AC) and using it to power your applications. The +5-V power supply PCB is an application that allows you to build a simple power supply for digital logic applications.

I must caution you that the power coming out of your wall socket can conceivably destroy your application, cause a fire, burn you, or even electrocute you. Despite the fact that it is commonly used for appliances, light, and electronic devices in the home, it is not to be trifled with. In this book I have provided a circuit that can be built safely and inexpensively. I highly recommend that this circuit or a commercial bench power supply be used for your digital electronic projects.

Power coming from your wall sockets (the mains) comes in as either a 110- or 220-V peak-to-peak sine wave with a frequency of 50 or 60 cycles per second [or hertz (Hz)], as is shown by Fig. 5.2. In North America, power is provided at 110 V or more peak-to-peak voltage (typically 115 V) at 60 Hz. Different

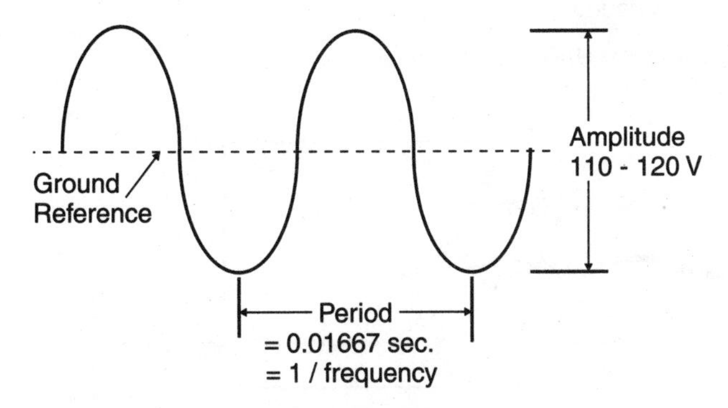

Figure 5.2 North American 110-V, 60-Hz power.

countries around the world will use different peak-to-peak voltage levels and operating frequencies. If you are going to design a power supply for a specific country's use, make sure you understand what are the characteristics of the power supply. The +5-V power supply PCB provided in this book may not be appropriate for your country.

Also, make sure you understand what are the legal requirements for circuits that plug into mains power for specific countries. The information provided here is strictly rule of thumb to make sure that the circuit is safe for use in North America. The information contained here may be incorrect or illegal for different jurisdictions within North America. These safety and legal issues are the primary reasons why I recommend that you avoid building your own mains-interfacing power supplies unless there are compelling reasons to do so.

This power coming in is normally provided by a socket, which is built into your walls. Figure 5.3 shows the layout of the socket and labels the individual connections. *Power* is the alternating voltage sine wave shown in Fig. 5.2. *Neutral* is the return path for this current, while *ground* is a shunt to *earth ground* if the circuit gets damaged and the live voltage is passed to the neutral connection.

This power input is called *alternating current* (AC) and works somewhat differently from the basic direct current (DC) circuits discussed elsewhere in this book. Note that the *ground* is not used as a return path as it is in the DC circuit; the *neutral* line should be considered as the only return path for current. Ground, as mentioned before, is an emergency return path if the circuit or power cord running from the socket to the device is damaged. Normally ground is connected to the device's case to prevent the user from getting a shock from the circuit, as shown in Fig. 5.4.

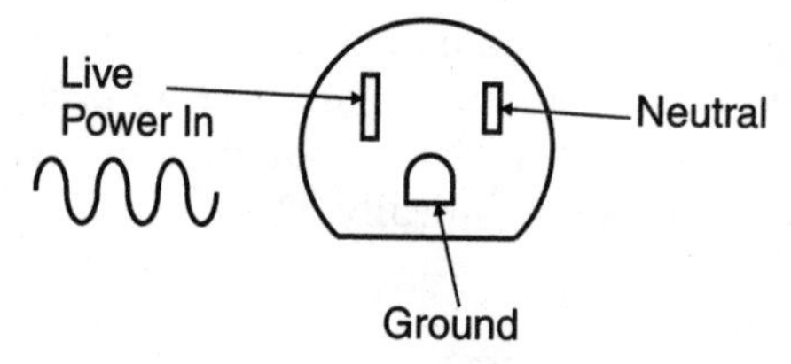

Figure 5.3 North American 110-V, 60-Hz wall plug.

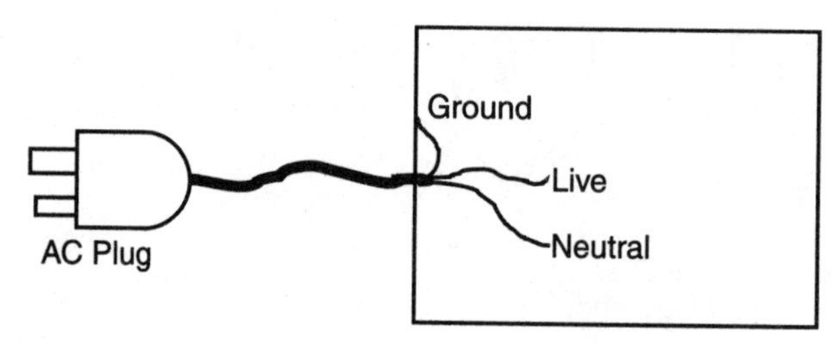

Figure 5.4 AC plug connection.

Because the voltage coming from the mains is so high, it has to be converted into a lower DC voltage for the digital electronic application circuit. This is done in three stages. The first is reducing the voltage from more than 100 V to 10 V or less by a *transformer.* A transformer is a device made up of two coils that share their magnetic field. When current is passed through one coil, the second coil will produce an *induced* voltage and current that can be used to power the circuit.

As is shown in Fig. 5.5, the transformer can simply consist of a magnetic toroidal core with power supplied at the primary-side coil and the transformed AC voltage out provided at the secondary-side coil. Figure 5.5 also gives the relationship between the voltage and current on the secondary-side coil based on the number of turns for each coil. Note that the current is inversely proportional to the turns ratio. In North America (which has 110 V AC), an 8:1 transformer is commonly used. This means that with 110 V in, there will be 14 V out. For 220 V, a 16:1 transformer should be used for the same voltage output.

While the voltage has been lowered, it is still AC and it is still going positive and negative. This voltage has to be *rectified* into a straight DC voltage. This is done by diodes in either a *half-wave* or *full-wave* rectifier, as shown in Fig. 5.6.

Full-wave rectifiers transform the positive and negative "lobes" of the AC circuit into a positive voltage where the half-wave rectifier "clips" the negative wave (resulting in half the total power available to the circuit). For this reason

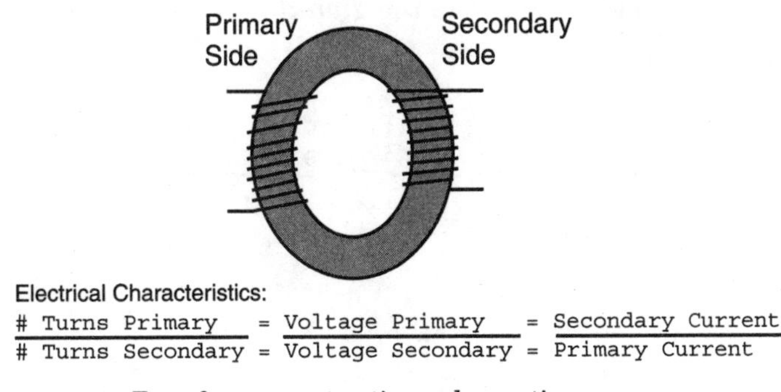

Figure 5.5 Transformer construction and operation.

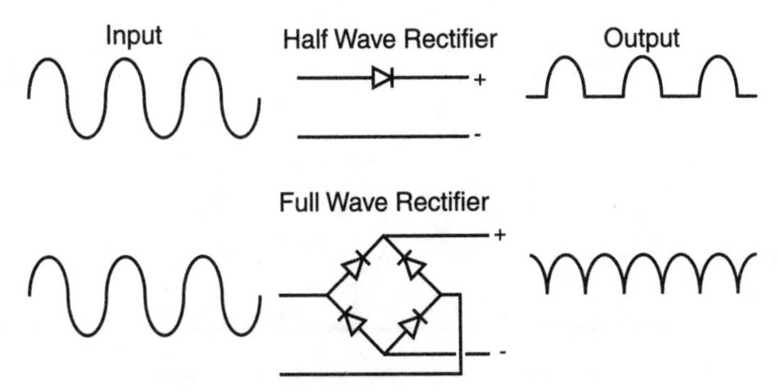

Figure 5.6 AC current rectifiers.

(as well as the fact that full-wave rectifiers can be cheaply purchased) I prefer to use the full-wave rectifier.

Inputting the rectified signal directly from the diodes into a voltage regulator should not be attempted; instead a filtering electrolytic capacitor of a few tens of microfarads should be used. The filtered signal output from the full-wave rectifier will look like Fig. 5.7.

As long as the rectified signal does not drop below the minimum voltage of the voltage regulator, the regulated DC voltage output will be constant. The filtering cap should be a minimum of 10 μF, with a good rule of thumb being that for digital circuits: A 20-μF capacitor is required for each ampere of current drawn. For DC electric motors, this value increases to 100 μF per ampere drawn to help prevent inductive "kick-back spikes" from being driven back through the transformer to the mains circuit. For the +5-V power supplies presented in this book, a 10-μF electrolytic cap is used to provide 500 mA to your digital electronics projects.

Using the transformer, full-wave rectifier, an electrolytic filter capacitor, and a 7805 voltage regulator, a +5-V, 0.5-amp power supply for digital logic applications could be created as shown in Fig. 5.8. The *voltage regulator* converts the rectified transformer-reduced AC voltage into a voltage that can be used by the digital logic.

There are a few things to note in this circuit. The first is, the mains ground is connected to the case and not to the *digital ground.* In any DC-powered cir-

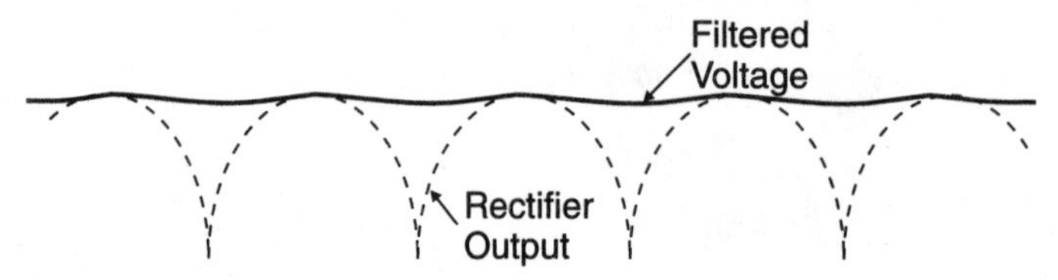

Figure 5.7 Filtered and rectified AC current.

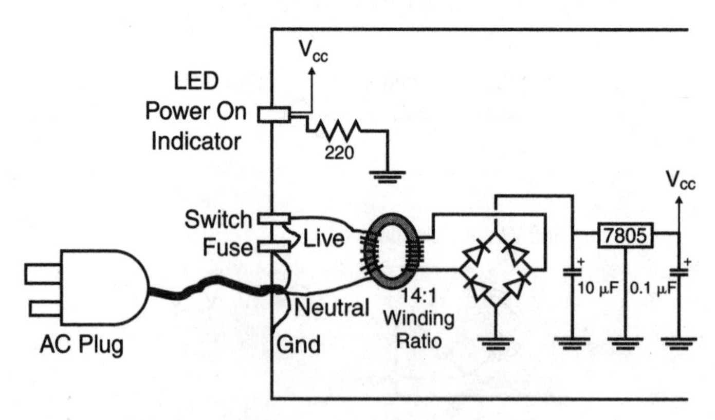

Figure 5.8 Complete voltage regulator from mains.

cuit, the negative terminal of the full wave rectifier can be called *digital ground,* but should be left floating relative to earth ground, which is provided by the AC plug. In this case, *digital ground* is simply a common negative terminal for the circuit.

I have put a fuse in the power line, which will cut out in high-current-draw situations (like short circuits). It is rated at 0.1 amp, which may seem low, but remember that current output is inversely proportional to the turns ratio of the transformer; 0.1 amps at 110 V translates into 1.4 amps at 8 V at the output. Without this fuse, very large (and very dangerous) currents could build up inside the circuit. For example, 2 amps at 110 V translates to 28 amps at 8 V or 224 W of power.

Along with the fuse, the *switch* in the circuit should be one that is certified for switching AC voltages. AC switches usually have a mechanical assembly inside them that snaps the switch contacts on and off. This minimizes arcing within the switch. This may seem hard to believe, but if you look inside an AC switch while it is opening or closing, you will see a blue spark and sometimes hear a pop. This is caused by high inductive voltages produced by the transformer coils that kick back when the AC power to the transformer is shut off.

Note that the switch is connected to the "live" wire after the fuse. A switch should be on the live terminal of the circuit to prevent active voltages when the switch is opened. You should be working on the circuit *only* when the power cord is unplugged from the wall socket.

If you do build mains power supply circuits, like this one, I recommend that you use number 14 gauge stranded wire for all connections. Connections should consist of soldered connections (not household Marette connectors) for safety. Heat-shrink tubing should be placed over all solder joints and bare wire. As well, only Underwriters Laboratories/Canadian Standards Association (UL/CSA) (or the local country testing organization) approved plugs, wires, switches, fuse holders, and transformers should be used in a properly grounded metal case. If any of these terms are unfamiliar to you or you doubt your ability to build the circuit safely, then don't build it!

After going through the theory and practice of a mains power supply, I want to present to you what I consider to be a superior method for providing mains power to your digital logic application. Instead of the plug, wire transformer, case, fuse, switch, and rectifier combination, I suggest that you use a wall-socket-mounted AC to DC power supply (a wall-wart).

This device encapsulates many of the elements listed above, is UL/CSA approved, is much easier to work with, and will be cheaper than if you bought the parts listed above. Chances are, you already have a few wall-warts in your home that are used for toys, personal tape and CD sound systems, or other electronic devices. As long as you have a few volts greater than the regulator's output (I use 3 V above the regulated output as a rule of thumb), then you don't have to buy anything at all. The regulator circuit shown in Fig. 5.8 can be simplified to Fig. 5.9, which is much cheaper and safer than the do-it-yourself transformer solution. This is the basis of the power supply PCB included with the book.

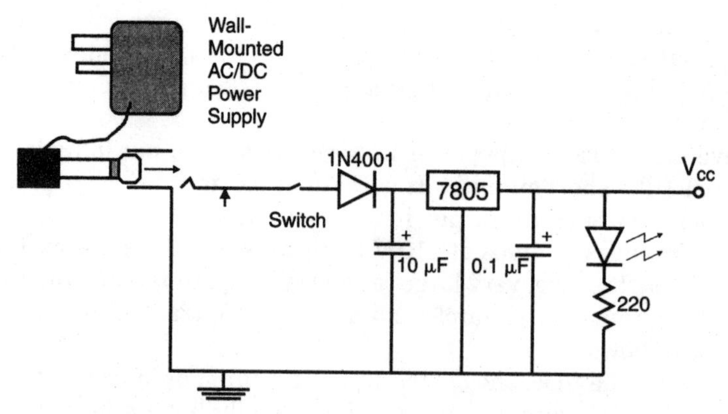

Figure 5.9 Digital power supply using a wall-mounted AC/DC supply.

Figure 5.10 Power plug types.

When using a wall-wart power supply, make sure that you use a power plug instead of a phono plug for connecting to your application. A power plug is different from a phono plug in that the positive terminal is a barrel inside the outside negative terminal, as shown in Fig. 5.10.

The advantage of the power plug is that the positive and negative terminals can never be shorted together (as they can in the phono plug). The socket for the power plug has a pole that is inserted into the power plug. Like the phono plug, power plug sockets are available in designs that will disable alternative power sources (i.e., batteries) when the plug is inserted. I realize that I use a phono plug later in the chapter to describe the action of power source isolation, but this was done because the operation of the phono plug is easier to visualize (and draw) than the power plug.

A good idea for your applications should be to provide a power-on indicator to your circuit. For this function, I normally use an LED and current-limiting resistor.

Power Requirements

Before specifying the power supply for an application, you should understand what is the total current required by the application when it is executing in different modes. Looking at the different modes of operation is usually not intuitively obvious for new designers, but you can find that your application does not work properly because of issues with how the application executes at different times.

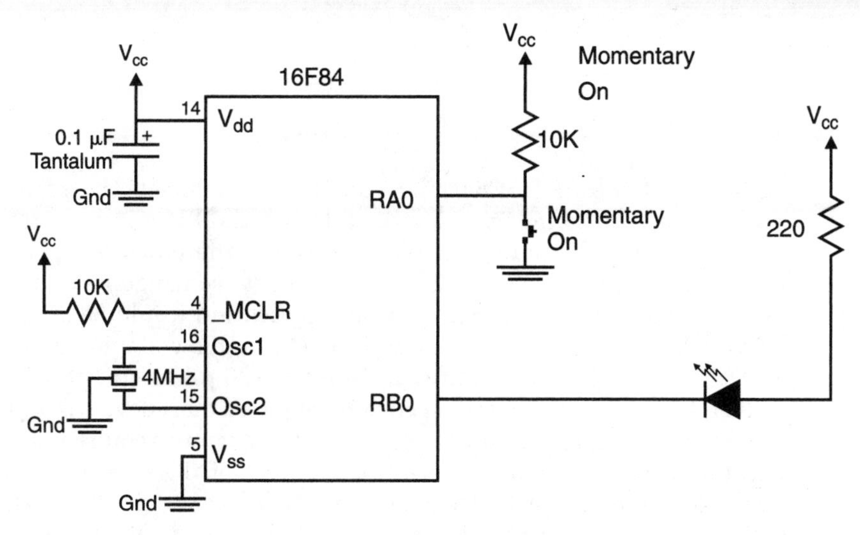

Figure 5.11 Turning on an LED by using a button as a control.

What I normally do when figuring out the power requirements for an application is to first list all the parts used in an application. For example, for the flashing LED circuit shown in Fig. 5.11, the bill of materials is:

Part	Description
PICmicro	PIC16F84-04/P
V_{dd}/V_{ss} decoupling capacitor	0.1-μF tantalum
_MCLR pull-up	10k, ¼ W
4 MHz ceramic resonator	4 MHz with built-in capacitors
RA0 pull-up	10k, ¼ W
RA0 switch	Momentary on
RB0 LED	Red LED
RB0 LED resistor	220 Ω, ¼ W

For this application, the operating modes are:

1. Power off (application not connected to the battery)
2. PICmicro microcontroller running; switch open, LED off
3. Application driving one display and SW1 or SW2 closed

At any time power is applied to the circuit, the oscillator in PICmicro MCU is running. I tend to look at the worst case for the current requirement calculations, which in this case means that the switch is closed and the LED is lit. Assuming that the LED has a voltage drop of 2.0 V (which is typical for an LED), the current available to each LED is:

$$I = V/R$$
$$= 2.0/220$$
$$= 9.1 \text{ mA}$$

Looking at the PIC16F84 microcontroller datasheet, the power consumed at 4 MHz is about 4 mA, making the total for the PICmicro microcontroller to drive the LEDs about 13 mA. When I specify the power supply required for an application, I like to add 25 to 50 percent "contingency" current. In this case, the total current specified for an application would be 30 mA.

The 78L05 is able to source up to 100 mA, so this voltage regulator is appropriate for this application (it requires about 100 μA to operate).

When looking up the specified current required for a device from its manufacturer's datasheet, make sure you look up the current required at the expected operating frequency and temperature. Because of their design, CMOS circuits use more current the faster they are run. Operating temperatures can also affect the amount of current required by a device. If you are unsure what the correct current requirements are for a device, use the maximum listed in the datasheet. It is always better to plan for more current draw than is actually required than finding out you planned for too little.

Linear Voltage Regulators

When I first started being interested in electronics, power supply design consisted of choosing and arranging a number of discrete parts so that the circuit would meet the requirements of the project. Designing the power supply and wiring the parts needed for it together was often a difficult chore for many people just using designs they had used for a long time and hoping they would be appropriate for the application at hand. I remember approaching a number of projects with trepidation because I had had some power supplies burn out on me, and in one case a specified electrolytic capacitor exploded and its replacement exploded as well. As a teenager, I developed a healthy respect for mains current as it melted the wiring in the power supply I had built.

Today, the situation has changed drastically, with the availability of cheap and simple integrated circuits that can carry out the regulation function. When these chips are used with a commercial wall-wart transformer, they are very safe as well. However, I still see a lot of circuits that do not use what I would consider an appropriate power supply. In this section, I want to introduce you to the issues of regulating power and go through my favorite circuits with an explanation of where I feel they should be used.

Voltage regulators are circuits that lower an incoming voltage to a specific level that can be used by another circuit (often referred to as the *load*). Along with lowering this voltage, sufficient current must be produced to drive all the devices in the load without affecting the regulated voltage.

Earlier in this book, I showed how water could be used as an analogy to electrical voltage and current. This is also a useful way of describing how a voltage regulator works.

In the water example, water from a higher-pressure source must be provided at a lower pressure. A common way of doing this is using a bowl, which contains a floating block that is connected to a valve that regulates the flow of water into the bowl. This is used in automobile carburetors, as shown in Fig. 5.12. As liquid is drawn out of the bowl, it is replaced when the float block (most commonly known as the *float*) drops and allows more liquid in from the high-pressure source, as shown in Fig. 5.13.

While this device is very simple and easy to understand, the actual implementation could be quite "fiddly." Issues that have to be accounted for are the ability of float to provide enough force to close off the valve as well as to respond quickly when a volume of liquid is drawn from the bowl.

A similar situation exists with voltage regulators, which work in almost the same way as a carburetor. A linear voltage regulator has the block diagram shown in Fig. 5.14. The *comparator* behaves like the float, providing the input to the transistor (which acts like the valve in the carburetor). As the current draw from the load increases, the comparator will sense the increased load, which causes an output voltage drop and allows the transistor to pass more power through.

While this seems very simple, as I have explained it, in practice this circuit is very hard to correctly implement. Depending on the characteristics of the different parts used in the application, load variations can result in *oscillations,* which are discussed below.

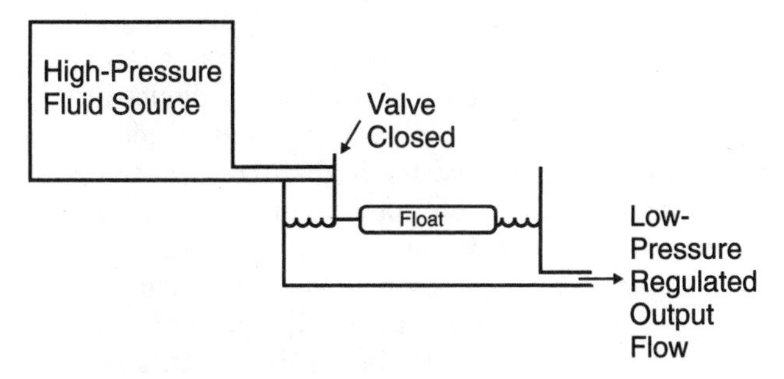

Figure 5.12 Car carburetor as a flow regulator.

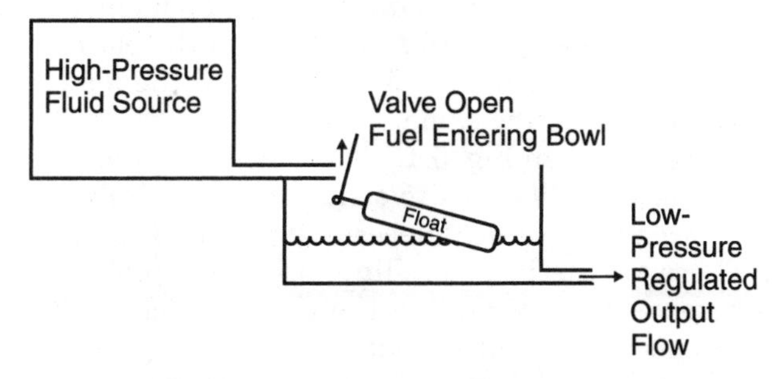

Figure 5.13 Car carburetor allowing in fuel.

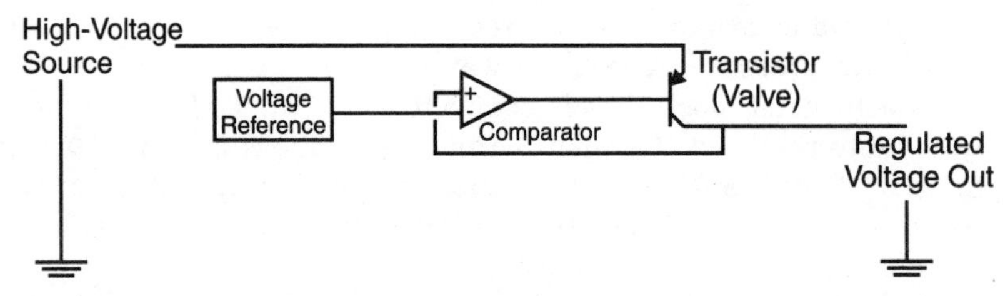

Figure 5.14 Simple regulator controlling voltage.

One of the problems of this circuit is dissipating the heat that is generated by the transistor switch. When the circuit is operating, there is a voltage drop across this transistor, along with the current being drawn across it. This results in power dissipation that can be expressed by the formula:

$$P = V \times I$$

Where V is the voltage drop across the transistor and
I is the current drawn by the load.

For a voltage regulator that supplies 1 amp of current at 5 V from a 12-V supply, there will be a voltage drop of 7 V across it. The actual power generated will be 7 W (7 V times 1 amp) and will require some kind of heat sink to dissipate it to prevent the transistor from being damaged.

This power loss (in the example above, higher than what the load dissipates) gives the accurate impression that the linear voltage regulator is inefficient. For relatively low power applications (requiring less than 1 amp), they are more than adequate and the low cost of integrated devices makes them very attractive. In the following section, I will present high-efficiency *switched-mode* power supplies that are better suited for high-current applications or applications that require the voltage to be stepped up or made into a negative voltage.

Linear voltage regulators have to be designed in such a way that they will track the load's demands and not begin to oscillate when the load changes. Oscillation is a very easy trap to fall into in this type of circuit (which is why the power supply I talked about at the beginning of this section caused electrolytic capacitors to blow up). Take the case where the load requires instantaneous increased power, and the comparator and transistor switch cannot keep up. This potentially translates to the power supply providing more power when it is not required. The comparator will recognize that the power is too high and cut it down, but the change will not take place until the load power requirement has increased or stayed the same.

This is shown graphically in Fig. 5.15. The output lags the comparator output, which causes these oscillations to continue indefinitely. This is actually how an oscillator circuit works as a necessary part in a radio and other electronic circuits. There are methods for eliminating this problem in power supplies, but this involves understanding the characteristics of the individual parts from the perspective of control theory, and is really not necessary when there are small, easy-to-use integrated circuits that provide this function for much less cost than the parts required for a circuit using discrete parts.

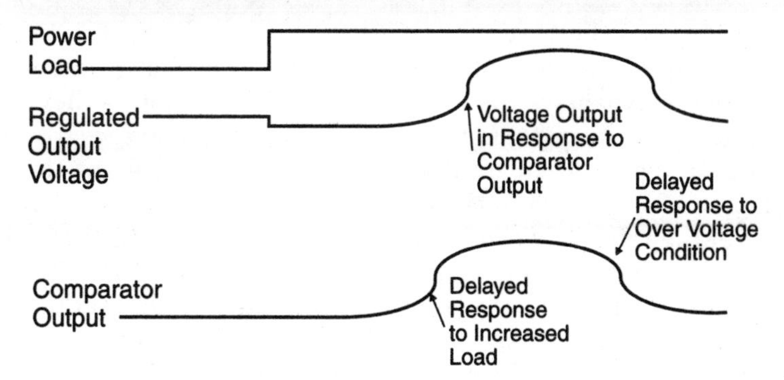

Figure 5.15 Power supply oscillations.

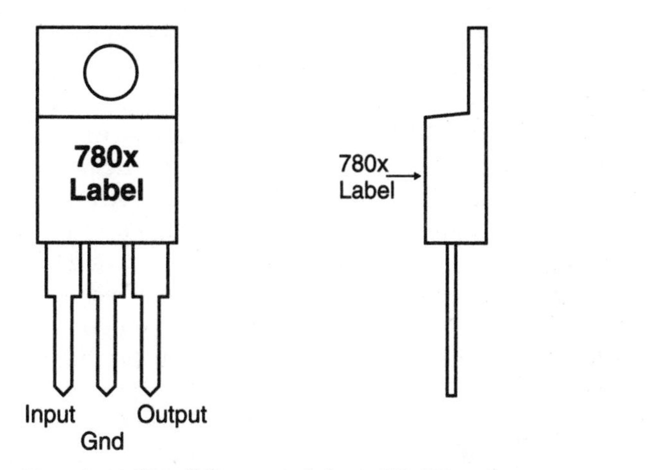

Figure 5.16 780x Voltage regulator in TO-220 package.

For my own applications requiring between 50 mA and 500 mA, I like to use the 78xx and 78Lxx series of integrated voltage regulators. These three-pin devices can provide up to 1 amp and 100 mA of current, respectively, at a specific voltage. They are designed to avoid the oscillation problems caused by varying loads and have automatic shutdown capabilities if their internal temperatures get too high or if the load is too great for their driver circuits.

For people starting out in designing electronic applications, I suggest you invest in a few 7805s. This 5-V regulator, which can be found for 40 cents each or less, is packaged in a TO-220 transistor package and looks like Fig. 5.16. I always remember the pin-out by the convention that inputs come into a schematic from the left and outputs are on the right.

The 7805 is capable of sourcing up to 1 amp and should have a heat sink associated with it (although I often don't follow this rule). A good rule of thumb for heat sinks is that 1 square centimeter of surface area is required to dissipate the heat from 1 W of power. For the example above of 7 W being generated by a transistor with a 7-V drop across it and passing 1 amp of current, the necessary dissipation area is 7 square centimeters.

The 78L05 (and the 12-V output 78L12) can supply up to 100 mA of current and is packaged in a transistor TO-92 package, as shown in Fig. 5.17. 78Lxx parts are more expensive than the straight 78xx versions—they can cost up to sixty cents each. Even though they are more expensive, you would be hard pressed to design a circuit that could be built as cheaply as the 78Lxx. Note that the pin-out, relative to the label, is reversed from that of the 78xx parts. Because of the relatively small output current (and proportionately lower dissipated power) heat sinking is usually not required for the 78Lxx parts.

There are other voltage regulators available on the market with similar output capabilities, and they are cheaper, but they lack the current and temperature overload capabilities of the 78(L)xx parts. Personally, I feel these parts should be avoided because the overcurrent/temperature features of the 78(L)xx parts have saved me from having to replace the voltage regulators on numerous occasions. Not having this feature might save you a few cents (literally just a few cents), but will expose you and your customers to the possibility of having to replace the voltage regulator.

A good example of this is the Parallax Basic Stamp—the number one complaint for the product is burned out power regulators. The most recommended fix for the problem is to replace the original regulator with a 78L05.

The 78(L)xx voltage regulator's output can be shifted upward to provide different voltages by providing a different ground (which acts as a voltage reference) to the rest of the application. I'm mentioning this because EPROM/flash programming requires a +13- to +14-V supply, and you don't want to design a unique power supply for this application unless you can help it.

The easiest way to provide this capability is to use silicon diodes to shift the ground reference by the 0.7 V of the forward junction. Adding two silicon diodes as shown in Fig. 5.18 will create a 13.4-V power supply, which can be used for programming the PICmicro.

The inputs and outputs to voltage regulators should have *filtering* capacitors on their inputs and outputs. The typical filter capacitor I use is a 10-μF electrolytic across the input and 0.1-μF tantalum across the output. The basic

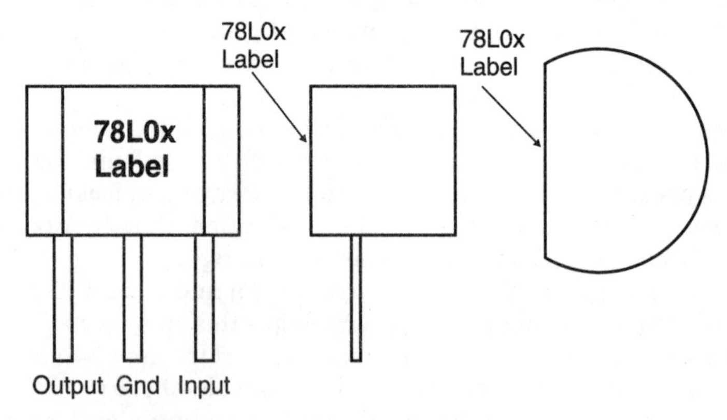

Figure 5.17 78L0x voltage regulator in TO-92 package.

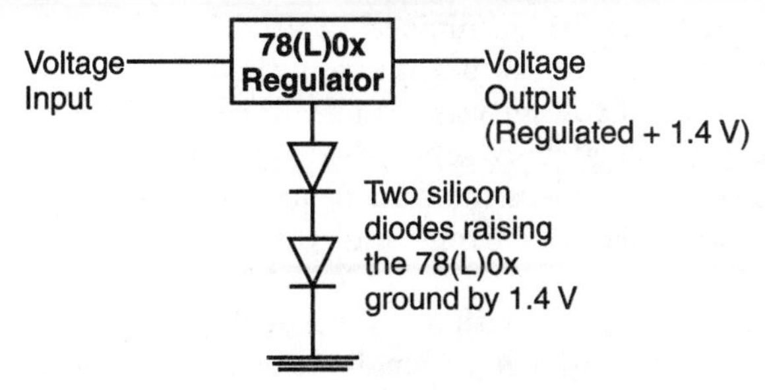

Figure 5.18 Raising a voltage regulator output.

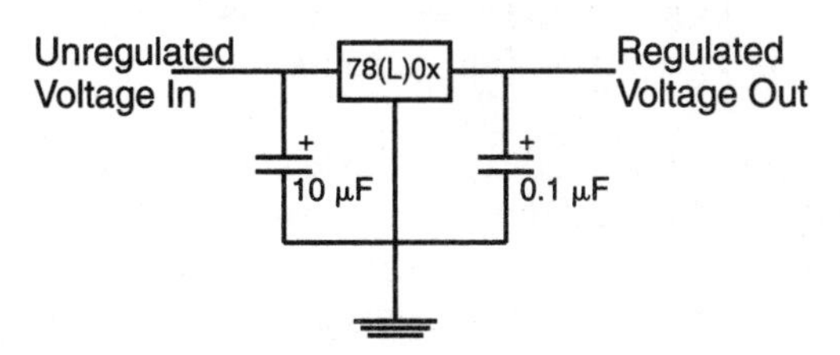

Figure 5.19 Using a 780(L)x as a voltage regulator.

circuit for using these parts is shown in Fig. 5.19. Both the 10-μF electrolytic and 0.1-μF tantalum capacitors should be *derated* by 50 percent or more to eliminate any chance they will burn out or blow up. This is to say the actual voltage rating should be twice or more the voltage levels the parts are subjected to in the application.

These capacitors will eliminate any noise from the input voltage and help provide a very clean supply to your application circuit. They are also required for proper operation of the voltage regulator chips—you may find they work properly without these capacitors for small loads, but for larger loads, the capacitors are definitely required.

The last point I want to make about voltage regulator circuits is that if a switch is to be added to turn the circuit on and off, then it should be *upstream* of the input filter capacitor. I'm mentioning this because sometimes it will be easier to wire in the switch between the capacitor and the voltage regulator. In this case, the switch could be opened and input voltage removed with a charge left on the capacitor. In this case, you may find yourself getting shocks for no apparent reason.

There is another type of voltage regulator circuit that is useful, quite inexpensive, and well suited to applications where parts may be inserted or removed while power is applied. Going back to the carburetor analogy to the voltage regulator, there is another, simpler way to provide liquid at a specific pressure and that is to let higher pressure liquid fill the bowl and allow anything extra to spill over. The fuel in the bowl will provide a constant pressure output at the bottom of the bowl, as shown in Fig. 5.20. This type of pressure regulator is obviously very inefficient because of the overflow spilling out of the bowl. The overflow can be minimized by limiting the amount of liquid falling into the bowl to the amount output from the bowl.

The analogous method of voltage regulation can be provided by a *zener diode* and a current-limiting resistor. A zener diode is an interesting beast that behaves like a regular diode unless the voltage across it is greater than its rated voltage, at which point it passes current until the voltage across it is the rated value.

A +5-V regulator can be built with a 5-V zener diode and a resistor, as I've shown in Fig. 5.21. I normally use zener diodes only in circuits that require less than 50 mA. A zener diode voltage regulator can be built for around 25 cents. In this circuit, the voltage across the zener diode's anode will always be at 5.1 V, unless the load is so high that it exceeds the current available through the resistor. This is a confusing statement and needs some explanation.

The current-limiting resistor is used to regulate the amount of current that is available to the load. Without it, the zener would pass as much current as the V_{in} power supply could source (essentially being a short circuit). The current-limiting resistor provides a voltage drop and current limit to avoid this problem.

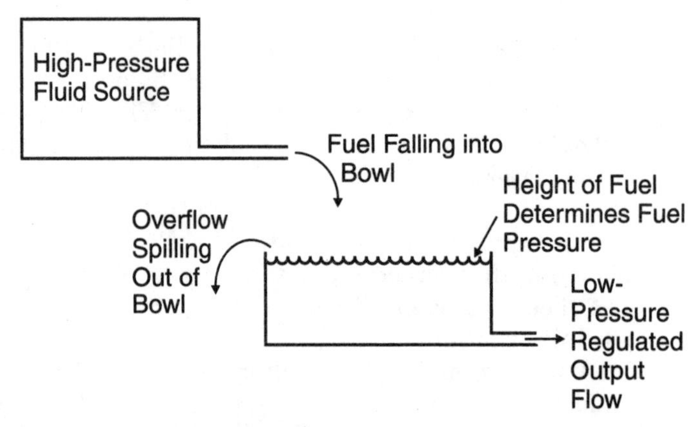

Figure 5.20 Simple carburetor operation.

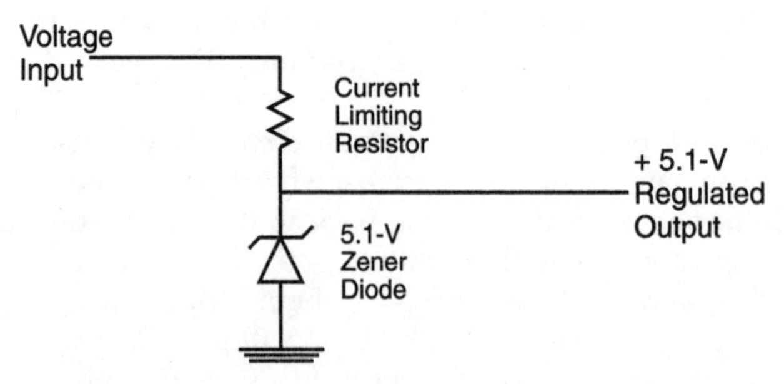

Figure 5.21 Zener diode voltage regulator.

The value of the current-limiting resistor is calculated from the expected voltage input, the zener voltage, and the current drawn by the load. For example, if 10 V is supplied by the voltage input supply and 5 V and 25 mA is required by the load, the resistance value would be calculated by first finding the voltage across the resistor and the current through it.

In this example, the voltage across the resistor is 5 V (10 V − 5 V) and the current through it is 25 mA. Using Ohm's and Kirchhoff's laws, we can calculate the resistance.

$$\begin{aligned} R &= V/I \\ &= (10\ \text{V} - 5\ \text{V})\ /25\ \text{mA} \\ &= 5\ \text{V}/25\ \text{mA} \\ &= 200\ \Omega \end{aligned}$$

So a 200-Ω current-limiting resistor is required for this power supply in this application. In addition to the resistor value, you should know the power being dissipated through the resistor and zener diode. For the resistor, power is calculated as:

$$\begin{aligned} P &= V \times I \\ &= (10\ \text{V} - 5\ \text{V}) \times 25\ \text{mA} \\ &= 5\ \text{V} \times 25\ \text{mA} \\ &= 125\ \text{mW} \end{aligned}$$

which means a ¼-W resistor can be used safely.

For the zener diode's power rating, I always go with the worst case, which is when there is no load. In this case, because the voltage diode is the resistor, the power dissipated is also the same (125 mW). A ¼-W zener diode can be used safely in this circuit.

This circuit is best suited for situations where low power is required. The circuit, essentially a voltage divider, is well suited to applications where chips with the load may be pulled out and replaced with power active. (This is known as *hot plugging* and *hot unplugging*).

When a load is placed in parallel with the zener diode, the resulting current through the zener diode will drop. In this case, while the current passing through the zener diode drops, the voltage across it remains constant. Remember that the current passing through the current-limiting resistor remains constant in this circuit regardless of the current passing through the load.

Like the linear voltage regulator, the zener diode voltage regulator should have filtering caps provided with it to prevent noise from the voltage input power supply from being passed to the load circuit and vice versa.

Switch-Mode Power Supplies

The voltage regulator (or *linear regulator*) is used to provide a regulated voltage from a higher, often unregulated, voltage. As I showed above, voltage regulation is accomplished by limiting power to the load. This method is quite inefficient in terms of power and is not always the best way of providing battery power in an application.

For example, if a 9-V battery is used to drive a 100-mA load at 5 V, stepping down the 9 V to 5 V will have a 4-V drop and 100 mA current and consume a total of 0.4 W.

Since the application consumes a total of 0.5 W, the application is only 55 percent efficient with the power being supplied to it. This isn't a problem for small circuits which are powered from an AC main line because the total current is quite small and the lost power is insignificant, but it is an issue for battery-powered applications or applications that require a lot of current.

The *switch-mode power supply* (SWPS) is a device that uses the characteristic of coil "fly back" after current passing through it is turned off. By rapidly switching on and off a small coil, a higher voltage can be produced by the circuit shown in Fig. 5.22. This is one of the most popular of four or five switch-mode power supply designs available.

In the circuit diagram, the N-channel MOSFET is used as a switch and should be capable of handling the peak current loads (calculated below). A pulse-width modulation (PWM) signal (*switch control*) turns the transistor on and off. The purpose of the diode is to prevent current from sliding back down into the switch after being driven from the coil. Finally, the capacitor is used to filter out as much voltage ripple as possible.

This circuit will transform V_{in} in to V_{out} for a given load based on a simple set of equations:

$$I_{peak} = 2 \times I_{out} \times (V_{out}/V_{in})$$

$$T_{off} = L \times \frac{I_{peak}}{V_{out} - V_{in}}$$

$$T_{on} = (V_{out}/V_{in}) - 1$$

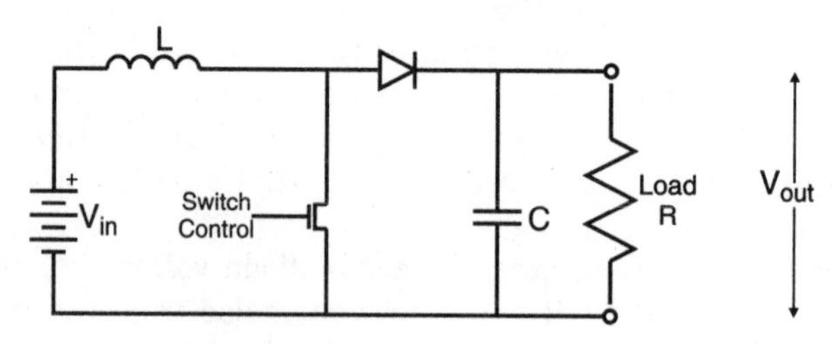

Figure 5.22 Switch-mode power supply circuit.

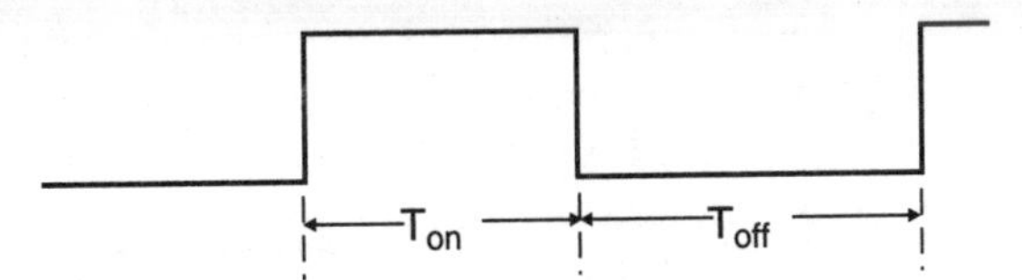

Figure 5.23 Switch-mode power supply PWM signal.

These formulas are recursive. That is to say, they are based on each other, but this reflects one of the basic aspect of switch-mode power supply design—the process is iterative, based on the parts available for the power supply and the requirements of the load.

The T_{on} and T_{off} values are used to determine the PWM's characteristics. T_{on} is the time the switch is on and current is passing through it. T_{off} (also known as T_{don} in some references) is the time the switch is off and current is flowing through the diode. These two values are combined to produce the switch PWM signal shown in Fig. 5.23.

I have looked at using on-board microcontrollers to drive a switch-mode power supply that would produce different voltages for programming parts. For this case, I wanted to raise 5 V to 13 V for a 50-mA load using a 50-kHz PWM frequency generated by the microcontroller. Using the formulas above, I wanted to calculate T_{off} and T_{on} using a 22-μH coil (which I happened to have on hand). To generate the *switch control* PWM signal, I want to use a microcontroller running at a 1-μsec/instruction cycle execution rate. The formulas above could then be applied:

$$
\begin{aligned}
I_{peak} &= 2 \times I_{out} \times (V_{out}/V_{in}) \\
&= 2 \times 50 \text{ mA} \times (13/5) \\
&= 260 \text{ mA} \\
T_{off} &= L \times \frac{I_{peak}}{V_{out} - V_{in}} \\
&= 22 \text{ μH} \times \frac{260 \text{ mA}}{13 - 5} \\
&= 22\,(10^{-6}) \times \frac{260\,(10^{-3})}{8} \\
&= 0.715 \text{ μsec} \\
T_{on} &= T_{off} \times [\,(V_{out}/V_{in}) - 1] \\
&= 0.715 \text{ msec} \times [\,(13/5) - 1] \\
&= 1.144 \text{ μsec} \\
T_{period} &= T_{on} + T_{off} \\
&= 1.859 \text{ μsec}
\end{aligned}
$$

For this case—and this is where I mean determining the values is an iterative process—the T_{on} and T_{off} specified are too fast to be produced by the microcontroller. So, to implement this PWM, I would have had to change the parameters and parts until I got components that met the needs of the application.

For example, if the coil is changed to 5 μh, the values for T_{on} and T_{off} become 0.1625 msec and 0.26 msec, respectively. This is much faster than what I had previously (and even more inappropriate for the microcontroller to drive the PWM used in the switch-mode power supply). Instead the coil size has to be increased: Going up to a 220 mH coil, the values for T_{on} and T_{off} are 7 μsec and 11.4 μsec, which are easier to output from the microcontroller running at the 1 μsec/instruction cycle execution rate.

When this PWM is being output by a microcontroller that is driving the output by "bit banging" code, it will be very difficult (if not impossible) for the microcontroller to execute main line code and do other work. This really shows that the microcontroller is not well suited to being a switch-mode power supply controller unless the built-in PWM output available in some microcontrollers is used.

Instead of a microcontroller, there are many other chips available on the market that will control a switch-mode power supply cheaply and reliably. Many of these chips will also monitor and regulate the output voltage level (something that can't be done easily with a microcontroller) in addition to providing a method of producing negative voltages and step-down switch-mode power supplies.

If you are just looking for a clocking circuit to drive the switch control, then I recommend looking at the venerable 555 timer. This chip is cheap and there is a lot of information written about it (including using it for this type of application).

Battery Power

In our society, battery-powered electronic devices can be found just about everywhere. When I look around my office, I can see a Palm Pilot, a digital clock, a digital multimeter, a radio, and a couple of calculators, all of which run on alkaline "radio" batteries. It is probably a big surprise then to find out that I don't recommend that you use battery power for your own electronic projects, unless you follow some specific rules.

These rules are based on an understanding of how batteries operate. You are probably familiar with three types of batteries: alkaline disposable dry cells, or radio, batteries (which I usually refer to as *alkaline batteries*), and rechargeable nickel-cadmium cells ("Ni-cads"). The last type of battery is the lithium cell that is normally used for clocks and retaining data in CMOS devices. These batteries may appear similar, but they each work differently and have different operating characteristics.

All types of batteries are given a *rated life* in the units of ampere-hours (A-h). This rating is used to determine how long a battery will provide its rated operating characteristics with a given drain. For example, a battery with a 2 A-h rating will supply 2 amps of current for 1 hour—or 200 mA of current for

10 hours. A typical value for a AA alkaline radio battery is in the range of 100 to 300 mA-h. Of the three types of batteries discussed in this section, alkaline batteries tend to have the greatest ampere-hour rating, with lithium having the smallest. Lithium batteries are designed for providing a *trickle* (very small) current for a very long time.

When you use different types of batteries in an application, there is one suggestion and warning that I should make. When prototyping a circuit, *never* use Ni-cad batteries. Ni-cads can provide a very large amount of current when short-circuited (alkaline batteries provide a maximum amount of current usually in the range of hundreds of mA). If you are prototyping with Ni-cads and a short circuit is encountered, the full charge within the batteries can be released. This will result in the batteries getting very hot and potentially exploding. This will not happen with alkaline batteries.

One of the biggest differences between the battery types is the voltage output. For alkaline batteries, the voltage output on new batteries is 1.7 to 2.0 V per cell. This voltage tends to drop linearly over use and 1.5 V per cell is output when half the charge has been used up. Ni-cads tend to output around 1.2 V per cell, and lithium provides much higher voltage at about 2.2 V or higher per cell. To provide higher voltages, cells are connected in series and their voltage outputs summed together. For example, 9-V "square" batteries have six 1.5-V cells built into them to provide the nominal 9-V output.

This leads to the question of how many cells are required to provide a stable V_{cc} (normally 5.0 V) for the PICmicro and digital logic of your circuits. If alkaline batteries are used, then three are required; the number is four for Ni-cad and two for lithium. But, as I will explain in this section, this is not an optimal arrangement for powering applications.

The next major concern about using batteries is the voltage output over time as current is drawn. This is different for the three different types of cells, as shown in Fig. 5.24.

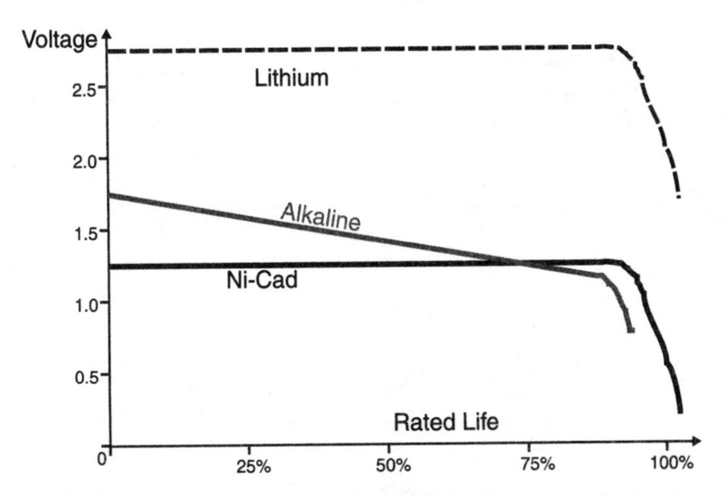

Figure 5.24 Operation of different battery types.

Ni-cads and lithium batteries tend to give constant voltage while they are being drained, but die off suddenly when their charge is depleted. Alkaline batteries, as I noted above, tend to have their output voltage linearly drop as the charge within the cell is depleted. This characteristic is what separates the alkaline cell from the other two types.

Another characteristic that is different between the types of battery cells is how long a charge can be retained in them. Lithium and Ni-cads will keep their charge for a very long time, if there is no drain on them. As you have probably found, alkaline and other dry cells will leak and lose their charge over time. Ni-Cad cells are particularly good at maintaining charge and will stay charged for 10 years or more. Lithium cells also have good charge-retention capabilities, although they generally need replacement after 5 or so years, even if their rated capacity hasn't been used.

Lithium batteries, with their approximately 2.0-V output, are not enough for digital logic electronic operation, but can be used to maintain a device in "sleep" or standby modes. Ideally the lithium battery should only power the circuit when V_{cc} (or regulated power) to the circuit is lost. The circuit I normally use for this function is shown in Fig. 5.25.

In this circuit, as long as power is being applied to the circuit, no current will flow from the lithium battery because the voltage applied is greater than what is being applied by the battery. When V_{cc} is lost, the lithium cell supplies current to the circuit. Ideally, there should be some kind of sensor to detect the low-voltage condition (when the lithium cell is powering the circuit because the primary power has been lost). The low-voltage sensor should cause the circuit to enter a sleep mode to minimize the required power—this can be accomplished by using a comparator, or the *brown-out reset* function of some circuits.

The voltage-over-time characteristic of alkaline batteries means that battery power should be regulated in some manner to provide the nominal 5.0 V required for the logic. This can be done either by providing a small voltage and increasing it by a switch-mode supply or by providing more than 5.0 V and regulating it downward. While are many digital circuits that can run from lower than 5.0-V sources, and some that have on-board regulators that can regulate the voltage to an appropriate level for the circuits, you generally cannot run your applications directly from batteries.

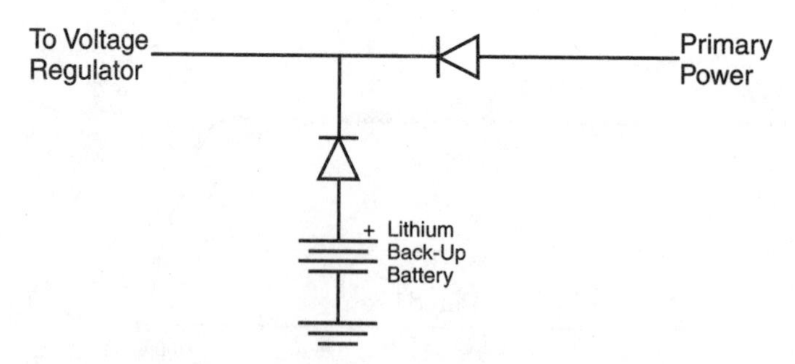

Figure 5.25 Isolating a battery in a supply.

For "quick and dirty" circuits, I will often use a 9-V alkaline battery connected to a 78L05 voltage regulator, as I have discussed in this chapter.

To get the most out of the battery, I will add a wall power source socket to the circuit in such a way that the battery will provide power if the socket is not available. This can be done by using the diode circuit above, or by using a socket that connects a battery when the wall source is no longer available. The power connection operation looks like Figs. 5.26 and 5.27.

In this circuit, when the phono plug is not plugged in, the switch built into the socket is closed and power is drawn from the battery. When the plug is installed, the battery is isolated from the circuit. As I have discussed earlier in this chapter, although I used a phono plug for this example, I normally use a power plug because it will not arc when the plug is slid in and out.

While Ni-Cad cells can be recharged, they cannot be recharged by a straight DC power source (and some variants may be damaged by a straight DC power source). Placing the battery downstream of the plug, as shown in Fig. 5.28, should *never* be done.

Warning! Lithium cells and alkaline cells cannot be recharged by Ni-cad recharging equipment, and attempting to recharge the batteries this way can

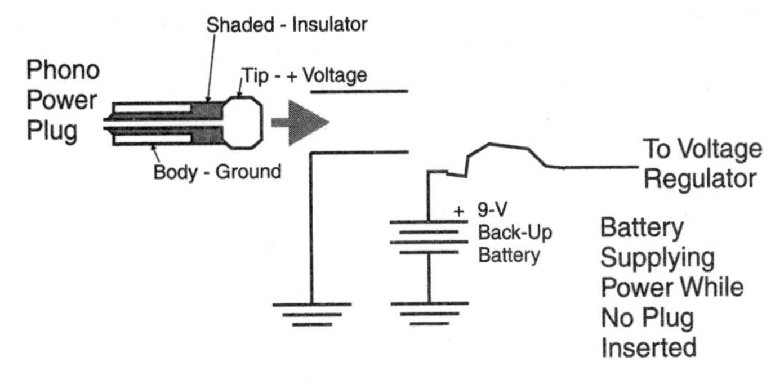

Figure 5.26 Battery in a multiple power source circuit.

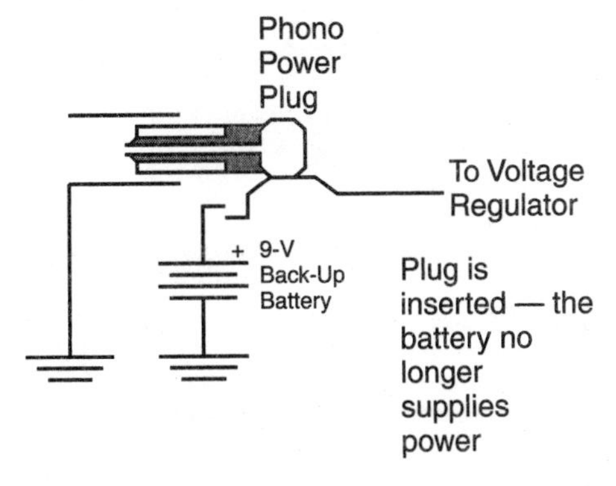

Figure 5.27 Battery disabled in a multiple power source circuit.

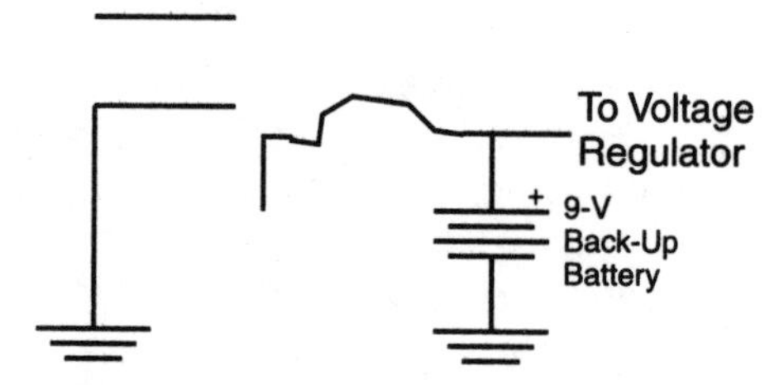

Figure 5.28 Invalid battery charging in a multiple power source circuit.

result in their exploding. While there are some alkaline batteries that can be recharged, use the manufacturer's recharging equipment only. When a lithium cell is depleted—throw it out!

Ni-cad batteries have a potential problem that you should be aware of. If they are used for the same length of time before being recharged, they can develop what is known as *memory* and will stop producing current after this length of time. To avoid this problem, Ni-cads should be used for varying lengths of time and periodically deep-discharged (run to almost zero volts output). This will break up the dendrites that build up over time and are the cause of memory. For many people, memory is not a problem with their applications, and some of the newer Ni-cad cells are designed in such a way that memory is not a problem.

Chapter 6

Test Equipment

In this book, I try to emphasize the real world and actual operation of the digital circuits. To help you see what is happening, I recommend that you use at least a digital multimeter for looking at signal voltages and measure actual application current requirements. It is possible to work through virtually all the projects in this book without any type of test equipment, but I recommend that you spend $20 to $40 on the basic instruments to observe what is happening and help you with your own applications later.

Digital Multimeters

Earlier in the book, I introduced the concepts of how ammeters and voltmeters work. These electromechanical devices have been replaced by small hand-held *digital multimeters* (DMMs), which are cheaper, more accurate, and easier to read and can perform a number of other useful measurements as well (Fig. 6.1). Digital multimeters can be bought for as little as $20 U.S.

The output of a digital multimeter is a 3- or 4-digit numerical display. In many devices, the measurement is selected via a switch on the DMM, and the display's decimal point moves over. If the value is too large for the display, something like a 1 in the left most digit will be displayed with the other digits blanked out. The table below lists what would be shown on a 4-digit display for different resistor measurements:

Resistance	Range	Output	Comments
220 Ω	200	1	Value too large for display
220 Ω	2k	0.220	220 Ω = 0.22k
220 Ω	20k	0.022	

Figure 6.1 Digital multimeter.

Note that the ranges start at 2, rather than 1 as you would probably expect. This is a common convention, left over from when displays were more expensive and the single digit 1 minimized extra cost.

Some digital multimeters have up to 6 digits, but I must stress that, as you work with the digital multimeters for your own digital electronics projects, never use more than 3 digits. The extra accuracy is not needed and adds a lot to the cost of the device.

Each digit represents a power of 10. For a 3-digit display, the value is supposedly accurate to 1 part in 1000. Greater than 1 part per 1000 accuracy is required only very rarely, in very specialized cases. When this level of accuracy is required, precision power supplies, crystals, and components would be used with a specially calibrated digital multimeter. It may be interesting to see differences between devices at 10 millionths of a volt, but this accuracy is not practical for any typical digital electronic applications that I can think of.

Digital multimeters are not *fast response* devices. You may find that it can take as long as 10 seconds before the display stabilizes on a value. This time to stabilize means that the digital multimeter is not capable of measuring changing digital signals, unless the signal changes once every few seconds. If this is not possible, you would be better off using one of the other tools that are described in the following sections of this chapter.

Along with the ability to measure current, voltage, and resistors, digital multimeters are available with the following features and capabilities:

1. Perform autoranging while measuring a parameter
2. Measure capacitance
3. Measure temperature
4. Measure a bipolar transistor's beta
5. Perform diode checks
6. Measure frequency

These features are nice to have, but not critical for the projects in this book. If you don't currently have a digital multimeter but an analog meter, I recommend that you scrap the analog meter and buy yourself a digital multimeter to help you with understanding what is happening in your circuit.

If you don't currently have a voltmeter or ammeter of any type, then you shouldn't be reading this, you should be going out and buying one of the cheaper ones and learning what you can do with it.

Logic Probes

Along with a digital multimeter, I consider a *logic probe* to be a basic tool for anyone developing digital circuits. A logic probe (Fig. 6.2) is a pencil-like device that can be used to check different parts of a board and return a visual/audio signal that indicates whether a signal is logic high, low, or tristate (not connected to a driver or a pulled-up or pulled-down input).

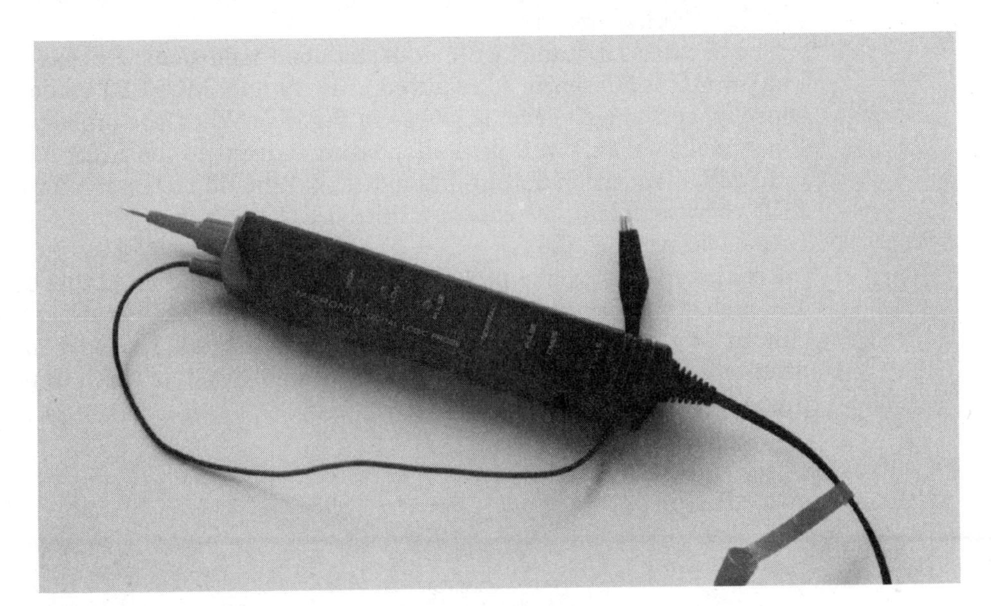

Figure 6.2 Logic probe.

There are usually two controls on a logic probe. The *pulse/continuous* switch will cause an LED to flash when a changing signal is encountered, or a single LED to light for a continuous level. Along with an LED, many logic probes have a speaker in them that will provide beeps and tones so you can tell what is going on without looking at the LEDs.

The speaker audio output is a feature I highly recommend having on any logic probe that you use. This output will allow you to focus on probing the circuit, rather than looking at the instrument.

The other switch is *TTL/CMOS* and selects which voltage threshold should be used. Normally TTL is 1.4 V and CMOS is 2.5 V.

Good-quality logic probes can be bought for as little as $20, and I recommend you buy one rather than try to build one yourself. You may think you could build one yourself with a resistor and LED like the circuit shown in Fig. 6.3 but this circuit has three problems:

1. It doesn't detect low outputs, just high with enough current drive capabilities to light the LED.
2. If a circuit doesn't have enough current source capability, the LED won't light.
3. It gives no indication of level transition.

To detect high, low, and tristate/no connection, another LED will have to be added to the circuit. If the single LED is not lit, then the pin could be attached to a low, tristate, or low-current high output. To fix this, voltage comparators, capable of driving an LED, should be used in the circuit as shown in Fig. 6.4. In this circuit, if the probe is driven high or low by a pin, one of the LEDs will light. A switch-selectable *threshold reference* voltage divider could be put in for a TTL/CMOS selection.

The circuit as it stands will not detect open conditions; for that feature some kind of MOSFET sensor is required. A P-channel MOSFET could be put in to sense when the logic level is low as in Fig. 6.5. When the probe pin is low, the P-channel MOSFET will turn on, passing current to the LED.

To add a transition output a "single shot" should be triggered by each of the LED drivers. This will cause a third LED to flash when the LED drivers become active (Fig. 6.6).

Looking at this, you're probably confident you could build this circuit easily. You probably could, although there may be some problems with getting the circuit to fit in a small enough package to be useful. In addition, if you did attempt to build one yourself, it would probably cost as much in parts as buying a prebuilt and tested unit.

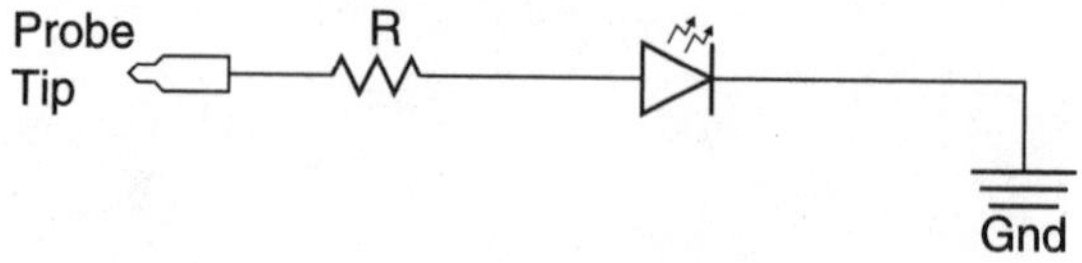

Probe Tip
R
Gnd

Figure 6.3 Basic logic probe.

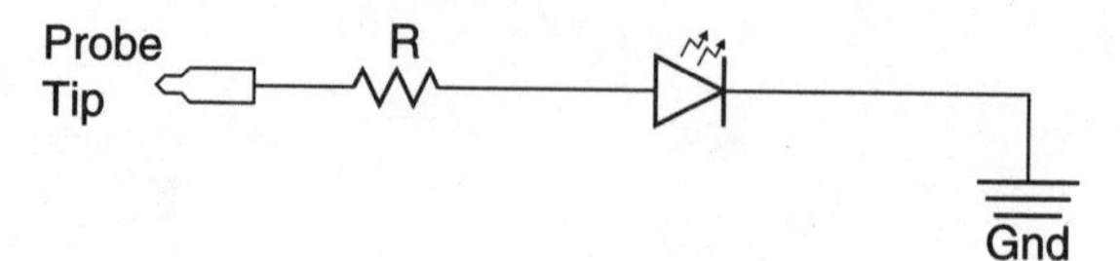

Figure 6.3 Basic logic probe.

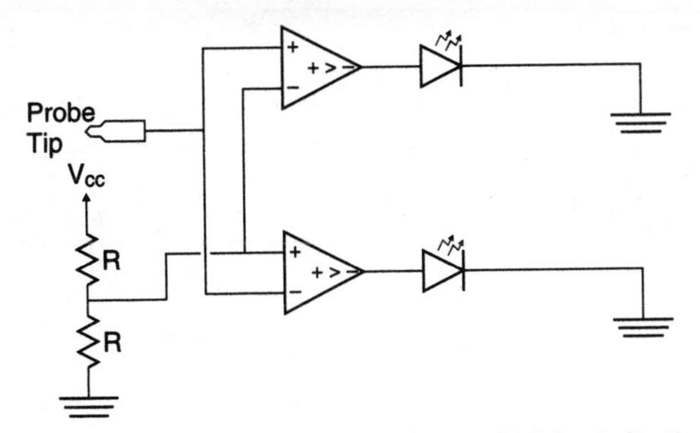

Figure 6.4 Logic probe capable of returning high/low indication.

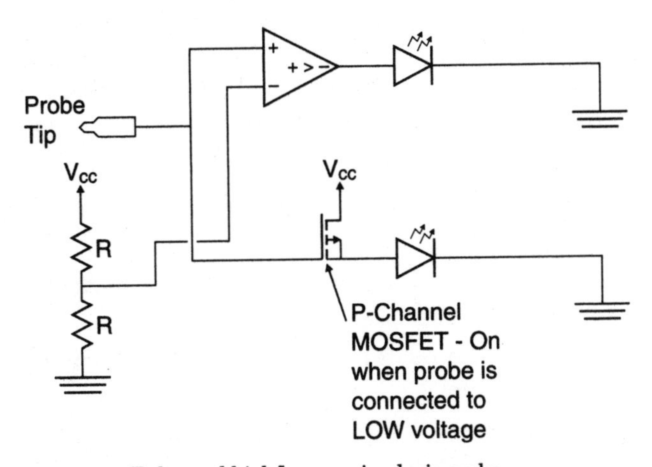

Figure 6.5 Enhanced high/low sensing logic probe.

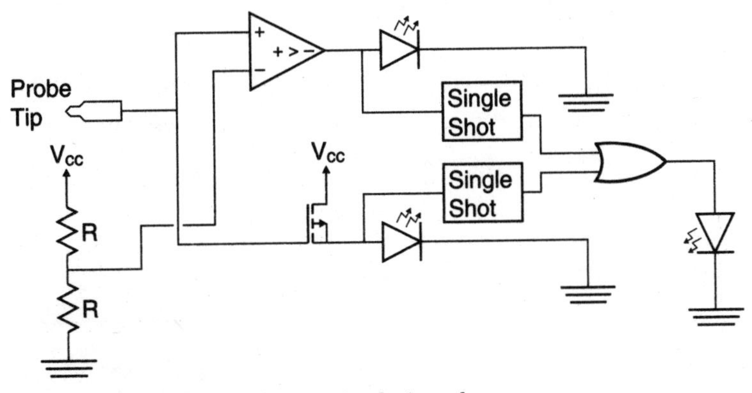

Figure 6.6 Complete high/low sensing logic probe.

For these reasons, I recommend that you avoid the headaches of building one and just buy one ready made. As unlikely as this sounds, the Radio Shack logic probe (catalog number 22-203) contains all the features I've listed here and is quite inexpensive (it costs $20), making it the device you should look at first. I've owned three of them over the years—they are as good or better than other logic probes costing twice as much or more from more "prestigious" manufacturers.

Examining the Binary Counter

Before going on, I would like to examine the last project circuit that was built: the 74LS193-based counter. This circuit incremented a counter when a button was pressed. Using the digital multimeter and the logic probe that have just been discussed, I would like to take a look at a number of different areas of the circuit in order to show you how you can characterize a digital electronics circuit.

Later in the book, when I discuss *failure analyzing* or *debugging* a circuit, one of the most important tasks that you can do with the circuit is to *characterize* it. Along with showing you how a circuit is characterized, you should get some ideas on the voltage/current levels you should see on your own circuits.

The parameters that I am going to examine in the binary counter are:

1. The voltage levels output by the push button (which is the same as the switch inputs) when it is driving out a 1 and a 0 and is not connected to a TTL input
2. The voltage levels output by the push button when it is driving out a 1 and a 0 and is connected to a TTL input
3. The current drawn by the push button when it is driving out a 1 and a 0 and it is not connected to a TTL input
4. The current drawn by the push button when it is driving out a 1 and a 0 and is connected to a TTL input
5. The current drawn by the 74LS193 when the push button is pressed and released and the 74LS193 is driving out 0 to the four LEDs
6. The current drawn by the 74LS193 when it is driving one, two, three, and four LEDs
7. The voltage at an LED when it is being driven with a 1 and a 0
8. The current passing through an LED when it is being driven with a 1 and a 0

Along with measuring these parameters, I will be discussing my predictions and commenting on what you will be seeing with the instruments as well as your own senses. As I noted in the binary counter project, when I used the 74HC193, I found that I could affect the output by just *touching* the chip. Touch can be a surprisingly powerful tool when you are looking for problems in electronic circuits.

In order to help you visualize how the tests are going to be implemented, I have taken the binary counter circuit and indicated with numbers where the

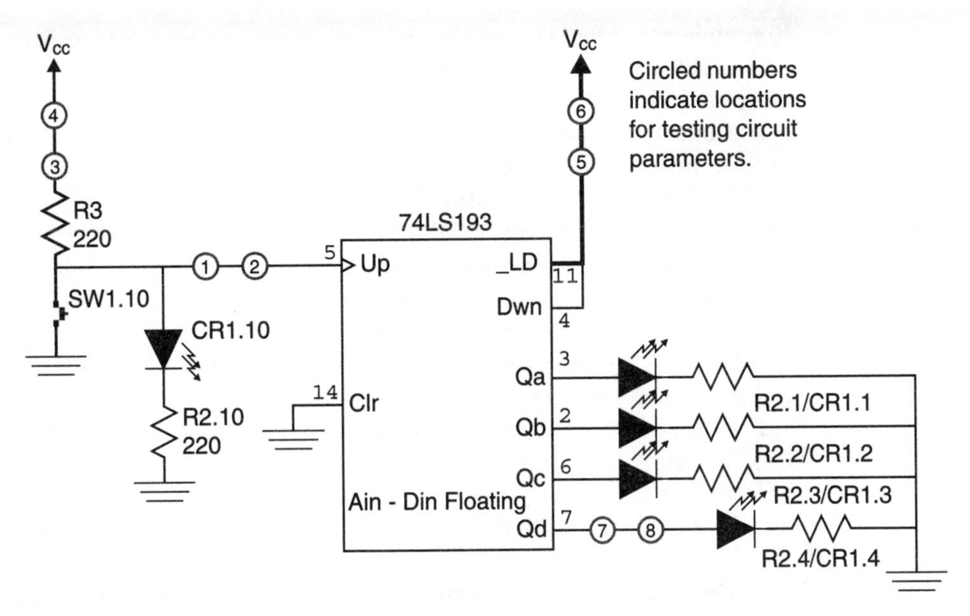

Figure 6.7 Binary counter parameter measurement points.

measurements are to be taken. Figure 6.7 should be used to guide you to where the measurements are taking place.

For the first set of measurements that I would like you to make, disconnect the push button from the 74LS193 and measure the voltage coming from the switch. The easiest way of doing this is to turn off the power to the circuit and simply remove the 74LS193 from the breadboard.

Once this is done, connect the ground connection of your digital multimeter to the ground of the circuit (I use the ground screw terminal on the +5-V power supply) and then probe the switch at the point where the 74LS193's pin 5 would be. In Chap. 3, "Setting up Your Own Digital Electronics Lab," I discussed how the expected voltage when the switch was 1 (the LED on) was calculated by Kirchhoff's law (that the sum of the voltages in a circuit is equal to the voltage applied). Assuming that the LED had a voltage drop of 2.0 V, I could calculate the voltage across each of the two resistors:

$$V_{\text{applied}} = V_{\text{R3}} + V_{\text{R2}} + V_{\text{LED}}$$

$$5 \text{ V} = V_{\text{R3}} + V_{\text{R2}} + 2.0 \text{ V}$$

$$3 \text{ V} = V_{\text{R3}} + V_{\text{R2}}$$

Since, R3 and R2 are both 220 Ω, I can rewrite the equation to:

$$3 \text{ V} = 2 \times V_{\text{R}}$$

$$V_{\text{R}} = 1.5 \text{ V}$$

Remembering that this is the voltage across R3 and R2, to find the voltage across R2 and the LED, I sum V_R and V_{LED} to expect 3.5 V being driven from the circuit.

Measuring to confirm this, I found that the actual voltage is 3.60 V (on my digital multimeter) which has an error of 2.8 percent. Places where the error could creep in are the resistors (I used 5 percent tolerance parts) and the actual voltage applied (I measured it to be 5.07 V).

When the switch is pressed, I measured 0.05 V at point 1 on Fig. 6.7, which is reasonably close to the 0.0 V I expected. This error could be caused by resistive voltage drops in the breadboard, power supply, and wiring, which could result in the ground at the switch being different from the point I set the digital multimeter at.

I checked the voltage values with a logic probe set to TTL (rather than CMOS) and normal (instead of pulse), and I found that the logic probe was able to differentiate between 1 and 0 without any problems. This may seem like a simplistic test, but it is useful for making sure you can see what is actually being driven within the circuit.

Turning off the binary counter, I want to put the 74LS193 back in the circuit for checking the button voltage measurements for test point 2. So far in the book, I have talked about voltage thresholds for the input circuit, and you are probably under the impression that the input pins are infinite impedances (resistances) and no current flows through them. As will be discussed later in the book, this is an erroneous assumption and can lead to problems. In this section, I will introduce you to some of the issues of the circuit that I will expand on in a few chapters.

When I measured the 1 push button voltage with the 74LS193 installed, I found that the voltage was actually 3.61 V. The 0 voltage was 0.05 V again. The difference doesn't seem to merit the previous paragraph, but as I work through TTL chips, you will start to see where simply approximating a TTL input to an infinite resistance will lead to problems.

For parameter measurement 3, turn off the +5-V power supply, remove the 74LS193, and wire your digital multimeter between R3 and V_{cc}. You can use the V_{cc} connection of the terminal block to connect the red lead of the digital multimeter, and I find wrapping a wire around the black lead will give you quite a good connection. The red lead can remain in place for the following three parameter tests. When you are connecting your digital multimeter, make sure you have the red lead in the correct socket—many inexpensive digital multimeters (mine included) have multiple sockets for the leads, depending on the function that is being used.

With the TTL chip taken out, you should be expecting about 6.8 mA. This value was found from the information from the previous tests and Ohm's law. Since I know the voltage across each of the two resistors (as well as their value), I calculated the expected current as:

$$\begin{aligned} I &= V/R \\ &= 1.5 \text{ V}/220 \ \Omega \\ &= 6.8 \text{ mA} \end{aligned}$$

The actual value measured was 7.2 mA (5.8 percent difference).

When the button is pressed, the expected current jumps to 22.7 mA because the 5 V applied is passed through the single 220-Ω R3 resistor. The actual measured value I got was 25.2 mA.

Turning off the 5-V power supply and plugging in the 74LS193 (for parameter measurements 4), I found that I had a slight difference in the current drawn through the circuit for a 1 (it changed to 7.3 mA). The interesting and unexpected result was when the button was pushed (0), the current draw decreased to 24.8 mA. 0.4 mA may not seem like a lot (it is a change of 1.6 percent), but it indicates that the resistance of the circuit actually dropped by 12.5 Ω.

There is one thing I would like to bring to your attention that will probably not seem so trivial. When you first created your circuit, chances are that you, like me, moved all the unused switches to the off position so the LEDs wouldn't distract you. After a while of running like this, you probably noticed that the 220-Ω R1 resistor got quite hot and the 7805 in the voltage regulator got *very* hot.

Previously, I discussed the concept of power and that it was defined as:

$$\begin{aligned} P &= V \times I \\ &= V^2/R \\ &= I^2 \times R \end{aligned}$$

Using the last formula, we can calculate the power being dissipated in the 220-Ω resistor that is connected to ground as:

$$\begin{aligned} P &= (25.2 \text{ mA})^2 \times 220 \ \Omega \\ &= 0.14 \text{ W} \end{aligned}$$

When you multiply this by 8 (for the eight LEDs), the total current jumps to 201.6 mA and the power dissipated jumps to a full watt! If you are powering the power supply with significantly more than 9 V, you will find that you will be dissipating 2 or 3 additional watts through the 7805. If you think about this for a second, you will realize that 3 or 4 watts of power is being used to light eight LEDs. This will probably seem quite inefficient to you (and you would be right).

The purpose of the power supply and the circuits that I have shown you so far has been to educate you in the basic operation of TTL circuits. As you become more familiar with designing your own circuits, you will be able to look for cases (such as this) where a lot of power is being dissipated for a seemingly very simple function and design an application to reduce the amount of power required.

The fifth parameter measurement is to look at the current drawn by the 74LS193 when the button is off (1 driven) and the button is on (0 driven). Turn the power off from the circuit and connect the push button back to V_{cc}. Insert the digital multimeter between the 74LS193's V_{cc} pin (pin 16) and then turn back on the power. If any of the LEDs are on when you apply power, click the button until they are all off. When you make the measurements, make sure all the LEDs are off—the current drawn through the 74LS193 will make the results that you see confusing.

With all the LEDs off and the button driving a 1 into the chip, I measured 19.6 mA drawn by the 74LS193 against a typical value of 19.0 mA from the 74LS193's datasheet.

The surprising result was when I looked at the current draw (with all LEDs off) and the push button pressed (0 input to the 74LS193). Where I expected there to be a current draw increase, I found that there was actually a current draw *decrease*. Later in the book, I will discuss the reason why there is a decrease.

The next parameter (6) to be checked is the current through the 74LS193 while driving a differing number of LEDs. For this experiment, I only checked the current through the 74LS193 when the button was up (1 being input into the 74LS193). My expectation for this parameter measurement was that the current drawn by the 74LS193 would increase by the amount of current drawn by the LED.

Going back the 74LS193 datasheet, I can see that the typical high output voltage is 3.4 V. Going back to the assumption that there is a 2.0-V drop across the LED, this means that there is 1.4 V across the 220-Ω resistor. Using Ohm's law, I calculate that there will be 6.3 mA passing from the 74LS193 to the LED/resistor, and this value can be added to the 19-mA typical 74LS193 current draw. In the table below, I have listed what I expect for the current draw and what I actually see:

LEDs on	Expected	Actual	Difference
0	19.0 mA	19.6 mA	0.6 mA
1	25.3 mA	24.0 mA	1.3 mA
2	31.6 mA	28.3 mA	3.3 mA
3	37.9 mA	33.0 mA	4.9 mA
4	44.2 mA	37.5 mA	6.7 mA

From this table, you will see that there is a difference of roughly 1.5 mA for each actual LED turned on against the expected value. This probably seems like a very small value, but it works out to 15 percent—a value that can cause

$$I = V/R$$

$$= 6.8\ \text{mA}$$

significant problems for large circuits when power supplies are being specified and the actual current requirements are different by this amount.

The next parameter measurement (7) should help to rectify some of the confusion. Turn off the 5-V power supply, disconnect the red digital multimeter lead from the 5-V power supply terminal, and reconnect the 74LS193 to V_{cc}. Set up the digital multimeter to read voltages and set the red lead to 74LS193 pin 7 and the black lead to gnd.

Turn the power back on and measure the voltage being driven from the 74LS193 for one, two, three, and four LEDs being lit. As I indicated, the 74LS193's datasheet specifies that the typical high-level output is 3.4 V. This voltage should remain steady, as the 74LS193 is capable of supplying at least 20 mA from each pin.

When I measured the voltage being driven to the LED, I found it fairly steady at 3.13 V regardless of how many LEDs were lit. The 0.27-V output is lower than the typical output, but within the minimum to maximum range of 2.7 to 3.4 V. If the expected current through the LED/resistor is calculated, you will find that it is about 1.2 mA—which is very close to the amount of current that is missing from the expected to actual calculations above.

The last parameter measurement (8) is the current through an LED/resistor when it is on (a 1 being driven to it). To set this up, turn off the 5-V power supply. Break the connection between one of the 74LS193 pins and an LED, and in its place insert the digital multimeter leads and set the digital multimeter to read current. When this is done, turn on the digital multimeter and the 5-V power supply and push the button until the LED being measured lights.

I found that the actual current through the LED/resistor was 5.1 mA. If this is multiplied by 220 Ω, 1.1 V is calculated to drop through the resistor. When the 2-V drop of the LED is added to this value, the 3.1-V drop matches extremely well to the voltage that was measured at the output pin of the 74LS193.

Reading back over this section, I notice two things. The first is that I have given a lot of attention to discrepancies between expected (calculated) and actual values. I realize that I have gone on ad nauseum, comparing actual values to what I expected from basic laws and the chip's datasheet. I felt it was important to do this so you get an idea of what you are going to see when you design and debug your own circuits. This section should have also given you some techniques that can be used to check basic circuit and TTL chip parameters in circuit.

The second point that I hope you picked up on is that you can demonstrate Ohm's law and other basic electrical laws by using very simple and, at first glance, inappropriate circuits. A recurring theme of this book is that digital electronics actually uses electronic devices that follow the same basic rules as the basic electronic parts presented at the start of the book. The better you can conceptualize this point, the better you will be able to design circuits that will best meet your requirements with minimum power, maximum speed, and lowest cost.

Oscilloscopes

If you are expecting a tax return, a grant, or a stock dividend, put off that purchase of a wide-screen TV or a new PC and consider instead buying yourself an oscilloscope to help you with working with electronics. No single tool will give you the capabilities of an oscilloscope to understand what is going on in your application. I have owned the Tektronix TDS-210 shown in Fig. 6.8 for the past 4 years, and can honestly say that the projects in this book would not have been possible without it.

There are two different types of oscilloscopes. The most basic one is the analog oscilloscope that starts a sweep generator when a voltage trigger level is reached. The sweep generator causes a cathode-ray-tube (CRT) electron beam to move across the screen. The continuing signal is "drawn" on the circuit screen, deflected up and down proportionately to the voltage of the input signal. This operation is shown in Fig. 6.9.

The circuit screen is marked off in a *graticule,* which indicates the time between features on the screen and their magnitudes. The oscilloscope itself is calibrated so these values can be read off the screen simply by counting the divisions on the graticule.

Multiple signals can be displayed by alternating the position of the single electron beam. One way is to alternate the source, displayed each time the oscilloscope is triggered. The second method is to change which source is being displayed as the beam sweeps (this is known as *chopping*). Figure 6.10 shows how these methods work and appear on an oscilloscope display.

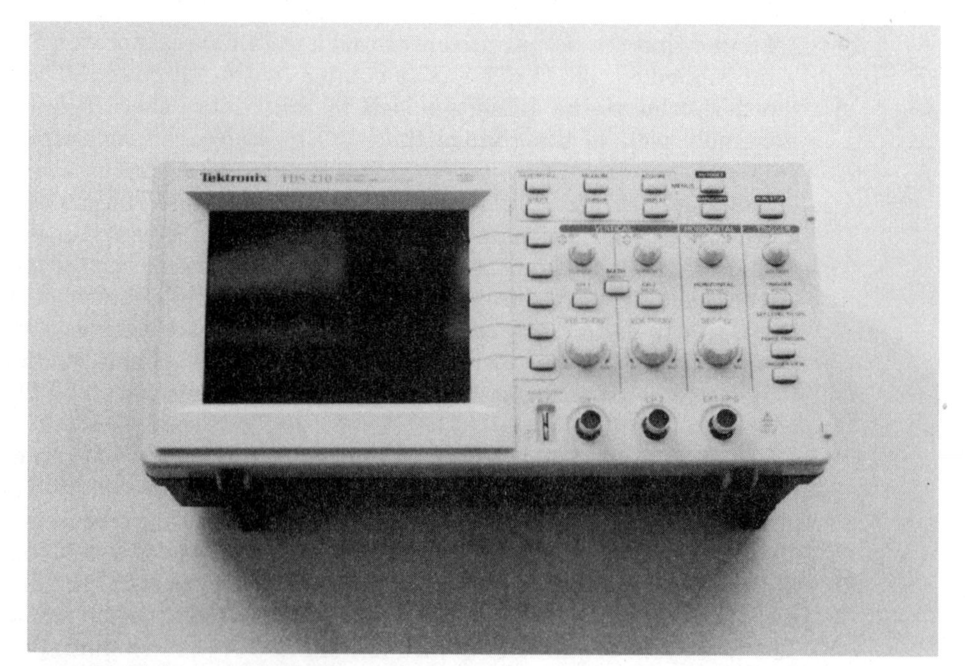

Figure 6.8 Digitizing oscilloscope.

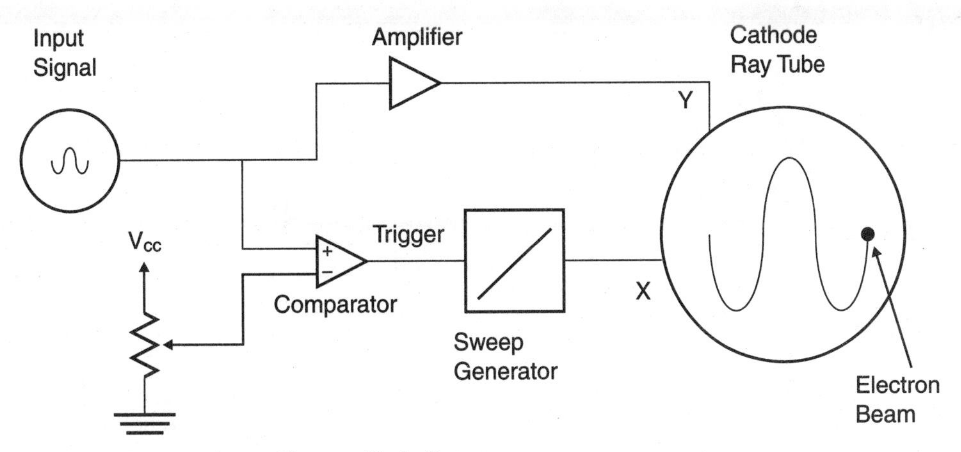

Figure 6.9 Basic analog oscilloscope block diagram.

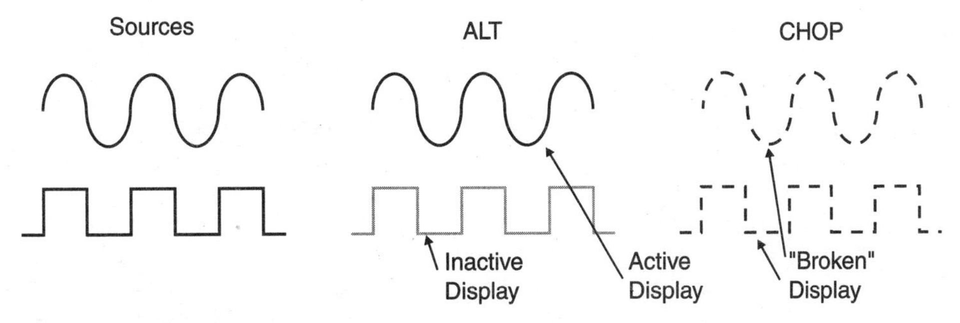

Figure 6.10 Basic analog oscilloscope operating modes.

Neither method is perfect for looking at multiple signals, and in either case, important data could be lost or not visible. There are two problems with the analog oscilloscope when it comes to working with digital circuits. The first is, each time the electron beam sweeps, the signal on the display is very dim and fades quickly. If the intensity is turned up, the phosphors on the backside of the CRT could be damaged. The second problem is, it is very easy to miss single events in multiple sources because the "alt" and "chop" displays miss the changes. These problems get worse with faster sweep speeds and multiple signals.

Over the years, some manufacturers have come up with ways to store signals and improve phosphor performance, but there are still a lot of problems working with an analog oscilloscope with digital signals. Analog oscilloscopes are excellent for very low speed, repeating signals, such as you would find in your stereo with a reference signal, but they are not that helpful for digital signals.

Instead of using an analog oscilloscope for digital applications, I would recommend using a digital oscilloscope, more properly known as *digital storage oscilloscopes* (DSOs) or *digitizers*. These oscilloscopes save incoming signal values digitally and display them in the same formats as an analog oscilloscope. The block diagram for a digital oscilloscope is shown in Fig. 6.11. In this

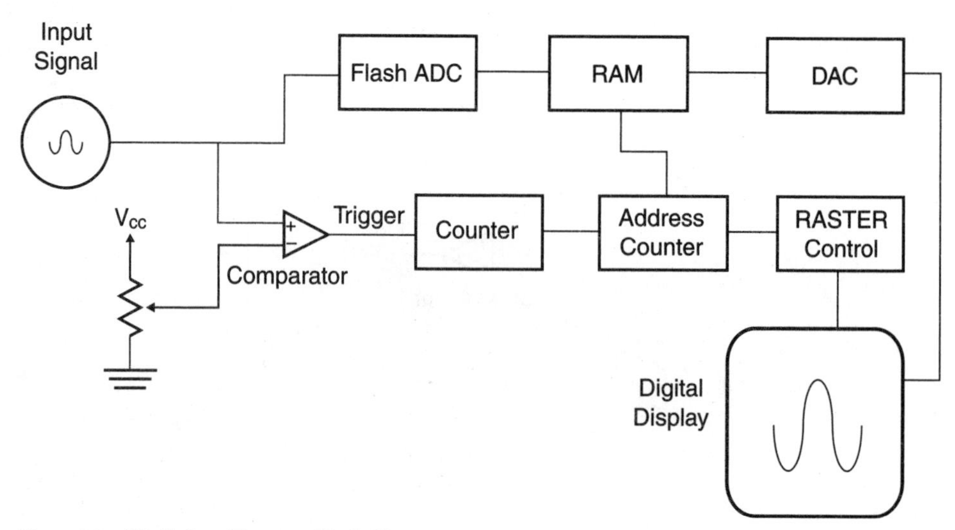

Figure 6.11 Digital oscilloscope block diagram.

circuit, the incoming circuit is digitized by the flash ADC and stored into memory. When the triggered event has finished, the data is read from memory and displayed on the oscilloscope's display.

The obvious advantage to a digital oscilloscope is its ability to capture and display an event's multiple signals without any data being lost or difficult to observe. There is also the added advantage of being able to transfer the data from the digital oscilloscope to a PC without having to take a picture of the screen and then digitizing it (as you would with an analog oscilloscope). The oscilloscope pictures shown in this book were transferred via RS-232 from my TDS-210 oscilloscope and stored into .tif format files for publishing.

Events can also be displayed more easily in a digital oscilloscope because many models have the capability to sample continuously and then, when the trigger point is reached, the counter just continues to the end of the sample. This is shown throughout the book in the various oscilloscope pictures.

The drawback of a digital oscilloscope is its cost. When I was in university, all oscilloscopes were analog, and once, in a demonstration, a radar lab was opened up and the class was shown a digital oscilloscope. It was a big event and the professor responsible for it had barriers put up to keep inquisitive fingers from damaging the instrument. This was probably a reasonable precaution for a $150,000 piece of equipment. A couple of years later, when I first started working, the cost of digitizing storage oscilloscopes was still several times that of an analog oscilloscope—high enough to make it difficult to justify their purchase.

Today, a digitizing oscilloscope is about twice the cost of a comparable analog oscilloscope, and relatively small devices like my hand-held oscilloscope (Fig. 6.12) can be bought for a hundred dollars or so.

Despite the higher costs, I highly recommend that you get a digital oscilloscope rather than an analog one. The digital oscilloscope can make virtually

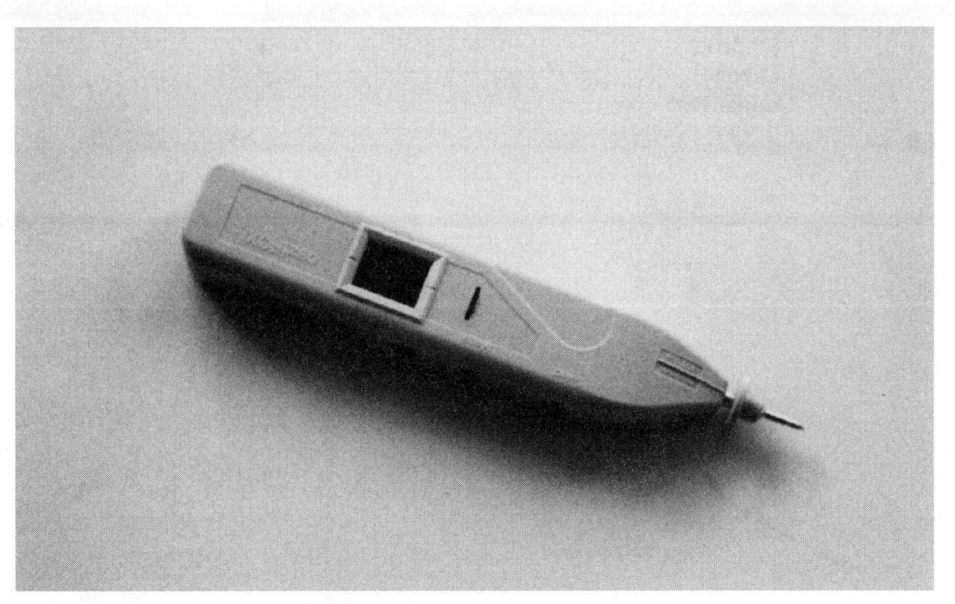

Figure 6.12 Hand-held digital oscilloscope.

all the same measurements of the analog oscilloscope, but the ability to monitor individual events will make it worth its weight in gold. Even reasonably slow digital oscilloscopes (the OsziFOX hand-held oscilloscope shown in Fig. 6.12 can capture signals at a frequency up to 5 MHz) are more useful when working with digital circuits than analog oscilloscopes that can run 10 times faster.

There are a few issues that you will have to work through and understand before you are comfortable with an oscilloscope. The first is understanding how probes work. You might expect that oscilloscope probes are actually straight-through wires into the oscilloscope circuitry. While this is possible, most oscilloscopes have 10× probes, which provide a high input impedance to the circuit. Built into the probe is a 10:1 voltage divider, as shown in Fig. 6.13. The nominal impedance is 1-MΩ resistance and probably 10- to 15-pF capacitance. This low probe impedance will allow an oscilloscope to monitor a low impedance circuit without affecting its operation significantly.

Depending on the equipment, you may not have a compensated display (i.e., instead of the signals coming at 0 to 5 V for digital logic, they are displayed at 0 to 0.5 V).

Along with 10× probes, there are cases where any kind of load at all on the circuit can affect its operation. To avoid this, MOSFET transistor probes (known as *FET probes*) like that in Fig. 6.14 are available. These probes do not drain any current, and the two MOSFET transistors are very close to the probe time to avoid any "stub" effects. FET probes are quite expensive (usually starting at a few hundred dollars), and are designed for monitoring very high speed, impedance-controlled circuits and nets. None of the projects presented in this book come close to requiring them.

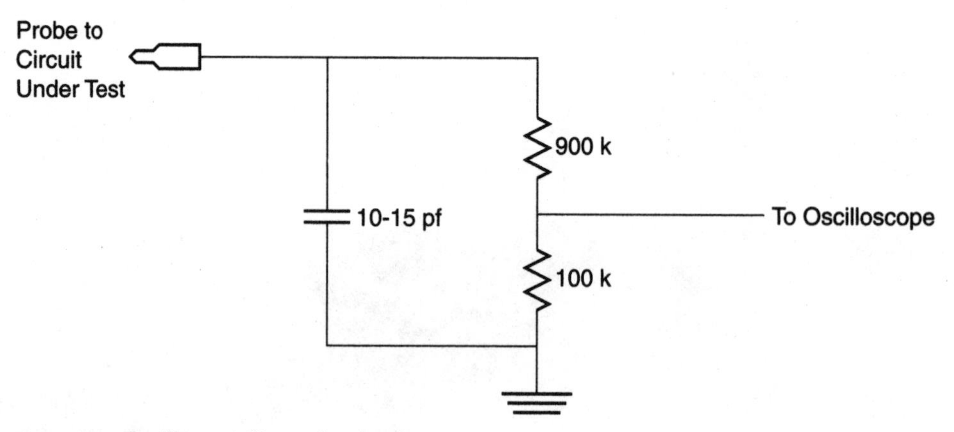

Figure 6.13 Oscilloscope 10× probe circuit.

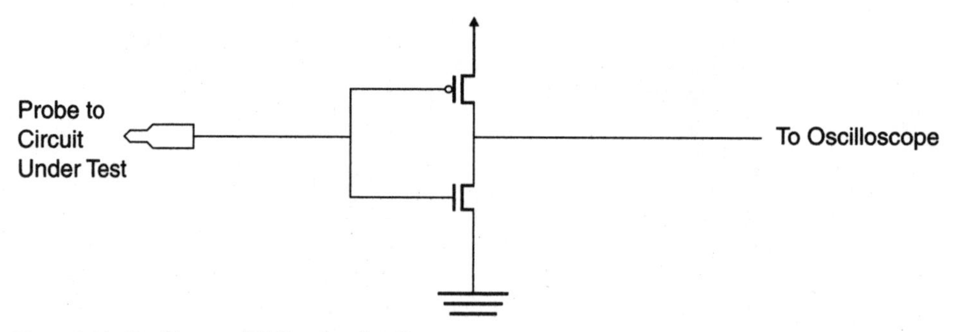

Figure 6.14 Oscilloscope FET probe circuit.

The last issue regarding oscilloscopes is *framing* the event you want to observe. It is probably obvious to say that it is important that the entire event should be shown on the oscilloscope display, but this is easier said than done when you first start working with the oscilloscope. To make it easier on yourself, work with repeating signals as much as possible and play with the oscilloscope's controls until you are happy. As time goes on, you will find that you are better able to figure out how to set up the oscilloscope so you can see exactly what is happening very quickly. When you first start working with an oscilloscope, don't be surprised if it takes 15 minutes or more for setting up each scope shot. This time will go down to 10 seconds or less as you become more familiar with using the oscilloscope.

To make measurements a little more comfortable and avoid cramping your arms trying to hold the probe while resetting the project or pressing a button to initiate an event, you should buy some "chip clips." This tool will allow you to fasten the probes to the circuit rather than having to hold onto them.

Logic Analyzers

A logic analyzer borrows a lot of the technology from the digital oscilloscope discussed in the previous section. The logic analyzer, shown in Fig. 6.15, uses the incoming multiple line data pattern for the trigger and then loads its memory with the bit data until it is full. Logic analyzers tend to be quite expensive, although there is a trend today by some manufacturers to include logic analyzer capabilities within a digital oscilloscope.

Instead of recording digital representations of analog data, digital logic values are recorded at various points in the circuit. Logic analyzers are much more difficult to use within a circuit because of the amount of connections to be made and the need to label signals on the display and come up with a pattern to compare against to trigger the sample. To help cut down on the need to reenter data, most logic analyzers have a disk or nonvolatile memory built into them to save specific setups. Logic analyzers can present data in two different ways. A graphical logic display, like that in Fig. 6.16, is useful in looking at signals and comparing them to simulator output or even circuit drawings.

The second type of display is a *state display,* in which a clock is used to strobe data into the memory rather than save the data according to a logic analyzer internal clock. This type of display is best suited for monitoring the execution of an application. If the data and addresses are captured, it can be a fast and inexpensive way of creating a processor emulator/tracer.

The two most critical parameters of the logic analyzer are the speed at which data can be processed at and the depth of memory behind each pin. The need for fast memory should be obvious.

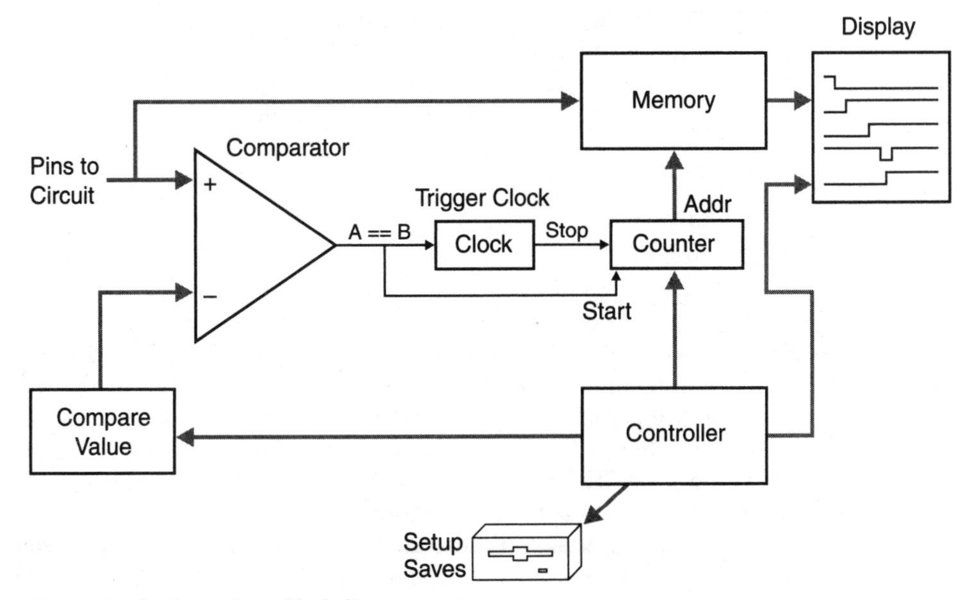

Figure 6.15 Logic analyzer block diagram.

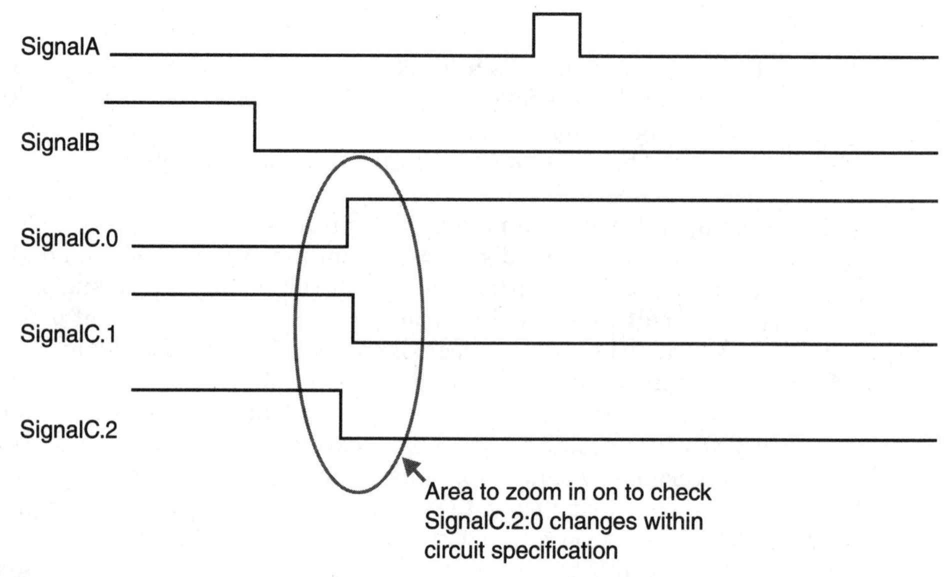

Figure 6.16 Logic analyzer display.

The *depth* relates to how much data can be stored and, more importantly, how much you can zoom in on a problem. In Fig. 6.16, channels SignalC1, SignalC2, and SignalC3 are supposed to all change at the same point. To confirm this, you would want to check these transitions by zooming in to see what is actually happening. A logic analyzer with insufficient depth would not be able to display any data except at the current resolution. To go to a higher resolution, another picture would have to be taken.

If you are looking at what types of test equipment you should buy, may I suggest the following list of items in priority order:

1. Digital multimeter
2. Logic probe
3. Oscilloscope
4. Logic analyzer

From this list, the first one or two tools (the digital multimeter and logic probe) will be sufficient to work the projects presented in this book and help you to develop your own. As you gain more experience in electronics, you will be able to choose the tools that are best suited for your particular style of development. With this knowledge, you should be able to come up with a suite of tools that meets all your requirements.

With this list, please note that there are other tools that you can use to debug a circuit. For example, you may find that a continuity tester from a hardware store is an excellent tool to use for finding shorts and open circuits.

I find a magnifying glass, flashlight, and a dentist's mirror to be excellent tools to help find soldering and wiring problems.

Your fingers are one of your most basic tools. As I indicated when I built the binary counter circuit (Project 7), I found that by tapping on the 74HC193 with my finger, I was able to generate spurious increments/decrements in the circuit. It never hurts to tap a CMOS chip with your finger to make sure that all the inputs are in the correct state and not floating (disconnected).

Your finger is also a good temperature probe. You will find that TTL chips will get warm to the touch when they are operating. If the chip isn't warm after a few moments, you may have a power problem. When you are checking a circuit for hot or cold spots with your finger, I recommend that you use the top of your finger and not the pad. There are two reasons for suggesting this. The pad of the finger tends to have calluses, which makes the tip less sensitive to hot and cold. This can be a big problem if the circuit is very hot—your finger can be burned before your brain registers the pain. Another reason for using the top of your finger to check for hot spots is that if you burn it, it will be painful, but it will not be as uncomfortable as a burn on the surface you hold a pen or pencil in (or click a mouse with).

Now, having said this, I add *be careful*!

A small resistor dissipating 1 W of heat can give you a nasty burn. If you have any reason to suspect that a chip or part of a circuit could be hot, turn off the power and wait a few seconds to see if it takes a much longer time to cool down than other parts. It should also be obvious that a finger should never be used to probe a circuit where there are, or there is the possibility of, live AC currents.

Chapter

7

Digital Electronics in an Analog World

I have introduced the concept that digital devices are actually constructed of analog devices, and you can visualize that when you are playing with gates and simple applications, as I have shown so far in this book. It wasn't until I was seeing the results of failure analysis on 4-Mbit memory chips, that I realized that the operation of the transistors and other analog parts in a silicon chip circuit could affect other parts that are located on and off the same chip.

Along with circuits within a chip affecting other ones, you will be interfacing your applications to devices of various types. In order to do this successfully, you will have to understand more about how the chips work from the analog perspective rather than that of straight 1s and 0s.

In this chapter, I will expand on many of the concepts I presented earlier in the book and discuss how digital circuits and operations are implemented by using analog devices like transistors. The actual capabilities of the circuits will be discussed so you have a better idea of what changes have to be made in order to successfully interface TTL chips to other devices. Along with this, I will look at the issues of matching circuits so that circuits can pass information without falling into analog electronics traps that corrupt the data or render it unusable.

Decoupling Chips

When the transistors within a chip change state, there is a change in current flow within the chips associated with the change. As I have indicated earlier in the book, silicon semiconductors are *current-based* devices, which must pass current in order to affect the operation of other devices. Depending on where the transistor is located and what type of device it is driving, these current changes could be in the range of microamperes to tens of milliamperes.

Current changes coupled with the current being drawn through resistive devices means that the voltage across these devices will fluctuate. When the voltage level fluctuates within the chip, the current within the chip changes, meaning that transistors that should be operating well away from their transition points can actually be at a point very close to their transition points. I use the term *transition point* to indicate where a logic input is perceived to have changed from a 0 to a 1 or vice versa. In this chapter, I will show how TTL logic actually works and what is the actual transition point for it. I will also discuss CMOS logic and how it differs from TTL.

When the transistors reach their transition points, the chip can output an invalid value or a spurious bit can change data that is stored within a flip-flop in the chip. Another possible scenario is that the chip enters a state that prevents it from responding to changes in inputs. The latter case is usually known as *locking up*.

To stop these changing currents from affecting the operation of the chip, a current filter, known as a *decoupling capacitor* is placed across the V_{cc} and ground pins of the TTL chip. The decoupling capacitor helps to average out the current by supplying some extra current when the transistor state change requires it or by absorbing current when the chip's current requirements decrease. To decouple a TTL (or basically any digital device), I use a 0.1-μF capacitor across the V_{cc} and Gnd pins of a chip. To decouple a CMOS chip, I wire the 0.1-μF capacitor across the V_{pp} and V_{ss} pins. A typical power connection is shown in Fig. 7.1.

The most-used values for decoupling capacitors are in the range of 0.01 to 0.1 μF. When a PCB board is laid out or a prototype card is wired, the critical parameter of the decoupling capacitor is that its positive (+) terminal is as close

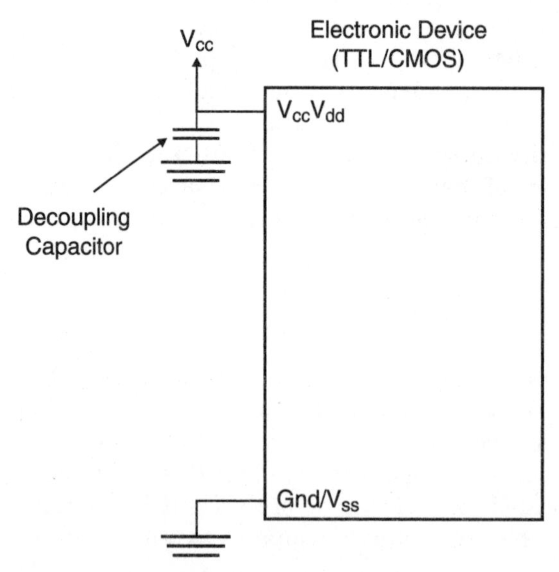

Figure 7.1 Digital electronic chip power connections.

as possible to the V_{cc} (or V_{dd} pin in CMOS logic) pin of the chip. If there are multiple V_{cc}/V_{dd} pins on a chip, then the manufacturer's datasheet must be consulted to find out which pins should have decoupling capacitors wired to them.

Decoupling capacitors should be placed across V_{cc} and Gnd of all chips, which can develop significant current transients. In Fig. 7.1, I show the decoupling capacitor to be a polarized 0.1-μF part. This complicates the wiring somewhat, but it makes for much more reliable applications. Note that for many chips with V_{cc} and Gnd at opposite corners (such as standard TTL chips), there are sockets with capacitors built in.

The capacitor is shown as polarized because I normally use tantalum capacitors for decoupling, although ceramic disk or polyester caps can be used for this purpose. Electrolytic capacitors should not be used for this purpose because they cannot react quickly enough to quickly fluctuating voltage levels because of their high *equivalent series resistance* (ESR), which turns the capacitor into an *RC* network that responds according to the value of the resistance and capacitance in the circuit. Later in this chapter, I will discuss *RC* networks and note that the higher the resistance or capacitance, the slower the response to changes. Ideally, the resistance parameter should be changed because the higher the capacitance, the better the capacitor will be at filtering out the current changes and preventing the chip from changing state or locking up.

Tantalum capacitors are the best because they have the lowest ESR and excellent capacitance per unit volume, but there are a few things to watch for. The first is the polarity of the capacitor; tantalum capacitors inserted backward, like electrolytic capacitors, can catch fire or explode. This is not an issue with ceramic disk or polyester caps.

The second thing to watch out for is to make sure you derate the specified voltage of the part to 40 percent or less for your application. *Derating* capacitors means that instead of using them at their rated voltage, you should choose capacitors that are rated at 3 times or more of the voltage in the application you plan to use them in. For my applications, which use tantalum decoupling capacitors at 5 V, I will use parts that are rated at 16 V (which corresponds to a derated value less than one-third of the rated value).

When tantalum capacitors are first powered up in an application, voltage spikes (which are caused by power supply start-up and current transients within chips) can cause them to develop pinhole breakthroughs in the dielectric, which can cause the plates to touch. When the plates touch, heat is generated by the short circuit, which boils away more dielectric. This process snowballs until the part explodes or catches fire. Derating the cap values ensures that the dielectric layer is thicker, minimizing the opportunity for pinholes to develop.

The price of a tantalum capacitor is proportional to its voltage rating. For a hobbyist or for a prototype circuit, a 16-V tantalum capacitor is not prohibitively expensive, but for a large, complex design, you might find that the decoupling capacitors are the single largest bill of material cost item in a product. In these cases, you will find yourself under a lot of pressure to specify lower-rated-voltage parts or use ceramic or other types of capacitors with higher ESRs to save money.

You can use lower-rated tantalum decoupling capacitors in production, but I suggest that you contact the part manufacturer first and understand what are the issues surrounding the capacitor's behavior when V_{cc} is close to the rated voltage. I would expect that with more modern parts, the need for significant derating of the voltage will become less important.

How Digital Circuits Are Implemented

When I work with new engineers on problems with electronic circuits, one of the biggest misconceptions that I find is that the engineers think of digital circuits as being different from analog components. This misconception leads people to think that digital electronics does not have any of the same issues of analog components and that the concepts and application of digital electronic circuits are completely different. As I will show in this section and the next, this idea is incorrect and is often a dangerous one to make, as it will lead you into trying things that cannot work and avoiding opportunities that could.

So far in this book, I have tried to present to you the different operating characteristics of TTL logic circuits. In this section and the next, I want to expand on this to show how TTL circuits are actually implemented and to go through their operation. In doing this, I hope to dispel assumptions that you have made, and this will probably save you a lot of time trying to figure out a problem in the future.

Before going into the TTL gate circuitry, there are four common misconceptions that I would like to change right away. One misconception is that TTL gates are controlled by voltage. Most people, when quoting TTL specifications, will quote the input switching voltage that is provided in the TTL part's datasheet. As I will show in these two sections, the critical parameter is actually current—the threshold voltage level is normally quoted because at this voltage, the TTL gate changes between being a current sink and a current source, and it can be easily measured.

The second and third misconceptions about digital electronics are that they are switches and that they can be used only for digital logic functions. If you look at the gate circuits from an analog circuit analysis perspective, you will discover that they are all basically very high gain amplifiers. While I don't take advantage of this aspect of digital electronic gates in the projects shown in this book, you will see many different circuits that use digital chips in situations where they do not seem to be appropriate or where they might not be expected to even work at all.

The last misconception is that the basic TTL gate is based on a simple inverter, like the one shown in Fig. 7.2. While this circuit *will* work as an inverter and I will recommend its use in "quick and dirty" situations where an inverter is required, it will not function efficiently in a circuit. The output pull-up is capable of sourcing only a set amount of current; if this circuit is used to drive several other gates, you will find that the current available to each one will diminish quickly. An important feature of all digital electronic gate circuits is their ability to provide (*source*) and take away (*sink*) relatively large amounts of current. The ability to sink and source reasonable currents is important to avoid the situation where there is a large capacitance on the output. In these cases, the capacitor will have to charge, which results in the circuit response slowing down and not being able to operate very quickly.

When the first logic gates were designed, the initial circuit (and standard) developers had to work through these misconceptions, and discovered other problems as well that would have to be accounted for when they were specifying basic gates. While I do not present the circuit shown in Fig. 7.2 in a very good light, it was actually used for many years as the basis for resistor-transistor logic (RTL).

While the circuit shown in Fig. 7.2 works and can be used to develop more complex functions, it is not optimal for the reasons that I have mentioned, as well as a few others. After many years (and I'm sure what seemed like endless meetings), the transistor-transistor logic standard was created; it specified the inverter circuit shown in Fig. 7.3 as the basis for the logic technology family.

When you first look at the circuit in Fig. 7.3, I'm sure that it will appear to be strange. The circuit actually works quite well, although its operation will not be immediately obvious.

For example, the input to the inverter probably seems like a strange way to use an NPN transistor. The example RTL inverter circuit shown above probably makes more sense, with current coming through the input resistor turning on the transistor and pulling the output line low. The use of the NPN transistor is confusing, although its operation is quite straightforward and very clever.

When I see this circuit, I mentally replace the input NPN transistor with two diodes, as I have shown in Fig. 7.4. As can be seen in the modeled gate, current can flow only from the gate, and if the input is driven high (or not driven at all and left floating), then the current through the resistor passing to the NPN transistor's base will flow to the remainder of the circuit. If the Input is pulled low, then the current will have a path of less resistance to ground and it will follow it rather than go through the rest of the inverter circuit.

If you put a voltmeter on a TTL input pin, you may or may not find that it is at some value above 1 V, but if you put a logic probe on it, you will find that

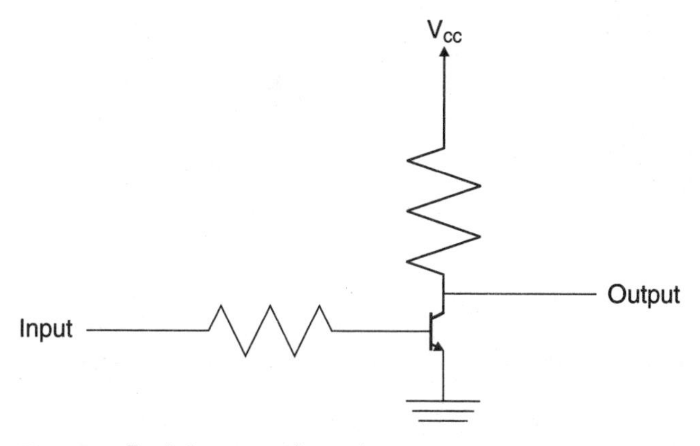

Figure 7.2 Basic inverter schematic.

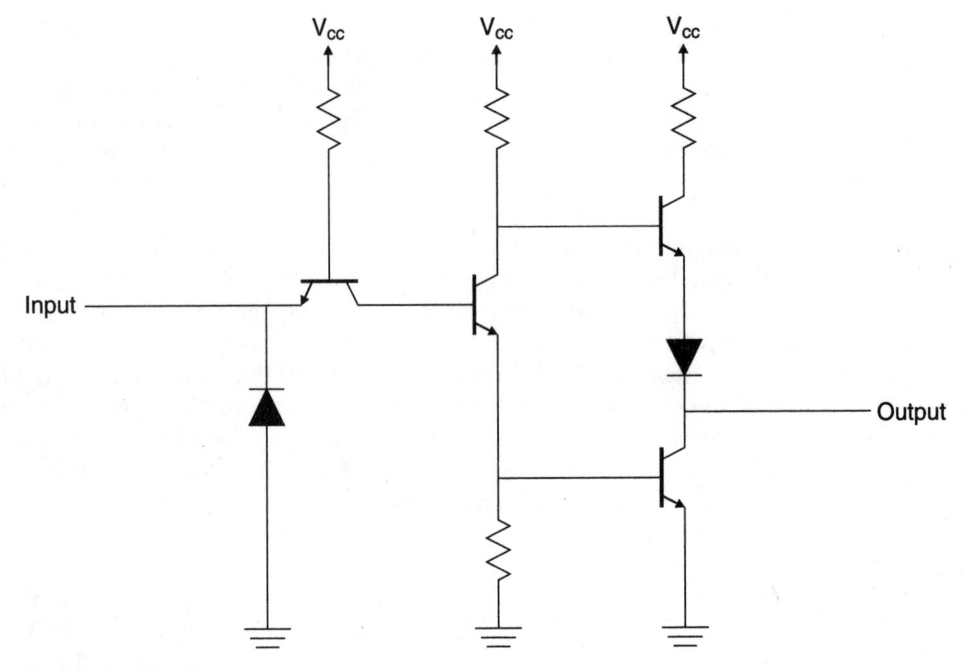

Figure 7.3 TTL inverter circuit.

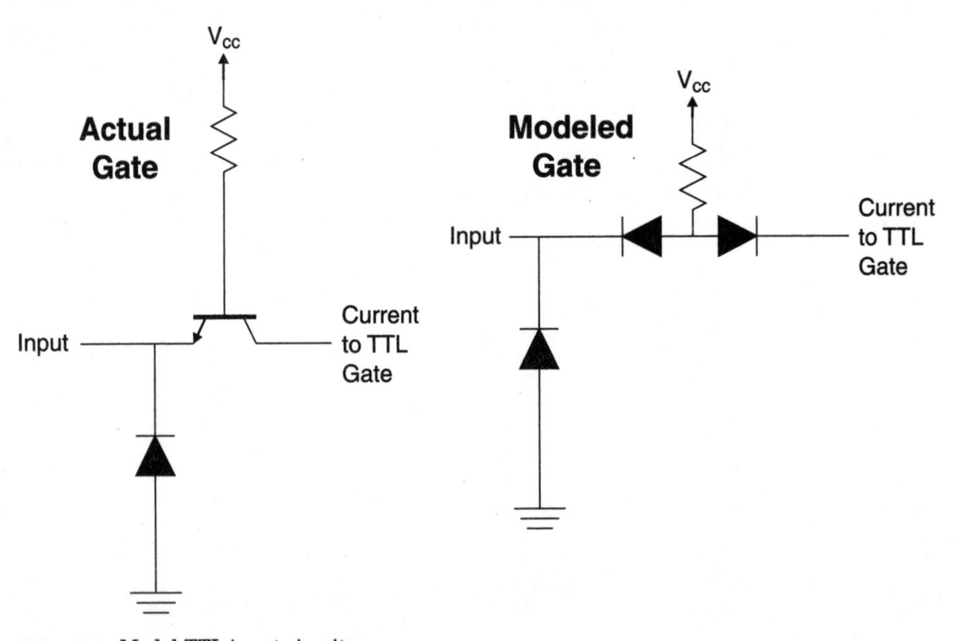

Figure 7.4 Model TTL input circuit.

it does not identify the pin as being high or low. The reason for these strange observations lies in the isolation the pin has been given and how the circuit is wired. I am pointing this out because I don't want you to think that the input pin is pulled up and you can take advantage of this in your applications.

The diode connected to the input is there for clamping protection. If a device drives a negative voltage on the pin, then the operation of the logic gate could be affected and incorrect values could be recognized within it. With this clamping diode, negative voltages will cause current to pass through ground, hopefully equalizing the circuit.

For positive voltages, note that the input NPN transistor performs the same function. Looking at Fig. 7.4, you should see that if a voltage greater than V_{cc} is applied to the gate, it will be blocked by the diodic function of the NPN transistor. This clamping function allows a TTL gate to be used in various applications where the voltages applied to them cannot be guaranteed to be perfect; it avoids situations where the internal circuitry can be damaged or binary values being passed within the circuit can be inadvertently changed.

If the Input is left floating or is driven with a high voltage, the circuit will respond in exactly the same way with the current through the input current limiting resistor being passed to the inverter portion of the chip as shown in Fig. 7.5. When the current flows from the input section of the gate, it turns on the transistor connected to the input NPN transistor's collector. This transistor

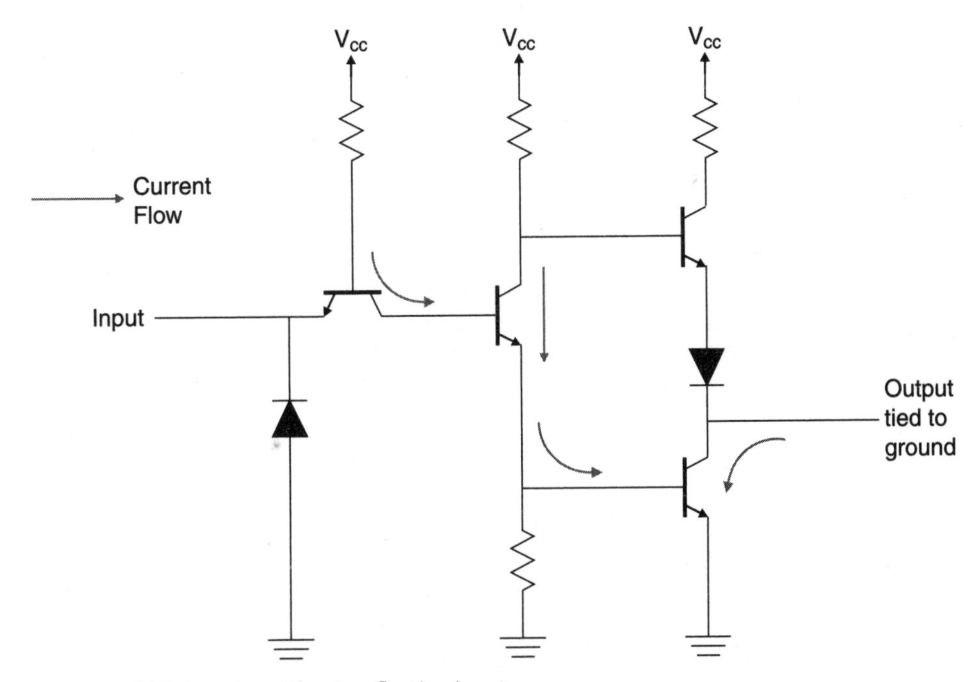

Figure 7.5 TTL inverter with a 1 or floating input.

allows current to pass through the circuit, part of which turns "on" the bottom right transistor. When the bottom right transistor is turned on, the output is essentially "tied" to ground, as I have shown in Fig. 7.5. Note that there is no current available at the base of the upper transistor, which is turned off.

When current is drawn through the input pin, there is no current passed from the input NPN transistor's collector to the next transistor's base. In this case, the current passing through the middle resistor will pass to the top right transistor, as shown in Fig. 7.6. When this transistor is turned on, the output is tied to V_{cc} through the small current-limiting resistor on the right of the circuit. This current-limiting resistor is used to limit the maximum amount of current available (or sourced) by the gate.

There are two things I would like to point out to you in the three transistors that make up the body of the inverter circuit. The first is the use of the diode above the output. The reason for this diode being there is to protect the gate from back-driving voltages. *Back driving* is the term used for forcing an output into a different state. While there are circuits built into the TTL gate for protection, you should always avoid this situation, as you will end up in a push-pull fight between devices that will result in an indeterminate voltage/current being available at the output. This is known as *contention,* and I will discuss it in more detail elsewhere in the book.

The second thing to notice about this three NPN transistor circuit is that it is really an amplifier. For straight TTL logic, the current available when the

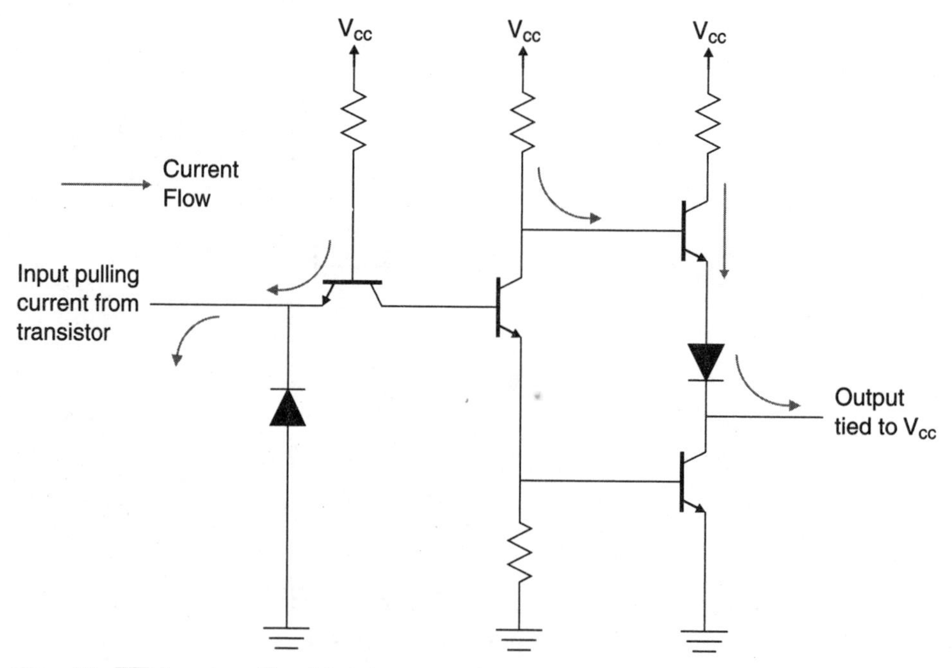

Figure 7.6 TTL inverter with a 0 input.

input pin is high is very small (on the order of 1 mA for most TTL logic families). This current is amplified into a much higher output current (in the range of 20 to 50 mA).

As I pointed out above, you will find a number of circuits that use TTL (and CMOS) logic in ways that you would not expect. The reason why these circuits work is the implementation of the gate circuitry as an amplifier rather than a simple switch, as digital logic is normally portrayed.

While the basic TTL gate that I have shown here works quite well and is quite fast, there are a number of aspects to the circuitry that make it suboptimal for many applications. This is also true for the different versions of TTL, including:

- *High speed* or *Fast* (74F), which is faster and has more current sink/source capabilities and requires a moderate amount of power
- *Shottky* (74S), which is very fast and has more current sink/source capabilities, which require quite a bit of power
- *Low-power* (74L), which is, relatively speaking, very slow, but consumes minimal power
- *Low-power Shottky* (74LS), which has approximately the same speed as regular TTL, about the same source/sink capabilities, and much-reduced power consumption

The problems with TTL (bipolar) digital logic centers around the difficulty in manufacturing the different components used in the gates. When bipolar transistors are built on silicon, they must first have their own alternating N-type and P-type "tubs" that they are built into—these tubs prevent current from the devices leaking out and potentially affecting other devices. This means that the complexity of building the transistors is much higher than you would expect.

Along with building transistors being much more difficult than expected, it is extremely difficult to build physically small resistors of specific values on silicon. For relatively simple integrated circuits, like the ones used in this book, physically small resistors are not required, but if very large integrated circuits were built with the technology, then there would be serious manufacturing concerns.

The last problem with bipolar TTL is the amount of current and power that is consumed. This is especially true when devices like LEDs and buttons are included in circuits. For basic TTL circuits, each inverter requires somewhere between 1 and 1.5 and a half mA to operate. When this is translated into millions of gates, this value climbs to thousands of amps.

To solve these problems, an entirely new class of logic was required. Earlier in the book, I discussed *field-effect transistors* (FETs), which use a voltage potential to change the conductivity of silicon. These devices do not require currents to operate and can be implemented much more simply than TTL devices. The first microprocessors and dense memory chip devices used NMOS technology, which employed an N-channel MOSFET as a switch to control whether a pull-up voltage was passed along to the next device in the chain. NMOS technology was very prevalent in the late '70s and early '80s, but it was never considered to be an optimal technology for very complex integrated circuits because of the need for the resistor in the circuit.

The successor to NMOS was CMOS, which avoided the need for resistors in gate circuitry altogether. In this technology, N-channel MOSFETs and P-Channel MOSFETS are used together to control current flow according to digital logic conditions. The circuit for a CMOS inverter is shown in Fig. 7.7. In the CMOS inverter, when the input voltage is high, the P-channel MOSFET at the top is turned off and the N-channel MOSFET at the bottom is turned on, pulling the output to ground (or V_{ss}). When the input voltage is at ground potential, then the operation of the transistors is reversed and the output is at V_{dd} potential and the P-channel MOSFET can source current.

The advantages of the CMOS circuit are its very miniscule power requirements and minimal silicon real estate requirements. Because MOSFETs are controlled by voltage, the gate uses current only when it changes state and a small amount of charge is passed from one transistor to the other. The amount of charge passed through the two transistors increases with the frequency of the operation of the gate, which means that CMOS devices that are run by battery power operate at the slowest possible speeds.

P-channel and N-channel MOSFETs can be built in much less space than their bipolar analogs. Neither device requires the complex tubs of the bipolar devices, and they are cheaper to manufacture. For these reasons, CMOS has allowed the highly complex, several-hundred-million-transistor computer processor chips that we are familiar with to be possible.

The most significant drawback to CMOS is the speed at which the transistors turn on and off. The gates of the MOSFET transistors are essentially capacitors, and charging the gates can take more time than the current flow in bipolar transistors. Over the last 5 years, the speed of MOSFET devices has been improved to the point where they are approaching the speed of bipolar

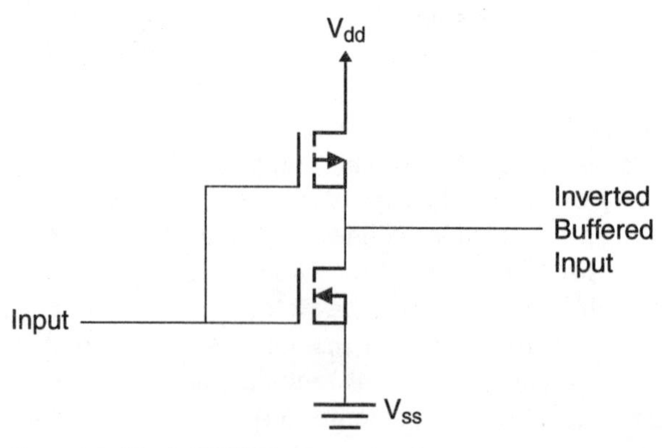

Figure 7.7 Basic CMOS inverter circuit.

transistors, but this comes at a high development cost and additional space on the silicon chip substrate.

It is important to note that CMOS logic gates do not have the built-in pull-up of the TTL gates. As I noted in Project 7, when I used a CMOS part instead of a TTL part, I found that if I left the inputs high, they behaved as if they were actually pulled to ground. This can be quite disconcerting if you are familiar with TTL logic, and it could lead you to believe a chip is defective because it does not behave as if unconnected inputs were actually tied to V_{cc}. If you have a TTL chip that is behaving strangely, always remember to check the part number to see if it is CMOS and not TTL. CMOS parts are labeled *74HCxx* or *74HCTxx*.

Going back to the start of this section, you might think that CMOS logic actually works like the mythical device, about which I tried to correct some common misconceptions. While CMOS logic does operate from voltage potentials rather than current flow, you have to remember that the devices primarily control current flows (like TTL chips), and many of the same rules governing interconnecting them still apply.

Note that the CMOS inverter circuit in Fig. 7.7 does not have the same clamping diodes and back-driving protection of the TTL gates. You will find that CMOS is much more easily damaged by electrostatic discharge (ESD) and wiring errors. When I am designing and building circuits, I try to take the problem areas of both technologies into account. The result of this care is circuits that are much more robust and less likely to fail from reliability problems.

In the logic gates I have shown so far, the outputs have been able to source current or output a high voltage. These types of outputs are known as *totem pole* outputs because they resemble Native American totem poles. In some cases, such as dotted-AND buses, you will want to use outputs that cannot source current and only sink it. In these cases, you will want to use open collector or open drain (for CMOS logic) parts that are single transistors like the ones shown in Fig. 7.8.

One potential problem you will have with your applications is that, periodically, the chip you are using will seem "bad" because it does not seem to be able to drive positive voltages. While you might think the driver transistor is "blown," chances are you have an open collector or open drain part. I know

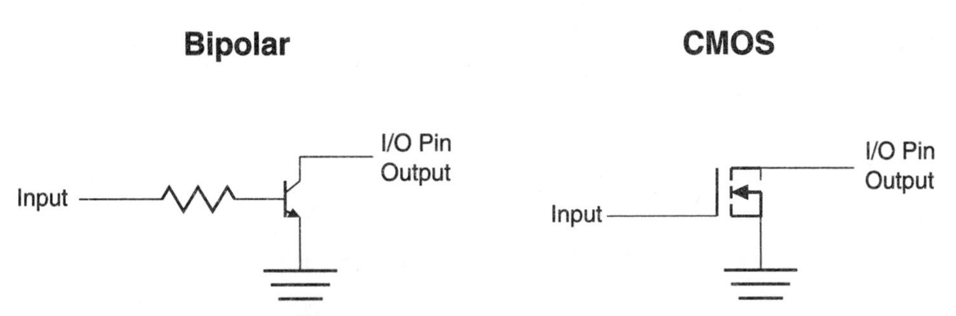

Figure 7.8 Basic open collector/drain output circuits.

that with the Microchip PICmicro® microcontroller, one of the most common questions concerns problems with one of the I/O pins—this pin is wired to behave as an open drain only, and it is a problem that trips many people up. If you find that you have an open collector output and you need a totem pole, you can put a pull-up resistor (1k to 10k) on the output, and it will behave like a totem pole for you.

I have discussed TTL and CMOS inverters almost exclusively in this section, and while they are interesting, they are not representative of the actual circuits used in TTL and CMOS circuits because they have only one input. For any kind of logic operation, multiple inputs are required to implement NAND gates in TTL and NAND or NOR gates in CMOS, which are the basis for creating more complex circuits, as I have discussed earlier in the book.

To implement a multiple-input TTL NAND gate, the input NPN transistor is given additional emitters, as is shown in Fig. 7.9. Going back to the diode analog I used at the start of the section, this is similar to adding a number of additional diodes to allow current flow away from the middle transistor of the circuit if any of the inputs are pulled low.

The practical limit for emitters in TTL is eight. This means that up to eight inputs can be built into a single gate, avoiding the need for multiple gate delays when a three- (to eight-) input gate is required in an application. When you look at some TTL datasheets, you will see that internal AND, OR, NAND, and NOR gates are often drawn with multiple inputs—they are implemented by using the multiple-emitter NPN transistor that is shown in Fig. 7.9.

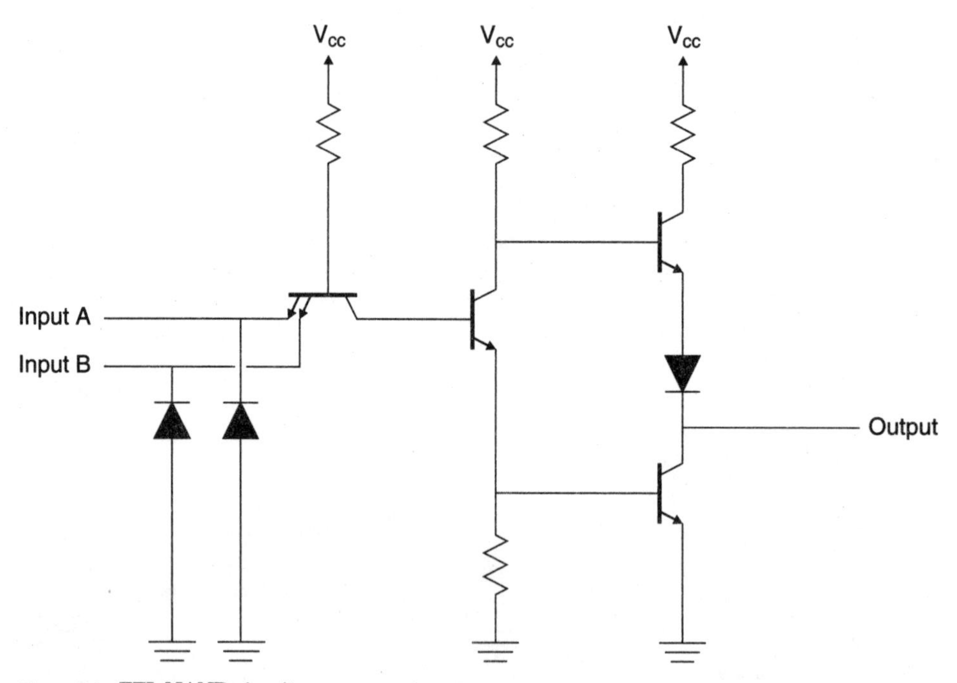

Figure 7.9 TTL NAND circuit.

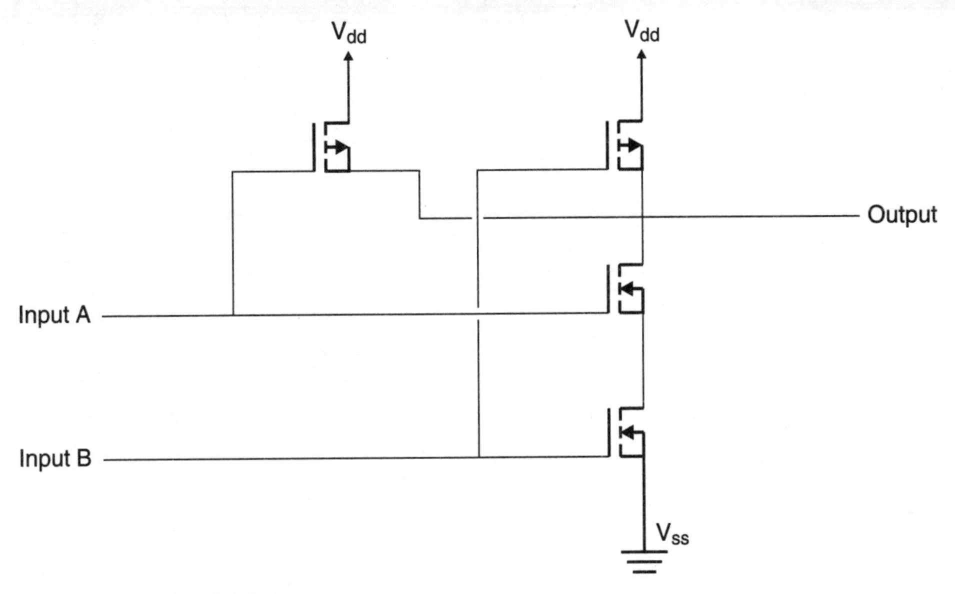

Figure 7.10 CMOS NAND Gate.

For CMOS circuits, implementing multiple input gates is relatively simple—multiple transistors must be added to the gate instead of modifying one that is already there. Figure 7.10 shows how two P-channel and two N-channel MOSFETs are interconnected to create a NAND gate.

In the CMOS NAND gate example, if either of the two inputs is high, then its corresponding P-channel MOSFET will be turned off and the N-channel MOSFET will be turned on. If both inputs are high, then both N-channel MOSFETs will be turned on and the output will be tied to ground. More than two inputs can be handled by this method, but depending on the application, it may be more efficient and faster (because of the extra capacitance of multiple transistors) to use multiple gates rather than one, as would be done in TTL.

Project 8—Discrete-Component TTL Gate

To help you understand and conceptualize how the TTL logic gate works, I thought it would be useful to have you build one. This might sound daunting, but with a few common transistors, resistors, and diodes, you can create an inverter circuit that is virtually identical to the one built in standard (74xx) TTL logic chips.

I built the circuit shown in Fig. 7.11 on a breadboard, with the +5-V power supply and interface PCBs that come with the book, and came up with the wiring diagram shown in Fig. 7.12. When you assemble the circuit, make sure you note the orientations of the four transistors and follow them as they are shown in Fig. 7.12. If you don't, the circuit will not work or will behave strangely and not like a commercially available TTL gate as you would expect.

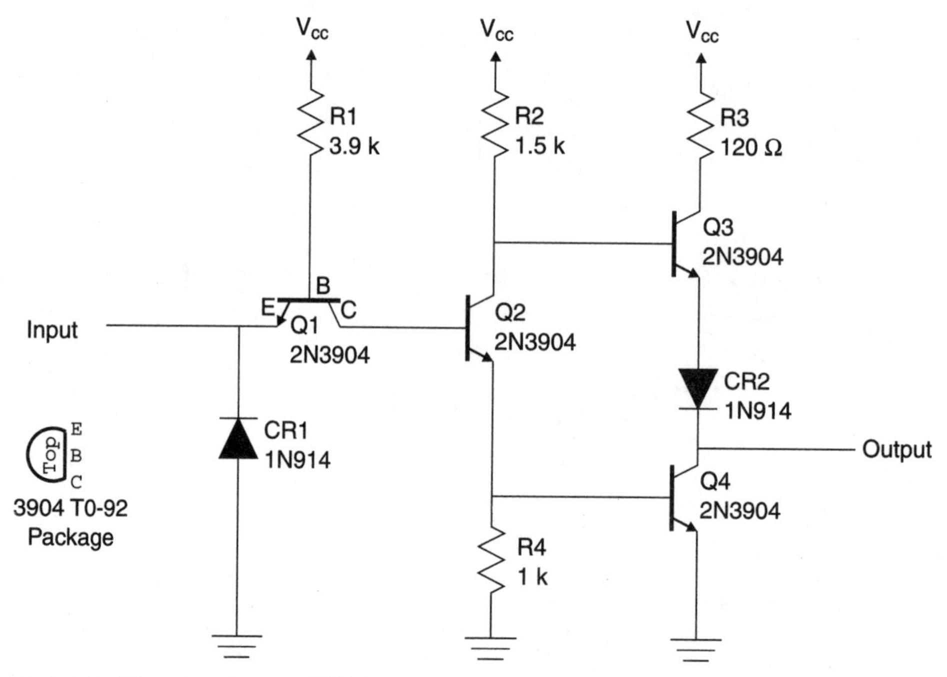

Figure 7.11 Discrete component TTL inverter circuit.

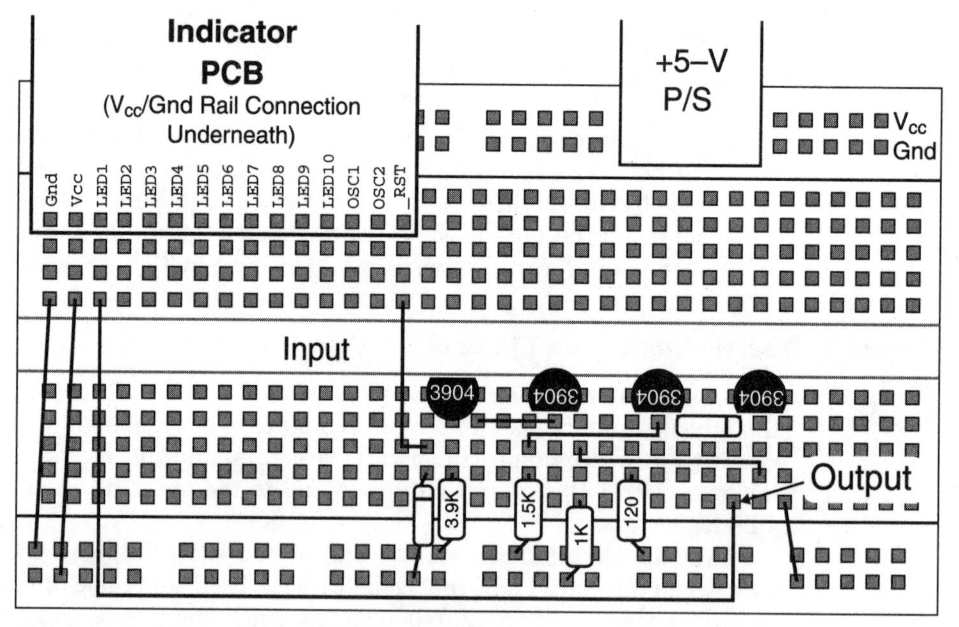

Figure 7.12 Project 8—discrete TTL inverter wiring.

If you were to compare all operating characteristics of this circuit with one inverter built into a 7404, you would find that the two circuits are essentially identical. After building the circuit, I looked at the circuit's operating voltages and currents as I will discuss below.

This circuit took me about five minutes to build (including putting all the parts together), using the following bill of materials:

Part	Description
Q1–Q4	2N3904 NPN transistor in TO-220 package
CR1–CR2	1N914 silicon diodes
R1	3.9k, ¼-W resistor
R2	1.5k, ¼-W resistor
R3	1k, ¼-W resistor
R4	120-Ω, ¼-W resistor
R5	10k, 1-turn PCB-mount potentiometer
Miscellaneous	Breadboard, interface PCB, +5-V power supply, wire

When you have the circuit together, turn on the +5-V power supply and press the _RST button. When the button is pressed, you should see that the far-left LED turns on (it is normally off when the button is released).

If you disconnect the input of the circuit from the _RST breadboard connection, you will discover that the LED stays off. If you have read through the previous section, you will know that this was to be expected because of the operation of the NPN transistor used in the input circuit.

With the input still disconnected, take a milliammeter and connect it across Q1's emitter and ground. When you do this, you should see the leftmost LED come on and see that 1 mA or thereabouts (I measured 1.06 mA) is passing through the milliammeter. This is also expected because the milliammeter has become the low-resistance path for the current passing through R1.

Assuming that the voltage drop across Q1's base/emitter junction is 0.7 V (as for a typical silicon diode), we can calculate the expected current when the emitter is shorted to ground (or pulled low by another TTL gate) using Kirchhoff's and Ohm's laws:

$$5\text{ V} = V_{be} + V_{\text{R1}}$$

$$V_{\text{R1}} = 5\text{ V} - V_{be}$$

$$= 5 - 0.7\text{ V}$$

$$= 4.3\text{ V}$$

$$\text{Current} = V_{R1}/\text{R1}$$

$$= 4.3\ \text{V}/3.9\text{k}$$

$$= 1.103\ \text{mA}$$

The 1.06 mA that I measured is only a few percent off from the calculated (expected) value, and this discrepancy can be accounted for by the voltage drop within my digital multimeter (set to measure milliamperes) as well as any internal resistances of the instrument.

Next, connect the input to one of the two buttons on the Interface card that have LEDs associated with them (LED9 or LED10). You will find that the circuit behaves in exactly the same way, except that the LED associated with the button (the input) never seems to light. If you look at it in a darkened room with the adjacent LEDs turned off, you may see a very faint glow in the LED.

The reason for this is that the input circuit is holding the voltage of the circuit at around 1.7 V. This is too low for the LED to light properly and will cause some current flow within the two resistors in the LED9 or LED10 circuits.

This is the reason why I use *only* 74LSxx (Low Power Shottky) parts with the interface card when I want to use the LEDs to indicate what the input state of the switch is. The 74LSxx gates have a different input section that does not affect the operation of the interface card's LEDs and diodes. As I indicated above, 74LSxx TTL operates at about the same speed as straight 74xx, so I did not consider this to be a hardship.

If you look at the output voltages, you will see that when the inverter's output is high, 4.25 V is driven out, and when the inverter's output is low, the output is around zero volts. This should not be a surprise to you, as the output works identically to the TTL parts you have been working with so far.

The last test I want you to perform on this gate is to find out what is the threshold voltage where the input changes to an output. To perform this test, connect a potentiometer, wired as a voltage divider to the circuit as shown in Fig. 7.13. This potentiometer allows you to find the point at which the voltage changes from a high to a low and back again and to measure it easily with a digital multimeter.

Using the potentiometer, I found that the threshold voltage (where the LED's brightness was about half the normal on level) was 1.44 V. This compares very favorably with the 1.4 V normally quoted as TTL's threshold voltage.

Next, disconnect the potentiometer from the inverter circuit and measure the voltage at the potentiometer's wiper without the load of the inverter. The voltage I got was 1.16 V, which is almost 20 percent different from the voltage measured when the potentiometer was connected to the inverter's input circuit. This should not be a surprising result. Using the basic DC laws, I can calculate the resistance of the potentiometer's wiper to ground:

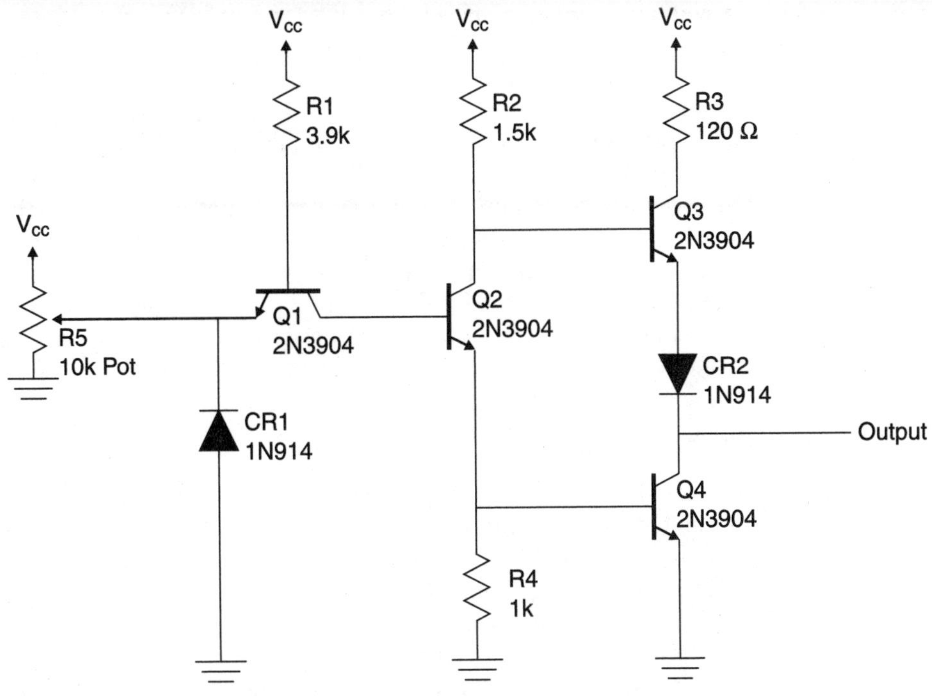

Figure 7.13 TTL inverter threshold test circuit.

$$V_{out}/V_{tot} = R_{wiper}/R_{tot}$$
$$1.16\text{ V}/5\text{ V} = R_{wiper}/10\text{k}$$
$$R_{wiper} = 10\text{k} \times 1.16/5$$
$$= 2.32\text{k}$$

From the earlier tests on the circuit, we know that the current flowing through the base-emitter junction of Q1 is 1.06 mA. This means that, if the resistance experienced by the current sourced by R1 is less than 2.32k, then the current will follow this path instead of the transistors/resistors inside the inverter. This is probably somewhat surprising, but the path inside the inverter circuit is actually quite complex and cannot be simply analyzed.

You can test this by placing a 2.2k resistor from the inverter's input to ground, and you will see that the circuit will behave as if the input was pulled to ground. Higher-value resistors will not behave the same way (or have the LED partially light).

By working through this project, you should have gotten a much better appreciation for what actually goes on in a TTL gate and how it works. It is important to remember that the gate is current controlled and really responds *only* when current is sourced from it to ground. You should also see that the internal pull-up of a TTL input pin cannot be used to advantage outside of the chip. Lastly, with the testing of the circuit with the potentiometer, you should get a better understanding that the TTL gate is not truly a "digital" device; that there are intermediate values and operations available.

This "analog" operation of the TTL gate is something that you should be aware of and give you an idea of why you should only operate these devices in the TTL voltage ranges without any marginal or noisy connections. TTL (and CMOS) logic chips will behave predictably and reliably when they are operated in the regions they are best designed for—taking them outside those regions or working around the "threshold" ranges can result in unpredictable and unreliable operation of the chips and your application.

Project 9—CMOS Touch Switch

In the previous sections, I have gone into quite a bit of detail about how a TTL inverter (NAND) circuit works, but I have not gone into the same level of detail for CMOS logic. CMOS logic has one fundamental difference between it and TTL: It is voltage controlled, while TTL is current controlled. This difference allows CMOS to work with very little current, resulting in very little power being dissipated by the circuitry.

This low input current can be demonstrated by using the circuit shown in Fig. 7.14. By connecting the input of a CMOS inverter to a bared wire, you can charge your body to V_{cc} (5 V) or Gnd (0 V) potential to change the input of the inverter. When your finger is touching both the V_{cc} bared wire and the 74HC04 inverter input bared wire, your body has the same electrical potential as V_{cc} and you will turn the LED at the output of the Inverter off. Similarly, when you touch Gnd and the inverter input bared wire, any charge in your body will be passed to ground and the inverter will turn the LED on.

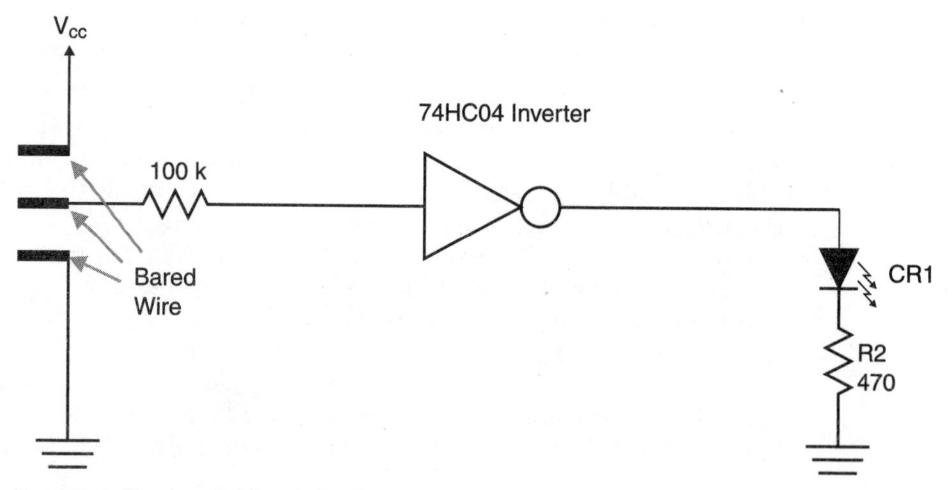

Figure 7.14 Touch switch input circuit.

You may have read that skin has a resistance of 1.5k. In case you feel that there is actually some current passing from you to the inverter input, I have included a 100k resistor in series for the 74HC04 input and the bared wire that you will touch. With this resistor, the maximum amount of current that can pass through you to the inverter input is 50 μAs.

This 100K resistor also will provide some measure of ESD protection to the CMOS input. I will discuss ESD protection later in the book, but when working with CMOS devices, you should always be wary of accidentally damaging the circuits.

I wired my prototype circuit as shown in Fig. 7.15. Note that the bared wires (I used the bare wire that is meant for straddling pins in a wiring kit) are connected in such a way that you can easily touch the center (inverter input) with either V_{cc} or Gnd. The bill of materials for the circuit is:

Part	Description
7404	74HC04 CMOS hex inverter
100k	100k, ¼-W resistor
0.1-μF	0.1-μF, 16-V tantalum capacitor
Miscellaneous	+5-V power supply PCB, interface PCB, breadboard, wire

After building the circuit, you can test it out and you will find that when you touch the leftmost wire (V_{cc}nd the center wire (inverter input), you will indeed

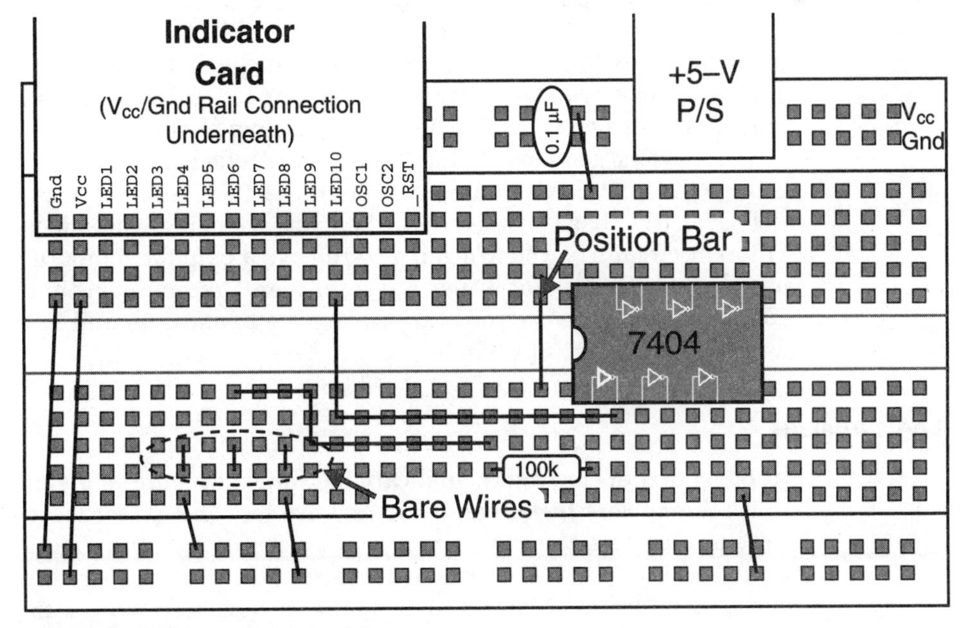

Figure 7.15 CMOS touch switch wiring.

turn off the LED. The same also goes for when you touch the rightmost wire (Gnd) and the center wire—the LED will turn on.

What will probably be surprising for you is that when you take your finger off the wire, the LED will stay in the same state for some time. This time ranges from a few seconds to many minutes. You may find that you can affect the time that the LED is in a specific state by placing your hand near the circuit and touching the breadboard the circuit is on—even touching the table the breadboard is on will change the LED state.

The reason that the circuit stays in the same state for an indefinite amount of time and changes on what seems to be a whim lies in the capacitive nature of the CMOS transistor gates. The N-channel and P-channel transistors are simply flat plates that do not allow current flow. When you take your finger away, you will leave a trapped charge in the wires and chips that will stay there until it leaks away in the silicon, or an induced voltage changes the state of the gate.

This property can make CMOS logic appear to be very unstable and is the reason for some of the problems I had when I first built a counter circuit with a CMOS part with floating inputs instead of a TTL part. As I have shown in the previous sections, TTL has built in pull-ups that provide a default high input if the pin is left unconnected.

CMOS does not have this feature and can be influenced by the slightest variation in electrical field around it. This may seem like an exaggeration, but remember that I said earlier that digital electronic gates can be thought of as very high gain amplifiers. The CMOS gate itself, while normally acting as if a low voltage is being input into it if nothing is connected to it, is a very high gain amplifier that reacts to any induced voltages in a very strong and often unpredictable manner.

In Fig. 7.15, I put a wire on the breadboard and labeled it the *position bar.* This wire is meant to be used to position a TTL chip if you want to try it out instead of the CMOS chip. You will find that the TTL chip is completely insensitive to your finger on either one of the inputs and will act as if the input is pulled high (which it is). The TTL chip does not change because, as I've shown in the previous sections, TTL is current controlled, and not enough current to change the input state can flow through the 100k resistor and to your body.

The position bar is a good tool when you are using a breadboard to test out different circuits or you are going to remove a chip for reprogramming. If you use the position bar for either purpose, then I recommend that you add it to your breadboard *before* you start wiring the application and find you have no way to get the wire in there.

Logic Analogs

In many digital electronics applications, you will require individual logic gates that take up very little real estate and are fast enough for most simple applications. *Fast enough* means circuits that can switch in less than a microsecond. Going back to the RTL (resistor-transistor logic) technology discussed previously, there are some simple analogs that can be used when a full logic chip is not required.

For example, the AND function can be re-created by using the dual diode circuit (Fig. 7.16) for a TTL Input. This circuit will pull down the input if either of the two diodes is driven low. If the input is to be CMOS, then I recommend using the resistor/diode circuit shown in Fig. 7.17. In this circuit, the CMOS input will be driven high only if both driving outputs are high. In the resistor/diode AND gate equivalent, if input A is low, the output will be low. If input A is high, then current can flow through the resistor to the output or through input B (if input B is driven low). When input B is high and input A is low, then no voltage will be driven to the output. Only if input A and input B are high, will the output then be high. A typical value for R in this circuit is 10k. Diodes for these circuits can be virtually any small-signal silicon diodes (I like to use 1N914's because they are very cheap).

In the circuit shown in Fig. 7.18, if either input is driven high, the output voltage will be high. The resistor will pull the output low if neither input A nor input B is driving high. Note that the circuit shown in Fig. 7.18 will work for either TTL or CMOS inputs, although for TTL, I will use a 330- to 470-Ω resistor pull-down. For CMOS, you can use up to a 10k resistor.

The last analog is one you're probably very familiar with: the inverter that can be cobbled together by using a pulled-up transistor as shown in Fig. 7.19.

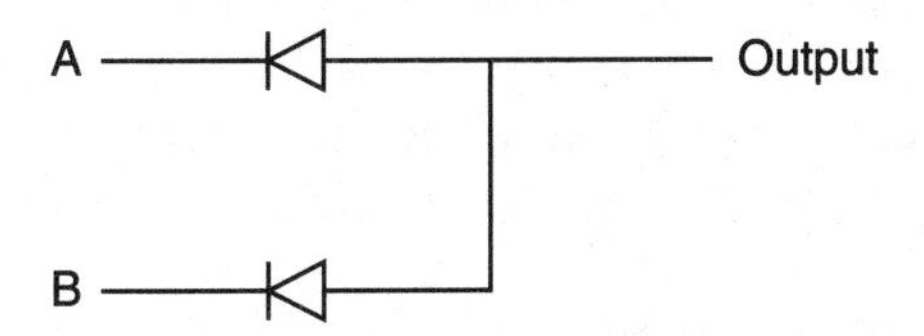

Figure 7.16 AND gate circuit for TTL circuits.

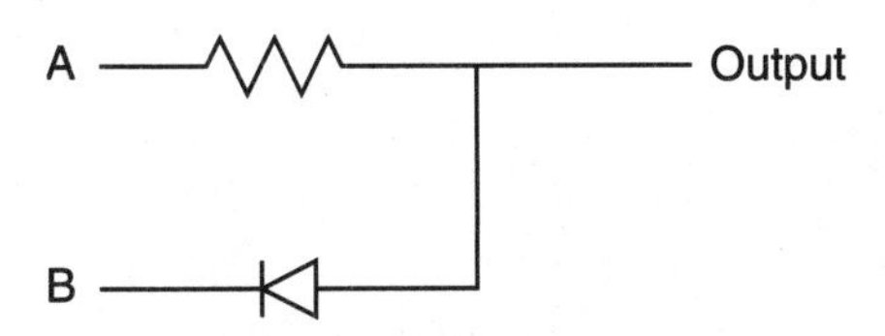

Figure 7.17 AND gate circuit for CMOS inputs.

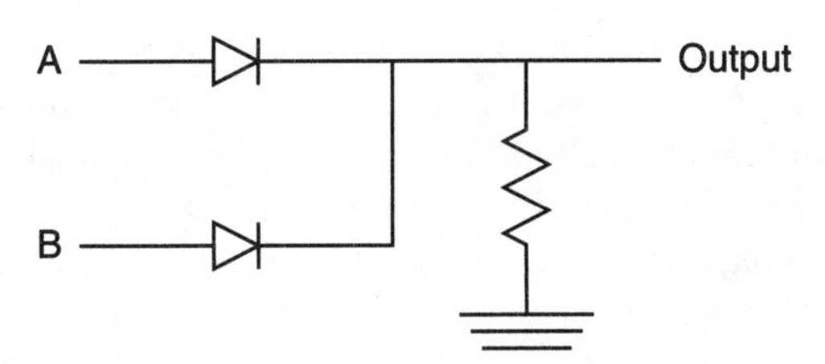

Figure 7.18 OR gate analog circuit.

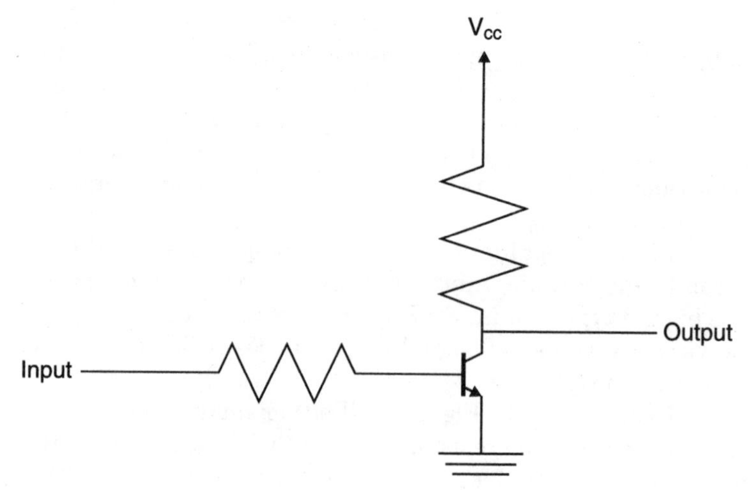

Figure 7.19 Basic inverter schematic.

In this circuit, if the input is high (current being supplied to the circuit), the NPN transistor will be turned on and pulling the line low. When input is low, the transistor will be off and the resistor will provide a high-voltage output. The base resistor value has to be chosen in such a way that the transistor will always be in saturation mode, but not passing excessive current. The pull-up resistor must provide sufficient current for the transistor, when it is turned off, to drive whatever circuit it is connected to.

A common term for these types of circuits is M**L or MML, which stands for *Mickey Mouse Logic.* This expression implies that the solution is not very high-tech or complex, but if designed right will eliminate the cost of adding a chip for just one gate.

Project 10—AND Gate in Use

The MML AND gate that I presented in the previous section can be very simply implemented by the circuit in Fig. 7.20. In this circuit, two of the switch inputs are buffered by TTL inverters and their outputs drive a pair of diodes, which are used to control the operation of another set of buffers, which control an output LED.

I use the two inverters to convert, or *buffer,* the voltage signals into TTL logic signals. This may seem like an unconventional use for the two inverters, but the circuit is really just converting the voltage-based digital signal into the current-based digital signal used by TTL logic. By this method, switches can be used to drive either TTL or CMOS logic.

The bill of material for the project is:

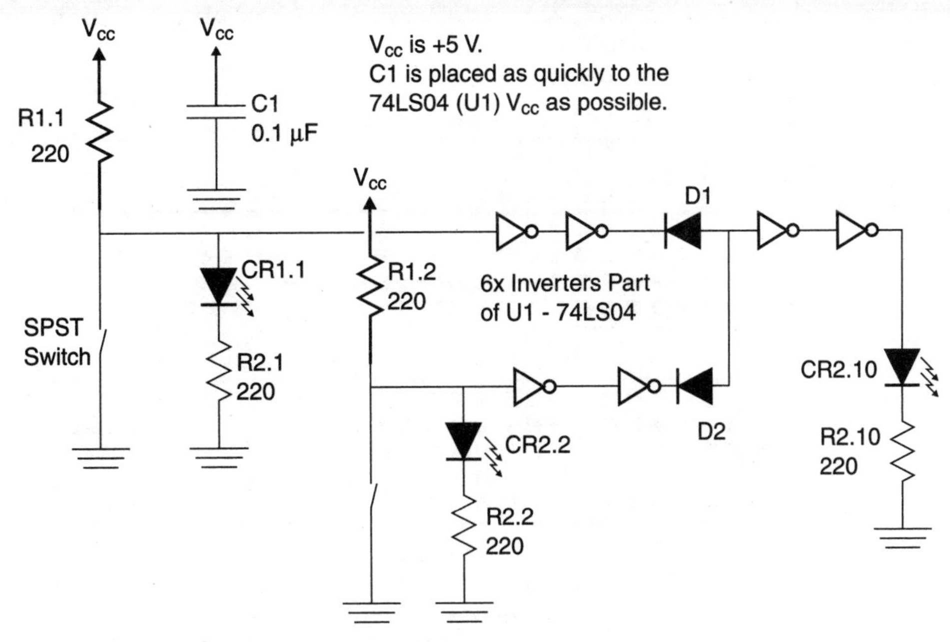

Figure 7.20 Circuit to test AND analog operation.

Part	Description
U1	74LS04
D1, D2	1N914 diode
C1	0.1-µF tantalum capacitor
Miscellaneous	+5-V power supply, interface PCB, breadboard, wire

The application's circuit could be wired as shown in Fig. 7.21 and should just take you a few minutes.

When you have completed wiring the circuit, build a truth table like:

CR1.1	CR1.2	CR1.10
0	0	
0	1	
1	1	
1	0	

and find out how the two diodes respond. You will find that they work exactly like a TTL AND gate.

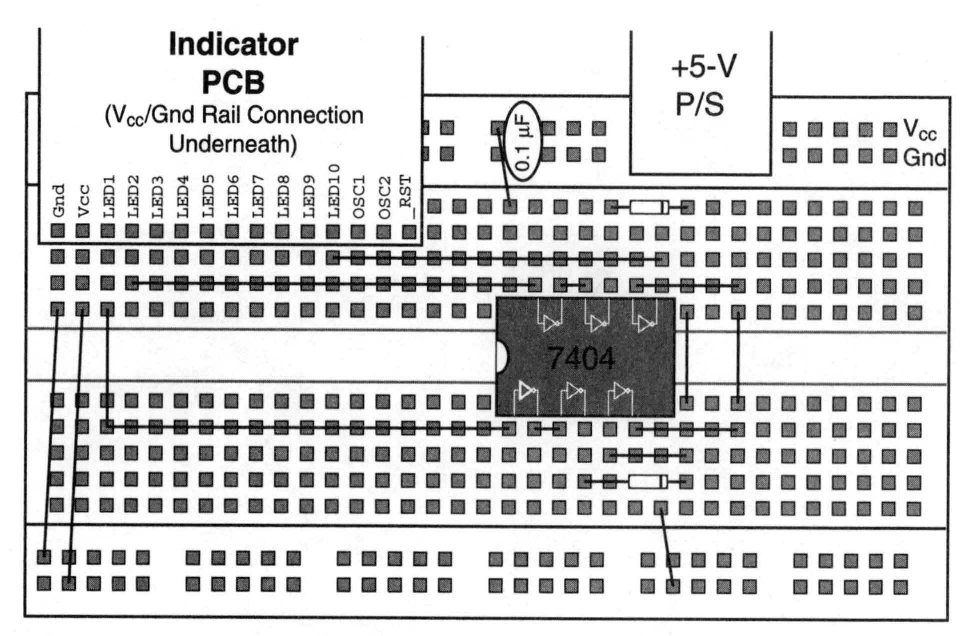

Figure 7.21 Project 9—TTL analog AND gate wiring.

Pull-ups/Pull-downs

How to handle unused digital inputs is not something that is explained well in school. I remember when I was in university and first started working, the need for resistors tying unused inputs to high or low voltages was never really explained, and while many example circuits used resistors for the task, very few designers really understood what the correct way of doing it was. It wasn't until I became experienced in specifying tests for products that I saw why pull-ups and pull-downs are really required and what kind of impact they have on a circuit.

In the previous sections, I have introduced you to the actual workings of the TTL chip, along with its inputs and outputs. So far in the book, when I have had an unused input for a chip, I have either left it unconnected (floating) or I have tied the input to V_{cc} or ground to set a specific operation in motion. This was done for convenience and to avoid making the circuits difficult for you to wire. In this section, I would like to discuss what is happening in the analog circuits in these cases and what I consider to be the correct way to tie TTL chip inputs to V_{cc} or ground.

In the circuit I showed in the previous sections, the bipolar TTL inverter chip has what amounts to a built-in pull-up; the NPN transistor behaves like two diodes, with the V_{cc} supply current within the chip passing through the part of the transistor that offered the least resistance. When the circuit connected to the input provided a path that was 500 Ω or less, then the current supplied to the input pin passed through this resistance (placing the pin at a low voltage level) and not through the transistors in the inverter. The resulting output was high.

I have discussed the idea that unused TTL pins can be left unconnected (or floating) and the chip will behave as if this pin has a high voltage input into it. The problem with doing this is that not all chips have this bipolar transistor input. As I found out to my embarrassment in Project 7, using an HC (CMOS input) chip instead of an LS chip with floating inputs resulted in a circuit that didn't work properly. Although there are CMOS versions of standard TTL logic chips available, most digital electronic chips that you will be working with in your career will be CMOS, and you will find that if any inputs are left floating, the chip will behave unpredictably. This is because the PMOS/NMOS transistor input gate behaves as a high-gain amplifier instead of the bipolar TTL input you are familiar with.

As a result of this, you should *always* tie unchanging pins to a high or low voltage source. Notice that I did not say V_{cc} or ground in the previous statement. The reason for this will be explained in this section.

The most common circuit for connecting TTL inputs to a set input is the pull-up, as shown in Fig. 7.22. In this circuit, a fairly high value resistor (1k to 10k normally) connects the input pin to the circuit's V_{cc}. The reason for doing this is to allow the input pin to be pulled down to a low voltage level by a circuit like the button input shown in Fig. 7.23.

In the button input circuit, when the button is pressed, the input pin's state can be changed without causing a serious drain on the circuit's power supply.

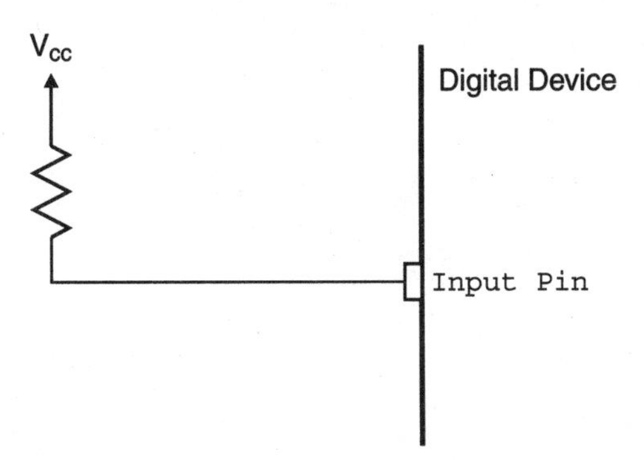

Figure 7.22 Pulled-up digital input.

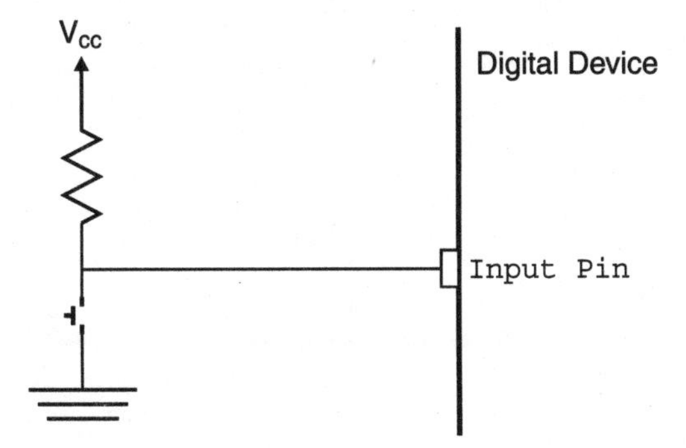

Figure 7.23 Pulled-up switch.

The resistor is known as a *current-limiting resistor* (this term is used throughout the book for resistors that prevent a circuit from drawing more current than desired). When the button is pressed, the path from the input pin to ground is made and current flows from the input pin to ground as well as from V_{cc} to ground through the current-limiting resistor.

The circuit shown in Fig. 7.23 is the circuit that I recommend using for digital inputs. Later in the book, I will discuss the concept of switch "bounce," but you should see that this circuit is a very simple way of getting button input from a user.

In addition to a pull-up, there is also a pull-down (Fig 7.24) with its associated pull-down switch (Fig. 7.25). In this case, a much lower level resistor is used to make the TTL input behave as if a low voltage is being applied to it.

In the pull-up case, understanding the circuit's operation is quite simple; the resistor connected to V_{cc} prevents a low-resistance path for current within the input, and the current supplied within the gate flows to the internal transistors. In the pull-down case, the resistor value must be carefully chosen to make sure it is the highest possible value to cause the gate to be pulled down.

With a pull-down, a TTL inverter circuit becomes the more complicated circuit (Fig. 7.26). The resistor must be a lower resistance path for the current

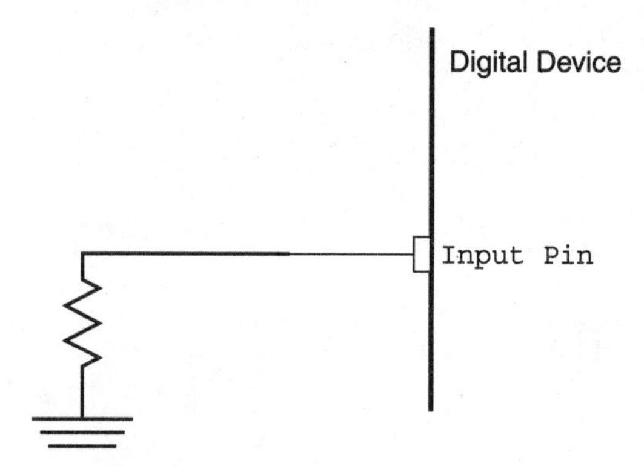

Figure 7.24 Pulled-down digital input.

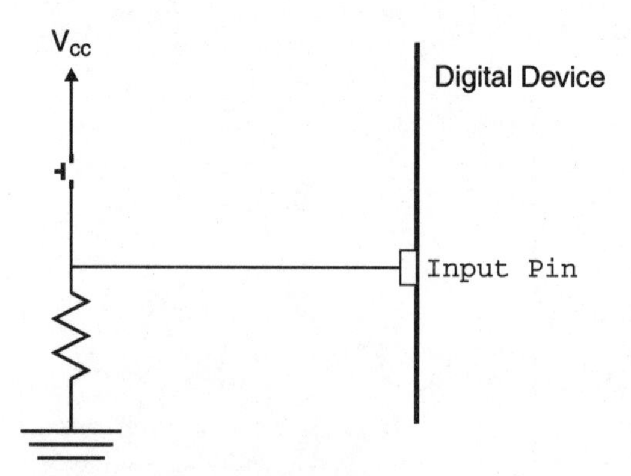

Figure 7.25 Pulled-down switch.

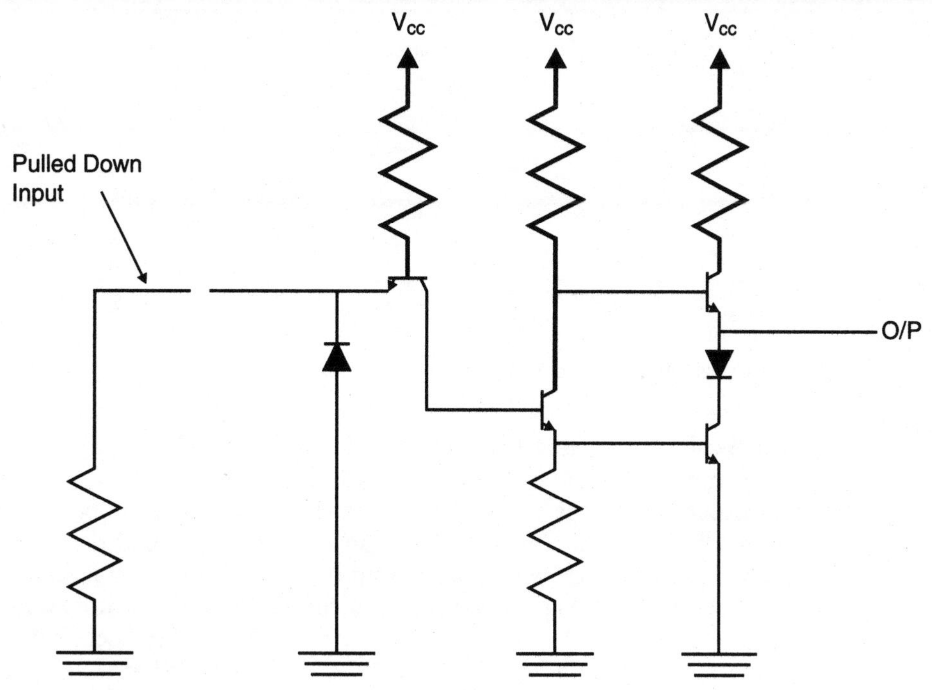

Figure 7.26 Pulled-down TTL inverter schematic.

than through the transistor/resistor combination for the pull down. Resistors in the range of 330 to 500 Ω are typically used, but different values are often required for different logic technologies.

Reading through the previous paragraphs, you are probably thinking that I have made the pull-down situation much more complex than it needs to be. Chances are, when you need a pull-down, you will just tie the TTL input pin to the circuit's ground. With the operation of most TTL chips that have negative active inputs, there won't be any problems when you simply wire them to ground, and only 1 mA will be passed to ground.

Problems do arise, however, when you are designing circuits that are going to be built and tested in a manufacturing setting. As I will discuss later in the book, many negative active inputs on TTL chips control the operation of tristate drivers, and when the product is being manufactured, these drivers have to be disabled to test the function of the chip that it is connected to them. The only way to do this is to drive the input pin high.

As I have shown, a simple resistor can be used to pull this line down, but the problem comes when something has to drive it high. Automated machines called *in-circuit testers* are often used to probe a circuit, driving lines both high and low to test the circuit's operation.

In-circuit testers can probe a product at many thousands of different circuit points, or *nets,* and are extremely complex machines. Part of the price for this function and complexity is that the probe circuits are not capable of driving a great deal of current into the circuit. They are usually designed as primarily

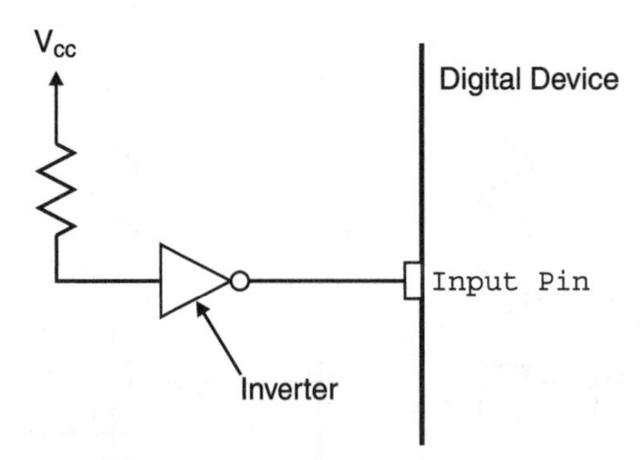

Figure 7.27 Pulled-down digital input using pull-up.

open-collector drivers with weak pull-ups. For a pull-down like the ones I have shown in this section to be overridden, 10 mA or more is usually required, and this amount cannot be provided by the tester. Buffers can be built into the test equipment, but this makes the tester more complex and more expensive.

In these cases, I always specify that an inverted pull-up, as shown in Fig. 7.27, be used. In this case, rather than try to "overpower" a pull-down, a traditional pull-up is controlled by the tester's pin. I have refused to declare boards ready for manufacturing because of this point, and I have sent them back to the designers for modification.

This probably sounds like a trivial point, but having pull-downs on TTL circuits can end up costing many thousands of dollars unnecessarily in manufacturing costs. If the TTL technology that the pull-down is connected to ever changes, or the part tolerance changes unacceptably, the product may behave unpredictably in the field. None of these potential problems are present with a pull-up.

To summarize how pull-ups and pull-downs are to be used in a circuit I present the following rules:

1. Never leave an input pin unconnected or floating.
2. Design for using only positive active unconnected inputs, and always pull them up with a resistor.
3. If negative active inputs are required, drive the pin with a pulled-up inverter.

Line Impedances

If you know anything about surfing, you'll know that all the best surfers come from the West Coast of North America, with very few coming from the East Coast. Have you ever wondered why?

Reading over this section, I realize that I should make something clear: When I ask the question above, I am discussing *wave surfing,* not Web or channel surfing. For this discussion, I am more interested in beach bums who are concerned with catching the perfect wave—not the couch potato who spends Sunday afternoons with a beer in one hand and a remote in the other trying to watch four football games at once.

The correct answer to the question of why the best surfers come from the West Coast is that bigger waves are present on the West Coast than on the East Coast.

Why are there larger waves on the West Coast?

The reason why there are larger waves on the west coast of North America is that the ocean bottom changes abruptly, instead of sloping upward at a fairly constant angle as it does on the east coast. Figure 7.28 shows a cross section of the different coasts. On the Atlantic (East) Coast, the bottom slopes quite regularly from the bottom of the ocean up to land. The energy of waves and current is dissipated along this slope, resulting in relatively calm and constant water from deep ocean to land. On the West Coast, the bottom angles up sharply as land is approached and the energy of the waves and currents is reflected; the only way the energy can be dissipated is in the form of large waves. In Fig. 7.28, I have marked the section of sea floor that angles up sharply as a discontinuity because the regular, continuous shape of the ocean floor changes abruptly.

After reading the previous paragraphs, if you are familiar with the sea floors, you'll probably think that I overly simplified the situation, and in any case, you are probably wondering what this has to do with digital electronics. Surprisingly enough, this example has a lot to do with digital electronics and how signals are carried through wires and printed circuit cards.

The wave example is exactly analogous to what happens to electrical signals when the medium they are traveling through changes significantly. It may be hard to visualize a large wave caused by a disruption in an electrical wire, but they do happen and can cause a lot of problems in a circuit that was not properly built.

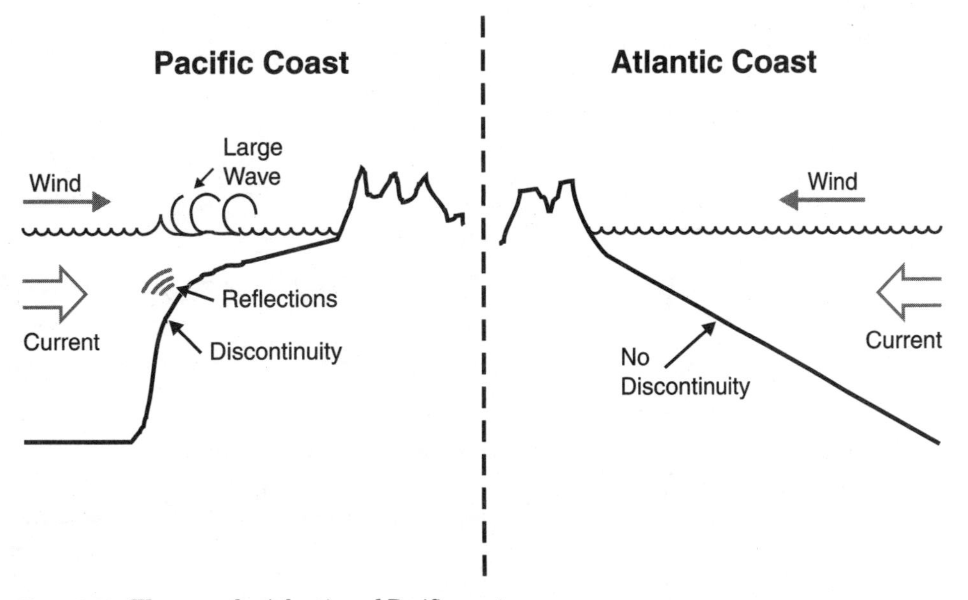

Figure 7.28 Waves on the Atlantic and Pacific coasts.

Any material that allows the movement of electrical signals has its own *characteristic impedance.* This is the resistance the material has to the passage of electrical signals. This is the same as the resistance of water to signals (i.e., waves).

While the resistance of water will seem to be very slight, it is nonetheless there. If you start a wave in a narrow pool of constant depth, you will see that the size (or *magnitude*) of the wave decreases the further it travels from the source. The process of decreasing is known as *attenuation.* I have specified a narrow pool because a wide pool, in which waves can travel (or *propagate*) in a circular motion, will have an additional decrease in magnitude over distance because the energy of the wave is spread over more area.

As shown in Fig. 7.29, if the bottom of the pool slopes up appropriately, the magnitude of the wave does not change—it has the same amount of energy at any point in its travel in the pool. Going back to electrical signals, this is exactly the situation that is desired. Electrical signals should not be attenuated as they travel in a conductor. This means that the characteristic impedance of the signal must be such that the signal is not attenuated as it passes through it.

Although it is not shown in Fig. 7.29, when the waves reach the end of the conductor, they should be terminated in such a way that the signal is not reflected back along the conductor. Reflections can cause spurious operations by input pins on the conductor. In Fig. 7.29, I have shown that the pool narrows to the point of the water level. In electrical systems, you may have to put on a terminator chip or resistor at the end of a signal line (net) to avoid any problems with reflections.

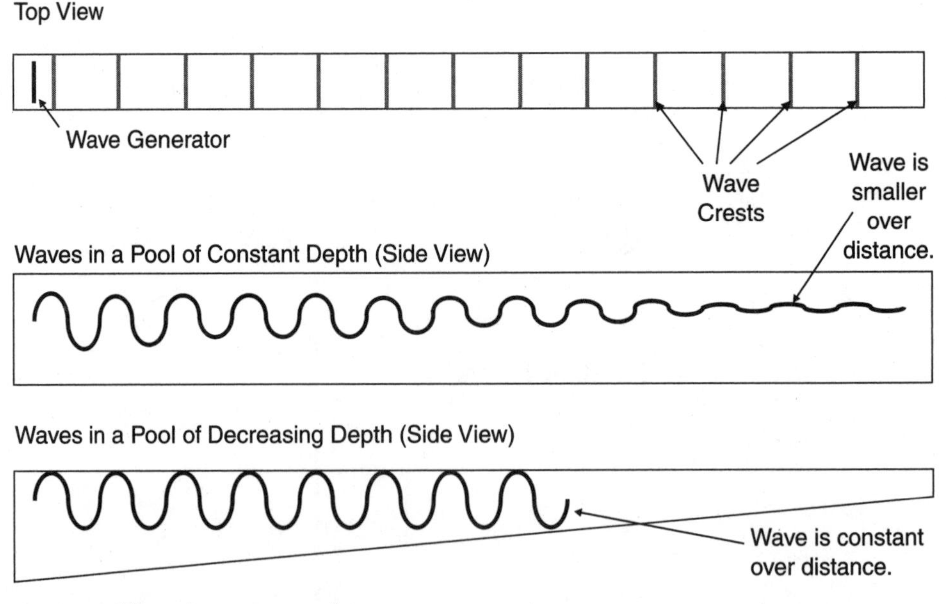

Figure 7.29 Waves in a narrow pool.

Before going on, I should make a very important comment: For the circuits and applications presented in this book, you do not have to worry about reflections of the characteristic impedance of the circuits. This is a problem only if you are running your signals at high speed. The faster the signals run, the more energy transitions there are, and this energy can be converted into significant voltages and currents in a circuit if the characteristic impedance is incorrect for the circuit or the circuit is improperly terminated and there is a reflection.

A very common improper termination of a circuit is a *stub,* similar to what I show in Fig. 7.30. When dealing with high speed signals, output pins are usually given the label *driver,* as they are driving a signal onto the circuit (or net). The *receiver* is the input pin and is usually matched to the driver to properly terminate the net and prevent reflections. Stubs are often the result of changing circuits during debugging and leaving a long conductor on the net. The stub can result in reflections back down the line because it is not properly terminated, or can result in radiated energy, which can cause problems with other circuits or with local radio and TV receivers. Stubs should be avoided at all costs, and later in the book, when I discuss reworking PCBs and circuits, I will discuss what should be done to avoid putting inadvertent stubs in your applications.

High speed is often defined as 20 MHz or faster. If you have looked through this book, you will see that none of the circuits presented here come close to this clock, or transition, speed. For the most part, you will not have to worry about the characteristic impedance of your circuit or any stubs on the nets. I would recommend that you try to keep your application wiring simple and avoid stubs, but this is good design and assembly practice and should be followed regardless of the speed the circuit will operate at.

There is one thing you will have to watch for in your circuits, and that is *discontinuities.* Going back to the West Coast/East Coast discussion, the great waves on the West Coast are caused by a discontinuity in the ocean floor. Changing wiring connections in a circuit can result in discontinuities and can result in reflections and prematurely attenuated signals. Problems caused by discontinuities can be much more severe at lower speeds than characteristic impedance issues. In the example circuit shown in Fig. 7.31, I have shown a signal being driven from a totem pole driver into a number of different forms of circuit wiring.

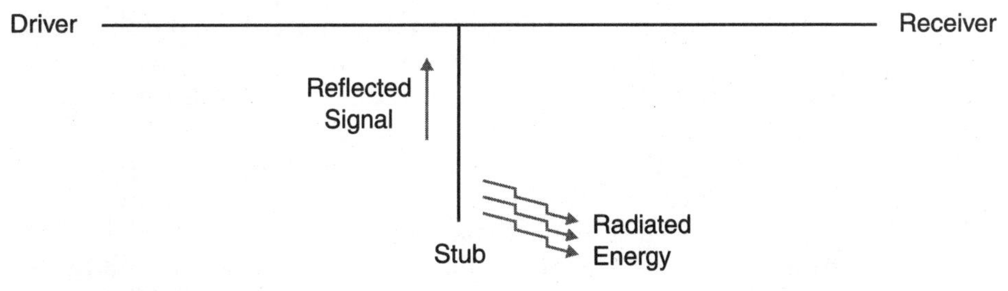

Figure 7.30 Stubs in a net.

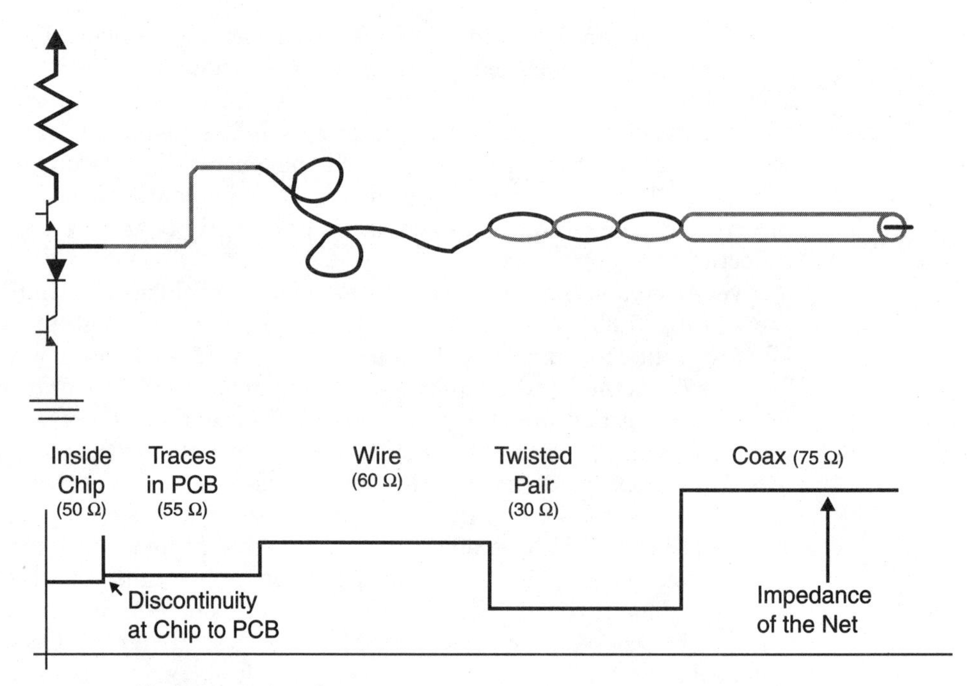

Figure 7.31 Variations in impedance over a net.

Each portion of the circuit shown in Fig. 7.31 has a different characteristic impedance (the ones given are strictly examples). If you were ever to build a circuit like this, you would find the signal at the end of the coax (coaxial cable) would be badly degraded at 20 kHz and totally unusable at 20 MHz, where you should start to expect to see high-speed problems.

Going back to the water analogy, you may know that to avoid detection, submarines will dive below changes in water temperature or salinity. In these cases, the characteristic impedance of the water to sound waves changes. Each change results in a progressively more garbled signal from the submarine to the pursuer, resulting in the pursuer having more problems hearing the submarine in the first place, and if it does find it, finding the correct bearing to it.

The solution to distortion of electrical signals due to impedance discontinuities is simple—don't change the wiring in a circuit. If you keep the wiring method constant, you will find that the signal will propagate quite cleanly without any significant distortion. If you are changing the way a signal is carried, then a new driver and receiver should be used (ideally designed for the wiring) to avoid the problems with the characteristic discontinuities.

The characteristic impedance and the presence of discontinuities can be measured by the use of a *time-delay reflectometer* (TDR). Figure 7.32 shows how this device works. A pulse generator drives a signal into a net. An oscilloscope connected to the signal monitors the pulse and the reflections returned from it. The TDR signal will indicate the initial pulse, and then the level of the oscilloscope signal will show you the characteristic impedance of different sections of the circuit. If the length of the characteristic impedance is measured

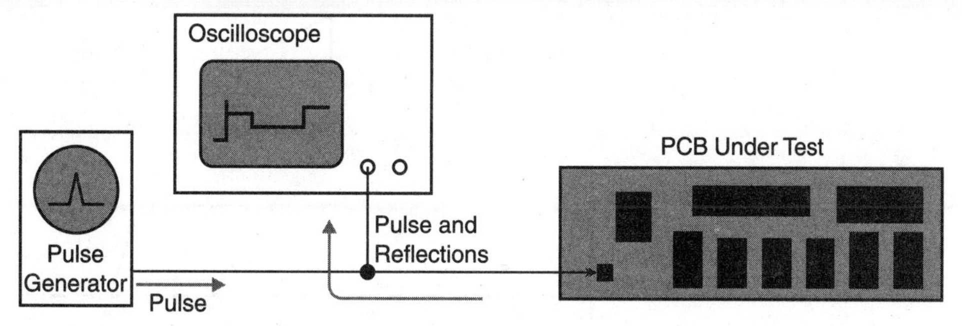

Figure 7.32 Time-delay reflectometer (TDR) operation.

on the oscilloscope screen (which is calibrated in units of time), the length of the section of characteristic impedance can be calculated (assume a signal travels 12 inches (30 cm) in 1 nsec).

A TDR can be an extremely useful tool; for example, they are used to look for breaks and shorts (discontinuities) in aircraft wiring. By connecting a TDR to one end of wire in an aircraft, it can be tested without having to open up the structure to visually inspect the wire, and the test is nondestructive. The wire is just as good after a TDR test as before. In some circuits, a TDR can be used to test the values of capacitors, inductors, and resistors built into a circuit. It can also test the quality of a PCB.

Despite its usefulness, a TDR can be a very difficult tool to work with. This is especially true for digital electronic circuits where distances are very short and it is impossible to place the TDR probe at the end of a net. If the probe is not placed at the end of a net, then there will be reflections coming into the oscilloscope from two sources, which will result in a very confusing display.

Passing Data between Digital and Analog Devices

When you start working with digital electronics, you tend to try to fit everything you are doing into a digital perspective. That is to say, digital controls and I/O are used in situations where they aren't warranted (and are definitely suboptimal). I am always reminded of the saying "When all you have is a hammer, everything looks like a nail" when I see how digital electronic devices are usually interfaced with real-world devices.

My pet peeve and favorite example of this is modern radio controls. I love having a digital frequency and volume readout, but buttons for tuning and volume control just don't work for me. This is especially true in a car where the lack of a knob means I miss the station or I can't "feel" the volume and set it either too loud or too soft.

Comparators

The most basic way of checking an analog voltage is to compare it against a known *reference* voltage. The aptly named *comparator* circuit performs this comparison action. Figure 7.33 shows the comparator's electronic symbol as well as its response to a fluctuating input.

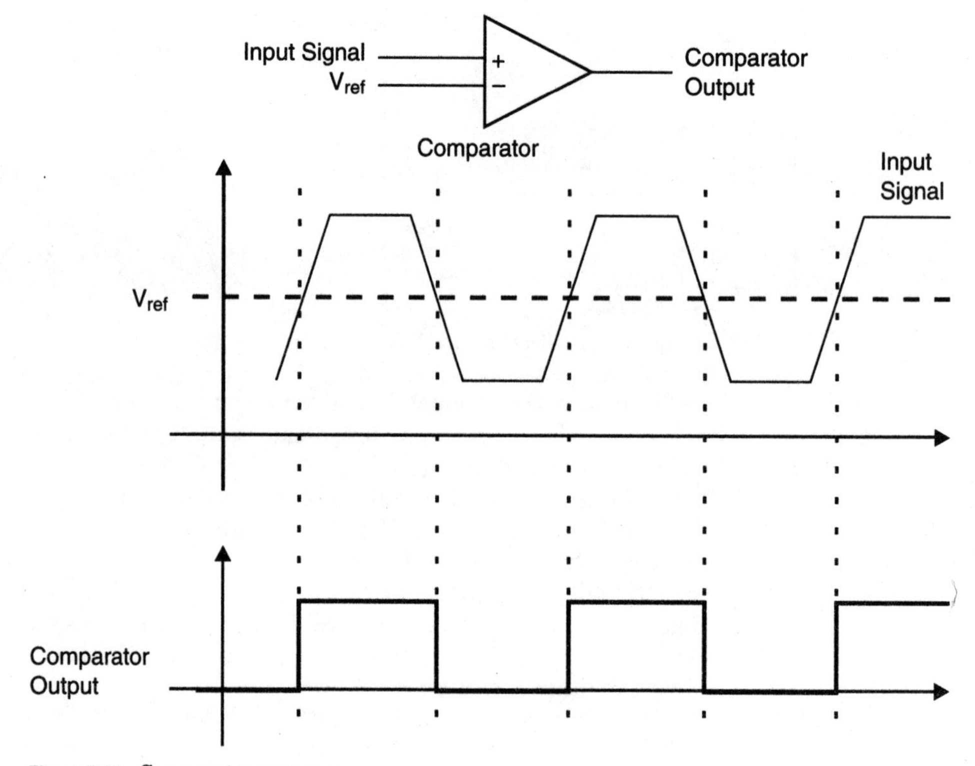

Figure 7.33 Comparator response.

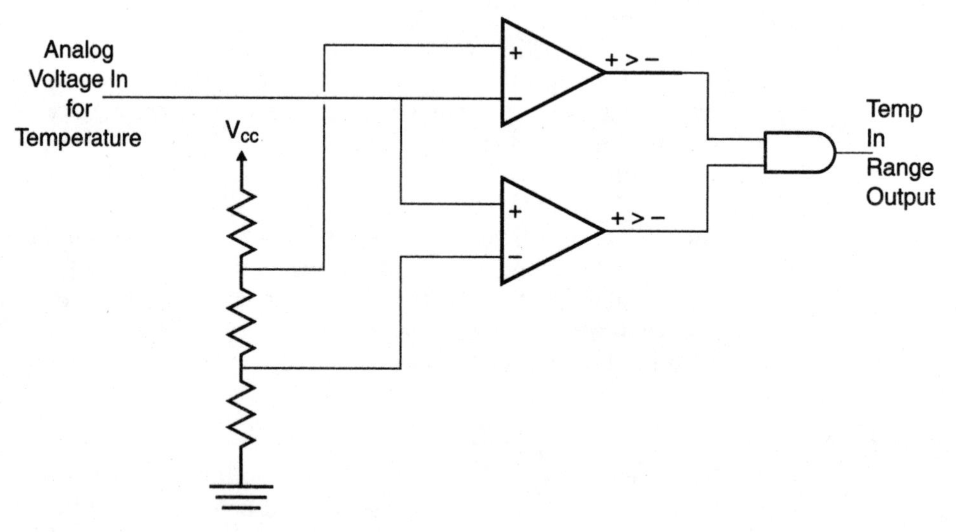

Figure 7.34 Thermostat using comparators.

In the comparator, the output is a 1 if the voltage is greater than the reference voltage. If a second comparator is used, with a voltage "window" (provided by a voltage divider circuit) as shown in Fig. 7.34, a test for a signal being within a window can be produced. In the thermostat circuit of Fig. 7.34, *Temperature in Range Output* is active if the analog voltage in is between high and low and could be used to implement an electronic thermostat.

Another application of the comparator is to "gang" a number together to test multiple conditions and return a digital representation for the voltage. This circuit is shown in Fig. 7.35. The *priority encoder* will return the binary number for the most significant comparator where the most significant comparator's "+ > –" output is true.

This circuit is known as a *flash* analog-to-digital converter because it is very fast, but it is also very expensive to implement. This is especially true for a flash ADC that has a large number of bits. This is because of the difficulty in building the precision resistors needed for the reference voltages and the need to "redrive" the analog voltage to be accurate across all the comparators. For an 8-bit-accuracy ADC, 256 resistors and comparators are required at a minimum.

Conversion between Analog and Digital

While it may seem that converting an analog voltage to an actual digital value can be somewhat difficult, there are many single chips dedicated to providing analog-to-digital conversion (ADC) that can be easily interfaced to digital circuits. When I discuss ADC in this book, I am primarily discussing measuring

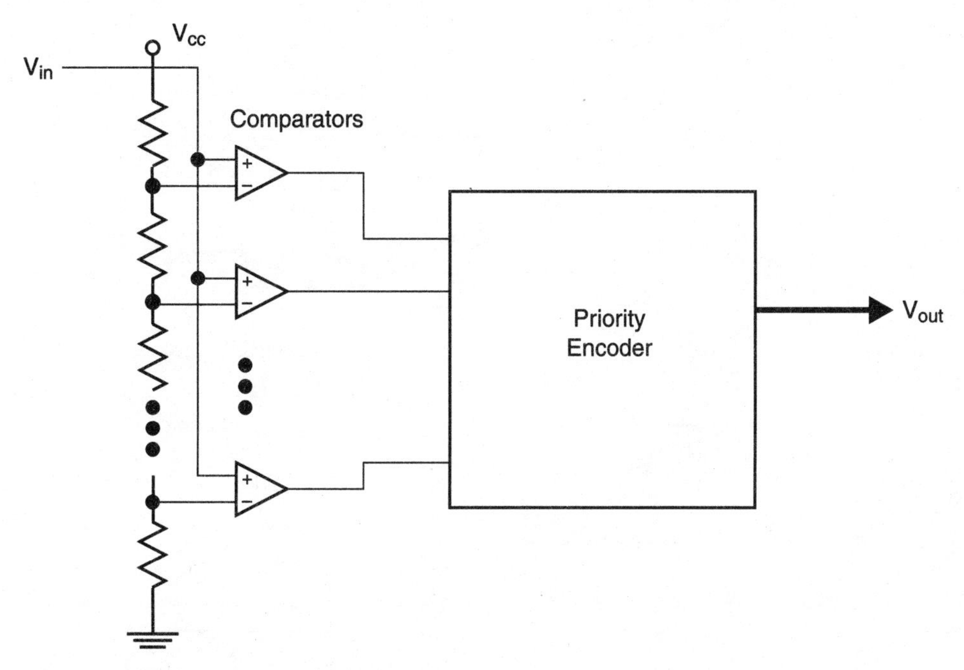

Figure 7.35 Flash analog-to-digital comparator.

steady-state voltages instead of rapidly changing signals. As I will show later in this section, it is difficult to accurately measure rapidly changing waveforms.

Besides the comparator, another common type of ADC is the *integrating* ADC. This device uses a *sweep generator,* which provides a voltage ramp that is compared against the incoming voltage. When the ramp is started, a timer is also started and stops when the comparator output is more than the voltage in (V_{in}). This circuit is shown in Fig. 7.36.

When using the integrating ADC in your applications, you should always ignore one or two least significant bits of the result. This is because of "jitter" in the operation of the input signal, comparator, timer (and its driver circuit), and sweep generator. If 8-bit accuracy is required in an ADC, then a 10-bit ADC or more should be used to avoid this problem; the least significant bits of the result can be ignored, and an 8-bit result is still produced.

The integrating ADC circuit is quite simple and much more cost-effective than the flash ADC, but does take a relatively large amount of time to execute. Often 20 μsec or more is required for a voltage conversion. This time leads to some interesting problems in sampling changing voltages. For example, in Fig. 7.37, I show an arbitrary frequency triangle wave going into an integrating ADC, which samples the signal at a much lower frequency. As can be seen in Fig. 7.37, the nature of the sawtooth is completely lost and the values read are essentially meaningless. There are methods for interpreting this data, but these fall into the realm of digital signal processing (DSP), which is beyond the material of this book. This inability to process high-speed analog signals is the reason why I suggest you work only with steady-state analog signals.

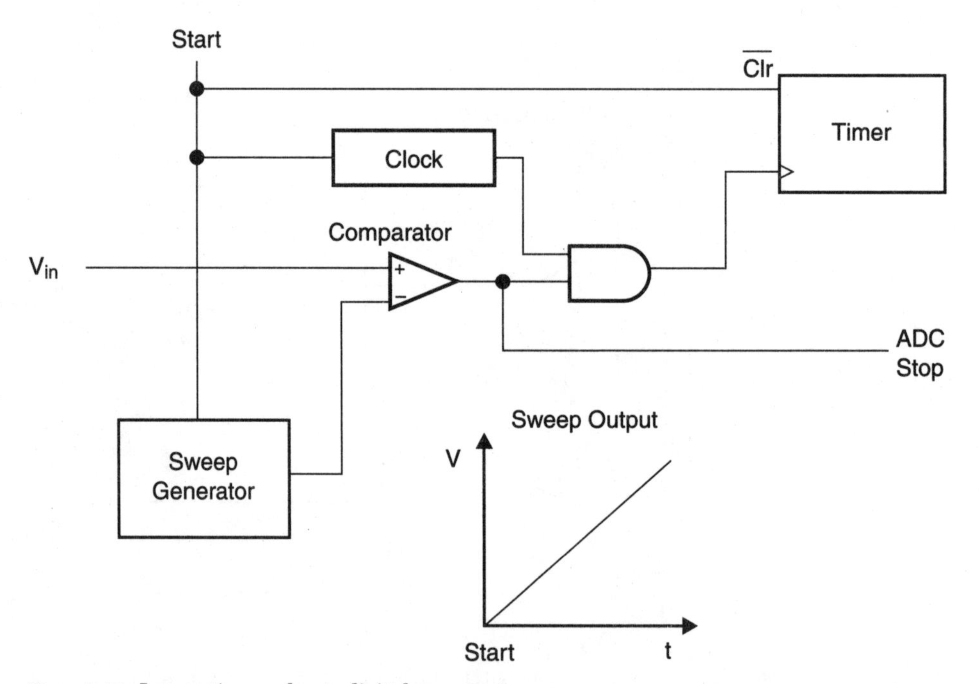

Figure 7.36 Integrating analog-to-digital converter.

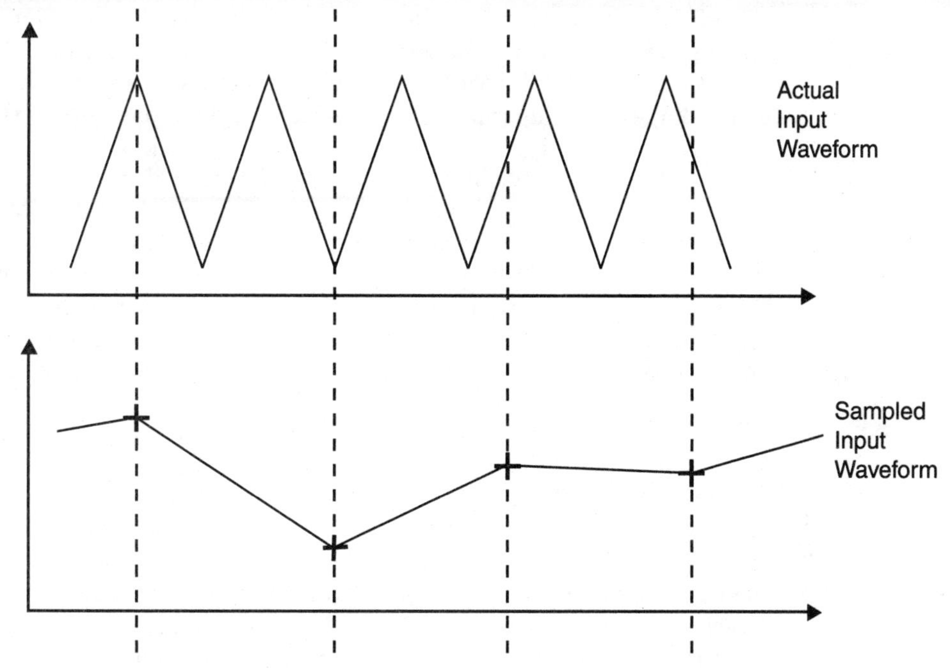

Figure 7.37 Missed sample waveform.

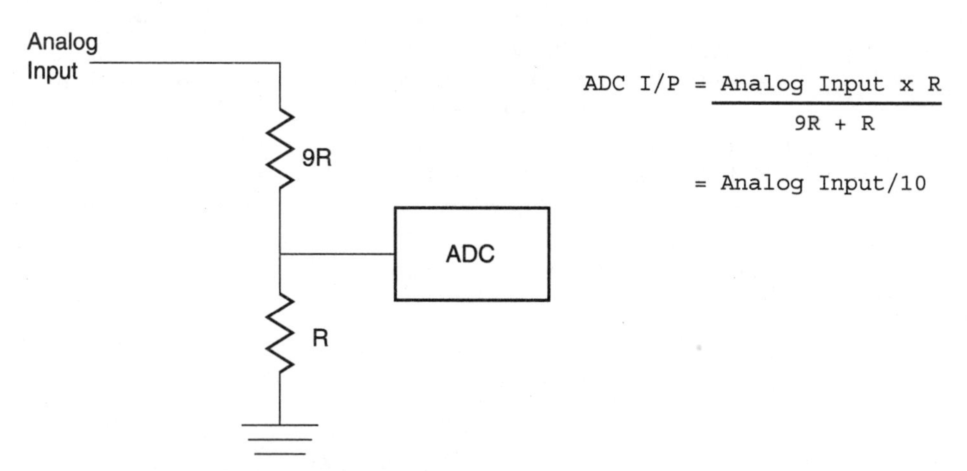

Figure 7.38 10-to-1 voltage divider ADC input.

You may get into situations where you want to measure voltages that are higher than the ADC is capable of measuring. Normally an ADC is capable of measuring only from ground to its input powering voltage (+5 V normally). In these cases, a voltage divider, as shown in Fig. 7.38, is used. Precision resistors (1 percent tolerance) should be used in the voltage divider to ensure there is no significant change in the accuracy of the ADC.

The final point I want to make about ADCs is that you should make sure you have a good common ground between the voltage input and the ADC. Not having a good ground will result in noise or ground shifts in the ADC input circuitry compared to the input source, which can lead to errors in the measured voltage.

One of the most obvious needs for interfacing a digital circuit to the real world is the need to provide an analog voltage output. Analog voltages are often seen as the best way to control meters, motors, and lights. In reality, *digital-to-analog converters* (DACs), the circuits that are used to convert digital signals into analog voltages, are quite difficult to implement for high-current devices. As I show here and elsewhere in this book, there are digital means for performing the same functions that are much simpler and better suited for device control.

The most basic element for analog voltage output is the *voltage divider*. This can consist of two resistors or a potentiometer (digital or otherwise) connected between a high voltage and ground with the wiper (or connection between the two resistors) outputting the analog voltage. The voltage divider circuit shown in Fig. 7.39 illustrates three aspects of DACs that I want to bring to your attention.

The first aspect is the V_{ref}. This is a known voltage of a given accuracy. Having a precise voltage reference is critical in all DACs (and analog-to-digital converters) to ensure that the output can be precisely predicted or measured.

Along with having a precision voltage, the resistor values (R1 and R2 in Fig. 7.39) also have to be precision parts. A digital potentiometer may give this accuracy, but at increased cost and complexity. Now, there are some situations where standard 5 percent tolerance resistors are good enough, but for most applications, a precision reference voltage and precision resistors are required.

In Fig. 7.39, I have also included an analog buffer to provide drive current for the voltage divider output. The graphic is for a *differential input* amplifier, which uses the output voltage as negative feedback to ensure the output doesn't change as a result of changes in the load. In all the DAC circuits shown in this section, I have included this buffer circuit because, while the DAC circuits can provide a set voltage, they cannot drive significant current loads.

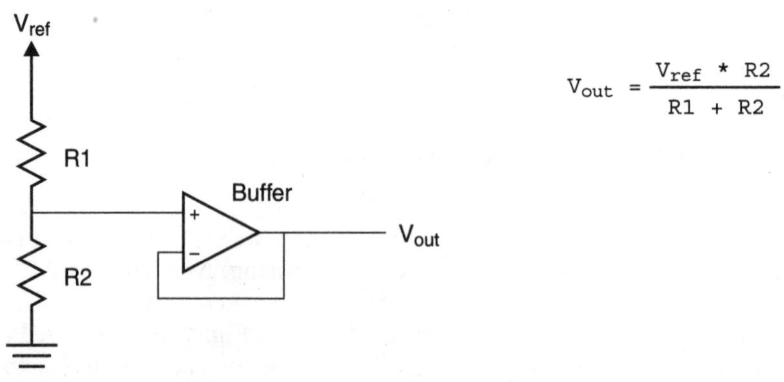

Figure 7.39 Voltage divider.

In the voltage divider circuit, any kind of current load will change the characteristic impedance of the voltage divider, and the voltage output will change with it. To avoid any loading problems, the buffer should have as high an input impedance as possible.

While the buffer circuit I have shown is very simple, in actuality it can be very complex for high current loads, especially for motors and other magnetic devices. In these cases filtering is required to prevent inductive loading and "kickback" from affecting the output and possibly causing oscillations within the buffer and the device it is driving.

Despite these dire warnings, simple voltage dividers are useful circuits in many applications. One that you will see in many projects is a voltage reference for LCD display contrast. In this application, the current required by the device is very small and a voltage divider can be used without the buffer. In addition the reference voltage does not need to be precise because the contrast is specific to the device and the user and is qualitative rather than quantitative.

The voltage divider circuit in Fig. 7.39 is capable of driving out only one voltage, which isn't very useful in many applications. In Fig. 7.40, I have drawn a circuit which I use when I need to output different voltages in a circuit. In this circuit, when a transistor is turned on, the resistor above it is pulled to ground. This truncates the voltage divider circuit and outputs a different voltage.

For example, if the transistor at control 1 is turned on, the output voltage is 0 V. When the transistor at control 2 is turned on, the output voltage is $V_{ref} \times R2/(R1 + R2)$. For the transistor at control 3, the voltage is characterized as $V_{ref} \times (R2 + R3)/(R1 + R2 + R3)$. If no transistors are turned on, then

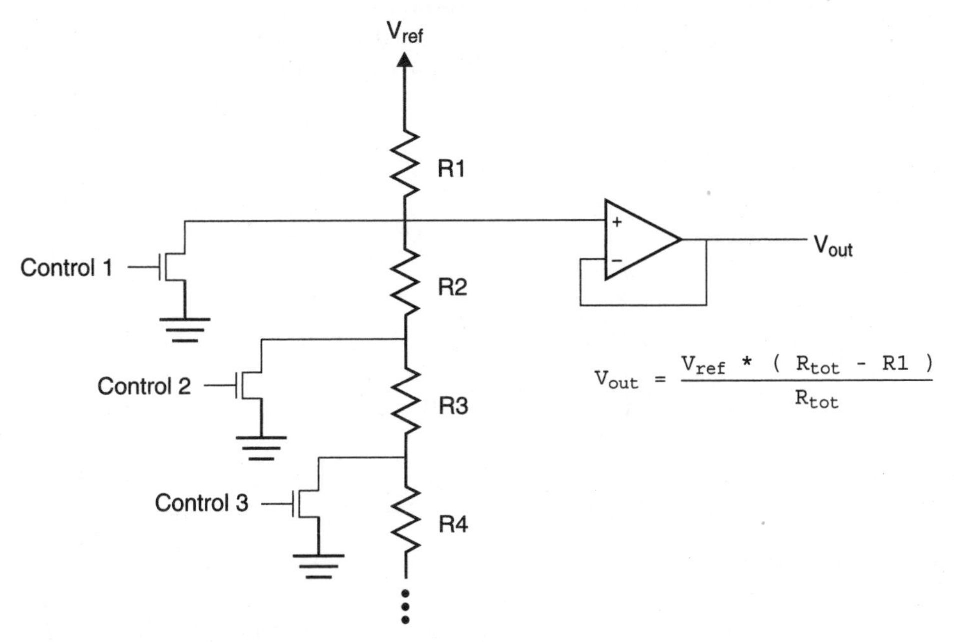

Figure 7.40 Multiple-control voltage divider.

the output is at V_{ref}. As more resistors become active, the output voltage approaches the general case of $V_{ref} \times (R_{total} - \mathrm{R1})/R_{total}$.

The advantage of this type of voltage divider is that it can be easily implemented with digital electronic circuits in which each I/O pin can be put in a high-impedance state or pulled to ground (which avoids the need for a separate transistor at each control).

There are some significant drawbacks to this method, however. The most obvious one is the difficulty in selecting resistor values for a large number of different voltage output levels. This is compounded by the circuit output only being a function of the highest turned-on transistor. Turning on multiple transistors will not provide an intermediate voltage output value. Despite these limitations, the circuit is useful when just a few analog voltages are required (such as providing composite video voltage levels).

Figure 7.41 shows a voltage divider circuit that can be used to provide a wide range of analog values (with intermediate values) quite easily. In this circuit, when the individual switches are closed, the "hex" fraction of V_{ref} is output. For example, closing SW3 and SW1 will result in V_{out} being equal to $V_{ref} \times 10/16$.

The biggest drawback of this circuit for use as a DAC is the difficulty in finding appropriate analog switches that can be controlled by a digital circuit and allow current to pass through without affecting the operation of the circuit. This type of voltage divider is the circuit most often used inside "canned" DAC devices.

The last type of DAC that I wanted to present to you is one that drives a pulse-width modulation (PWM) signal into a low-pass filter. In this circuit (shown in Fig. 7.42), the capacitor resists the extremes of the digital signal input and smooths it out into an analog voltage.

PWM signals are widely used in controlling and powering electronic devices—but not by this method. The problem with this circuit is that no mat-

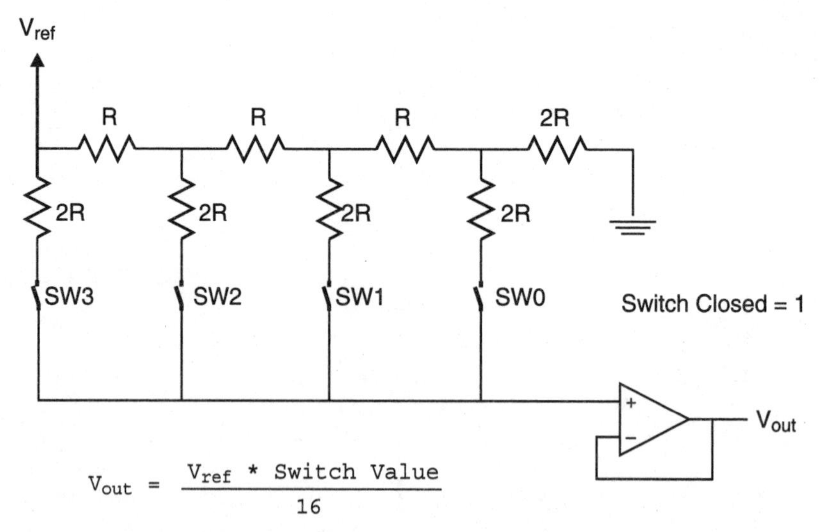

Figure 7.41 R/2R voltage divider ADC.

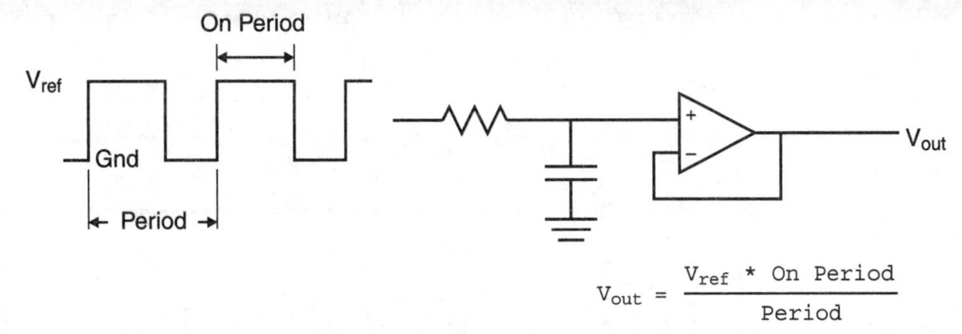

Figure 7.42 Pulse-width modulation analog voltage.

ter how good the filtering that is provided, there will always be some ripple in the output voltage that can cause problems. A much better way of using a PWM is to drive a control with a constant-voltage-level signal and use the PWM's duty cycle divided by the period to provide an average power level that is proportional to the desired analog voltage output.

Signal Integrity

In this chapter, I have discussed how digital circuits are implemented with analog components (including demonstrating how a simple gate can be built) as well as some of the issues surrounding them. To finish off the chapter, I wanted to leave you with a few thoughts on a subject that designers (and more important, companies) often forget about.

Signal integrity is the science of understanding how a logic (or other) electronic technology works and determining whether it is being used appropriately in an application. Chips, PCBs, prototyping methodologies, connectors, and interfaces are usually included in the initial signal integrity work. Once the various components are considered "qualified," then actual applications are checked over to make sure that they comply with the rules generated for the technologies and that the actual signals present in the application are at the correct levels for the technologies to work correctly.

This definition could be considered somewhat controversial. For many companies, ensuring "signal integrity" is simply accepting a technology and following the basic rules for the technology's use. Most companies do not check signals on the application (unless they have a failing unit) and are not familiar with how the board actually operates until the application reaches production. I have seen quite a few companies scrambling to try and fix problems with released products that don't work as expected in different situations.

To show what I mean, I want to take a closer look at the operation of the interface PCB and what the 1-kHz oscillator output looks like on a breadboard. Looking at 1-kHz output on an oscilloscope, you will see a nice square wave like the one shown in Fig. 7.43. But, if the timebase of the oscilloscope is turned up, you will see the surprisingly uneven signal shown in Fig. 7.44.

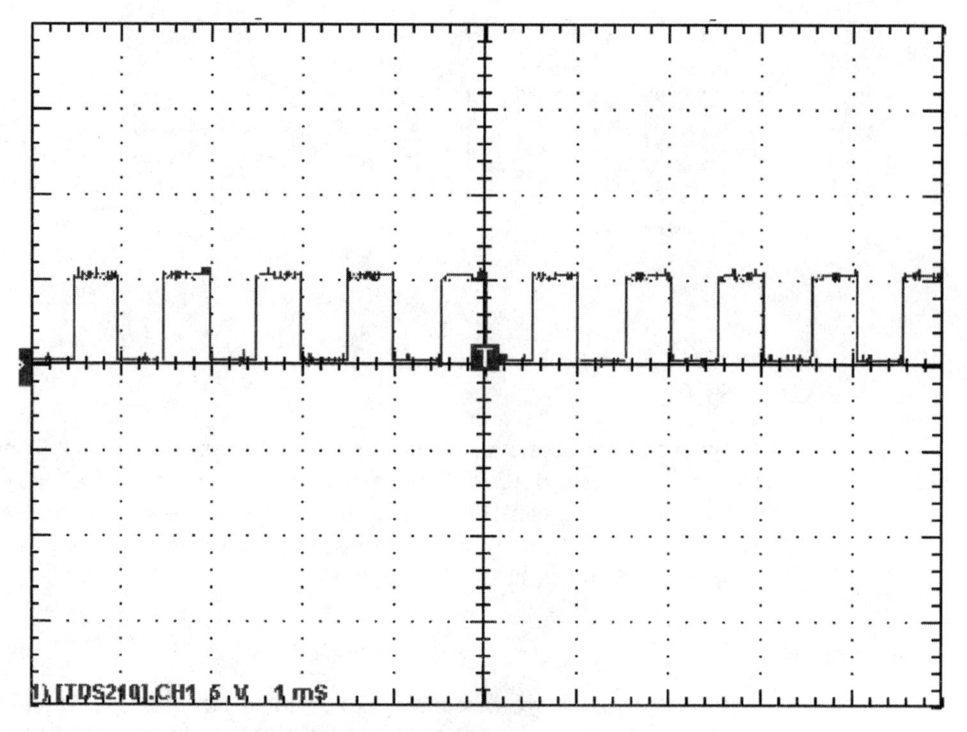

Figure 7.43 1-khz square wave from the interface PCB.

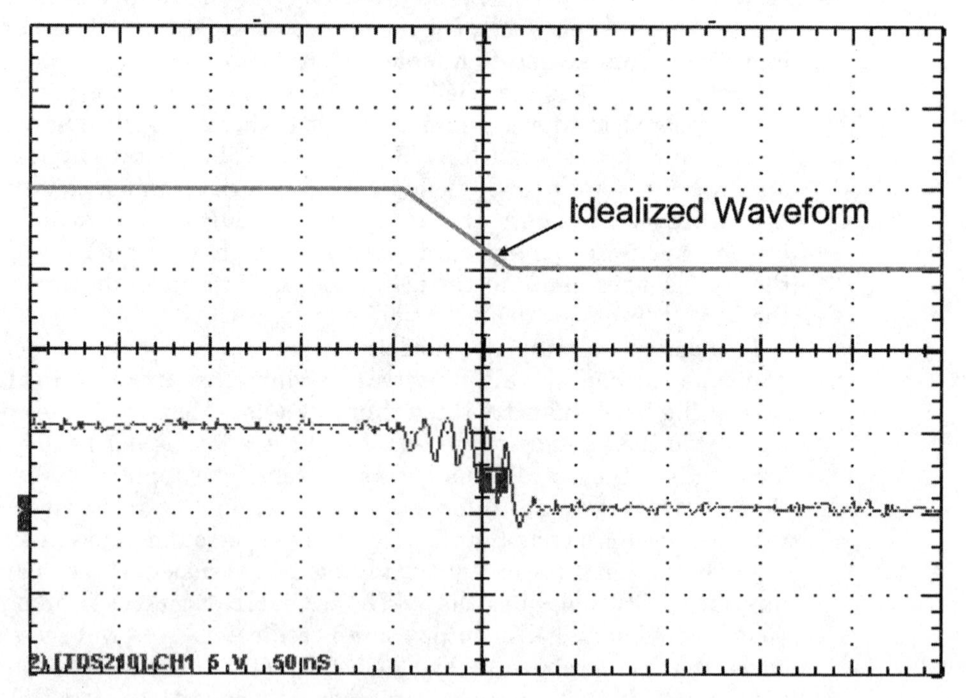

Figure 7.44 Edge on the 1-kHz square wave.

Depending on what this circuit is connected to, these bumps in the signal can cause a problem. Looking at Fig. 7.44, you will notice that the bumps become obvious at a 50-nsec/division setting on the oscilloscope, and for typical data speeds you work at for prototyping, this will not be a problem. If prototype circuits were the only location where this signal was used, I would probably agree with you. Where I would be uncomfortable is in the situation where this signal was present in a product.

The problem with this signal is that during transitions, other circuit operations could be triggered. If this is a 1-kHz clock, then there could be multiple clock signals "interpreted" by the logic receivers that would be treated as if the full period of the clock had gone by. This type of problem can be very difficult to find or even theorize on.

To minimize this type of problem, or if you seem to have problems with it in your applications, you can place a 30- to 100-pF capacitor across the output of the oscillator to ground. This will eliminate a lot of the "ringing."

The various unexpected features at the transition are a result of the analog aspects of the circuit (from both the internal gate transistors and the impedances of the line). As I discussed in the section "Line Impedances" in this chapter, an actual wire is not a perfect conductor; it consists of built-in resistances, capacitances, and inductances. These built-in impedances are the primary cause of the bumps and lumps in the actual signal.

There are a few comments that I should make about what you see in Fig. 7.44. The first is about the apparently gentle transition from the high to low voltage level that is caused by the interaction of resistive and capacitive elements of the signal. When a resistor and capacitor are wired together as shown in Fig. 7.45, an input signal is changed by the two components (which are part of

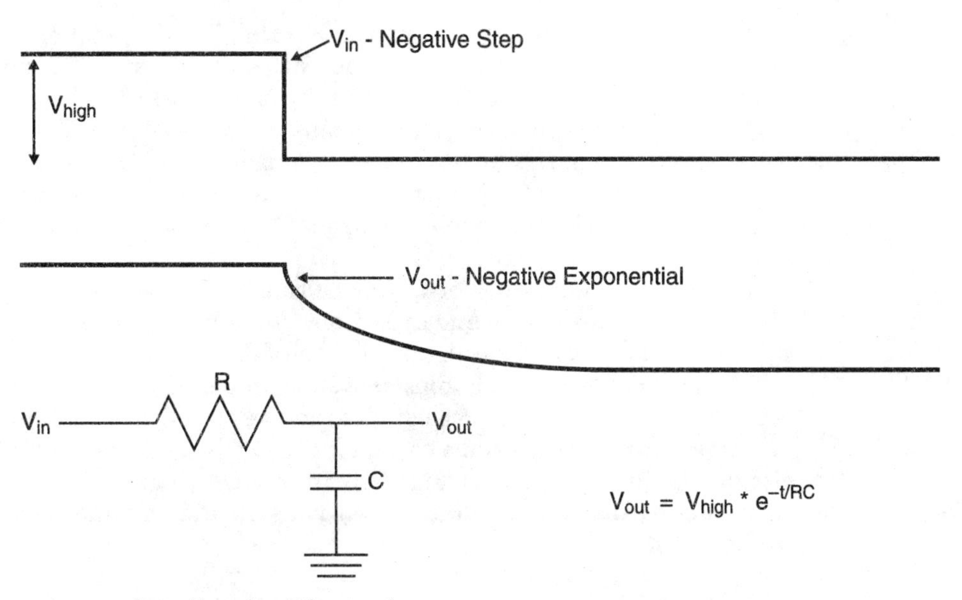

Figure 7.45 Operation of an *RC* network.

what makes up a wire) into the gentler exponential curve you see as output. A resistor and capacitor wired together like this are known as an *RC network,* which is used in many different circuits to provide a time delay. In the next chapter, I will show how the *RC* network is used to delay the start-up of a circuit's operation when it is powered up.

For an input that is going from 0 V to +5 V, the characteristic voltage for the output is:

$$V_{out}(t) = 5 \times (1 - e^{-t/RC}) \quad \text{V}$$

where *R* and *C* are the resistance and capacitance, respectively, of the circuit.

As noted in Fig. 7.45, when the output changes from +5 V to 0 V, the characteristic voltage for output is:

$$V_{out}(t) = 5 \times e^{-t/RC} \quad \text{V}$$

The *RC* term is known as the *RC* network *time constant* (and the product's units are seconds). The time constant is used to determine how long it takes for the circuit's output to respond to the change in input. The larger the value of resistance times capacitance, the longer it takes for the *RC* network to respond to changing inputs.

As I have pointed out in this chapter, TTL is not a voltage-controlled logic family. Despite this, you will find that placing an *RC* network between a TTL driver and a TTL receiver will result in a delay in the signal, although the actual delay cannot be calculated using the formulas above. When working with TTL, you can approximate the delay by the formula:

$$\text{Delay} = 2.2\,RC$$

This approximation is *very* rough, and you should only count on it being accurate to an order of magnitude—the operation of different TTL parts and different TTL inputs and outputs will change the actual delay.

The *RC* network is a form of a very simple low-pass filter. The *low-pass filter* is a device that removes high-frequency components of a signal. In the example here, where there is a "step function" from 0 to 5 V and back again, the *RC* network removes the high-frequency components of the signal, leaving just the low-frequency, "rounded" signal.

As a rule of thumb, the resistance of a connection between two gates should have minimum resistance and capacitance to minimize the time constant, and provide the fastest possible signal rise and fall.

The various lumps and bumps as well as undershoots and overshoots are a result of inductances and capacitances in the circuit and are known as *ringing.* Having an inductor and a capacitor together in a circuit creates a *resonant* circuit, like the one shown in Fig. 7.46. In this case, signals passed through the resistor and capacitor are alternatively resisted and amplified as I have shown in Fig. 7.46.

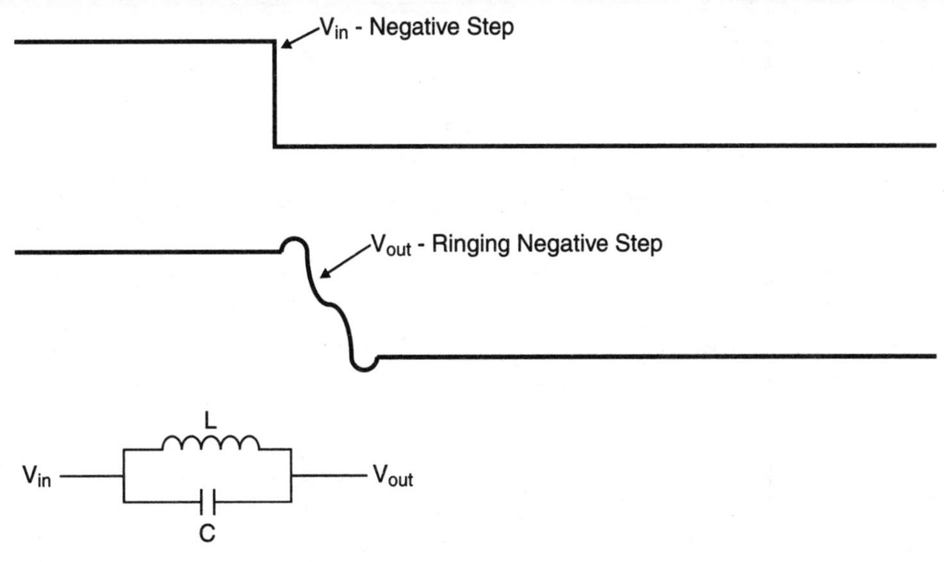

Figure 7.46 Operation of an *LC* network.

Designing circuits for use with inductors is quite complex and beyond the scope of this book. To minimize the potential for resonant circuits, connections in a circuit should be as short as possible and no wires should be hanging off into space. This is why I try to keep all my prototype circuit wiring as short as possible.

An ideal logic transition looks like Fig. 7.47. In this oscilloscope picture, notice that the signal does not have sharp edges, like the idealized logic transition, but there is no overshoot and undershoot that could affect the operation of other circuits or have them triggered at the incorrect times.

I should point out that to get a waveform in Fig. 7.47 with a similar fall time to the waveform in Fig. 7.44, I had to expand the oscilloscope's timebase from 50 nsec/division to 10 nsec/division. To put this in perspective, this signal transitions in around 66 *millionths* of a second. The point to learn from this is that when you have a good signal integrity, not only will the waveforms in the circuit be free of ringing, but they will operate at much faster speeds.

To get the type of waveform shown in Fig. 7.47 consistently, there are a number of rules that you should follow:

1. Avoid using CMOS (74HC) parts unless you really have to. The voltage control operation of the gates and the lack of internal pull-ups and pull-downs can make the circuits unreliable, and their high-gain operation can make their output very unpredictable, especially in the prototyping environment.
2. When prototyping, keep your wiring as short as possible and keep it close to the board. Also, spend some time thinking about how a circuit will be wired and try to avoid long wiring paths.

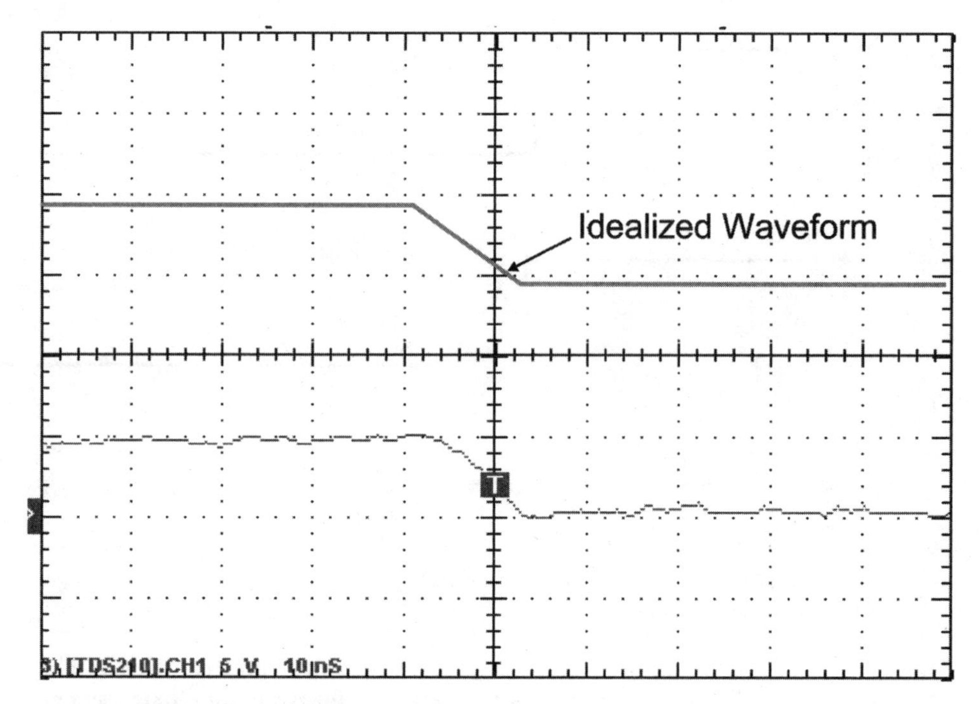

Figure 7.47 Edge on Good 1-kHz square wave.

3. Avoid discontinuities unless absolutely impossible. Connectors and unused or extra components will cause reflections in the line that can be amplified by the wiring in the circuit or by the components within the circuit.
4. Check the transitions for each output with an oscilloscope if possible. When doing this, use the shortest time base possible in the oscilloscope to ensure that any transient overshoots, undershoots, lumps and bumps, or slow response can be observed.
5. Make sure you have a decoupling capacitor on every component.
6. Check to see how susceptible your circuit is to outside interference. For a properly designed circuit, there should be no way another signal can affect the operation of the application.
7. Check your application at higher clock speeds. Sometimes you can discover if you have a resonant circuit that could potentially cause you problems later.
8. If a signal looks funny or is not exactly as you expect, try to understand why, using the techniques I outline in Chapter 13, "Debugging Projects."

Chapter

8

Simulators

The purpose of a digital electronic simulator is to allow a circuit designer to test out a design before going through the pain of building a prototype circuit or programming a programmable logic device (PLD). In a simulator, a set of inputs and outputs is defined, along with a circuit, and the operation of the circuit can be tested. Many modern simulator software packages include predefined components (matching 74 and 40 chip part numbers) and can be used with commercial "schematic capture" programs to eliminate the need for manually defining the circuit for the simulator software.

Digital electronic simulators can be found in a large range of prices, from free to several thousand dollars. The usefulness of the tool is a function of the predefined component models available, the ease with which data can be entered into the system, and how the output data is displayed. Often the inputs will be defined in a "script" file and the system will pass the input script parameters to the simulated circuit and record the results.

Having a graphical circuit input (or having a method of passing netlists from schematic capture tools) along with a good set of predefined models and graphical (waveform) output will result in a system that costs you $500 or more.

The irony with digital electronic simulators is that the cheaper they are, the faster they will appear to execute the simulation. You can buy simulators for less than $30 that run in *real time* (i.e., respond instantly to changes in the input conditions), while there are $1000 simulators that will take a half hour to complete the same operation. The difference in the simulators is the amount of *fidelity,* or accuracy, presented in the final result. A low-cost simulator will provide you with a basic idea of how the circuit works, while more complex simulators will give a very good prediction of how the circuit will actually operate and what its electrical signals look like.

To demonstrate the operation of the different simulator types available, I will use the example circuit shown in Fig. 8.1. This circuit adds two 2-bit numbers together and produces a 3-bit result. The circuit is assumed to be implemented

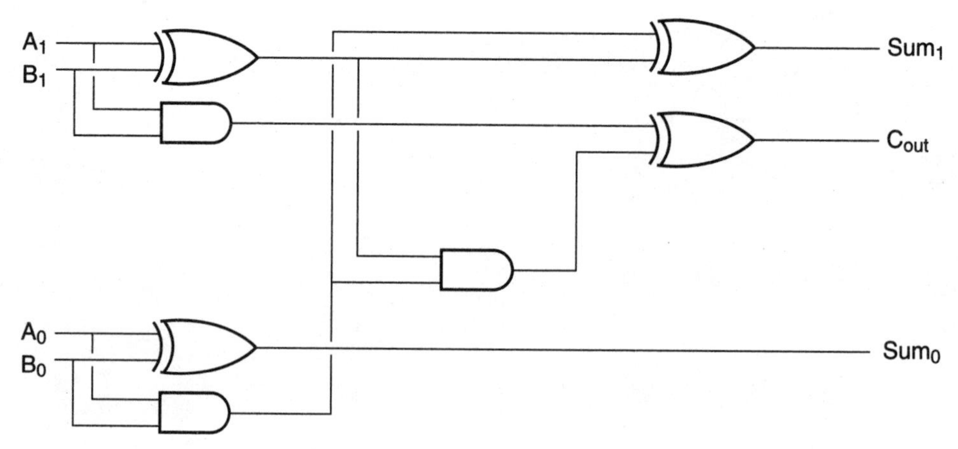

Figure 8.1 Two-bit adder circuit.

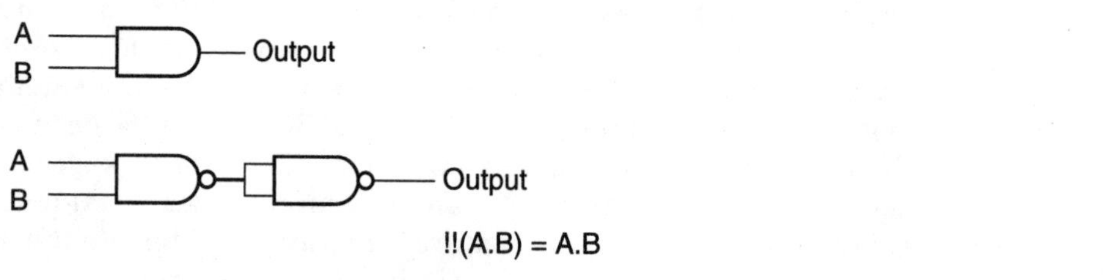

Figure 8.2 AND gate analog using NAND gates.

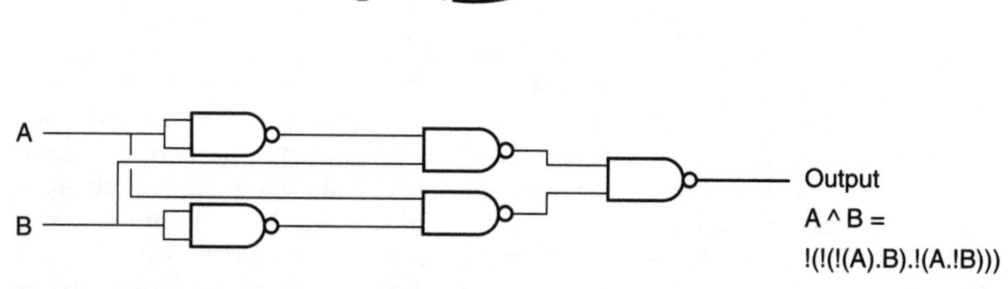

Figure 8.3 XOR gate analog using NAND gates.

in all two-input NAND-based TTL logic circuits (74LS00) as depicted in the AND analog shown in Fig. 8.2 and XOR analog shown Fig. 8.3.

Since both the AND and XOR gates in this circuit are assumed to be made from 74LS00 gates, which have an average gate delay of 10 nsec, it is also assumed that these circuits have gate delays of 20 nsec and 30 nsec, respectively.

Using this circuit with the gate delay assumptions will give you an idea of how the different methods of simulation work and what you can expect to see from them. Once I have discussed the different types of simulators, I will show the actual results I received using 74LS00 TTL chips.

I debated for a long time about whether I should include a chapter on digital electronics simulators. If you have read the books that I have written about microcontrollers, you will know that I feel very strongly about using software simulators to understand how an application works before it is programmed into the device. In the microcontroller books, I recommend that a low-end simulator downloaded from the Internet should be used rather than nothing at all.

For digital logic electronics, I do not recommend using a simulator except when a reasonably high performance simulator is available. The reason is that you want the simulated waveforms to be as accurate as possible; as I will show in this chapter, even with a high-performance simulator, it is unlikely that reality matches the simulated results. This is not to say that low-cost simulators cannot be used, just that you should understand what their limitations are and whether or not they will affect the operation of your circuit.

Low-Cost Simulators

A very inexpensive simulator works just through the output equations, sometimes with some kind of generic delay. The operation of the simulator is correct to a gross level, but subtle problems (*glitches*) will not be displayed, and corrections to the problems cannot be demonstrated on this type of simulator. For many beginner circuits, this is not a problem, but for circuits like the 2-bit adder I show above, potential concerns are masked and impossible to confirm for circuit debugging.

The simple simulator converts the circuit outputs into simple output equations and then calculates the number of gate delays and uses the worst case. For the 2-bit adder circuit, the output equations can be written as:

$$\mathrm{Sum}_0 = A_0 \wedge B_0$$

$$\mathrm{Sum}_1 = (A_1 \wedge B_1) \wedge (A_0 \,.\, B_0)$$

$$C_{out} = (A_1 \,.\, B_1) \wedge ((A_1 \wedge B_1) \,.\, (A_0 \,.\, B_0))$$

From the equations and the assumptions given it, the simulator will calculate the worst-case delays for each output. These delays are shown in the table below.

Output	Delay	Equation
Sum_0	30 nsec	$\mathrm{Sum}_0 = A_0 \wedge B_0$
Sum_1	60 nsec	$\mathrm{Sum}_1 = (A_1 \wedge B_1) \wedge (A_0 \,.\, B_0)$
C_{out}	80 nsec	$C_{out} = (A_1 \,.\, B_1) \wedge ((A_1 \wedge B_1) \,.\, (A_0 \,.\, B_0))$

For the 2-bit adder, I will use the same operational example; bits A_1 and A_0 will be initially driving 3 (both set) with B_1 and B_0 driving 0 (both bits reset). To

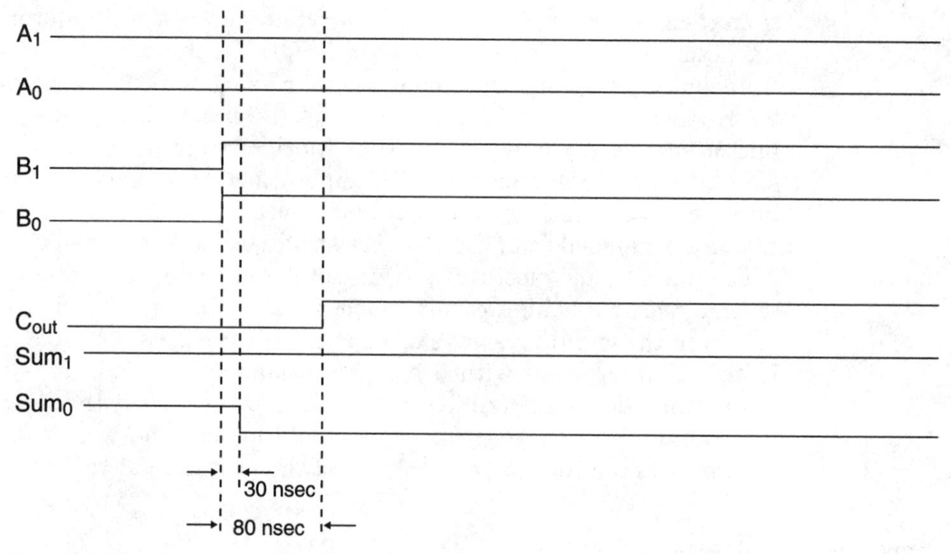

Figure 8.4 Waveform output from low-end simulator.

specify the operation of the two gates, I used the NAND analogs and assumed that TTL gates would make up the analogs. The resulting delays are 20 nsec for the AND gate and 30 nsec for the XOR.

At t_0, B_1 and B_0 will both be set, making the circuit perform the addition of 3 plus 3. If the low-cost simulator had a graphical output, it would look something like Fig. 8.4.

This waveform looks exactly as you would expect: The result of 3 added to 3 is 6, and the delays are as specified above. Unfortunately, some important characteristic features of the actual waveform are not displayed. As I indicated above, a simple digital electronics simulator can be used for learning about digital electronics, but it should not be used for design configuration or failure analysis.

Midrange Simulators

A higher-priced simulator works like the low-end simulator, but works through each of the gates in the circuit. This gives somewhat greater accuracy in representing the circuit's output waveform than the basic processors, but the simulated signals are not what I would consider accurate enough to rely on for building a circuit without any concerns.

To test out the 2-bit adder circuit, I used Electronic Workbench by Interactive Image Technologies Ltd. (http://www.interactiv.com). This tool will allow you to simulate digital, analog, and mixed digital/analog circuits. Electronics Workbench is an excellent tool for understanding how a circuit is working, although there are some issues with the operation of the simulator that you should be aware of.

To test out the operation of Electronics Workbench, I created the simple circuit shown in Fig. 8.5. This circuit uses custom-specified AND and XOR gates, which have the 20-nsec and 30-nsec gate delays, respectively. To drive the data input, I used the Word Generator, which can drive up to 16 inputs and then monitored the outputs with the 16-channel Logic Analyzer.

Electronic Workbench's Word Generator outputs a bit pattern according to a clock. As can be seen in Fig. 8.6, I use only the least 2 significant bits of the word generator, with the least significant bit always high and the second least significant bit being alternatively driven high and low. This simulates the adding of the 3 at the *B* inputs to the 3 at the *A* inputs.

The Word Generator sends data from the initial pattern address to the final. In this example, the Word Generator outputs 19 values before repeating. For the purposes of this simulation, to display how the 2-bit adder circuit works, I chose a relatively high simulation clock speed of 100 MHz. This speed was chosen so that I could observe the data flowing through each of the different gates.

The logic analyzer output (Fig. 8.7), is used to look at the operation of the circuit. For the AND and XOR timing parameters I specified above, you can see that there is a *glitch* in both the Sum_1 and C_{out} outputs. These glitches were not displayed in the low-cost simulator.

The output glitches are both 30 nsec in length, about one 30-billionth of a second in duration. This may seem very short, but the glitches can cause major

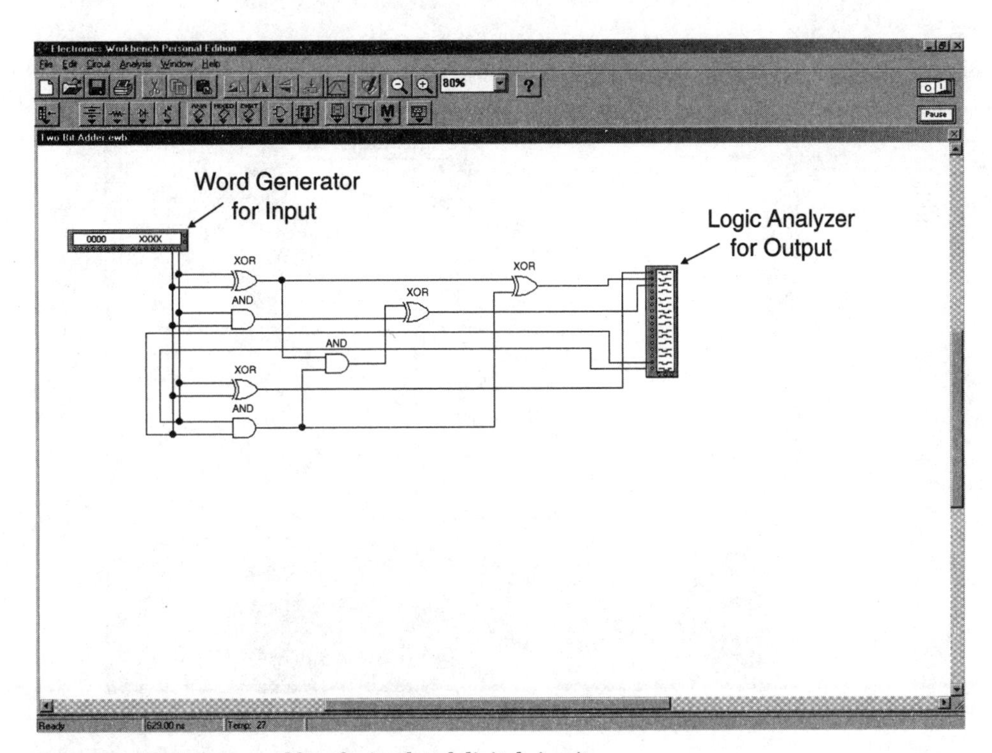

Figure 8.5 Electronics workbench simulated digital circuit.

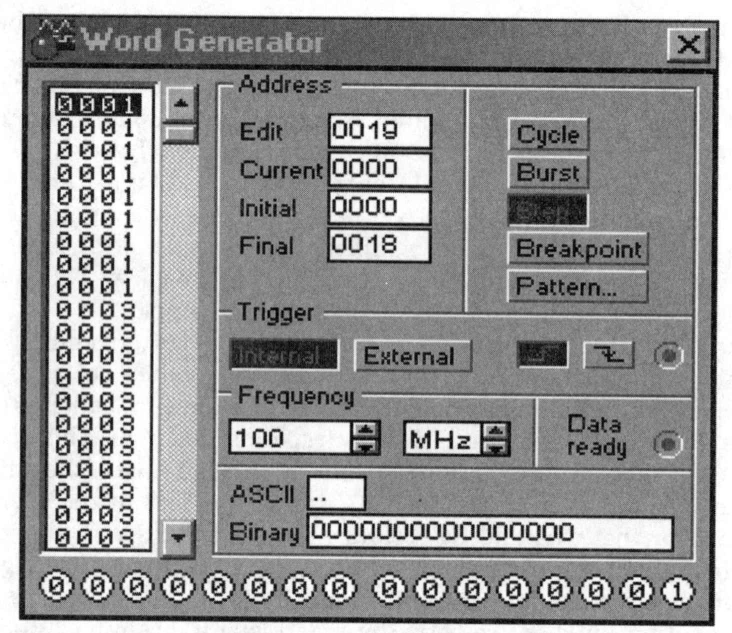

Figure 8.6 Electronic Workbench simulated digital input to 2-bit adder circuit.

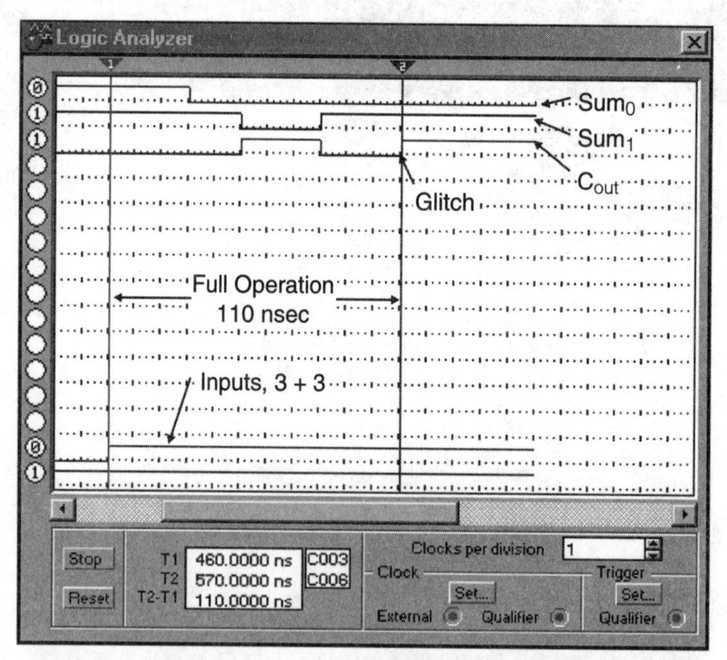

Figure 8.7 Electronic Workbench simulated digital output fron 2-bit adder circuit.

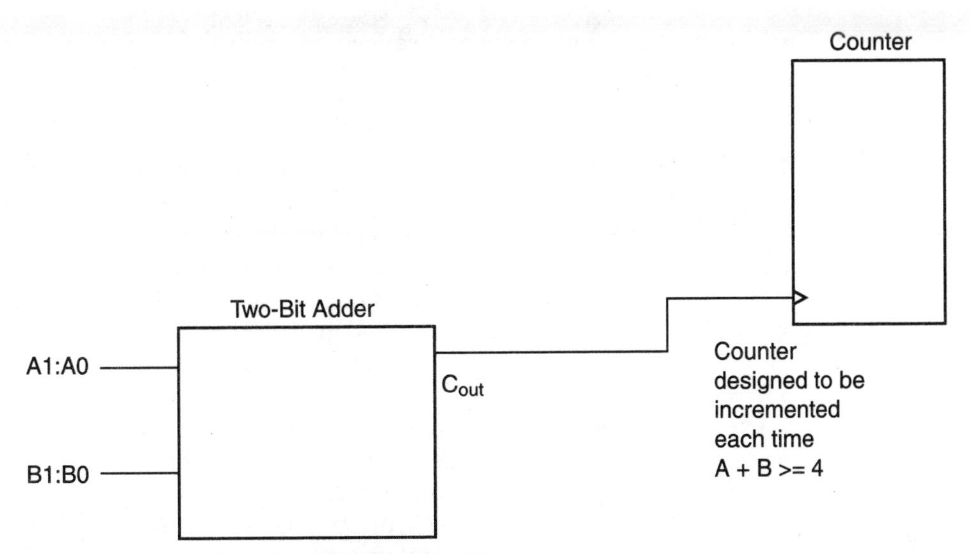

Figure 8.8 Counter driven by 2-bit adder circuit.

problems when you interface other circuits to it. For example, if the carry out was used to drive a counter (as shown in Fig. 8.8), every time B_1:B_0 is 2 or 3. In the example 2-bit adder circuit, the glitch on C_{out} can cause the counter to double-increment, resulting in an invalid result. This *glitch presentation* is a critical feature in digital logic simulators—if the simulator cannot display them, then you will find problems in the circuits that you will not be able to confirm. In addition to being unable to confirm the problems, you will not be able to understand whether a proposed fix will eliminate a problem.

High-End SPICE-Based Simulators

As I have discussed throughout this book and shown in more detail in the previous chapter, digital electronic circuits are composed of analog circuits and, as such, exhibit analog characteristics including resistance, capacitance, and inductance. It may sound ironic, but to properly simulate a digital circuit, you will have to use an analog simulation tool.

The most popular high-fidelity simulation tool is SPICE, which stands for Simulation Program for Integrated Circuits Emphasis. This tool simulates a circuit's output over time by discrete time-element modeling. That is, SPICE analyzes a circuit over very small incremental time periods. As SPICE works through each time period, it develops a picture of how the circuit works over time.

SPICE can carry out several different types of circuit *operational analyses* of a circuit's response to stimuli including:

- Nonlinear device (i.e., transistor) operational analysis. This capability is used for digital circuit analysis.

- Linear component (resistors, capacitors, inductors) operational analysis.
- Noise analysis and sensitivity.
- Distortion analysis (useful for audio amplifier designers).
- Fourier analysis (calculates and plots the frequency spectrum).
- Monte Carlo analysis (response to varying input).

To carry out this analysis, the circuit must be converted from a schematic diagram into a simple input file, as shown in Fig. 8.9, that shows how two resistors and a 5-V source are combined to form a voltage divider circuit. The simple input file shown in Fig. 8.9 lists the inputs for each of the circuit's three components.

These data/element statements are combined with four other statement types to make up a SPICE input file. The complete file consists of the title, data/element, command/control, output, and end statements.

SPICE, by itself, is very difficult to work with. Along with converting a schematic into a set of SPICE statements, the user will also have to specify the sampling period and interpret the output statements. Standard SPICE (including the variants you can download free from the Internet) simply prints out a list of the specified output values for the changing time intervals.

For most commercial SPICE implementations (like Electronic Workbench), the conversion of a circuit to a SPICE input file, as well as conversion of the list of outputs into a graphical display, is automated. This makes it much easier to set up and analyze circuits and allows relatively fast "what if" analysis of a circuit.

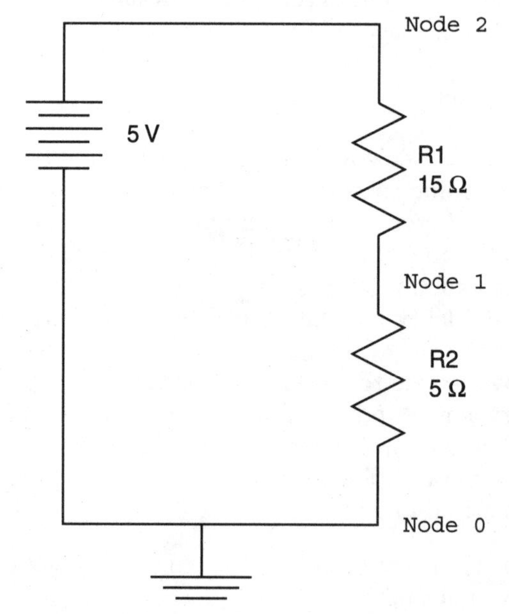

Spice Input File:

```
V1 2 0 5
R1 2 1 15
R2 0 1 5
```

Figure 8.9 Voltage divider and SPICE input.

Another important advantage of commercial SPICE implementations is the set of components that have been converted from standard basic components (a transistor can be made up from as many as 15 different basic components). In the appendices, I have listed Web pages from which you can download standard components, but these components are set up with different models that may or may not be appropriate for your application and the input they are being given.

To demonstrate how SPICE actually works, I went back to Project 8, the transistor-based TTL inverter, to compare its actual operation to an Electronic Workbench SPICE simulation. The Electronic Workbench circuit that I used is shown in Fig. 8.10. It processes a function generator's output and displays it on an oscilloscope. When I entered the circuit into Electronic Workbench, I used the TTL specified component values, rather than the actual values (which are different by less than 10 percent).

To drive this circuit, I used the function generator shown in Fig. 8.11 to drive a 0- to 5-V, 1-kHz square wave into the inverter circuit. Using the Electronic Workbench function generator, I set an amplitude of 2.5 V (which is from either side of *common*) and shifted the output up by 2.5 V. This gives a 0- to 5-V signal that simulates a digital output.

At the gross level, the inverter seems to follow the input signal very closely, as shown in Fig. 8.12. Turning up the oscilloscope's time base, the actual

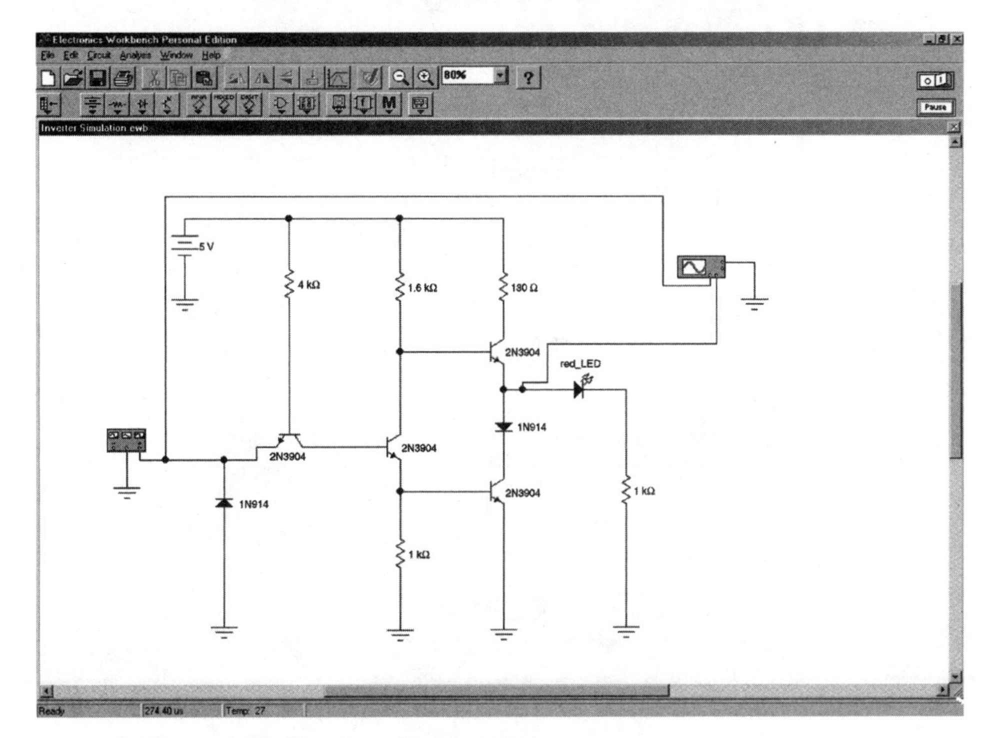

Figure 8.10 Electronic Workbench analog input screen.

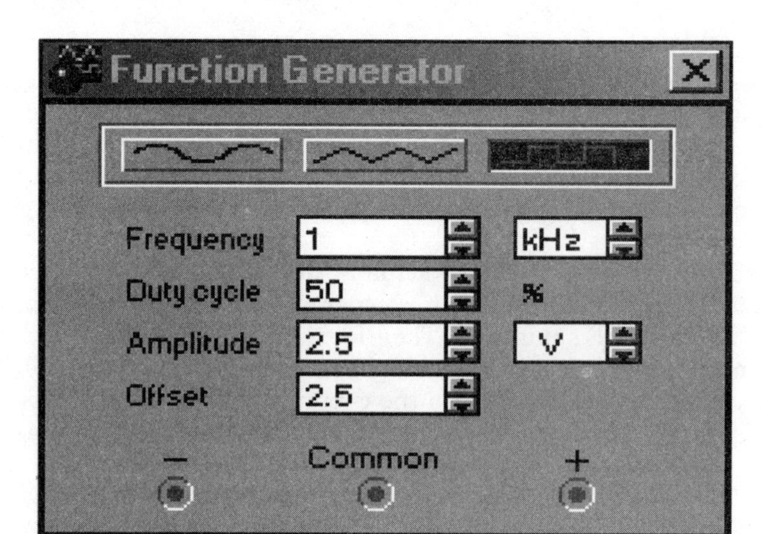

Figure 8.11 Electronic Workbench simulated 250-kHz, +5-V square-wave input.

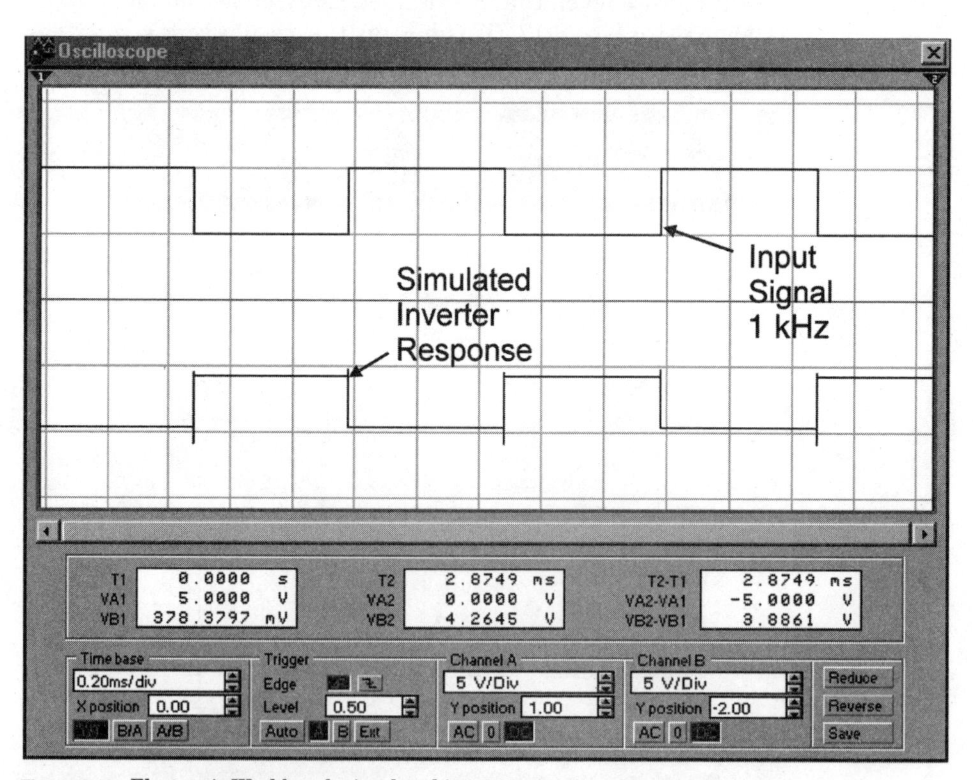

Figure 8.12 Electronic Workbench simulated inverter circuit response.

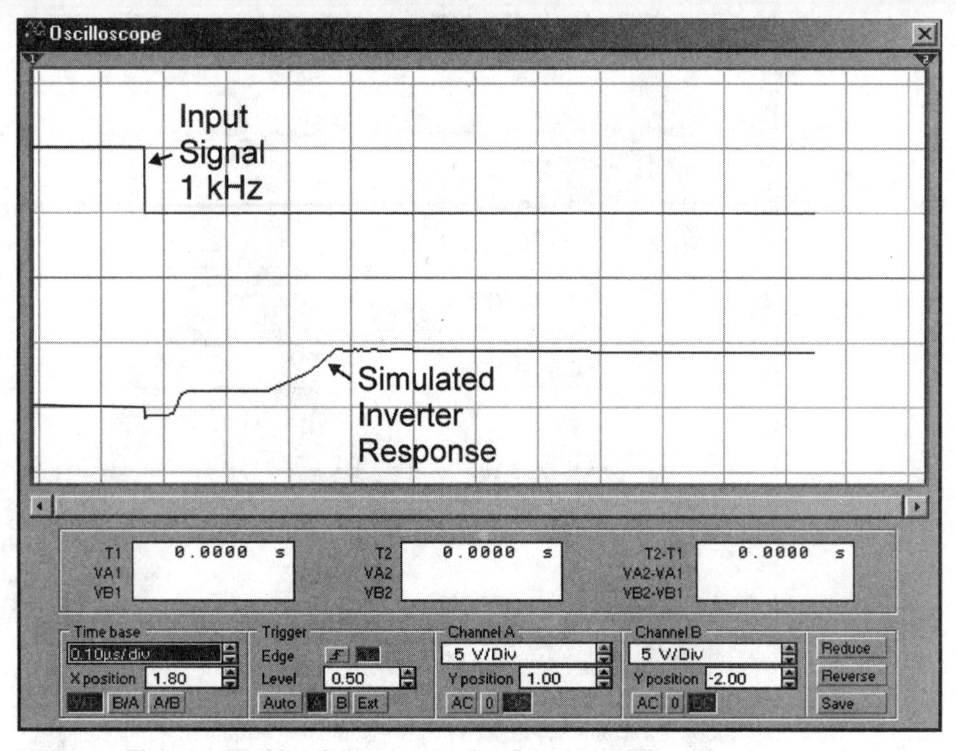

Figure 8.13 Electronic Workbench inverter simulated output on falling edge.

response to the falling edge shows that the output is not valid for approximately 300 nsec (this can be seen in Fig. 8.13). The inverter's response to a rising output is much faster, as can be seen in Fig. 8.14.

This response to rising and falling inputs is expected because of the current-limiting resistor on the top of the totem pole driver. What is interesting is the intermediate voltage level for about 100 nsec. The bottom transistor of the driver is not limited in any way, so the actual response is much more positive.

With this simulation completed, I rebuilt Project 8 as shown in Fig. 8.15. Note that the input to the inverter is wired directly into the 1-kHz oscillator output of the interface PCB (OSC2 pin). Looking at the input and output at a gross level, I observed the waveforms shown in Fig. 8.16.

In Fig. 8.16, the operation of the physical circuit appears better than the operation of the simulator circuit without the overshoots. When I zoom in on the edges of the circuit, I see in Fig. 8.17 that the actual response to the falling input is delayed by several hundred nanoseconds, just as in the simulation. In the response to the rising edge (Fig. 8.18), the output snaps down very quickly, just as it did in the simulated version.

If you compare the actual results to the simulated results, you'll see there are two main discrepancies. The first is the intermediate voltage predicted by

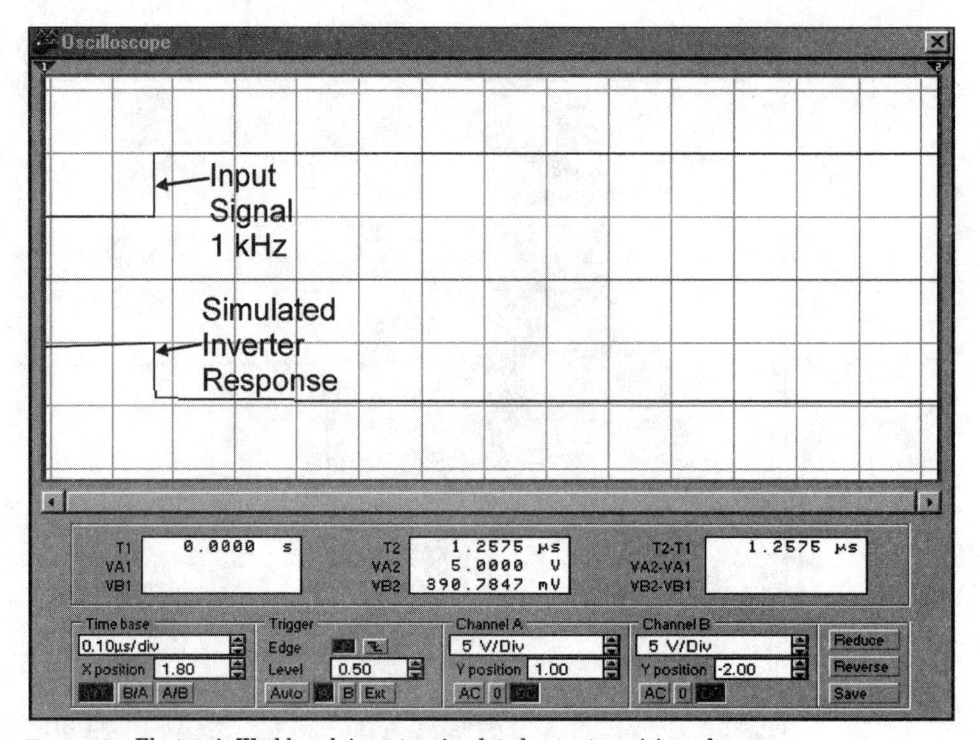

Figure 8.14 Electronic Workbench inverter simulated output on rising edge.

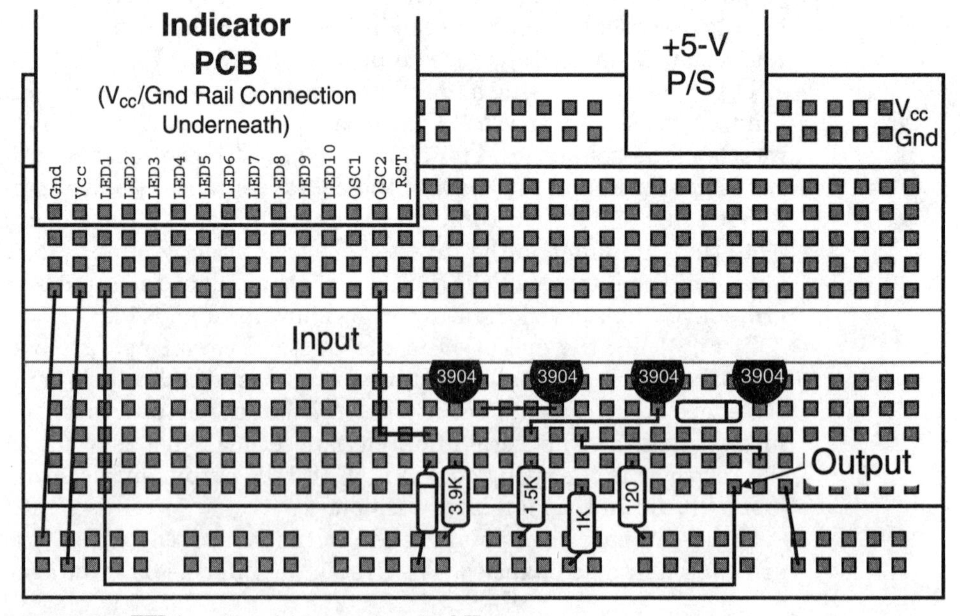

Figure 8.15 TTL inverter wiring running at 1 kHz.

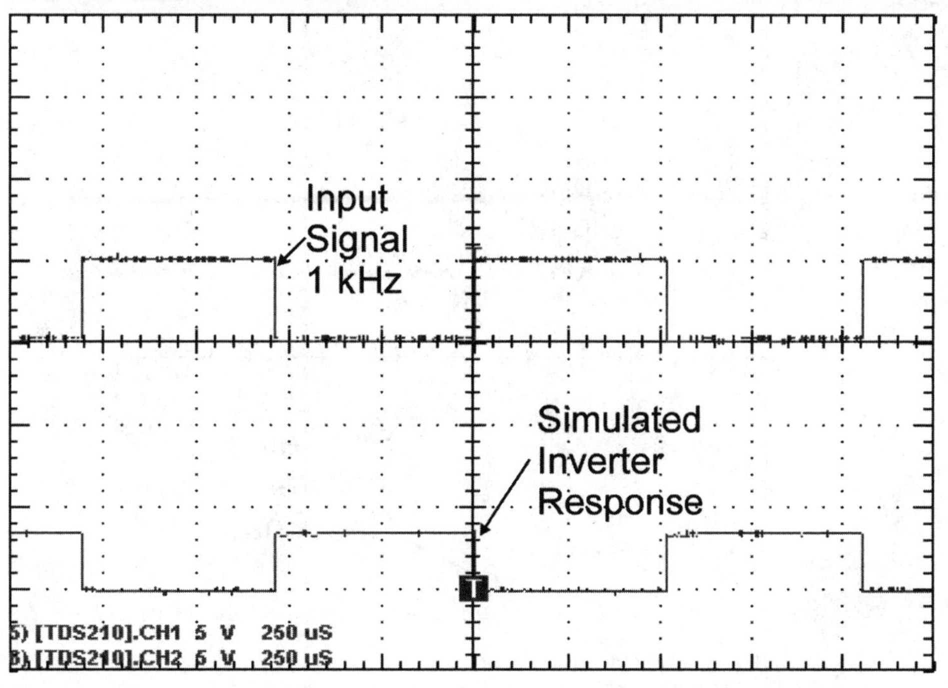

Figure 8.16 Actual inverter output from input.

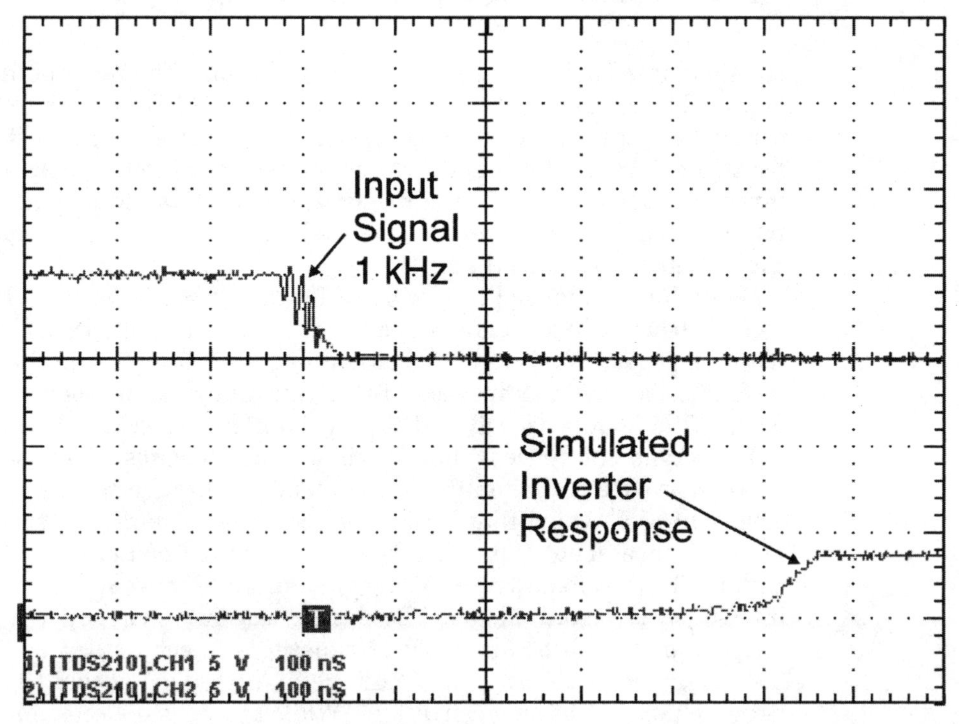

Figure 8.17 Actual inverter output on falling edge.

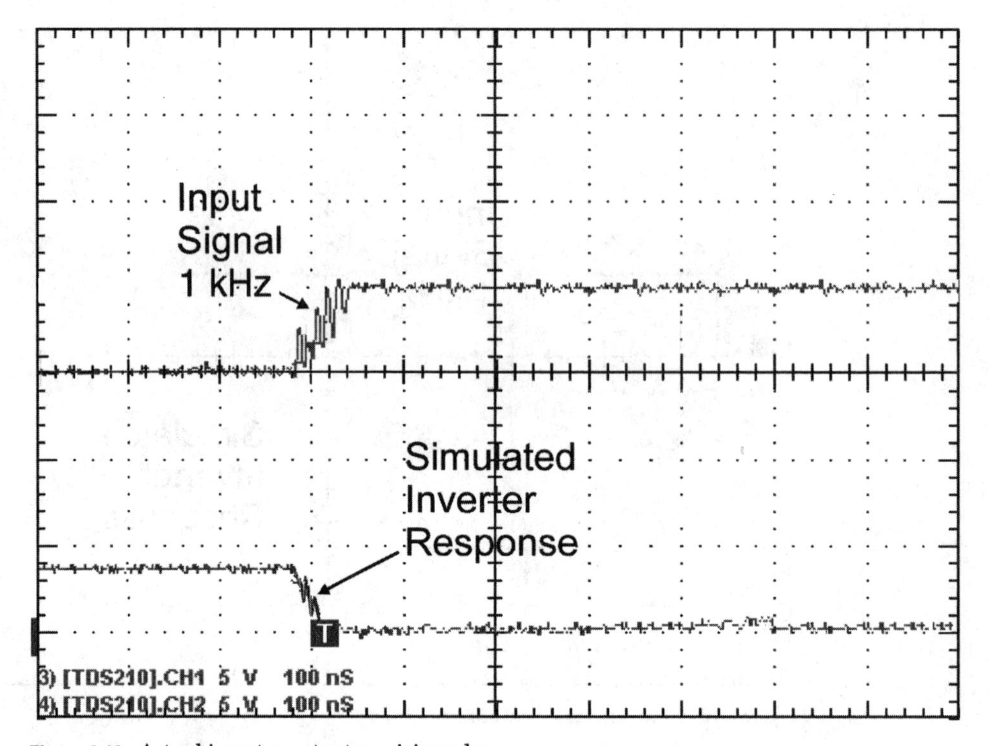

Figure 8.18 Actual inverter output on rising edge.

the simulator, but not present in the actual output. The cause of this difference is probably the parameter variances of the real LEDs and the 2N3904 transistors from the models used by Electronic Workbench. The most likely cause for this difference is the red LED connected to the totem pole output of the inverter. During the 300-nsec response period, I would expect that the two totem pole driver transistors are turned off and the voltage output is simply the voltage drop across the LED.

Depending on the application, this difference is something that I may be concerned about. To better understand the difference, I would probably experiment with the physical and simulated circuit to see if I can confirm that the voltage level is caused by the unpowered LED. This confirmation exercise may involve driving TTL inputs as well as diodes with different characteristic voltages.

The second difference is that the simulation indicates a 300-nsec delay from the falling edge to the output versus a 500-nsec delay in the actual circuit. This is not something that I would be concerned about. The reason why I am not concerned about this difference lies in the method of building the actual circuit: The breadboard prototyping system, while convenient, is not known for its high-speed-signal integrity. Remember that it is made up of many parallel strips of metal, all of which act as capacitors to each other. In doing simple measurements with my digital multimeter, I have found that the capacitance between pins can be as much as 5 pF. While this does not seem like a lot, when you think of a 63-row breadboard, the total capacitance is on the order of 630 pF (0.6 nF), which actually is significant and can greatly affect the operation circuits.

Project 11—Real World to Simulator Comparison

To finish off this chapter, I want to take a look at how good Electronic Workbench is at simulating the 2-bit adder circuit discussed in this chapter. Electronic Workbench has the ability to work in mixed mode, which means it can provide analog analysis of digital circuits. To use this mode, I used the function generator and oscilloscope instruments from the previous circuit to drive the 2-bit adder as shown in Fig. 8.19.

In this circuit, you'll notice that I have set the function generator to run at 1 MHz, to allow easier high-speed analysis of the circuit. I connected the output drivers to resistors/LEDs in the same way I would connect them when I built the actual circuit. The oscilloscope is triggered by the function generator which is driving a 1 or a 0 into B_1:B_0.

The oscilloscope output from the simulated circuit is shown in Fig. 8.20. Notice that the glitches, first presented on the logic analyzer screen of the full digital analysis, are shown in the mixed-mode analysis. I was surprised at the glitch on Sum_1 on the rising edge of B_1:B_0 (when C_{out} is falling). I had designed the circuit for producing glitches on data being added, and I neglected to consider what happened with the data.

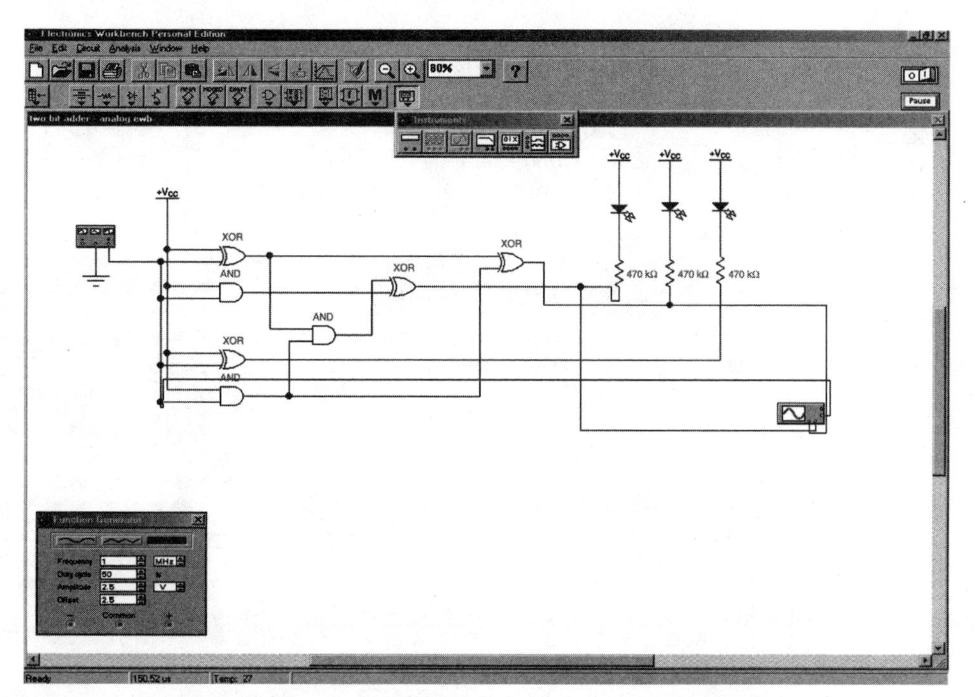

Figure 8.19 Electronic Workbench simulated digital circuit in an analog simulation.

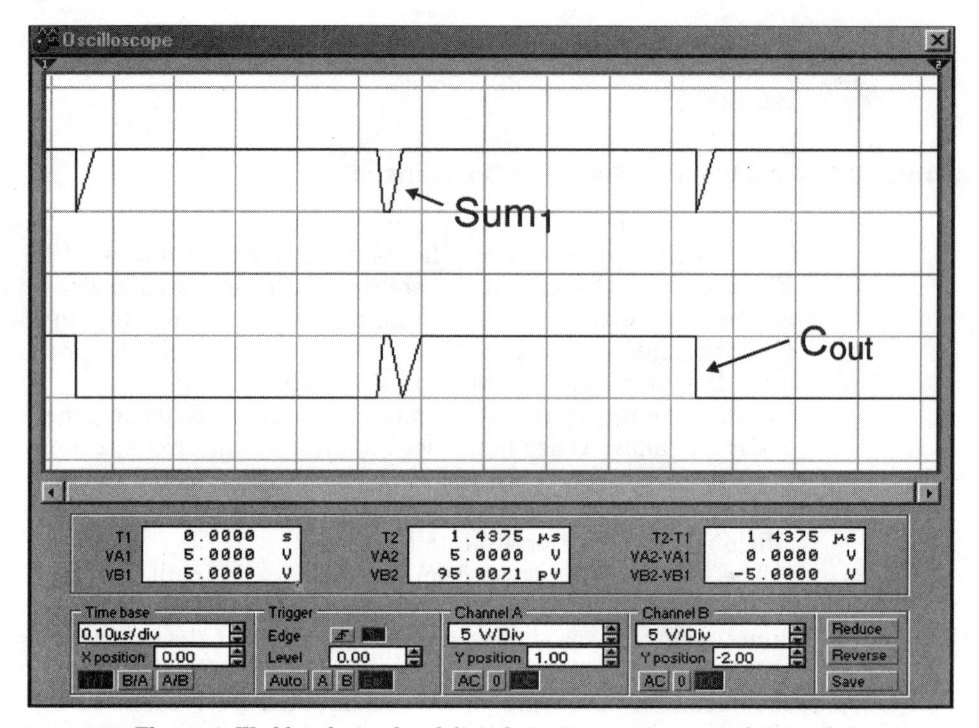

Figure 8.20 Electronic Workbench simulated digital circuit executing an analog simulation.

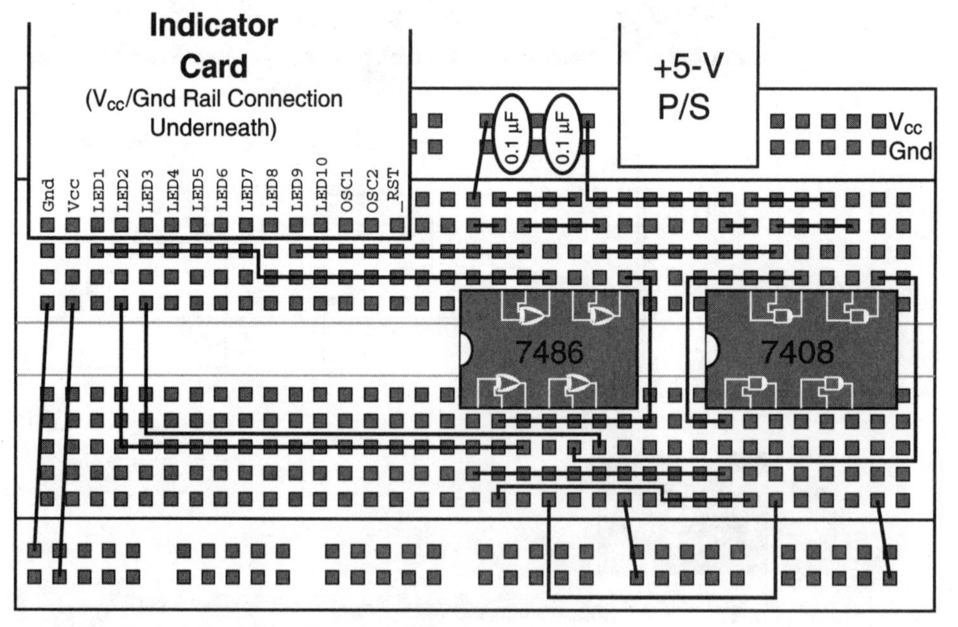

Figure 8.21 Project 11—2-bit adder test adder.

I built the 2-bit adder circuit with a 74LS86 and 74LS08 as shown in Fig. 8.21. The bill of materials for this circuit is as follows:

Part	Description
7486	74LS86 quad two-input XOR
7408	74LS08 quad two-input AND
0.1 μF	Two 0.1-μF, 16-V tantalum capacitors
Miscellaneous	Breadboard, +5-V power supply, indicator PCB, wiring

When I built my prototype, I used one of the push buttons on the Indicator PCB to pass the 0 or 1 to the 2-bit adder circuit. When the button is released, a 1 is being sent. I felt that this method would provide the cleanest edges and not generate the ringing signals that the oscillators can produce.

Looking at the output of the circuit, I saw some similar features to what was displayed on the Electronic Workbench oscilloscope. In Fig. 8.22, a small but definite glitch can be seen on the C_{out} output. A less-definite glitch can be seen on the Sum_1 output. The delay between the input and the correct output is very similar to that predicted by the simulation.

The Sum_1 glitches, being much less definite, practically speaking would not affect the operation of the circuit in any way. The glitch on the C_{out} output could cause problems with an application if it were used as part of the counter

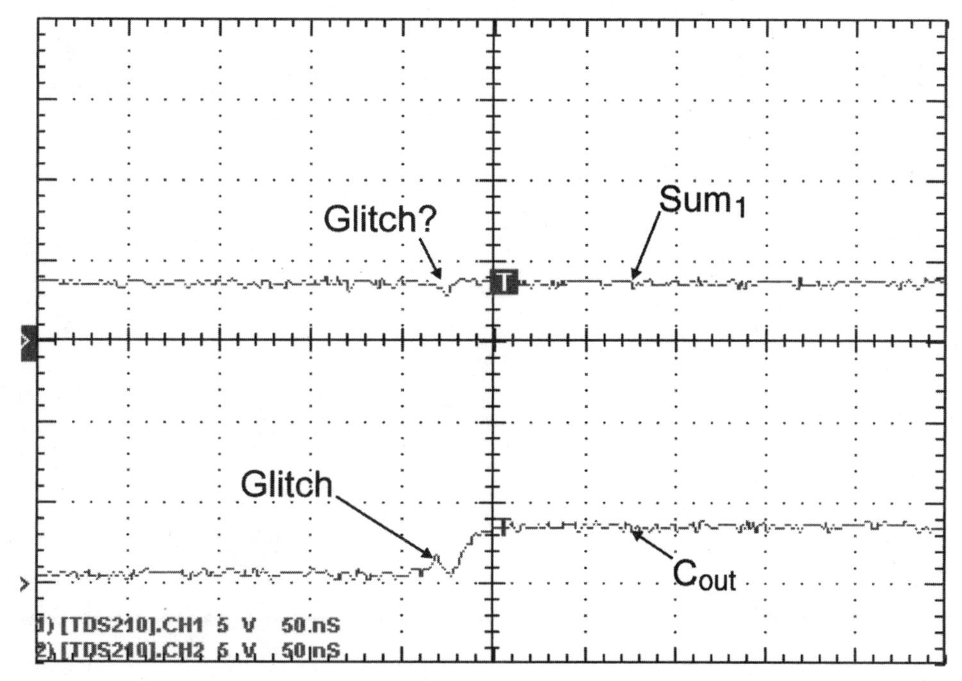

Figure 8.22 Actual output from 2-bit adder.

circuit shown earlier in the chapter. It could also cause a problem if the bit were polled during the transition/glitch period.

Looking at the simulator results versus the actual results, you may feel that a simulator could turn into Chicken Little, predicting disasters that are not likely to happen. Instead of thinking of a simulator in this way, and to continue to use a metaphor, I prefer thinking of the simulator as the little boy who cried wolf—sooner or later, it is going to predict something that will be a problem. If you design your circuits so that there are no glitches, or polling takes place outside of any window in which they could happen, you will never have problems with your designs, even if you have a marginal part.

This ability to predict possible glitch situations is the most valuable aspect of a simulator, and it is why I recommend that high performance tools that use SPICE, like Electronic Workbench, are used instead of simple simulators that do not use accurate device models.

Chapter

9

Common Digital Circuits

Before going on to some complete digital electronics projects in the next chapter, I want to discuss some of the practical aspects of wiring devices together. In this chapter, I will expand on the concepts that were presented earlier in the book and provide you with many of the subsystems that you will have to work with when you design your own digital electronics applications.

If you look at the topics presented in this chapter, I am sure that some of them will be surprising. These circuits are used to provide basic capabilities in your applications and avoid "reinventing the wheel" in order to develop your own applications. Many of the projects shown here are hip pocket applications that you should keep in the back of your mind in case you need to implement a specific function, you have run out of chips, or you do not have real estate on the board to use the full chips.

Where appropriate, I will also discuss CMOS chips, which differ from bipolar chips in how applications are wired to them. I realize that many companies describe their products as being 100 percent compatible with TTL, but you will find that there are many instances where a CMOS logic chip doesn't work with or work like a standard TTL chip.

As I have done previously, where possible I will present the functions with low-level TTL gates instead of integrated functions. After presenting the functions with low-level TTL gates, I will show how these functions are implemented with higher-level TTL chips that will provide these functions. As I work through the projects, I suggest that you look for your own opportunities for experimenting with the functions, so you can become more familiar with them.

Reset

When I am designing a complex digital electronics circuit, the first thing that I always worry about is reset and how the different circuits will start working when power is applied. When I introduced the concept of flip-flops, I noted that they could power up into *any* state—you probably read this and didn't realize the significance of this simple statement. In highly complex sequential digital circuits (such as a computer processor), it is imperative that every critical flip-flop in the system is set to a specific value before the processor "boots," or the computer will start executing at an unexpected location with unexpected data.

Many sequential chips have a *clear* pin that resets (sets to 0) all the flip-flops within them. To take advantage of this feature in a straight digital circuit, I want to hold this line active before allowing the clocks to start to run. The circuit that I typically use is shown in Fig 9.1. It consists of just a resistor, a capacitor, and an inverter.

When power is applied to this circuit, the capacitor acts as a short circuit and does not have any kind of voltage drop across it. As the charge builds up within the capacitor, the voltage across it increases. The resistor controls the rate of the charge buildup in the capacitor; the larger the value of resistor, the slower the charge buildup. This combination of resistor and capacitor is known as an *RC network.*

As I will discuss later in the book, the resistor-capacitor (*RC*) network connected to a CMOS input provides a delay from the time an initial voltage is input, expressed approximately by the formula:

$$\text{Delay} = 2.2\,RC \qquad \text{seconds}$$

For the circuit in Fig. 9.1, the approximate delay can be calculated as:

$$\text{Example delay} = 2.2 \times 10{,}000 \times 0.1\,(10^{-6}) \text{ seconds}$$

$$= 2.2 \text{ msec}$$

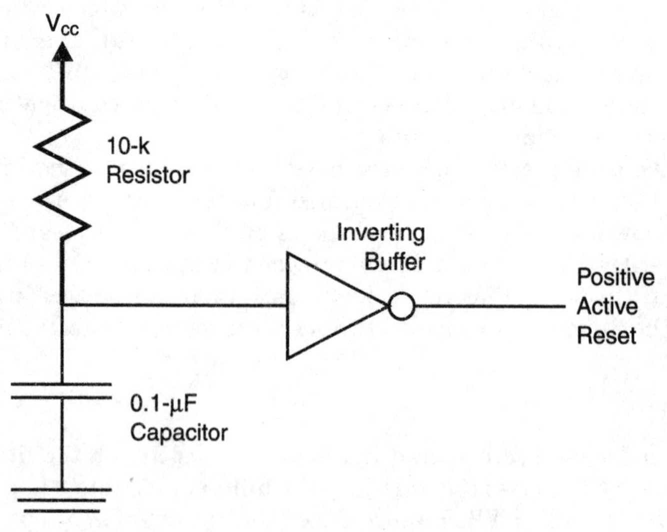

Figure 9.1 Basic reset circuit.

When the power-up operation of the power supply is compared to the characteristic rise time of the basic reset circuit (in Fig. 9.2), you can usually count on about a millisecond for the circuits to be reset and initialized before the application starts. While the *RC* network voltage is less than the threshold voltage for the inverter, the power supply is stabilizing through the circuit and, if necessary, memory circuits are being reset.

I normally prevent the oscillator output from being distributed throughout the circuit while the reset circuit is active. When the capacitor is charged to the point where the voltage across it is equal to or greater than the inverter's threshold voltage, I stop the resetting of flip-flops and allow the clock to be distributed throughout the circuit.

Depending on the circuit's oscillator, you may wish to use this circuit to start off a counter that counts down for a specific number of cycles to allow the oscillator to stabilize in frequency and waveform magnitude. This can be very important in computer processors and other very complex circuits with optimized timings, but for most circuits I do not find this necessary, and the 1 msec delay between power-up and operation is adequate for most applications.

If you are going to reset the application repeatedly, then I recommend using the circuit shown in Fig. 9.3. In this circuit, when the push button is pressed, the capacitor will be discharged through a 100-Ω resistor and held at a ground level until the button is released.

The purpose of the 100-Ω resistor is to limit the current passing through to the circuit board. In most cases the resistor is not absolutely required, but you will find circuits where the periodic influx of current will cause problems with the power supply. The power supply will detect the sudden influx of current, try to compensate, and have problems responding appropriately. To be on the safe side, always put in the 100-Ω resistor and you should not experience any problems with the circuit.

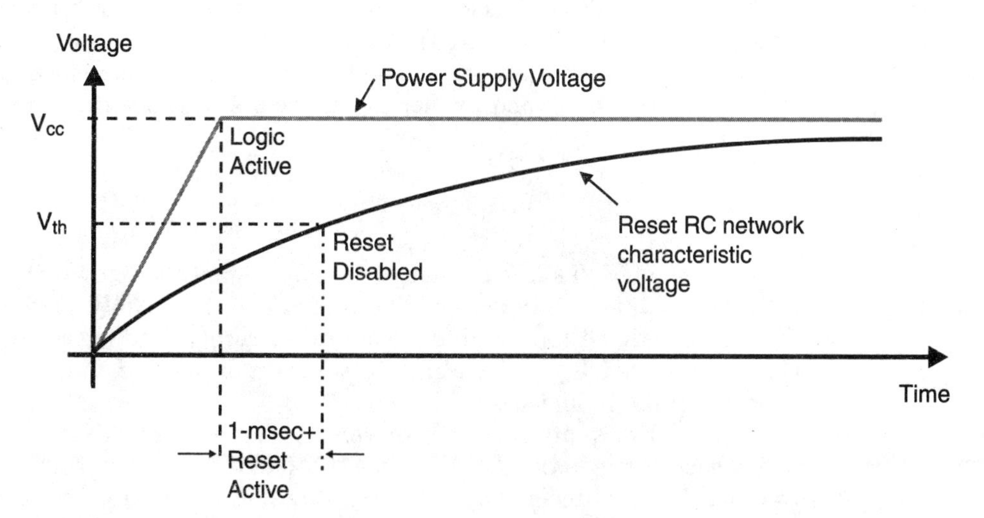

Figure 9.2 Operation of the reset circuit.

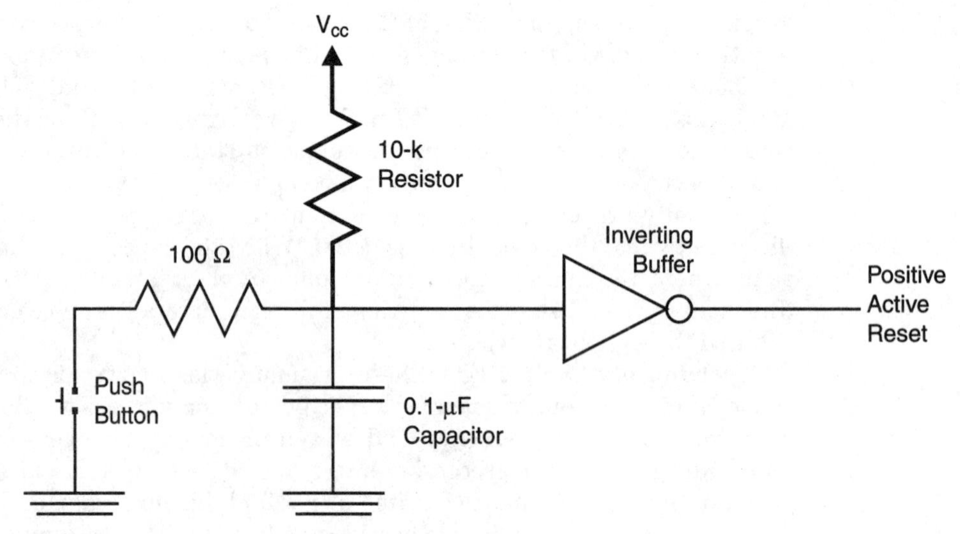

Figure 9.3 Reset circuit with push button control.

Oscillators

So far in this book, I have not discussed what is probably the most critical component needed to develop an advanced digital electronics application. So far, I have discussed digital components and how they are wired together, but I have not discussed the circuit that will be added to an application to make it run: the clock. Without a clock, many of the complex digital electronic circuits that we use (such as computers) would not be able to run through different operations without the user continually pressing a button to work through a sequence of operations.

The application's clock is a set of repeating pulses (usually with the same on and off time) that is input into the sequential circuits of an application to carry out the operations within them. The frequency output of a clock is measured in *hertz* (cycles per second), which is the reciprocal of the time the pulse is on and off:

$$\text{Frequency} = \frac{1}{\text{time on} + \text{time off}}$$

An example of this is a 1 pulse per second signal that drives the sequential circuits of a timekeeping clock. The term *clock* is probably confusing, as the digital clock that I have defined is only very rarely used to tell the time. The clock signal that I am discussing in this section is used to drive the digital counters of the timepiece.

The clock in a digital circuit is driven from an oscillator that uses some form of feedback to toggle the clock line in a consistent manner. In this book, you will find that I use the terms *clock* and *oscillator* interchangeably, with the clock (or oscillator) signal being responsible for the operation of the digital circuit.

I am continually amazed at how complex the subject of oscillators is. There are literally hundreds of different designs for oscillators, with different levels of complexity and output performance. In this book, I am concerned only with oscillators that work with TTL signals and do not provide variable output frequencies or varying output levels or waveforms. For digital circuits, the output of the oscillators ranges between 0 and 5 V and is a square wave, not sawtooth, triangle, sine, or any other non-digital form. This means that oscillators designed for radios and amplifiers [such as voltage-controlled oscillators (VCOs)] are not discussed in this book and are not appropriate for use with digital electronics.

The most basic type of oscillator is the *relaxation oscillator,* which feeds back the output of an inverter through a resistor/capacitor (*RC*) network, which delays their operation. The basic circuit and its defining output frequency equation is shown in Fig. 9.4.

In this circuit, the R1-*C* network is driven by the first inverter, and the characteristic *RC* response is fed back to the first inverter's input. When the voltage on the capacitor reaches the threshold voltage of the Schmidt-trigger input inverter, the inverter changes state and drives a new output voltage. This voltage is again passed through the R1-*C* network and delayed until the threshold voltage is reached again.

In Fig. 9.5, I have shown the voltage waveforms at the R1, R2, and *C* junction of the circuit as well as the output voltage on the interface card PCB's OSC2 output. Note that the R1, R2, and *C* junction voltage exceeds the V_{cc}/Gnd (+5 V and 0.0 V) limits. This is due to the capacitor being connected to the output driver.

Having the capacitor wired to the output driver moves the charge by 5 V each time the state changes. From this point, the output value changes the charge within the capacitor until it is back at the threshold voltage for the 74HC04, which is 2.5 V. You can see in Fig. 9.5 that the transitions take place every time the voltage across the capacitor is at 2.5 V relative to Gnd.

A CMOS inverter is used in this circuit to make sure that the oscillator cycles reliably. A TTL inverter cannot be used in this circuit because of the current drain of the input when a 0 is input. A Schmidt trigger input device (i.e., the 74HC14) could be used, but it is not necessary because the reference voltage of the capacitor is changing with every transition.

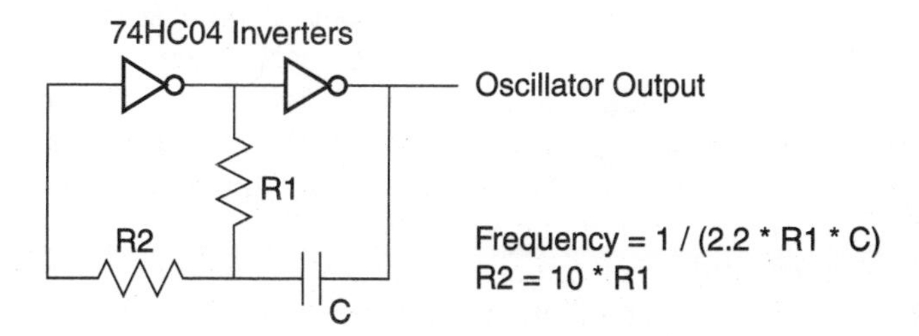

Figure 9.4 Basic relaxation oscillator.

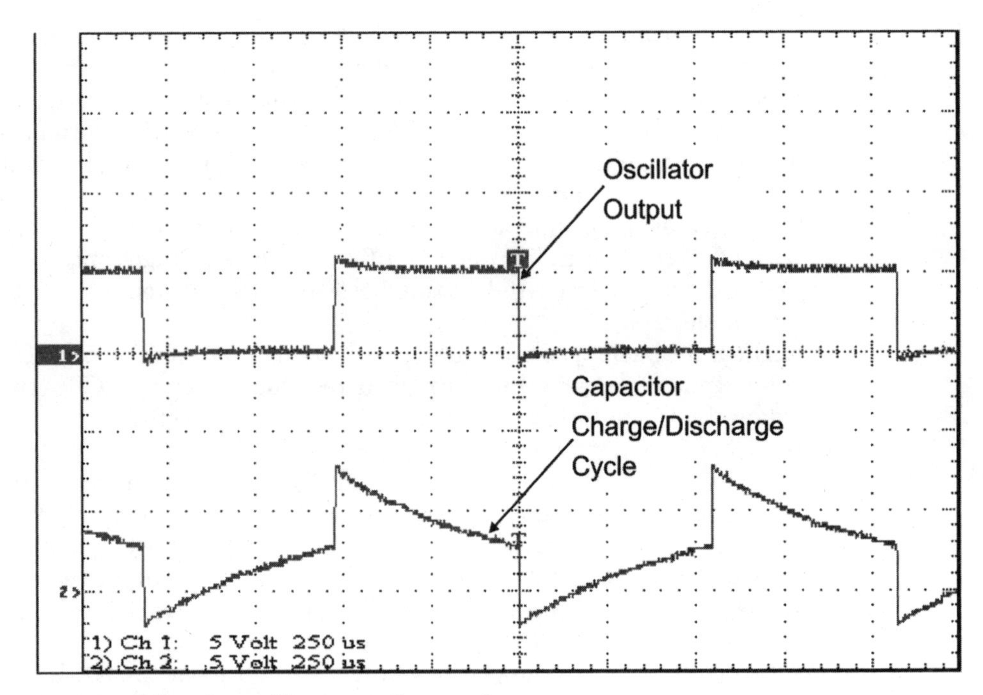

Figure 9.5 Relaxation oscillator operation waveforms.

I have included a basic 1000-Hz relaxation oscillator on the interface PCB board that comes with the book. When you assembled the PCB for the LEDs, you should have also soldered a 74HC04 onto the card along with a number of different components. Using the formula given in Fig. 9.5, I chose 4.7k and 47k resistors and a 0.1-μF capacitor to get the circuit running at approximately 1 kHz.

I say that the output is *approximately* 1 kHz because of the tolerances of the parts used in the circuit. For my prototypes, I used a 0.1-μF tantalum capacitor for the *C* in the relaxation oscillator circuit. This is probably not a "correct" use of a tantalum capacitor, as they can have tolerances approaching 30 percent of their rated value—I used it only because I tend to have a lot of them around. Along with the tolerance of the capacitor, there are also the tolerances of the resistors in the circuit to consider as well. These tolerances result in the opportunity for the actual clock signal to be off by 40 percent or more.

Your immediate response may be to add a potentiometer (variable resistor) into the circuit and tune it to the exact frequency that you want. Personally, I would discourage this practice as it involves a lot of work (especially if production parts are involved), which will drive up the cost of the product. If you are using a simple *RC* relaxation oscillator in your application, then additional costs are something that you would want to avoid.

The relaxation oscillator is adequate for many applications where a low-cost oscillator of an approximate value is required. I recommend that the circuit should not be used in any applications where any kind of precision is required.

Another aspect of this circuit that you must be aware of is the potential for large current transients within the chip as it charges and discharges the capacitor. For most circuits, this is not a problem, but if you have other sensitive circuits built into an application, you will want to keep the relaxation oscillator (as well as any other oscillators in the circuit) as electrically removed as possible from the other chips. As a rule of thumb, no other gates should be used in a chip if it is being used as an oscillator.

For the best clock accuracy, a *quartz crystal* should be used in an oscillator circuit like the one shown in Fig. 9.6. A quartz crystal is a *piezoelectric* device that provides a constant delay between one side of it and the other. The term *piezoelectric* refers to the property of quartz (and some other compounds) to mechanically deform when a current is applied to it or produce a voltage potential when it is mechanically deformed.

In an oscillator, the quartz crystal will have a voltage applied to one end of it, and this will cause the quartz crystal to deform. The rate at which this deformation takes place is known and will cause a voltage potential to be produced at the other end of the quartz crystal after a designed-in delay. This voltage is used as a feedback value to an inverter built into the oscillator circuit. The NPN bipolar-transistor-based inverter can be seen in Fig. 9.6.

The circuit in Fig. 9.6 is somewhat "fiddly" to build and to get working reliably. There are some formulas that can be used to specify the different resistor, capacitor, and inductor values, but personally, I would never use this circuit in my own applications.

This is why I did not put in any component values on the diagram; instead I would use the inverter-based oscillator shown in Fig. 9.7. In this circuit, instead of trying to understand a circuit well enough to specify the correct

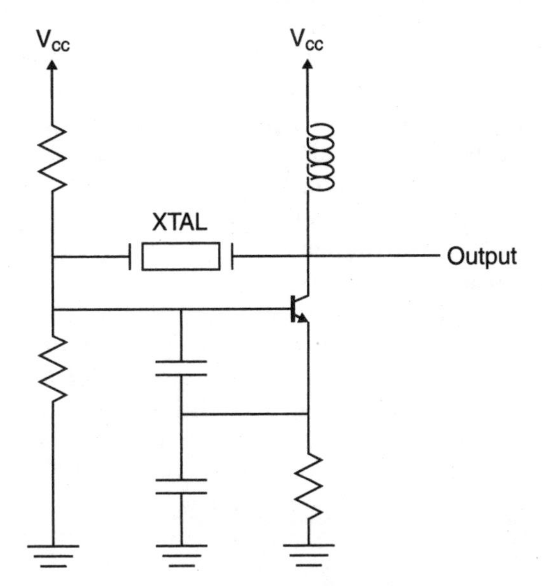

Figure 9.6 Basic crystal oscillator.

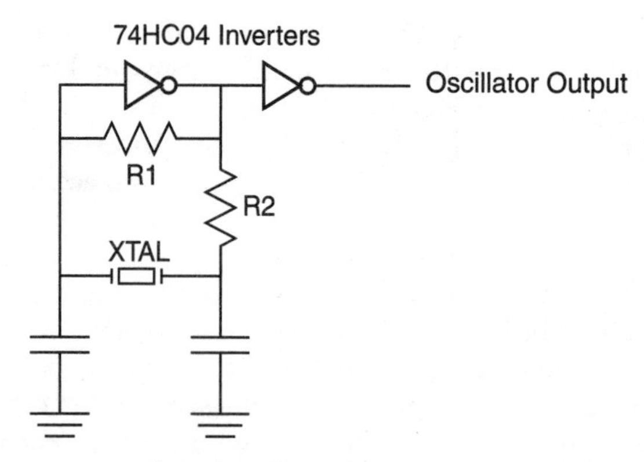

Figure 9.7 Crystal oscillator using inverters.

different analog values, you can simply put a crystal across the input and output of a CMOS inverter.

The capacitors and resistors are necessary to ensure that the oscillator runs reliably and there are not any large over- or undervoltage spikes (caused by the operation of the piezoelectric producing its own voltage output). For most megahertz-range oscillator circuits, 15- to 33-pF capacitors are adequate, and appropriate resistors are 1 to 10 MΩ for R1 and 10k to 100k for R2. You may find that, depending on the frequency of the crystal that you choose and how it is wired, you may have to vary these parts.

The second inverter in the circuit is not strictly required, but I like to have it in place to ensure the crystal is not loaded down by other devices and the operation of the oscillator doesn't change. Any large loads on the output side of the oscillator's primary inverter will affect the amount of current and voltage available to the crystal to pass the signal to the other side (and the oscillator's frequency will drop, or the oscillator won't work at all).

Changing the capacitance on the inverter output side of the primary inverter results in a small (1 or 2 percent) change to the output. I do not feel this is practical, and the nominal 0.01 percent or less error rate of the crystal should be accepted. I realize that there are applications (like digital clocks), where these changes are critical, but for the most part you should not have to tune the oscillator for the application.

Crystals work quite well, although there are two drawbacks that you should be aware of. Crystals are relatively expensive parts (especially compared to the *RC* network relaxation oscillator). You can pay up to $10 for a crystal (although about $1 is much more common). In addition, the oscillator is somewhat fragile and can be easily damaged by rough handling. You can also buy custom-built crystals that are designed to oscillate at a specified frequency.

A relatively new device that can be used in place of a crystal and does not have these shortcomings is the *ceramic resonator.* A ceramic resonator is used exactly the same way as a quartz crystal (see Fig. 9.8), but is usually much less expensive and is very rugged. If you have read my microcontroller books, you

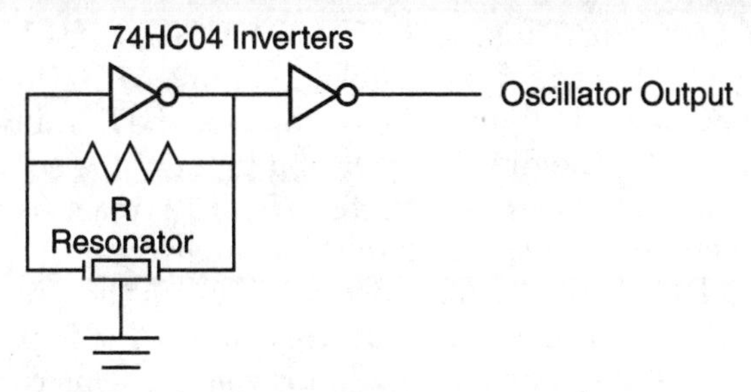

Figure 9.8 Oscillator using inverters and a ceramic resonator with built-in capacitators.

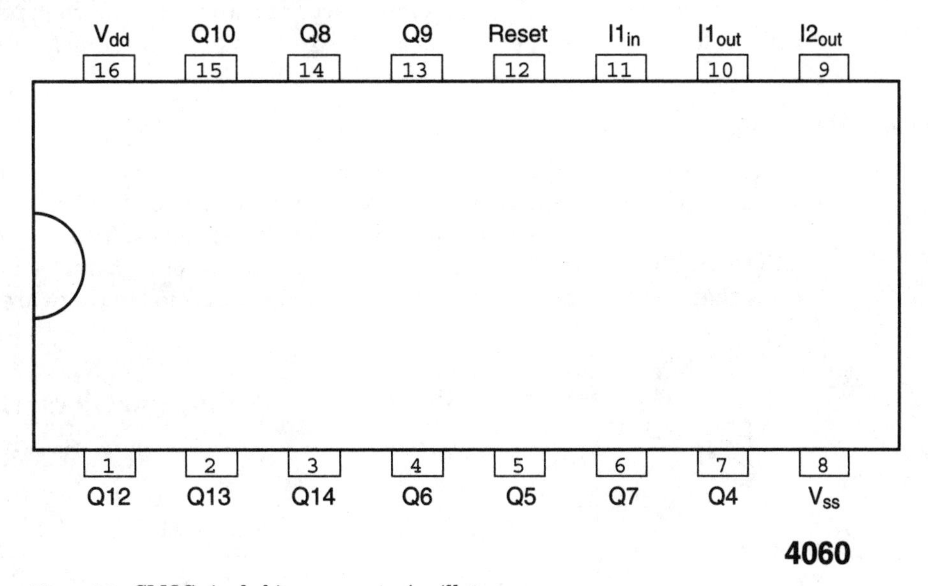

Figure 9.9 CMOS ripple binary counter/oscillator.

will know that I use ceramic resonators almost exclusively for clocking all my microcontroller applications. Despite the somewhat poorer accuracy of the parts (they are usually accurate to 0.5 percent), they really are the part of choice for most applications.

In this section, I have shown how simple CMOS inverters can be used to implement oscillators. In real life, very few people use these circuits because of the difficulty in specifying the correct parts for an application and the potentially large amount of real estate that they can take up. Instead, oscillators are usually implemented with some kind of premade, or canned, solution.

One of the most common chips that are used in digital logic applications is the CMOS 4060 (shown in Fig. 9.9). This chip can be used with the different oscillator types listed in this section and the divide-by outputs are very handy in many circuits (often eliminating the need for separate counters). The crystal, ceramic resonator, and relaxation oscillator circuits that I have shown in this section can be used with this part. The Q4 through Q13 outputs are divide-by counters (i.e., Q4 is the clock divided by 2 to 4 or 16 times).

When using the 4060, note that pin 11 is the input to the first inverter in the oscillator, pin 10 is output of the first inverter and input of the second inverter, while pin 9 is the output of the second inverter. If pin 12 (reset) is pulled high, the oscillator is stopped and the counters in the chip are reset.

Another option is a DIP oscillator. This device takes up the same amount of space as an 8- or 14-pin DIP package and usually consists of four pins (one at each corner). These parts may have an oscillator control. This control simply turns the output of the crystal oscillator on or off (like the ones shown in the 4060) built into it. For most applications, canned oscillators are the best solution, as they offer excellent frequency accuracy (over a very wide temperature range), low noise output (the packages are all metal), and moderate cost.

The 555 Timer Chip

When I was a teenager, the NE555 (Fig. 9.10; normally just called the 555) timer chip became available and very quickly became a standard for many different circuits because of its simplicity and robustness. The original bipolar 555 timer chip is not ideally suited to TTL applications (or really any applications that use any chips other than a single 555), and today there are a plethora

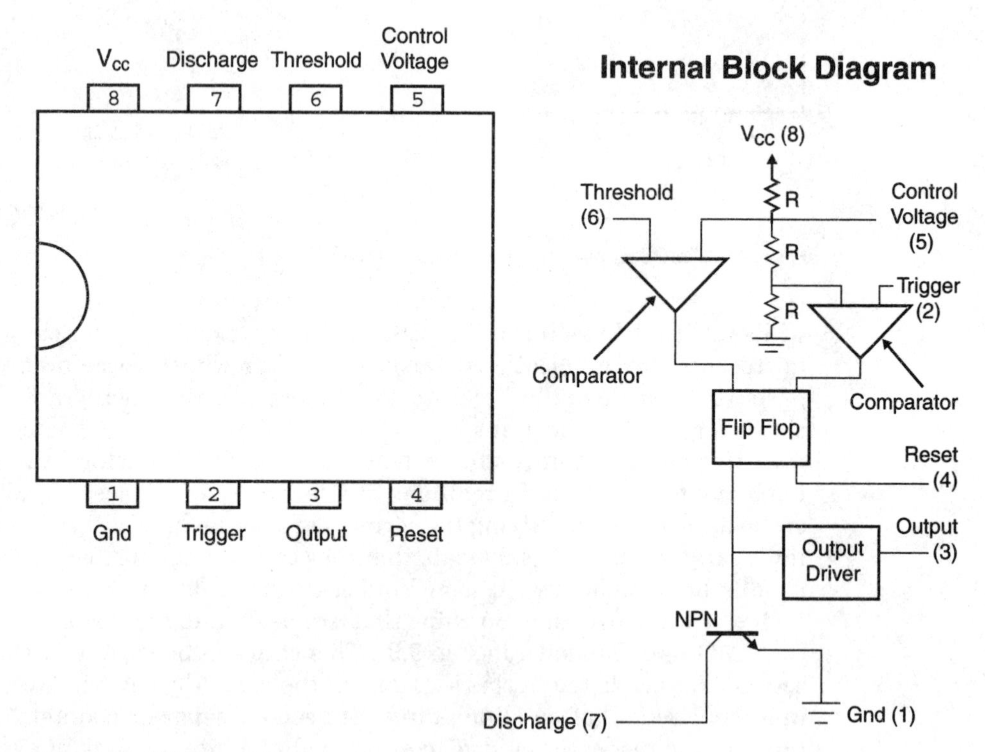

Figure 9.10 555 timer chip.

of chips available that provide functions and capabilities similar to those of the 555. Despite the problems with interfacing the 555 with other chips, it is a good idea to keep the 555 timer in your "hip pocket," as it can be simply added to a circuit or used as the basis for an application.

The 555 chip can be used to provide single pulses ranging from less than a microsecond to several seconds in length or to output a repeating signal with a duty cycle of 0 to 50 percent. The chip itself is highly tolerant of different input voltages and can drive much more than the TTL circuits described so far in this book. In looking through my old textbooks, data sheets, and notes, I found the following 555 timer–based circuits—they will give you an idea of the different applications built around this chip:

- Button debouncing circuit
- Delay driver
- Missing pulse detector
- Programmable timer
- Frequency divider
- Voltage-controlled oscillator
- Frequency meter
- Toy organ
- Lamp dimmer
- Light/dark detector
- Infrared transmitter/receiver
- Switch-mode power supply (SMPS) clock circuit

The most basic use for the 555 is as a simple oscillator. In Fig. 9.11, I show how two resistors along with a capacitor can be used with the 555 to generate a negative pulse train with the high and low time defined by the formulas shown in the figure.

In this configuration, the capacitor is charged through the R1 and R2 resistors. When the voltage across the capacitor (C) is equal to the threshold voltage (compared at pin 6 of the 555; it is two-thirds of the input voltage), the NPN transistor is turned on. When the NPN transistor is turned on, the charge accumulated in the capacitor is drained to ground through R2 and pin 7 of the 555. When the voltage across the capacitor reaches one-third of V_{cc}, the trigger voltage is reached and the NPN transistor is turned off, which allows the capacitor to start charging again toward the threshold voltage. This action repeats forever with the control to the NPN transistor driven out of the 555 on pin 3. The oscillator action described here is the basis for 10 of the 12 circuits listed above.

You see in Fig. 9.11 that there are two formulas that describe the operation of the output waveform. From these two formulas, it should be apparent that t_2 will never be greater than 50 percent of the total output waveform; this

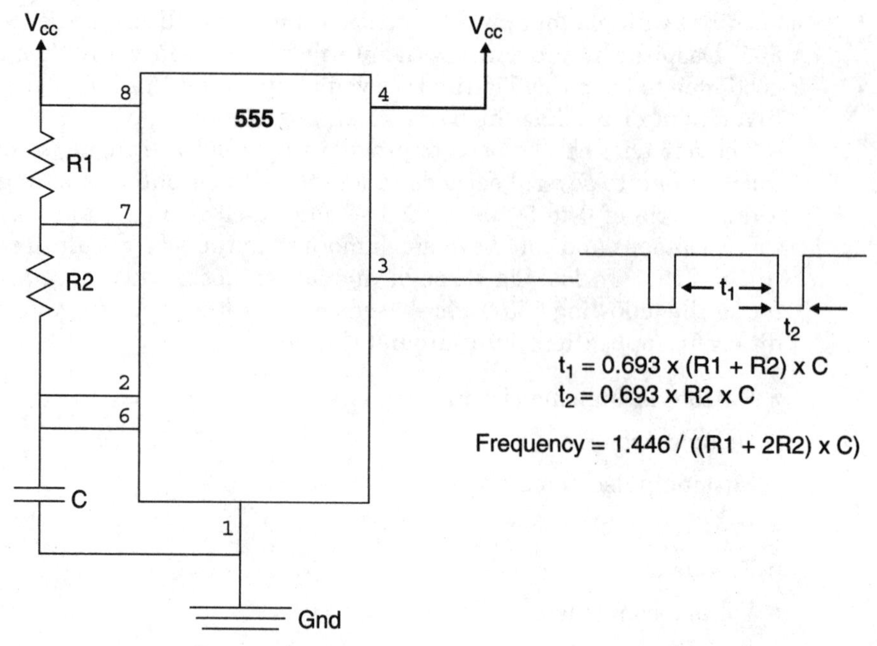

Figure 9.11 555 oscillator circuit.

means that the circuit cannot be used for a traditional PWM circuit. Note that R1 should *never* be reduced to 0 Ω—the reason for this warning is to prevent a circuit in which the discharge pin is connected directly to the circuit's V_{cc}. If the discharge pin is connected directly to V_{cc}, then, when the capacitor is discharging, all available current in the system will be passing through the 555.

In many simple applications, if all the current available to a circuit is passing through the 555, this is not a problem—but if there are additional TTL chips in the circuit, this will turn into a problem. If you want to have a 50 percent duty cycle (also called a *square wave*) output, I suggest running the 555 at twice the frequency you want for your circuit and then using a toggle flip-flop to divide the frequency by 2. The final signal will be at a 50 percent duty cycle and at the correct frequency.

The limited pulse width relative to the entire period is not a tremendous hardship, but you will have to be aware of it when you are designing a PWM output circuit. If you are designing a PWM output circuit, then I recommend that you do not use the 555 timer output directly; instead use the 555 timer to drive a PWM counter circuit.

The second commonly used configuration of the 555 is shown in Fig. 9.12. It will respond to a negative-going pulse by driving out a positive pulse for a given period. This circuit is of particular use in debouncing inputs, as I will show in the next chapter. Another very common use for this circuit is to delay an operation for a set amount of time.

This circuit is known as a *monostable* oscillator because it will respond to an input only once and then return a stable state. In normal operation, the capac-

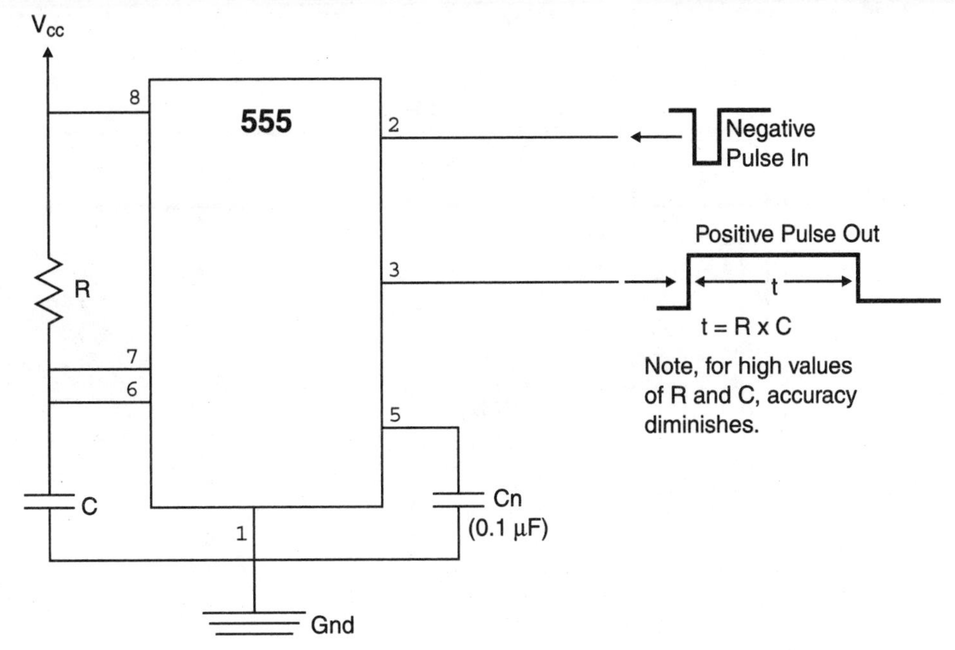

Figure 9.12 555 monostable circuit.

itor (C) is fully charged. When the trigger (pin 2) comparator detects the falling edge, the capacitor is discharged through the NPN transistor. Once the voltage across the capacitor has reached the threshold (pin 6) voltage, the transistor is turned off until the next triggering falling edge on pin 2. While the capacitor is recharging, the output is high.

Some circuits modify the control voltage (pin 5) of the 555 when the part is to work under specific voltage limits. By sourcing or sinking current through this pin, some circuits change the operating characteristics of the part. This is actually quite an advanced input method, as it changes the operating characteristics (and timing formulas) of the 555, and it is not recommended for most applications. The reset pin (4) is a positive active pin that will stop the 555 from oscillating, and it is usually used only in monostable output mode.

The 555 timer can be wired into a breadboard circuit in the two modes, as shown in Fig. 9.13. In either case, the 555 requires a minimum of work and time to add the circuit into a breadboard application.

As you might imagine, during the capacitor charge and discharge cycles for both oscillator and monostable operating modes, a great deal of current (on the order of 150 mA) is being diverted momentarily to the 555 timer circuit. This makes the 555 timer *very* disruptive to the operation of other circuits, including TTL chips. For this reason, I do not recommend that the 555 timer ever be used with any chips other than additional cascaded 555's. This is true even if large decoupling capacitors are placed on the 555 and other chips in the application circuit.

Having said this, I should point out there are CMOS versions of the 555 that do not have the same current transients as the standard bipolar part and coexists

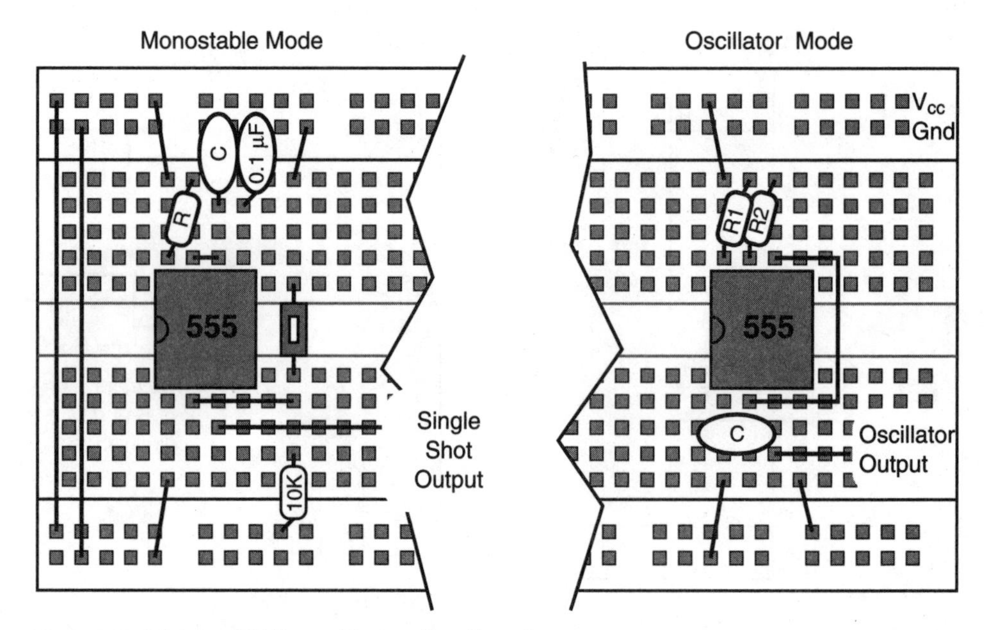

Figure 9.13 Wiring a 555 timer chip on a breadboard.

quite nicely with other parts. When I use 555 timers in the later parts of the book, I will be using the Texas Instruments TLC555, which can be powered by voltages between 1 and 18 V (the bipolar 555 can run from 4.5 to 18 V). The TLC555 has an internal transient current on the order of 60 μA while the standard 555 has an internal transient current of 3 mA. All other current requirements are proportionally lower in the TLC555 than in the standard 555, and the TLC555 is compatible with TTL and other digital logic technologies.

Project 12—Binary Clock

I was a young teenager when the first digital watches came out. At the time they were heralded as incredible feats of modernization; the transistors accepted by most people at this point in time took up about a quarter of a cubic inch (half a cubic centimeter). Computers were still "mainframes" that filled specially constructed rooms, and personal computers were still some years away. People were startled when a silicon chip, a fraction of the size of the gears, pendulums, and mainspring of a mechanical watch, could perform the same functions, often with much higher accuracy.

I have introduced you to all the constituent parts of a clock so far in this book. To summarize and demonstrate this information, I thought I would go through the design of a clock that uses these elements. The actual circuit will seem quite complex (especially when you wire it on a breadboard), but you will get an idea of what an accomplishment it was to come up with a digital watch 30 years ago.

The clock that you will build will have the following characteristics:

- It will accurately display hours (zero to 11) and minutes (zero to 59) in binary format using the bargraph LEDs used so far in the projects.
- The time base will be a 32.768-kHz watch crystal.
- The entire circuit will fit on three 6-in-long breadboards.
- The 5-V supply PCB supplied with this book will be sufficient to power this project.

These requirements probably seem quite simple, but as I go through this project (and the ones that follow), you will discover that it can be very difficult to squeeze together a circuit on a set project board. This project requires 12 chips, two bargraph LED displays, and a number of discrete components that can comfortably fit in the palm of your hand; you will be amazed at the size of the final circuit when you wire it on a breadboard circuit. Using a breadboard circuit is probably the least efficient method, in terms of space, of building a circuit, but it will pay off in minimizing the space used by the final circuit and the amount of time it takes to wire it.

Another important aspect of wiring a circuit on a breadboard is to make sure that space is left to allow you to experiment and add other components—this became an important issue for me when I was checking that the clock application would run properly. When I say that a circuit is to fit comfortably on three 6-in breadboards, what I am really looking for is the application to fit comfortably on two and a half, with enough space left over to add new circuits or redesign others that don't work properly. This application meets this requirement easily and is a requirement I will continue with for the later, more complex projects.

The output display consists of two 10-LED bargraph displays, with one devoted to minutes and the other to hours. While there is a second counter subsystem built into the circuit and you can monitor its behavior with an LED bargraph display, it is not usable when the clock is in operation, as it counts down over a minute. The minutes and hours counters count up and their values can be read directly when the clock is in operation.

The power requirement would be an issue only if I had a large number of LED displays. For the most part, TTL logic is not very power hungry, and you should see this application requiring less than 200 mA under all circumstances.

Before starting this project, I did something that I will do often for each complex project I work through in this book; I tried to break it up into separate smaller projects that all work together. This is very appropriate for most applications and will make the development of the entire application seem much less daunting. Figure 9.14 shows how I broke the application into four parts.

The first part of the circuit is the time base (or clock). To get the best possible accuracy, I wanted to use a commonly available 32.768-kHz watch crystal; 32,768 is actually 2 raised to the power of 15, so using this frequency, I knew

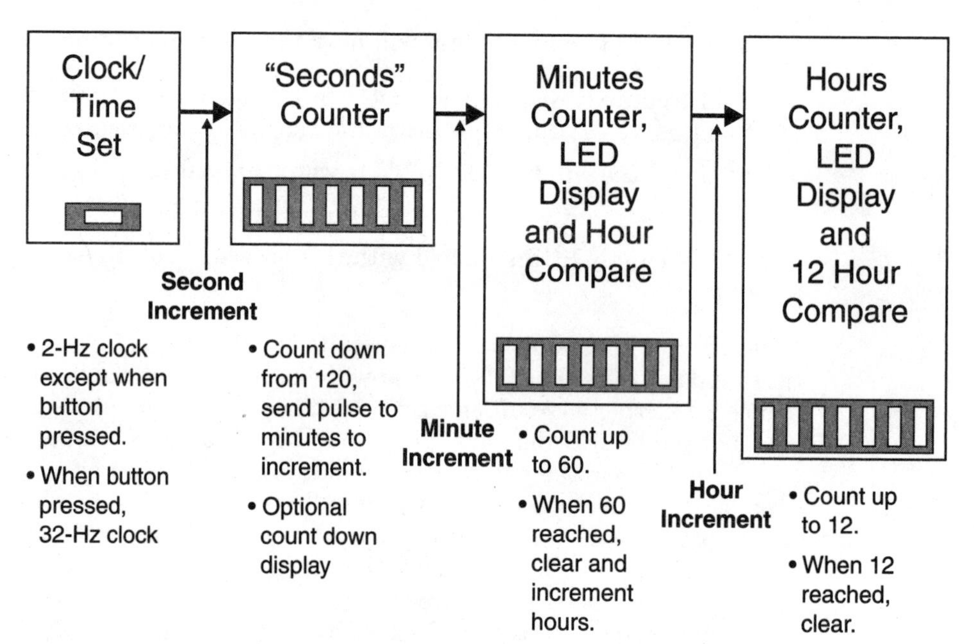

Figure 9.14 Project 12—block diagram.

that if I divided it down 15 times, I would have a stable, accurate 1-Hz clock that could be used to drive the circuit. For setting the clock, I wanted the user to be able to "speed up" this signal to drive the remaining counters faster to get to the desired time.

The second part of the circuit is a seconds counter that is loaded with an initial value and counts down to zero. When the clock counts down to zero, a clock pulse is sent on to the minutes counter. For the clock/time set part of this application, I wanted to use the CMOS 4060B oscillator/counter chip, but this can divide the oscillator clock only by 2 to the power 14, which results in a 2-Hz output from a 32.768-kHz crystal. To accommodate this situation (you could call it a feature), I start counting down from 120 rather than 60.

The minutes counter is the next part of the circuit and keeps track of and displays the current number of minutes. It is incremented each time the seconds counter underflows. The circuit consists of an 8-bit counter with a comparator that resets the counter and increments the hours counter. The 8-bit counter value is passed to LEDs and displayed as part of the current time. For the seconds counter, I loaded the counter with a set value and counted down. To ensure that the LEDs display the correct value, I reset to zero and display up to 59 (in binary, which has a value of 0b0111011).

Last, the hours counter increments each time the number of minutes reaches 60. To accommodate the operation of the counter used in this circuit (the 74LS193), I do not increment on the minutes being equal to 60. Instead, I increment on the minutes not being less than 60. The reason for doing this was to avoid having to use additional logic to convert the equals comparator output into a clock pulse for the hours counter. The hours counter has a comparator that checks the current value with 12 (0b01100) and resets the counter in this circuit when it is equal.

An important consideration in designing an application like this is to make sure that you can easily wire it on a breadboard. Breaking the application into four parts and minimizing the number of interconnections between the parts was important to simplify the wiring of the project on a breadboard. As you work through the circuit, you will see that there are only four lines connecting the four parts.

I will work through each part of this application as its own separate circuit. This allowed me to best debug the problems before they affected "downstream" circuits. I suggest that you build the circuit as I have, in four parts, and don't continue on until you are comfortable that each part works as you expect.

The bill of materials for the entire binary clock project is:

Part	Description
U1	4060B CMOS Oscillator/Counter
U2–U3, U5–U8, U11	74LS193 4-bit binary counter
U4	74LS86 quad two-input XOR gate
U9–U10, U12	74LS85 4-bit comparator
Y1	32.768-kHz crystal
D1–D20	Red LEDs; for wiring simplicity, use 10-LED bargraph display (*note:* D1–D8 are optional)
R1	10-MΩ, ¼-W resistor
R2, R5	10k, ¼-W resistor
R3, R4, R6	470 Ω, ¼-W, eight-resistor SIP (*note:* R3 is optional)
C1	100-pF capacitor (any type)
C2	15-pF capacitor (any type)
C3–C15	0.1 μF, 16-V tantalum capacitors
SW1	Momentary-on SPST push button switch
Miscellaneous	Breadboard, +5-V power supply, wire

With bargraph displays and SIP current-limiting resistors for them, there are a lot of components in this project. They also require a lot of wire. I used three boxes of breadboard wire for this project, so please make sure you have enough on hand.

When you are looking through this project's schematics, you will see that there are a lot of chip inputs tied to V_{cc} or Gnd. When you wire the application, I recommend that you wire in the connections to V_{cc} and Gnd before any of the logic connections. This will help make your final circuit a lot neater and will make it easier to go through.

I am really pleased with the clock circuit that I came up with for this application. It will output a nominal 2-Hz signal when the binary clock is in normal

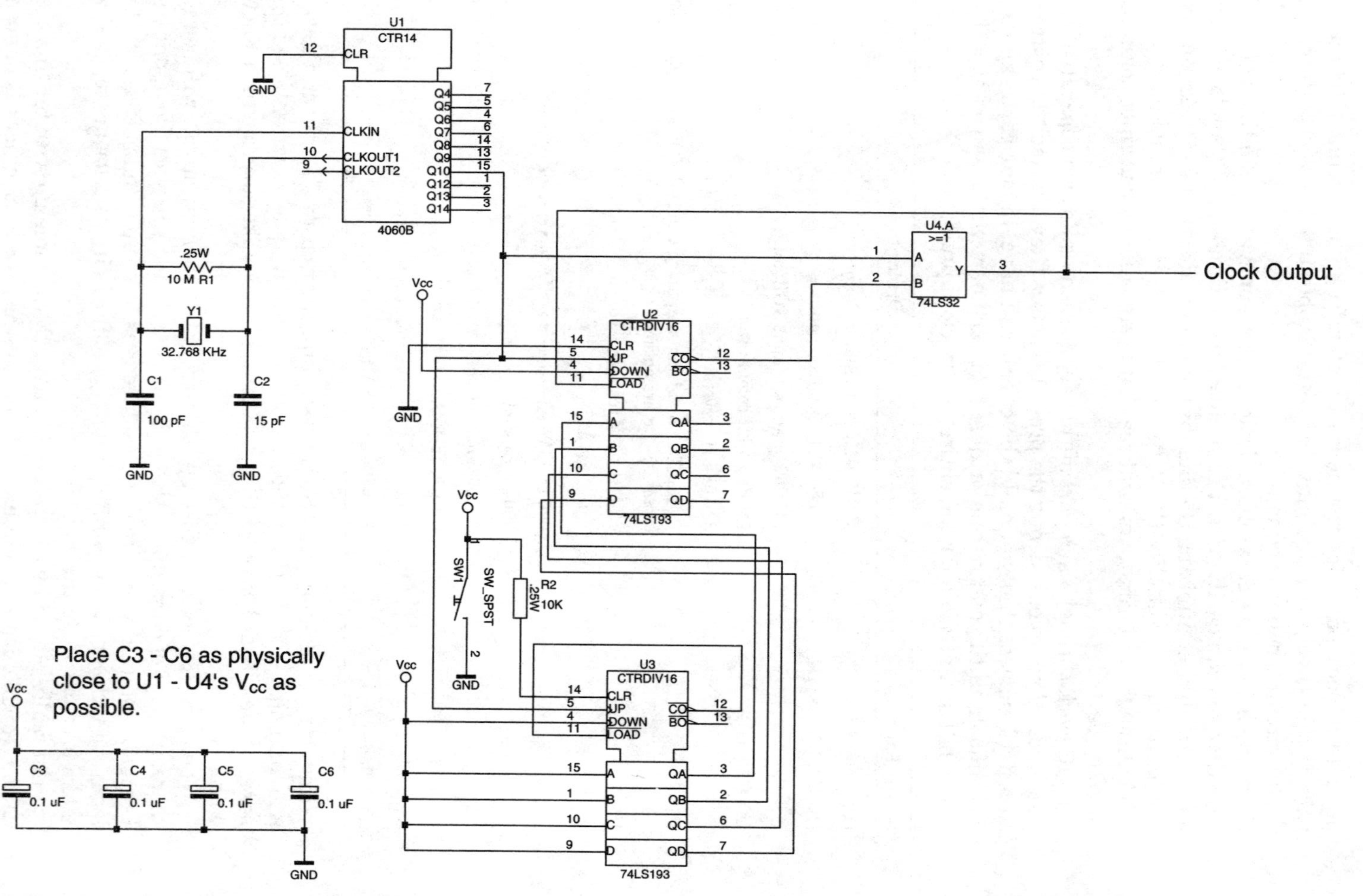

Figure 9.15 Project 8—binary clock clock/time set circuitry.

operation. When the momentary-on push button (SW1 in Fig. 9.15) is pressed, the update speed will increase to a 32-Hz signal that will increase the update rate of the clock 16 times and is used to set the time on the clock.

Looking at Fig. 9.15, you will probably not see how the time set operation works. The actual operation is quite simple and takes advantage of how the 74LS193's operation can be "manhandled" into doing something that it wasn't originally intended for. Looking over this project, it could be used as a tutorial in the different modes of operation of the 74LS193, as I use the seven chips built into this project in four different ways.

When the time set push button connected to U3 is released (circuit open), the 10k pull-up on the clear pin of U3 will cause it to *always* output all zeros. This is true even if one of the clock inputs is being toggled. These four U3 output bits are used as the reload values for U2. U2 is wired so that when the chip overflows (counts from 0b01111 to 0b00000), it will be reloaded with the values at the A, B, C, and D inputs. The U2 inputs are the same as the U3 outputs, and while the button that is released will always be 0b00000, which means that it will count from 0b00000 to 0b01111, reloading after it counts over 0b01111 and overflows. The U2 overflow is used as the time base counter for the rest of the binary clock circuit.

When the button on U3 is pressed, the clear pin is held low and the chip can start counting. The current count value is passed to U2 as its start counting value, as discussed above. When the count reaches 0b01111 and overflows, the load pin causes 0b01111 to be continually reloaded into U3, and the output value never varies from 0b01111, at least as far as U2 is concerned. When U2 overflows, it picks up the output value of U3 and starts counting to 0b01111 and overflows from there. This circuit allows the binary clock timer to run between 2 and 32 Hz.

When I originally designed this circuit, I used a different method of increasing the count frequency for setting the time and for the basic counts: I used the Q14 output of the 4060B (U1) chip. Q14 is the clock frequency divided by 2^{14}, which, in this case, is 2 Hz. This meant that to delay 1 minute, before incrementing the "minutes" counter, the circuit would have to delay 120 times (2×60). To count down from 120, I used the circuit shown in Fig. 9.16 to count down from 120, and when the count underflows, 120 is reloaded into the 74LS193s of the circuit and an increment pulse (minute clock) is passed to the minute counter circuit.

When I created my prototype circuit, I used the outputs of the two 74LS193s (U5 and U6) to drive eight LEDs in a 10-LED bargraph display. This let me monitor the operation of the clock circuit (and the speed-up time set) and predict when the minutes would be incremented. I have to warn you, this is a confusing display and you will probably have trouble figuring out absolute values of what is being presented. I found it confusing that each time the number counted down, *more* showed up in its place; this is probably a cryptic statement, but you will see what I mean when you build the circuit.

I suggest that you use the LEDs for debugging purposes, but if you build a display clock using this circuit, I recommend that you avoid these LEDs as they are quite confusing (for you as well as people who don't understand how binary numbers are represented).

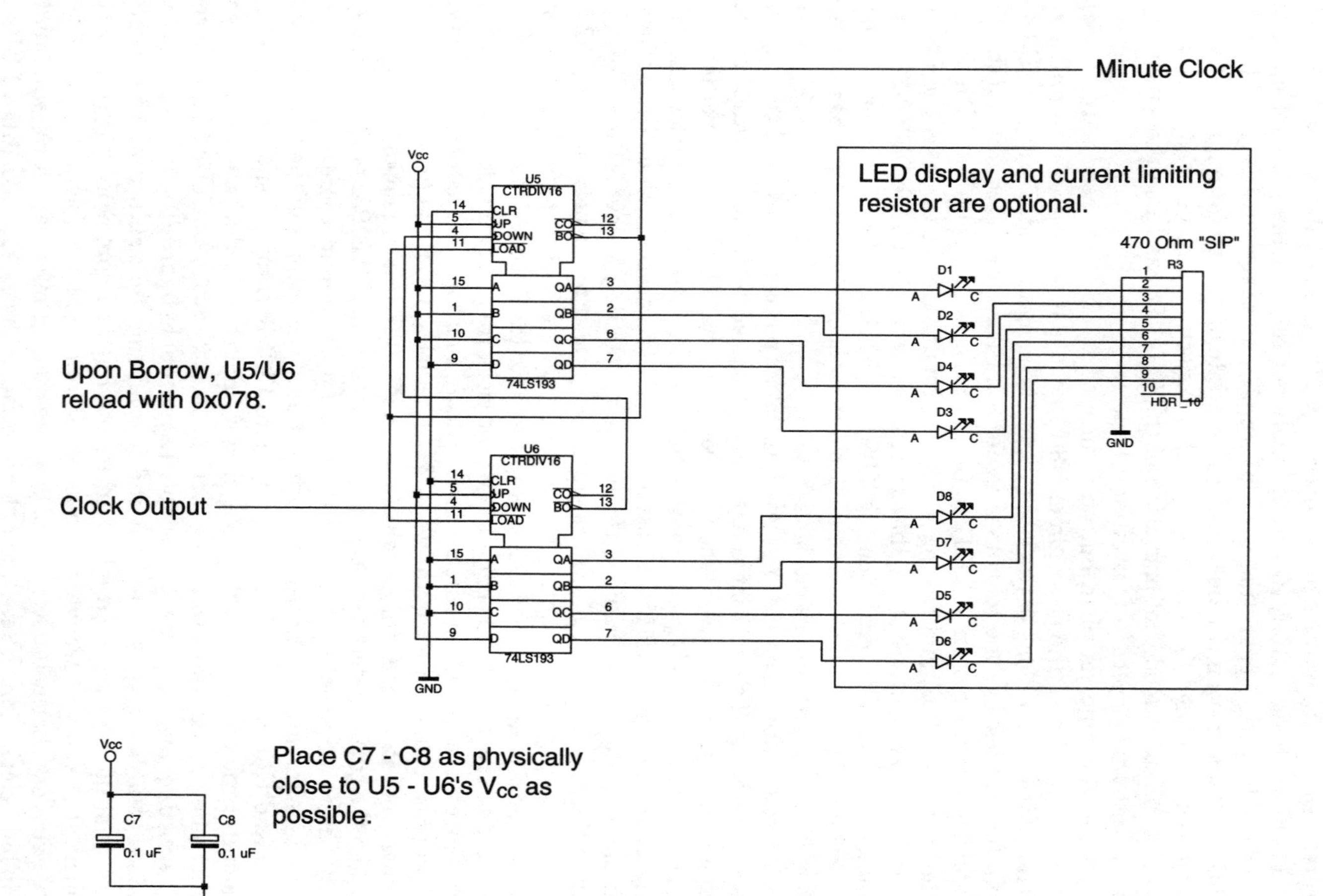

Figure 9.16 Project 8—binary clock seconds countdown circuitry.

The underflow output is used to increment the minutes counter, which has the circuit shown in Fig. 9.17. This circuit will increment the two 74LS193's each time the minute clock signal comes from the seconds counter until the value reaches 60 (0b000111100). The minute value is displayed on eight LEDs (although only six are necessary) and continuously compared.

When the minutes reach 60, the comparator equals value becomes active (positive) and is used to reset the minutes counter. To an observer, it will appear that the clock counts from 59 to zero.

The two 4-bit 74LS85 comparators in Fig. 9.17 are cascaded together to form a single 8-bit comparator. For the least significant 4-bit comparator, I have assumed that the comparisons of less significant (imaginary) bits are equal, and I have tied the inputs accordingly. The result values are passed to the next significant 4-bit comparator for use in its own comparison operations. The output of the most significant comparison is used to determine whether or not the minute value is less than, equal to, or greater than 60.

The minutes counter value is used to increment the hours value, but probably not in the way that you would expect. The 74LS193 counts up or down when it receives a *rising edge* on the up and down pins. Since the two 74LS193 minute counters never overflow or underflow, I could not use these outputs to toggle the hours. To get the correct signal to the hours counter, I used the less-than output of the comparators instead. For most of the binary clock's operation, the comparator less-than output is active. When the minute counter is equal to 60, the less-than comparator output is inactive (low) and then goes high again when the minute counter is reset to zero. This signal is passed to the hours counter (Fig. 9.18).

The hours counter is probably the most straightforward circuit of the entire application. The circuit simply consists of a 4-bit counter that is driving four LEDs and a comparator. The comparator resets the hours counter when it reaches 12 (0b01100).

When I built my prototype, I started with the clock circuit and worked my way up in terms of the amount of time that was being kept track of and displayed. This allowed me to debug the circuit as I went along, and when I was done, there was just some minor tuning that had to be done to the circuit. I want to go through the problems that I encountered and how I discovered what they were and how I fixed them.

The first problem I had was setting up the second counter reload value. When I first built the circuit, I discovered that the power supply "crowbarred" and turned itself off (indicated by the LED not lighting). Using my digital multimeter, I discovered that somewhere on the board, there was a short circuit that seemed to be between V_{cc} and Gnd. An intensive visual scan of the board didn't point me toward the problem (the second counter is quite a complex circuit, and it is difficult to see through all the wires to find the problem), so I used a trick that I am somewhat reluctant to share with you.

The reason for my reluctance to share this method is that it can be used only in circumstances like this, where a number of breadboards are used with

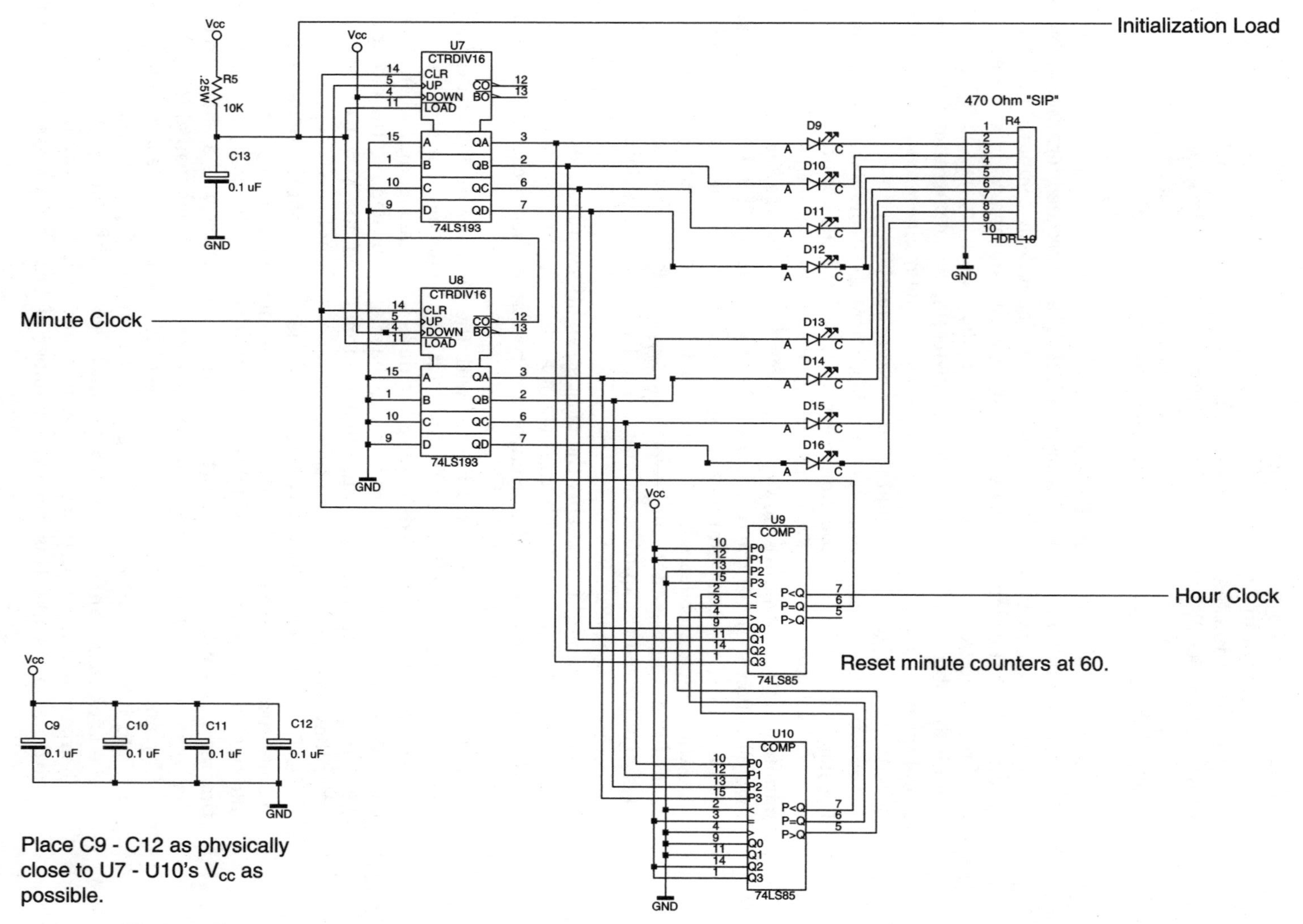

Figure 9.17 Project 8—binary clock minute display and compare.

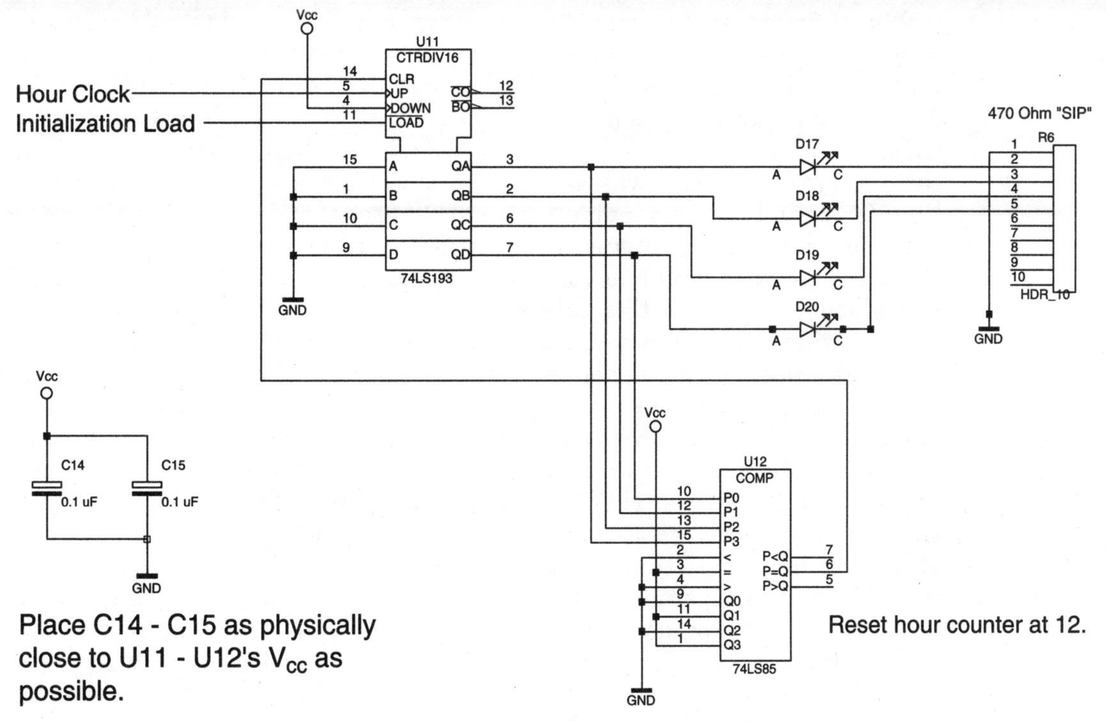

Figure 9.18 Project 8—binary clock hour display and compare.

multiple bus connections. Pulling the power supply off the breadboard, I found that the apparent power supply resistance of the circuit to be 1.3 Ω. I then went to each of the four sets of bus bars and measured the resistance between each one and discovered one set that had 0.1-Ω resistance while the others had 1.0-Ω resistance or greater. I went through all of the tie-ups and tie-downs on the bar until I found a V_{cc} and Gnd pair that were plugged into the same strip. Once the tie-up and tie-down were wired correctly, the problem went away.

Remember that this method—even though it helped me find my problem quite quickly—can be used *only* in this type of situation. Measuring the resistance of a PCB I happened to have lying around, I found that its resistance is (end-to-end) 0.3 Ω on the Gnd traces. The opportunity for finding a V_{cc}/Gnd short using an ohmmeter in this situation is very remote—chances are you would be better off looking for a "hot spot" in the board while it is powered up rather than for high/low-resistance situations.

I was successful in using an ohmmeter to find the short in this case because all of the bus bars were connected together by a series of wire connections. These wire connections provided extra resistance and allowed me to find the actual problem as well as separate it to specific parts of the breadboard. In most other situations, you will not be this lucky and trying to find a short by using this method will be an exercise in frustration.

When I first started getting the seconds counter working, I found that, occasionally, the upper 4 bits (the high-order 74LS193) would not reload when the counter underflowed. After searching through the wiring, I decided to see if a 74LS193 caused the problem that was out of specification. The reason for considering the chip (especially when I have said that there is virtually no such thing as problem silicon) was the very short clock pulse passed to the chip (less than 30 nsec). I tried swapping U5 and U6 to see if the problem would move to the lower 4 bits, and after exchanging the two chips the problem went away.

I tried a number of other chips in this position and they all worked perfectly. I was about to discard this chip when I decided to check one last thing—remembering Project 7, where I had a problem with the HC part, I decided to look at this one, and sure enough, it was a 74HC193. The reason why I didn't pick it out before was that it was built by the same manufacturer as all the other 74LS193s in the circuit and had a similar style of stamping on the top.

I don't understand why, but when I bought a tube of twenty 74LS193s, I received two 74HC193s. This is actually a common problem and one that you should always be aware of—somebody in a factory doesn't have quite enough of one part number, so they substitute in a few parts "that should be identical" that they have lying around. The lesson in all this is to always check the basic problem causes before assuming you have a bad chip or circuit design.

When the circuit was running, I discovered that, sometimes, the high 4 bits of the minutes counter would not be incremented at the correct time. When I traced through the circuit, I discovered that the clock increment pulse (from U2) was becoming unacceptably short as it (virtually) traveled through U6, U5, and U8 on its way to U7. A lot of this "characterization" had been done for the previous problem.

The correct fix to this problem would be to put in a *single-shot,* like the 555 timer in monostable mode (discussed in the previous section). I didn't want to because (1) I didn't have a CMOS 555 handy and I was afraid the 150-mA transient current would affect the 4060B oscillator operation and (2) it was 11 P.M. and I wanted to see the clock working properly before I went to bed.

Earlier in the section, I noted that when the 74LS193 clear pin is active, the 74LS193 stops counting and loads in 0b00000. I should also point out that load will stop the 74LS193 and set the output to the input value. When clear and load are active, the carry and borrow outputs (indicating an overflow and underflow condition, respectively) are disabled (made high). For applications like this one, the carry and borrow outputs are used to reload the 74LS193 and reset them according to where they are used.

The operation of the 74LS193, as I have outlined with respect to clear and load, means that if the carry or borrow outputs are used as downstream clocks, you will discover that the actual pulse will be very short.

The solution to this problem was to lengthen the pulse by combining it with the clock output from the 4060B in an OR gate. This is the single 74LS32 in the application and it lengthens the negative pulse to about 100 nsec (from 30 nsec) and is passed quite accurately through all the 74LS193's in the chain. This solution was found by a combination of experience and the knowledge that you can use gates to lengthen pulses in a similar manner to delaying them, as I show in the next section.

I guess you could say that this is somewhat of a kludgey solution to the problem, but it does work quite well in my prototype circuit. You may find that when you build this circuit, you don't need the gate inserted here [and in this case, U2's carry output (pin 12) can be used to reload U2 in addition to being passed directly to U6 as the clock output]. You may also find that it is not enough for the parts that you are using for your application. If I were designing this circuit for production, I would probably use single-shots (like the CMOS 555) for this type of application to make sure that the clock signals were always long enough to ensure the circuit will run reliably, no matter what parts are used.

As I write this, I have been running this clock for roughly a full day. Over this 24-hour period, the clock has run fast by about 4 minutes. This translates to an error of 0.278 percent and is somewhat more than I would expect for an application that uses a part designed as the time base for commercial clocks. I suspect the error lies in my choice of resistor and capacitors that are used along with the 32.768-kHz crystal. To get better accuracy, I should probably replace the two capacitors with variable 10- to 40-pF capacitors and adjust them until I find the best accuracy for the clock.

For most applications, this error is not horrendous (as I will show in the next chapter), but for a clock application, it is unacceptable. If I were designing a clock product, I would probably go back to a 4060B and ask crystal manufacturers what are the best support components to use with the crystal and what suggestions they have on PCB layout. The goal for production is to design products that do not have to be tweaked on the manufacturing floor.

A few weeks of talking to suppliers could result in a product that is extremely accurate and does not require any kind of frequency testing in order to know the exact frequency the product is running at before it goes out the door.

When you build the circuit, there are probably a few enhancements that you would want to consider to make it work better and make the display output somewhat more accessible to the average person.

The first suggestion that I have is to change the 4-bit clock set circuitry to 8 bits. Right now, with 4 bits, when you are setting the time, you will increment each minute in just under 4 seconds, which means that it will take you 3¾ minutes to increment an hour. It will take up to 45 minutes to set the time anywhere within 12 hours.

Extending the counter to 8 bits will reduce the maximum speed for updating a minute to about a quarter of a second, an hour to 14 seconds, and a full 12 hours to under 3 minutes. These times would be much more acceptable for people.

The second enhancement that you might consider is to use seven-segment LED displays instead of the bargraph displays I have used here. The change would not be as difficult as you might expect—the 74LS192 is pin- and operation-compatible with the 74LS193 but works with BCD (digits 0 through 9) instead of hex (digits 0 through F). In the minute circuit, U7 and U8 would be changed and the inputs to the comparator would be decimal 60 instead of hex 0x03C. The hour update would be a bit more substantial, with a second 74LS193 having to be added with an additional 74LS85 to be able to handle both potential hour digits. For the hours counter, I think I could avoid the second 74LS192 and 74LS85 by using a 74LS74 dual flip-flop to drive the hour's 10 digit (which is either off or 1).

To display the time on seven-segment LED displays, a 74LS47 could be used for each digit. I will be working with this chip in the next chapter, but it is quite a nice chip that will simplify your applications quite a bit.

If you perform the two suggested enhancements to this project, you will be adding either six or seven chips to it and increasing the wiring complexity accordingly. At the start of this section, I noted that the function provided by the first digital watch chips was startling for most people. With the enhancements I have suggested for this project, the binary clock will actually be very similar in capabilities to the original digital watch chips. I find that creating circuits like this, out of simple TTL chips, that simulate the operation of more complex chips really gives me an appreciation of how amazing modern digital electronics really is.

Time Delays

It seems like every week you can buy a new PC that is 50 to 100 MHz faster than what you could have bought the week before. Other electronic system vendors push the idea that their systems (such as video games) are faster and capable of doing more in a given amount of time than their competitors. Throughout this book, I have been talking about the issues of trying to minimize the number of gate delays in your circuits to allow them to execute as quickly as possible with a minimum of interference to the other circuits they are connected to. With this background, it is going to seem surprising that I am going to discuss some methods of *slowing down* your digital logic circuits.

With this introductory paragraph, you're probably thinking that for your product to be successful, you have to work at it being as fast as humanly (and electronically) possible. For some circuits, such as computer processors, this is a very reasonable perspective to look at when designing the circuits.

For most circuits, however, the need for speed is not as critical and can actually be detrimental in many cases (especially when interfacing to other devices and people). In these cases, the data signals will have to be slowed down in order to properly communicate to another device. An example of this is when a computer processor reads and writes to memory or I/O devices. As I will discuss later in the book, first a device address has to be passed to the peripheral devices followed by a control action (read or write), at which point data is transferred between the device and the computer processor. This process is shown in Fig. 9.19.

From the processor's perspective, the data transfer can take 1 clock cycle. For modern Pentium IV processors, this clock cycle can be as short as 500 psec or less. This clock period is much too short for virtually all external memory

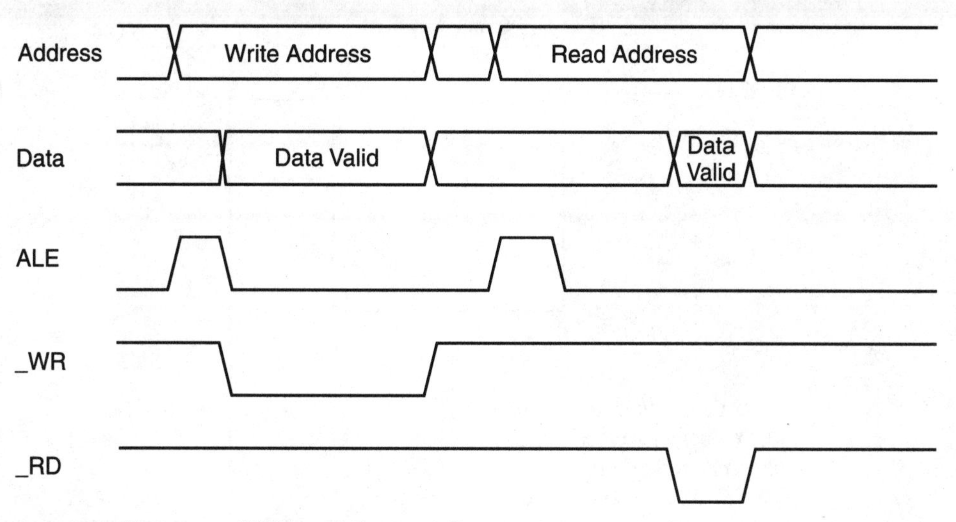

Figure 9.19 Memory and I/O bus timing waveforms.

and I/O devices on the market today. To allow processors to interface with much slower devices, the address and data are passed between the two devices on a bus using some type of protocol (like the one shown in Fig. 9.19).

As an example of bus data transmission protocol, 500 nsec is required for data to be transferred between the processor and the external device. The address latch enable (ALE) pulse (which indicates to the external devices that the address on the bus is valid) is 125 nsec long, with the data from the processor being present (and _RD line active) on the bus for the next 125 nsec of the cycle. For writing data, the data is output from the processor as soon as ALE has finished being asserted and is present on the bus for 312 nsec. The last 62.5 nsec of the read cycle is used as a "dead space" for the processor and peripherals to reset themselves for the next operation.

This seems like a very simple data protocol (and it really is). The question now is, how would this be implemented?

In the protocol that I have laid out, the ALE (address latch enable) line is used to indicate that the address on the bus is valid and should be stored and the decode process should be initiated. This may seem like an overly complex feature in a bus implementation, but it is a very common feature in many computer systems (the IBM PC's ISA bus has an ALE pin to indicate that the address is valid). Assuming that the processor initiates the read and write, the problem now arises as to how the 125-nsec ALE pulse is asserted and timed, along with the other features of the bus read/write.

One method of solving this problem is to create an overspeed clock that provides the correct timings for the bus. In Fig. 9.20, I have shown an 8× clock (16 MHz for this example), which drives a counter that outputs a number of signals that can be combined into the necessary signals. For example, to produce the write timings, the following equations for ALE and _WR could be used. (Along with these three timings, I will have to include another one,

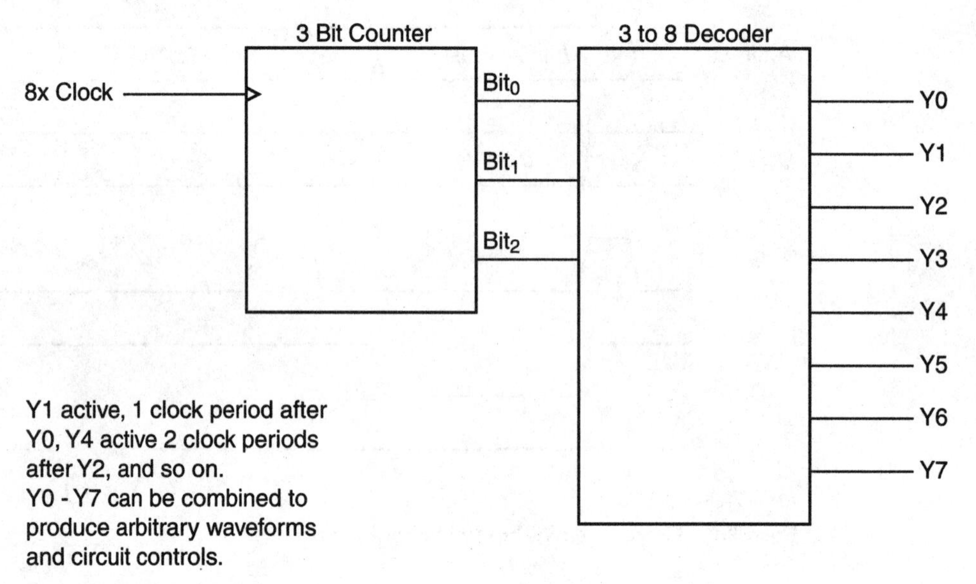

Figure 9.20 Clocked circuit delay.

_WRen, which enables the processor's data drivers to continue to drive out valid data when the _WR line becomes inactive in case the peripheral device stores data on the rising edge of the _WR line.)

$$ALE = (Y0.Y1) .! (_Write)$$

$$_WR = (Y2.Y3.Y4.Y5) + _Write$$

$$_WRen = (Y2.Y3.Y4.Y5.Y6) + _Write$$

For the read timings, the two bus signals would be implemented as:

$$ALE = (Y0.Y1) .! (_Read)$$

$$_RD = Y6 + _Read$$

In these equations, note that I have assumed that the operation takes place when the internal _Read and _Write signals are active (low or equal to zero). When these signals are active, the outputs can be properly asserted.

There are three problems with implementing the bus by this method. The first should be very obvious: The circuit would be very difficult to properly time (and you should realize that I have greatly simplified the operation of reading and writing by a computer processor). In the gate equations above, each of the signals, except for _WR and _WRen, can be governed by a 3-gate-delay equation. _WR and _WRen will require a minimum of 4 gate delays, which means that the timings of the other circuits have to be matched to provide accurate timings.

It is interesting to note that, although I originally stated that the read/write cycle was 500 nsec, I have taken away 187.5 nsec for overhead functions, leaving only about 310 nsec for the actual data transfer to take place. This shorter time period is the value that you should note when looking at microprocessor datasheets to ensure that the memory will work in your application.

The second problem is the increased internal complexity of the processor when this method is used. With this timing circuit, the 3-bit counter will have to be synchronized with the processor to ensure that the data signals are active at appropriate times. Depending on the processor, this can be a real problem and will probably restrict the operation of the processor to executing at the external bus speed, rather than the speed that the logic is capable of.

The last problem with this method is that, with an 8× clock, you will be increasing the current/power consumption of the circuit significantly. This is especially true for CMOS circuits. Using phased-locked loop (PLL) clock multipliers, the clock can be generated internally to the circuit, but this adds complexity to the circuit and still results in a higher-than-expected power dissipation level.

Despite these three problems, this method of producing timing delays is often used in microprocessors and other circuits for the simple reason that if the clock speed of the device is changed, then the external device timings change proportionally. This may not seem like an important point, but it can be very important for application designers who want to understand exactly how a chip will behave in a given set of circumstances.

Another method of providing delays is to use a "canned" delay line. These components usually consist of an inverting buffer driving a long copper line. At different points along the line, inverting taps are put in place to drive out the signal. Figure 9.21 shows how these components are implemented.

When you see an actual delay line component, you will probably refer to it as a *chip*. I hesitate to do so because I consider a chip to be simply a silicon

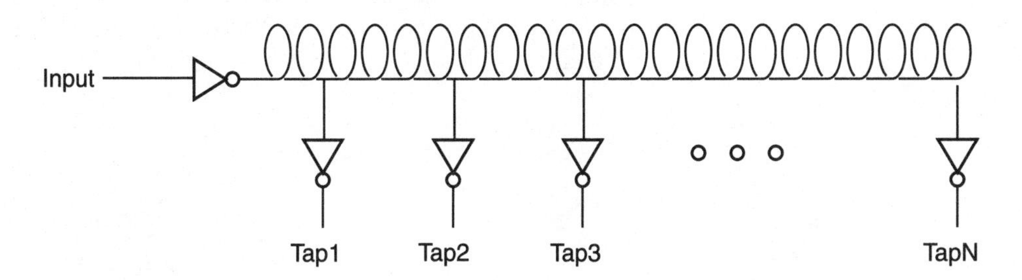

TapN delay = (2x inverter gate delay) + (1 nsec/ft * coil length)

Figure 9.21 Internal circuity of a delay module.

chip bonded to a lead frame and encapsulated in some manner. In a delay line component (or *module*), the wire delay is wound in a coil with the tap inverter inputs soldered to it at different intervals. The actual device requires very high precision mechanical assembly (more than a standard plastic encapsulated chip), as any errors in assembly or encapsulation will result in a useless part.

The wire in the part passes the digital signal to the various taps within the part with a delay of roughly 1 nsec per foot (30 cm) of wire. This rule of thumb is very important when you are designing high-speed digital electronic circuits. Chances are you will ask that the traces on PCBs be routed with 0.1-in (2.54-mm) precision to ensure that parallel signals all show up at the same place at the same time in the high-speed application circuit.

The advantage of the delay line module is that timing delay can be very precise. Custom-made delay line modules are available (the manufacturer solders the taps at specified points in the coil rather than at standard positions, which can be critical in some applications).

This high level of assembly/encapsulation precision has a price that you will have to pay. If you can buy a 74LS04 for less than a quarter in single units, you should not be surprised to discover that a delay line module will cost you over $10. The delay line provides you with the best control over different delays required in a circuit, but at quite a significant cost. Delay line modules should be considered only if no other options are available to you when you are designing a circuit.

In the original example of delaying signals by using an overspeed clock, substituting delay lines can have advantages as the delays do not change with the circuit's clock speed. This is important for peripherals that cannot tolerate a change in their interfaces.

The last method of delaying signals is to take advantage of the natural delays of digital electronic gates and simply chain a number of them together to get a needed delay. In Fig. 9.22, I have shown a 20-, 40-, and 60-nsec delay built out of a 74LS04 TTL chip.

The advantages of this method are that it is quite low in cost and reasonable precision can be built into the circuit. When you are designing delays for your applications, you should consult the chart on the next page, which shows the different gate (propagation) delays for various 74 logic families:

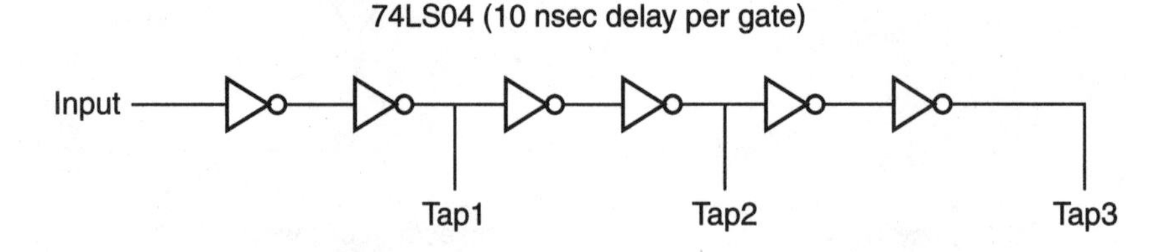

Figure 9.22 User-devised 20-nsec-per-tap delay line.

Family	Typical delay, nsec	Maximum delay, nsec
74AS	2	4.5
74S	3	4.5
74AC/74ACT	3	5.1
74F	3.5	5
74ALS	4	11
74H	5.9	10
74HC	9	18
74LS	10	15
74	11	22
74L	35	60
74C	50	90

Working with different technologies, you should be able to get quite accurate delays quite inexpensively. The disadvantage of this method is that it can take up a lot of space on a board (if the board gets too crowded, the canned delay line will have to be used).

Push Button Switches

As I present and discuss more complex applications, I will be relying on user input more and more. For most applications, this will consist of button input (i.e., press a button to initiate an action). If you are like most people, you will expect that each time the button is pressed, it will produce a clean "step" waveform like the one shown in Fig. 9.23.

Unfortunately, life is never this simple or clean. When a button is pressed, the contacts within the switch can literally bounce on and off each other several times before the button is in constant contact. Figure 9.24 shows an oscilloscope picture I took of a traditional button input circuit (a momentary-on push button with a 10k current-limiting pull up). This circuit is very similar to the two "LED9" and "LED10" that are built into the Interface PCB that comes with this book.

You should also note that releasing a push button could have the same bouncing problems as pressing it. Both bouncing problems can be reduced by using positive-click, self-cleaning buttons. You are probably familiar with these switches, as they produce an audible click when they are pressed or released and

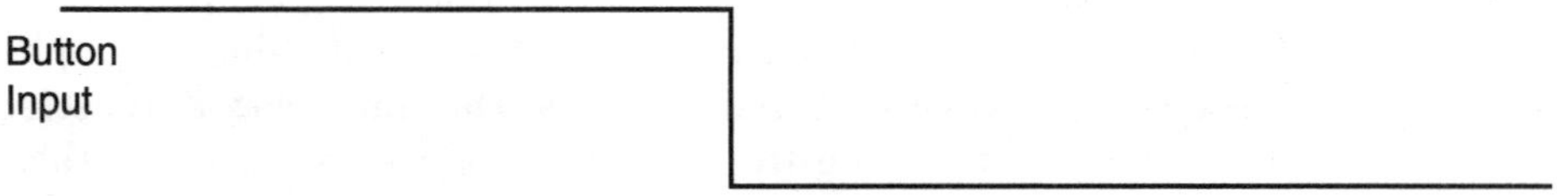

Button
Input

Figure 9.23 Idealized switch operation.

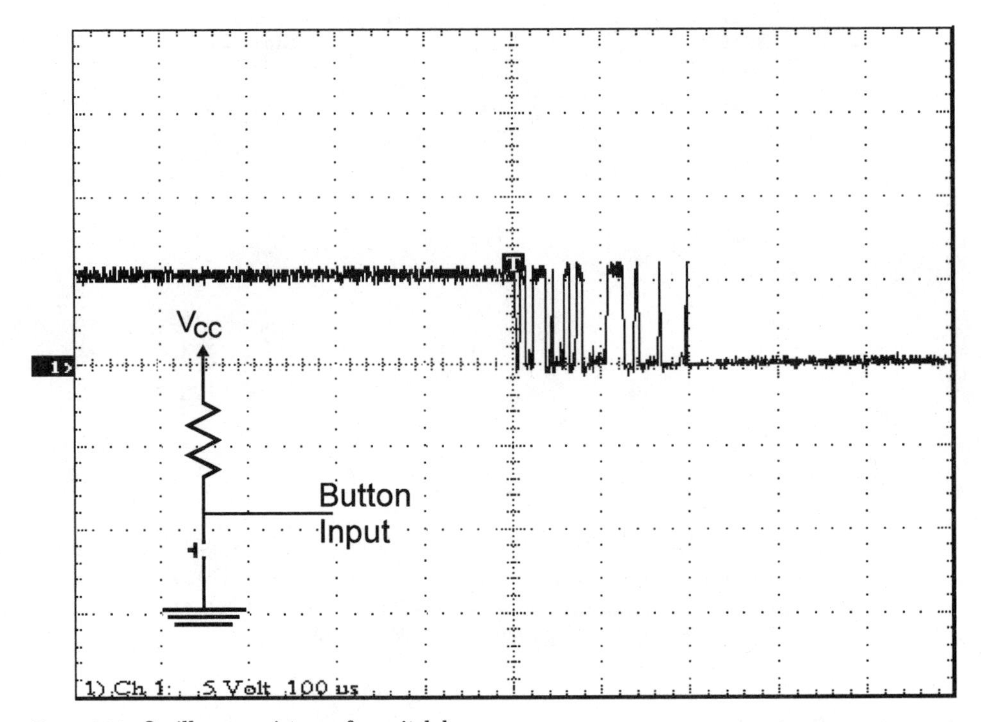

Figure 9.24 Oscilloscope picture of a switch bounce.

seem to almost break when you are pushing them. Even if you use these types of switches in your application, you will still have to come up with some way of debouncing the inputs, as you will not totally eliminate the problem.

In my microcontroller books, when I have discussed the issue of debouncing buttons, I did it almost exclusively from the perspective of implementing a *software debounce* in which the processor polls or monitors the state of the button. In the software debounce, the button state is declared "debounced" only when the button has not changed state for 20 msec. A *hardware debounce* is much more difficult to implement, as there is no processor to intelligently determine whether the button input is debounced and can be acted on. Over the years, I have found three methods of debouncing buttons that I feel comfortable with: (1) use a Schmitt trigger circuit, (2) use a 555 CMOS timer, and (3) avoid the need for debouncing entirely. These methods are quite simple and reliable.

When I looked around at different references to see what I haven't been using for button debouncing, I found a lot of circuits that involved "latching" the state of the button and then only "unlatching" the state when the action has completed. This is essentially implementing a set/reset (SR) flip-flop with the set input being the button and the reset input being the application circuit.

The problem with this method is that the time needed to perform the operation can take anywhere from microseconds to minutes. If the user has not released the button before the circuit has attempted to reset the flip-flop, the flip-flop will return to the set state and the circuit may repeat its operation as if another button press was made. Ideally, with any kind of flip-flop method, the button should be polled to determine whether the user has released the button before the next action is initiated. Unfortunately, adding the circuitry to poll the button adds to the complexity of the circuit and can be difficult to design.

The simplest circuit that I use for debouncing a button input is shown in Fig. 9.25. The Schmitt trigger inverter, coupled with the *RC* network, will filter out most of the bounces before the button press is registered as active. When the button is pressed, the capacitor will start to discharge through the 100-Ω resistor.

If there is any bouncing, then the discharge operation will stop and the capacitor will even start to recharge. The Schmitt trigger will provide the lowest possible voltage threshold for triggering on the capacitor's voltage. For most applications, this debouncing circuit is all that is required, and I will often put in a few 74LS14's (the Schmitt trigger version of the 74LS04 hex inverter) so that I can add debounced switches without a major impact on the application's circuit.

You will notice that the _RST output of the interface PCB that comes with this book has a button input that is similar to this one (but the 74HC04 that is specified to work with the application is not a Schmidt trigger device). This button input should work fairly well as debounced button input, but you may still find situations where you do not get a simple change to the input but multiple bounces instead.

This switch works best for buttons that do not have a lot of bouncing to begin with. If you are using buttons that do have quite a bit of bouncing, then I would use a (CMOS) 555 timer set up as a monostable output, as shown in Fig. 9.26. In this circuit, when the button is pressed, a 100-msec pulse is output from the 555 timer that can be used to initiate an action within the application. Instead of a 555 chip, you can also use the 556, which consists of two 555s in a single chip package.

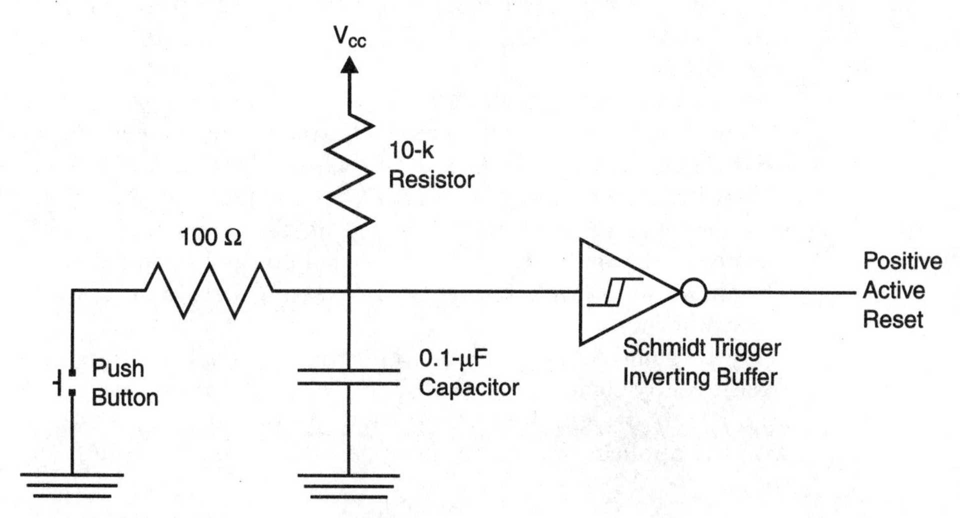

Figure 9.25 Simple push button debounce circuit.

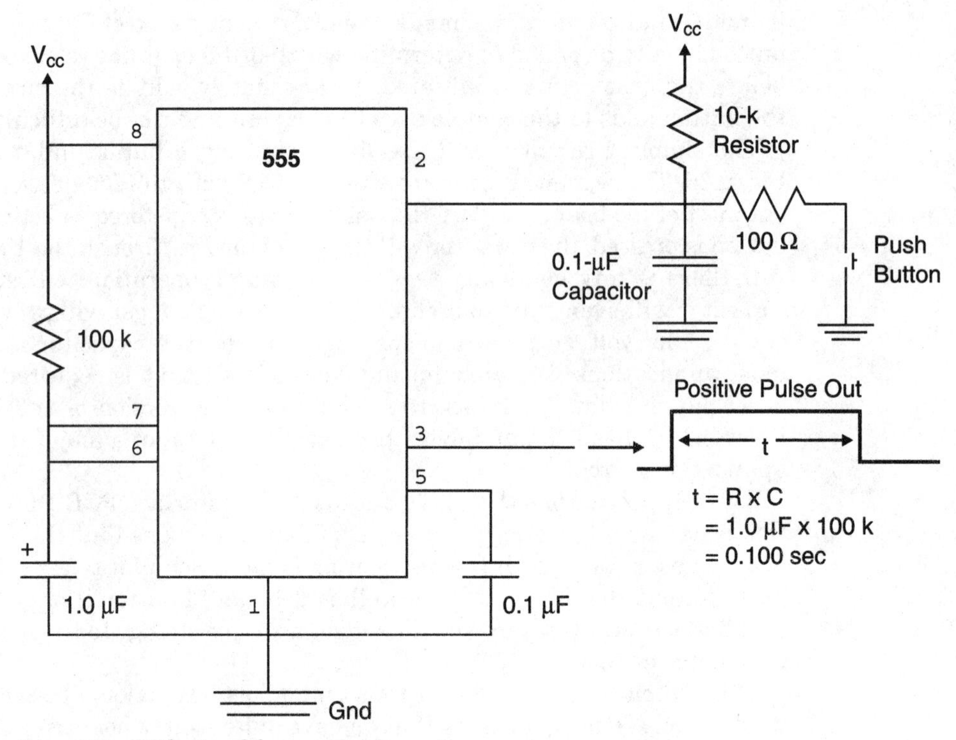

Figure 9.26 555 button debounce circuit.

Using the "single-shot," or monostable, output is probably the best way of implementing the debounced button input. Most TTL chips are designed with built-in Schmitt trigger inputs, so you can take advantage of the operation of the previous method of debouncing as well as have the single-shot for making sure that the button being held down is not interpreted as multiple button presses.

The most effective method I use for debouncing buttons is to design applications that do not need debounced buttons. This may sound more than a little strange (and "Zen"), but if you take a look at the binary clock project earlier in this chapter, you will see that I implemented a time change button that does not have to be debounced at all. When the button is pressed, a counter is allowed to count up and speed up the incrementing of the minutes. The circuit works well, has a very interesting feel about it, and was quite easy to get working.

My personal feelings about implementing push buttons in a digital logic circuit really come under the KISS ("Keep it simple, stupid") philosophy of application design. Remember that people are impressed with reliable, working applications—what goes on under the cover really doesn't concern them.

Project 13—Bounce Counter

With the theory of bounces, I wanted to take a look at a number of different ways of passing button input to a digital circuit and demonstrate bouncing actually happening. In doing this, I will show how button bouncing can affect applications and what you can do to prevent it. As I indicated in the previous section, the best way to avoid button bouncing problems is to design your applications in such a way that button bouncing is not a factor in the circuit's execution.

To show switch bouncing, I will use a 74LS193 counter along with a 74LS04 inverter to pass negative active button presses as positive active pulses to the 74LS193 counter chip. The 74LS193 counts up when it receives a rising edge, but by inverting the negative active button input, I should be counting once for each time the button is pressed. The basic circuit that I used is shown in Fig. 9.27.

In this circuit, I use one of the two momentary-on push buttons of the interface PCB to clear the 74LS193 when the experiment is finished by loading zeros into the 74LS193. The first four LEDs of the Interface PCB are used for displaying the current count of the 74LS193.

The 74LS193, 74LS04, and the interface PCB are wired on a breadboard as shown in Fig. 9.28. Remember to leave at least 10 rows of space for the button and the 555 debounce circuit that will be used later in the project.

The bill of materials for this project is:

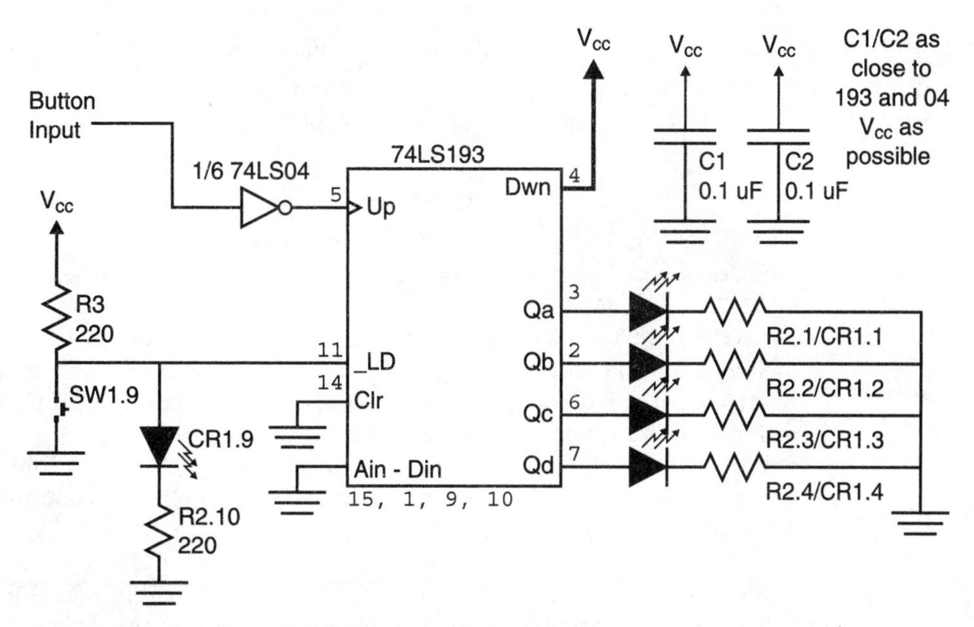

Figure 9.27 Project 13—basic 74193 bounce counter.

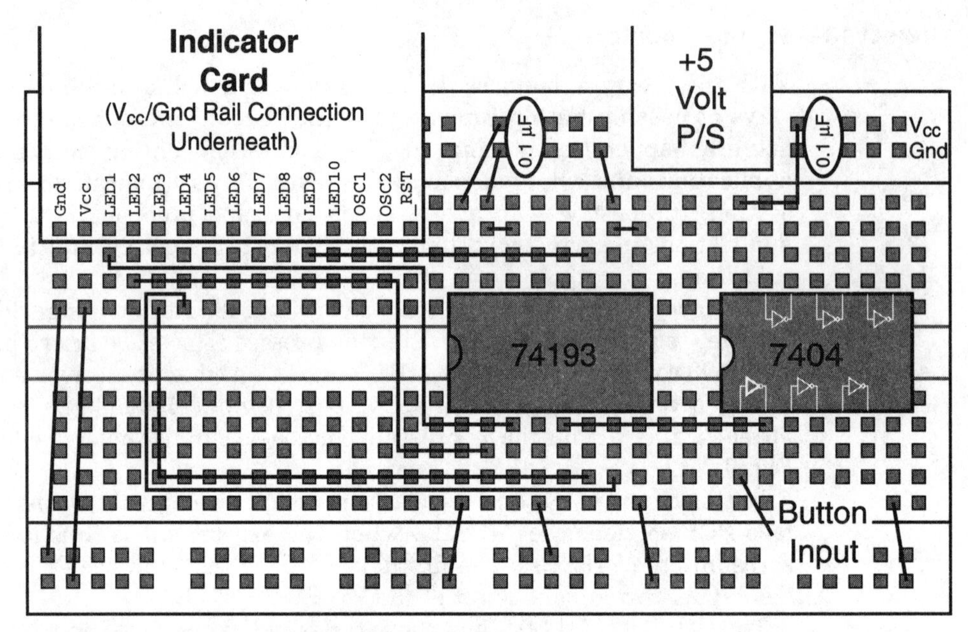

Figure 9.28 Project 13—bounce counter wiring layout.

Part	Description
U1	74LS193 counter
U2	74LS04 hex inverter
C1–C3	0.1-µF, 16-V tantalum capacitor
C4	1.0-µF, 35-V electrolytic capacitor
R1	10k, ¼-W resistor
R2	100k, ¼-W resistor
SW1	Push button momentary-on switch with no click (see text)
C1–C6	0.1-µF tantalum capacitors
Miscellaneous	Breadboard, interface PCB, +5-V power supply, wire

Once I built the basic 74LS04/74LS193 circuit, I added a 10k resistor for a switch input, as shown in Fig. 9.29. This is a standard method of providing button input to a digital circuit. With the counter circuit built, I then tried four different button input methods. I present the results below.

For the first test, I used an old panel-mounted push button that I had lying around. This button does not have the positive click (self-cleaning feature) I discussed in the previous section.

The action of the button seems to be firm, with little chance of bouncing, but when I connected it to the circuit, I found that it was impossible for me to get the counter to increment with a single press of the button. Often, I would find 7 or 8 added to the counter each time I pressed the button. With this high

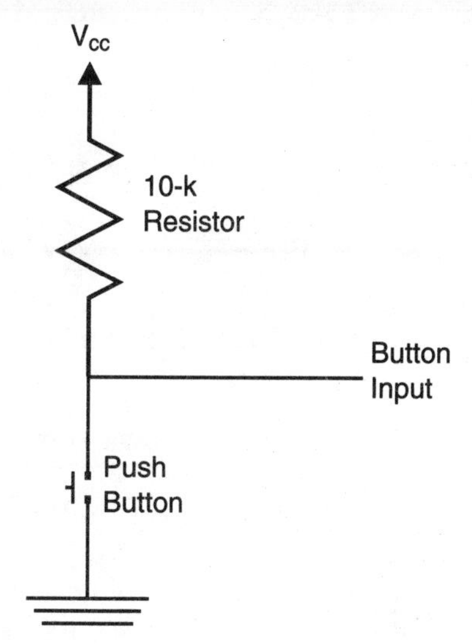

Figure 9.29 Button input circuit used for Project 13.

number of increments per press, it was impossible for me to follow through with pushing the button 25 times and checking the final counter value.

The initial conclusion I was left with was that a simple push button should *never* be used for input into a digital circuit.

Next, I used the remaining clicking, self-cleaning buttons (LED10) on the Interface PCB. While the buttons on the Interface PCB are not exactly the same as the one shown in Fig. 9.29, they do have a pull-up resistor, which they pull down to ground. When I used this button, I found that over the space of 25 button presses, the final result displayed on the LEDs was 0b01110, or 14, which indicated that over the 25 button presses, five extras (five bounces) were recorded. If there were no bounces, the LEDs would be reading 0b001001, or 9.

Next, I used the _RST button on the interface PCB. This button has a similar input, but goes through a low-pass *RC* filter and then through a dual CMOS inverter. The basic circuit is actually this circuit driving Schmitt trigger input instead of the CMOS inverter. I found that in 25 presses, there were two extra counts shown on the LEDs, which surprised me because I had expected there to be no bounces, because of the *RC* filter. If this type of button debouncing circuit is going to be used, then I recommend using a Schmitt input rather than a simple inverter, as I did here.

Last, I used a 555 timer to debounce the button input to a 0.1-sec pulse that was driven directly into the 74LS193 as the pulse is positive, rather than negative. The circuit that I used for this is shown in Fig. 9.30.

As would be expected, after 25 button presses, the final value output on the LEDs was 0b01001 (indicating 25 button presses were counted). I tried pressing the button 100 times, but found that the final counter value was 97, (0b0001, or 1, with an expected value of 0b00100, or 4).

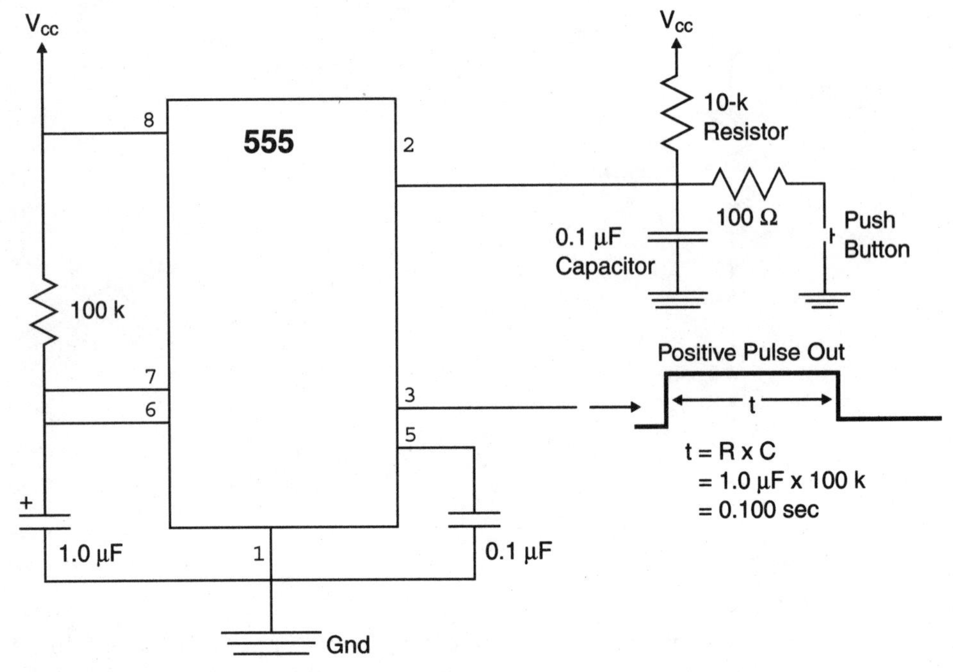

Figure 9.30 555 button debounce circuit.

The logical explanation for this is that in my haste to get through 100 button presses, I either pressed so fast that the output pulse masked the input falling edge and the button press was ignored, or I did not press hard enough for contact to be made.

In either case, I am happy. I would rather have a circuit that occasionally missed a button press instead of having extra ones. For most digital circuits this is not a problem because if I were entering in real data, I would probably have set up input parameters and then be checking the values before going on.

To summarize what I saw with the results, I created the chart below listing the button input type with the kind of results that were seen, along with summary comments. Also note that I sorted the button input types according to cost, with the least costly method being the simple no-click button and the most expensive being the 555 debounce input.

Cost	Method	Bounce results	Comments
Least	Simple button	Many counts per press	This method should *never* be used
	Click button	29 increments/25 presses	Reasonable method for hobbyist projects
	Click button with capacitor and CMOS buffer	27 increments/25 presses	Should have used buffer with Schmitt trigger input
Most	555 debounce	25 increments/25 presses	Best method when accuracy is required

Looking over the results, I want to repeat what I said earlier about the best way of adding button input to a digital circuit is to design the user interface in such a way that button debouncing is not required at all. Debouncing buttons is not something that can be done (cost-) effectively in digital circuits unless a dedicated chip (like the 74C922, which will be discussed later in the book) or an intelligent device (like a microcontroller) is used.

Shift Registers and Synchronous Serial Communications

Most intersystem (or intercomputer) communications are done serially. This means that a byte of data is sent over a single wire, 1 bit at a time, with the timing coordinated between the sender and the receiver. The obvious advantage of transmitting data serially is that fewer connections are required between devices. One of the disadvantages of transmitting data serially is the increased complexity of the two devices in order to convert parallel data to a serial data stream.

The basis for serial communications is the *shift register,* which converts a number of parallel bits into a time-dependant single string of bits, and converts the string of bits back into a set of parallel bits. Figure 9.31 shows this process with 8 parallel data bits being converted into a bit stream and transmitted to a receiver that re-creates the 8 bits back into their parallel data format.

The circuit that converts the parallel data into the serial stream is quite simple. Figure 9.32 shows a circuit (along with an operating waveform) for loading 4 bits into a series of four flip-flops.

These four flip-flops can be driven with data either from an external source or from the next significant bit, depending on the Ctrl bit state. If Ctrl is high, when the Clk (clock) is cycled, the data on the D_0 to D_3 bits is stored in the four flip-flops. If Ctrl is low, when Clk is cycled, each bit is updated with its next significant bit.

In Fig. 9.32, when the Clk is *falling* (transitioning from high to low), the output on Sdata changes. For flip-flops that change state on a rising clock

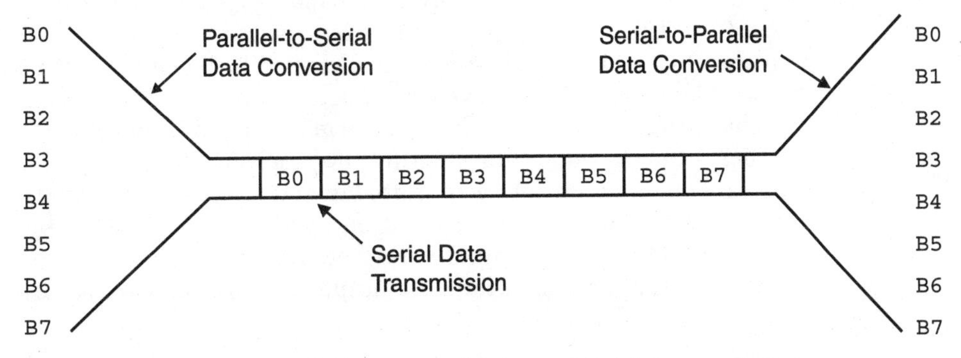

Figure 9.31 Converting a parallel byte into serial data and back into a parallel byte.

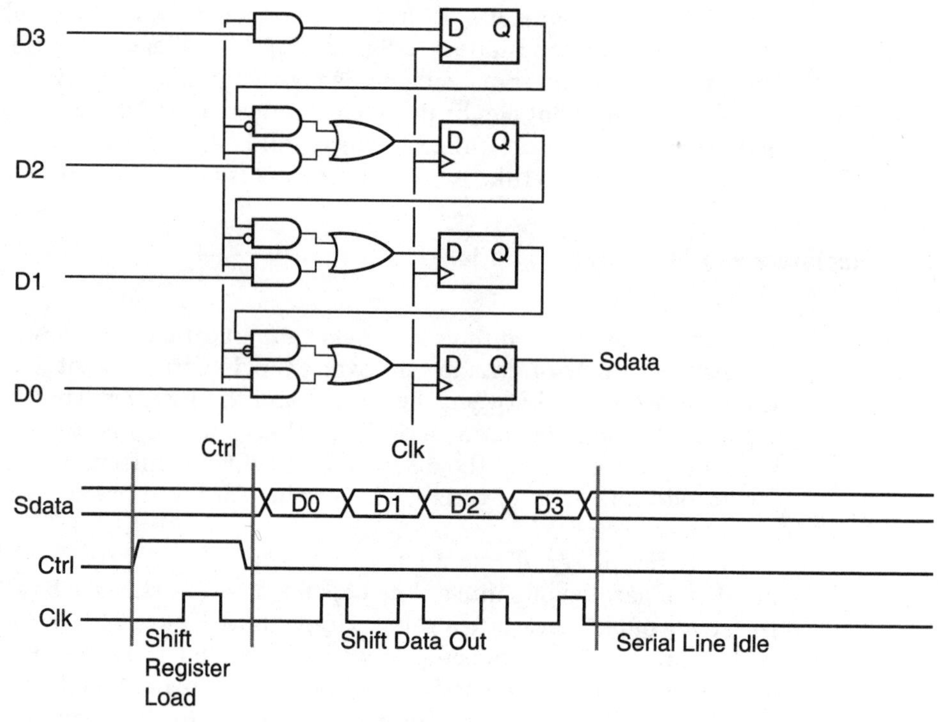

Figure 9.32 Parallel to serial data conversion.

edge, the data would change at a different point. The clock edge the data changes on can be specified, along with the order of the data that passes through the bits.

The process of each bit of data passing through the flip-flops is known as *shifting.* As can be seen in Fig. 9.32, the 4 data bits are shifted out on the Sdata line in ascending order, with the Clk line specifying when a new bit is to be shifted out. If this method were used to transmit data between two digital devices, it would be known as *synchronous serial data transmission.*

Sdata is received by simply using four flip-flops wired as I've shown in Fig. 9.33. The same clock that shifts out the data from the transmitter is used to shift in the data to the receiver. Along with the circuit used to shift in the data, I have included a waveform sheet for you to take a look at.

One confusing aspect of the waveforms is my use of the "*DoX*" convention to indicate the previous values within the receiver. These bits will be shifted out in the same way the data was shifted in and can cause problems if the circuitry attached to the receiver operates on specific values output by the receiver, which could be intermediate values occurring between two transmissions. If this were the case, then the data output to the circuit would be changed only after each bit has been shifted in.

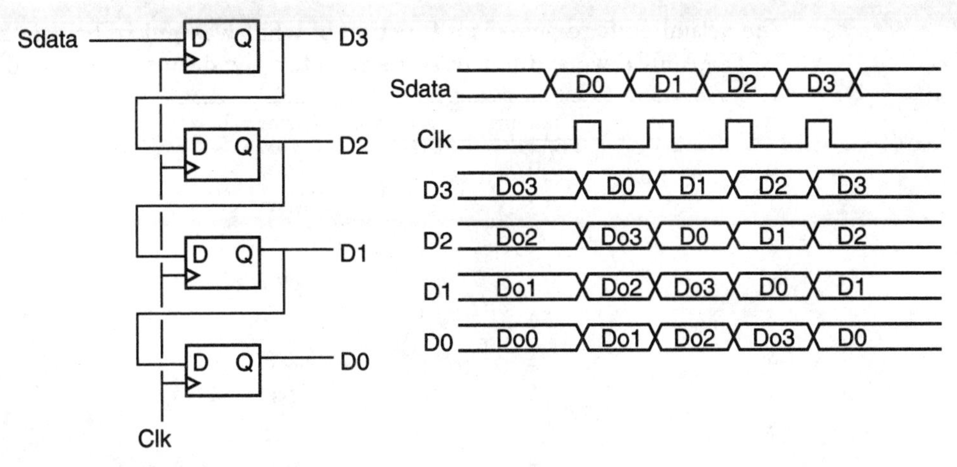

Figure 9.33 Serial to parallel data conversion.

Linear Feedback Shift Registers

One of the most interesting logic devices you can work with is the *linear feedback shift register* (LFSR). This circuit can be used to "pseudorandomize" data, encrypt and decrypt serial data, and provide very good serial data integrity checking. You may have heard the term *cyclical redundancy check* (CRC) applied to data transmission; this process uses a type of linear feedback shift register. Linear feedback shift registers can also be implemented fairly easily in software with a microcontroller or microprocessor, although it is in hardware where the function is most efficiently implemented.

The form of a linear feedback shift register is shown in Fig. 9.34. The XOR gates in between the shift register's inputs are known as *taps* and provide a *pseudorandom* output based on the serial inputs.

For example, if you had an 8-bit shift register with a single tap coming from bit 5, the circuit would look like the circuit shown in Fig. 9.35. The data going into the shift register would be characterized as:

Shift register in = data in ^ bit 5

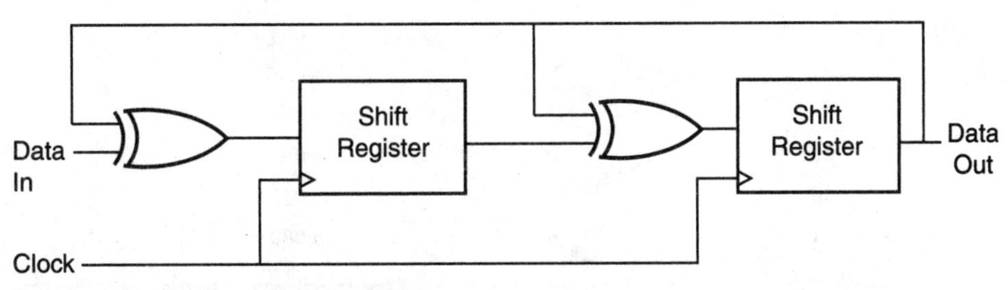

Figure 9.34 Linear feedback shift register.

The actual shift register is a function of what was put in before. If the 32 bits 0x055AA00FF were input into the register, the data states would be:

Cycle	Data in	Shift register values
0	1	0x000
1	1	0x001
2	1	0x003
3	1	0x007
4	1	0x00F
5	1	0x01F
6	1	0x03F
7	1	0x07E
8	0	0x0FD
9	0	0x0FB
10	0	0x0F7
11	0	0x0EF
12	0	0x0DF
13	0	0x0BE
14	0	0x07D
15	0	0x0FB
16	0	0x0F7
17	1	0x0EE
18	0	0x0DD
19	1	0x0BA
20	0	0x075
21	1	0x0EA
22	0	0x0D5
23	1	0x0AA
24	1	0x055
25	0	0x0AA
26	1	0x054
27	0	0x0A8
28	1	0x050
29	0	0x0A0
30	1	0x040
31	0	0x080

This might seem like a cumbersome method of getting 0x080, but the important thing is that this value is unique to the original string of bits. If a bit input into the linear feedback shift register were in error, then the result

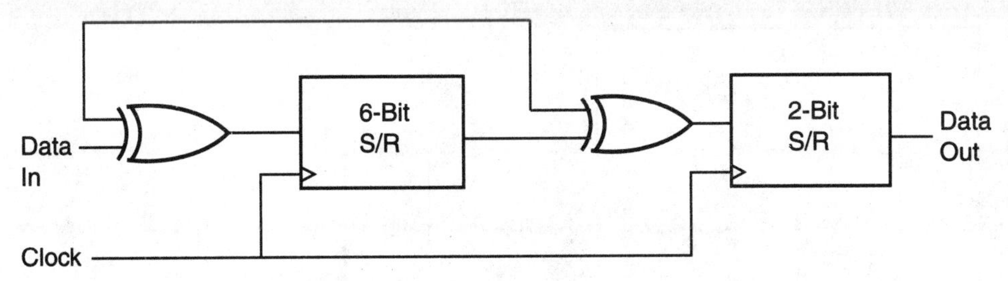

Figure 9.35 Linear feedback shift register application.

would not be 0x080; it would probably be something significantly different. In this mode, the LFSR is producing a control code for the data input to ensure the data is correct. The advantages of producing a control on a data string are that the hardware to create it is very simple and the control result will detect multiple bit errors (something that a checksum or parity checks may not do).

Linear feedback shift registers can also be used to generate pseudorandom displays. By using the output as the data-in circuit, I can produce a pseudorandom output that will change each time the data is shifted. This is actually the function of the circuit in Fig. 9.35. This circuit is useful for producing random numbers in an application or even displaying a seemingly random pattern of lights. I use this characteristic of linear feedback shift registers to make "Christmas lights" or random numbers for games.

Project 14—Random Light Generator

In the November/December issue of electronics magazines, you will find that there are a lot of simple Christmas decorations that you can build that involve flashing lights (usually LEDs). Many of these devices use microcontrollers or analog devices like the 555 timer to produce a random effect. With a few dollars worth of parts, you can create your own random light display for Christmas and at the same time demonstrate the pseudorandom operation of the linear feedback shift register.

For this application, I use the three different types of D flip-flop chips as a 16-bit-wide shift register. The use of the three types of shift registers is not to help acquaint you with the different devices, but instead to take advantage of the features of each device. In Fig. 9.36, you will see that I use the two D flip-flops of a 74LS74 along with a six D flip-flop 74LS174 and an eight D flip-flop 74LS374.

When the D flip-flops first become active, the stored values are *indeterminate,* which means that they cannot be predicted. Put another way, each flip-flop has an equally likely chance of being initialized with a 1 or a 0. For the linear feedback shift register, this is not a problem most of the time—as long as there is at least one bit set, then it will work properly.

The problem comes about when you get that one chance in 65,536 times when all 16 D flip-flops power up with 0's in them. When this happens, the XOR gate can never reload the least significant bit with a 1, and the application will

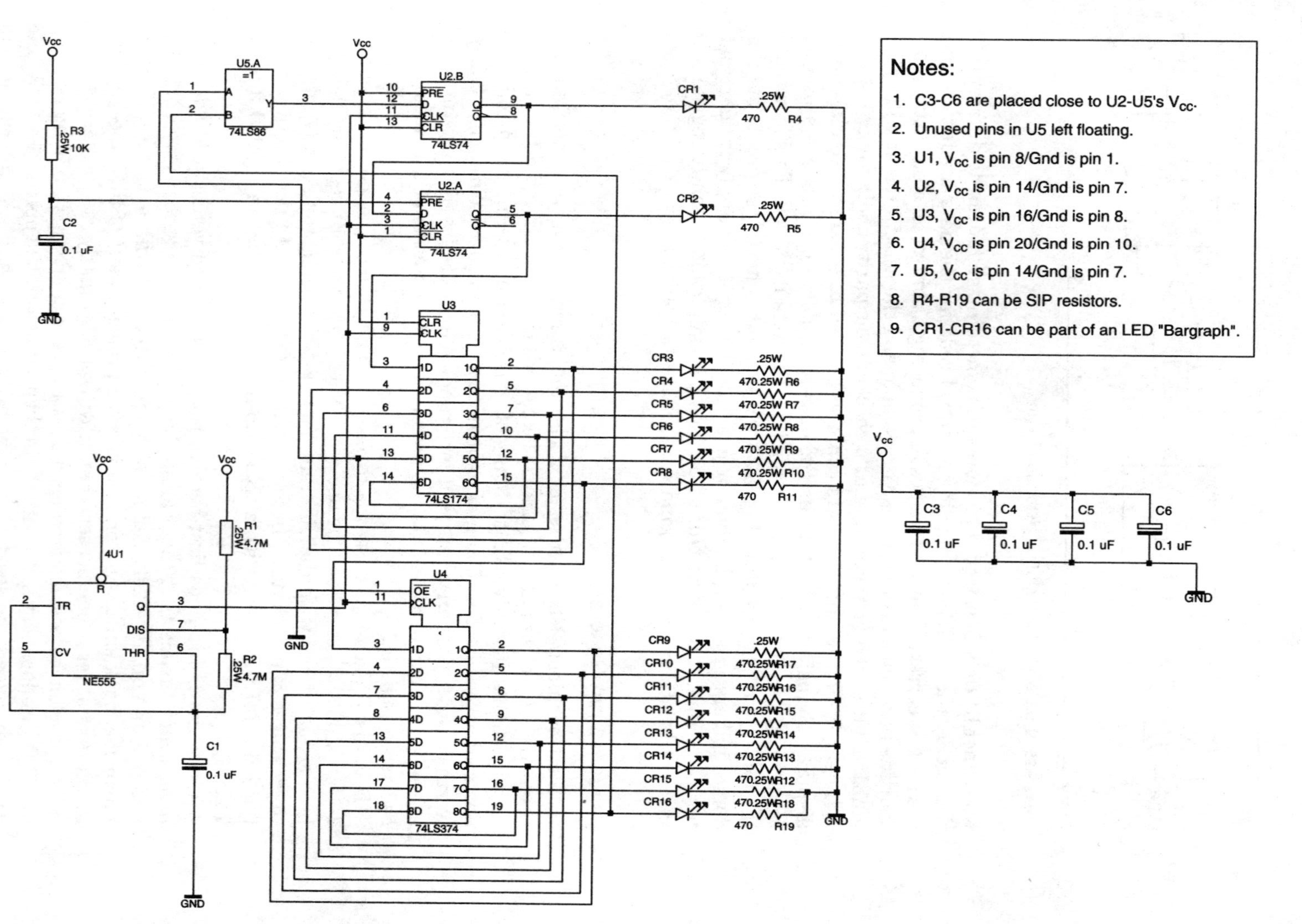

Figure 9.36 Project 12—Random light generator.

always have all the LEDs turned off. At first blush this may seem like it is a pretty unlikely occurrence, but I found with my testing that all the D flip-flops powered up with 0's about once every 10 times. As the saying goes, "your milage may vary," but the important point I want you to understand is that the probability that all the D flip-flops initializing with all zeros is not zero.

To avoid this problem from being a significant issue, I used the 74LS74 as one of the elements of the shift register because the bits can be asynchronously (i.e., without a clock) set or reset. To ensure that at least 1 bit was always set on power-up, I connected an *RC* network on the second least significant shift register bit's _PRE pin. This circuit will hold the _PRE input of this flip-flop low, forcing the register to store a 1 while the 555 timer is starting up. Once the 555 is active, the _PRE input should be at a valid 1 input, and it will behave as a normal shift register bit.

With the 74LS74 providing the necessary initialization conditions (or "protection"), the remaining 14 bits of the shift register were created by using a 74LS374 and a 74LS174. Note in the schematic diagram that I tie the 74LS174's _CLR line high (to make sure the register is never cleared) and the 74LS374's _OE line low (to enable the output drivers). The outputs (Q bits) are fed back to the inputs (D) bits of the D flip-flops to create the shift register.

Earlier in the chapter, when I introduced the concept of shift registers, I noted that I prefer making my own from flip-flops, as I have done in this application. The reason for doing this is that I can come up with much more appropriate circuits in an application instead of making do with a canned set of options. In this application, note that I have created a 16-bit shift register that has a power-up bit 1 set with only three chips. Creating a similar function with four bit shift registers (like the 74LS95) would require at least five chips (the extra for the preset function) and may be much more difficult to set up.

To drive this circuit, I used a 555 timer setup with a very long delay. Using the 4.7-MΩ resistors and 0.1-μF capacitor, I calculated the operational frequency as:

$$\begin{aligned}\text{Frequency} &= \frac{1.44}{(\text{R1} + 2 \times \text{R2}) \times \text{C1}} \\ &= \frac{1.44}{(4.7\ \text{M}\Omega + 2 \times 4.7\ \text{M}\Omega) \times 0.1\ \mu\text{F}} \\ &= \frac{1.44}{(3 \times 4.7\ \text{M}\Omega) \times 0.1\ \mu\text{F}} \\ &= \frac{1.44}{1.41}\ \text{Hz} \\ &= 1.02\ \text{Hz}\end{aligned}$$

By choosing these component values, I have come up with an oscillator frequency that allows the operation of the circuit to be easily observed.

That's really all there is to this circuit. I have not mentioned the LEDs, but if you have worked through all the projects so far, they should not really be a concern to you. At the end of this section, I will comment on how I would arrange the LEDs in a decoration.

The bill of materials for this project is:

Part	Description
U1	NE555 timer
U2	74LS74 dual D flip-flop
U3	74LS174 hex D flip-flop
U4	74LS374 octal D flip-flop
U5	74LS86 quad XOR gates
R1–R2	4.7-MΩ, ¼-W resistors
R3	10k, ¼-W resistor
R4–R19	470-Ω, ¼-W resistors; instead of individual resistors, two 8-resistor SIP packages can be used
CR1–CR16	16 LEDs; used as a bargraph display or Christmas decoration (suggestions for wiring decorations are in the text)
C1–C6	0.1-μF tantalum capacitors
Miscellaneous	Breadboard, +5-V power supply, wire, prototyping board

When you first look at the circuit, you probably feel that it can very easily be built into a standard 63-row breadboard. I found that it took two tries for me—the second time I created the prototype, I put two 10-LED bargraphs side by side and each chip with one row between it and the next one.

To build the application on a breadboard, I recommend following the steps listed below, which outlines the order in which the circuit should be built and little tests you can try out along the way.

1. Plan how the circuit is to be wired. I placed two 10-LED bargraph displays starting at row 2 of the breadboard with two 8-resistor 470-Ω SIPs tying the cathode of the LEDs to ground. From here, I placed the 74LS374, 74LS174, 74LS74, 74LS86, and NE555 on the breadboard with one row in between. You will find that you have one row left over.
2. Add the power supply to the breadboard and make the V_{cc}/Gnd buses. Once it is installed, short the anode of one of the LEDs to V_{cc} to make sure the power supply and LEDs are wired properly. Connect pin 4 of the 74LS74 to Gnd. This wiring will be used to test the wiring of the circuit and will be converted later into the *RC* network that is used to set one of the bits of the shift register to 1 on power up.
3. Wire the V_{cc}, Gnd, and necessary pull-ups and pull-downs of the chips. The schematic includes information on where the V_{cc} and Gnd pins are located on the different chips (the design system I use does not add these pins to the schematic).
4. Wire R1, R2, and C1 to the NE555 and complete the wiring needed for it to run. Power-up the breadboard and check the output of the timer. If you

use a logic probe, you will notice that the NE555 cycles about once every second, with the 1 time twice that of the 0.

5. Wire the output of the NE555 to the clock inputs of the 74LS74, 74LS174, and 74LS374.
6. Wire each output (Q pin) of the flip-flops to the next input (D pin) in the shift register chain. Leave the first input (pin 12 of the 74LS74) floating.
7. Wire each output (Q pin) of the flip-flops to the anodes of the LEDs. When this is done, turn on the power. You should see each LED turn on in sequence—although this may be confusing, depending on which values the various D flip-flops have on power up.
8. Disconnect pin 4 of the 74LS74 from ground and connect it to the *RC* network.
9. Connect the two inputs of the XOR (74LS86) gate and tie the output of the XOR gate to pin 12 of the 74LS74.
10. Power-up the circuit and you should see data shifting through and becoming pseudorandom on the basis of its initial state.

Once you have built this circuit, you will find that it runs quite well, although the pseudorandom aspect of it is not that readily apparent and the display is really not that interesting. To get a better idea of how this application works, disconnect the LEDs from the D flip-flops and reconnect them again at random. With the LEDs wired randomly to the D flip-flop outputs, you will suddenly discover that the display is actually quite a visually random event and quite attractive.

The appearance of the circuit's output can be enhanced by wiring different color LEDs into the circuit as well as moving them around randomly on a two-dimensional backdrop. One Christmas, I made a number of a "Frosty the Snowman" decorations with different color LEDs lighting up Frosty's hat—the effect was quite nice.

If you try to determine all the different LED combinations that can be displayed by this circuit, you will find that there is somewhere about 61,500. To go through each possible combination it will take slightly more than 17 hours. If you are interested in seeing the operation of the linear feedback shift register, I would suggest that you build the circuit without the 74LS374 and just use the 74LS74 and 74LS174 to show you the 240 or more combinations that are possible with an 8-bit circuit.

If you create a display like this, I do have a few suggestions for you. The first is to build the circuit on as permanent a platform as possible—the breadboard used for the prototype will not be acceptable and will not survive knocks and punishments of children or the passage of time. If you look through the book, you will see that I have included some discussions on more permanent prototyping methods, including designing your own PCB. This circuit is an excellent place to start learning this skill.

Second, look for LEDs that already have leads attached to them. Even if you are experienced at soldering, you will find that each LED will be a lot of work and take up to 5 minutes of your time. This time may seem extreme, but you will have to come up with some way of clamping the LED to the lead for soldering as well as some method for protecting the lead joints (such as heat-shrink tubing). This work will take a surprisingly long time. For 16 LEDs, this will be more than an hour and a quarter—if you are doing multiple displays, you will discover that the better part of a day can be lost soldering leads onto LEDs. I realize that LEDs with leads already attached to them cost quite a bit more (25 cents and up), but the time you save will keep you from going insane.

Last, when I first built the application, I discovered that the circuit would periodically blank out, and none of the LEDs would be lit. Theoretically, this should be impossible, so I spent a number of hours trying to figure out what happened. The problem seemed to take place when the power supply's V_{cc} dropped momentarily to about 2 V when the furnace fan started. It seems that even though the TTL D flip-flop gates cannot drive a 1 output, they will latch a 0 if the inputs are tied to the outputs.

The fix to this problem was simple—I just used a higher-capacity AC/DC power converter. The first power converter that I used (and used successfully for all the projects up to this point) was able to supply 150 mA at 9 V. This seemed to be insufficient for some cases—especially when other loads became active on the circuit. I changed the power supply input to a 250-mA, 12-V power converter and the problem has not reappeared.

I did note above that I used the same AC/DC power converter for all of the applications up to this point. For the problem that I saw, three things must have been true:

1. I was running this circuit for a very long period of time. I discovered the problem originally when I left the circuit running overnight.
2. The circuit's outputs are fed back in as inputs to provide the shift register function. This seemed to be a factor in the problem with the inputs.
3. The problem seems to be more likely when all of the LEDs are on (and the circuit is drawing a maximum amount of current).

With using the higher-capacity AC/DC power converter, I have run the circuit for up to 3 days without any problems. I just wanted to point out that I did have these problems and there was a fairly easy way to resolve them.

Buses

When I use the term *bus* in digital electronics, I use it to describe a circuit (or *net*) in an application in which there can be more than one driving circuit. Normally in a digital circuit, there is only one *driver,* or output pin that is driving a value onto the net. With a bus, there can be multiple drivers on a net—but it is important to note that there can only be one *signal* being driven in the circuit at any one time. Buses are important in digital circuits, such as computer processors, because they allow different devices to be interfaced to the same hardware.

The basic circuit needed to implement a bus is the *tristate driver,* or *buffer* (shown in Fig. 9.37). This circuit can either drive a high or low signal onto the bus (through the output pin), or no signal can be driven at all. If the pin is not driving, it is said to be in *high-impedance, high-Z,* or *tristate* mode, and it will not affect any data value that is on the bus.

The enable control bit is used to determine whether or not the output driver can place data onto the bus. In Fig. 9.37, if the enable line is low, then the input value will not be passed to the bus. When enable is low, then neither output transistor can be turned on and the bus will not be tied to V_{cc} or Gnd.

When enable is high, the value of input (positive and negative) is passed through the AND gates to the output transistors. If input is high and enable is high, then the top AND will output a 1 and the top output transistor will be turned on. If input is high and enable is high, then the inverted input will be low and will result in the output of the bottom AND being a 0 and the lower output transistor will be off. When input is low, then the upper transistor will be turned off and the lower one enabled.

The tristate driver is actually a very simple circuit. It is built into many different TTL and other digital electronics chips to control data being placed on a bus. As I have shown in Fig. 9.38, there are three different ways tristate drivers can be used in a bus circuit.

The simple driver (A in Fig. 9.38) will drive a value onto the bus when its Ctrl is enabled. The receiver (B) will pass information from the bus when its Ctrl is enabled. Last, the transceiver (C) can pass information to or from the bus according to the state of the two Ctrls. Note that with the transceiver, the two controls are never active at the same time.

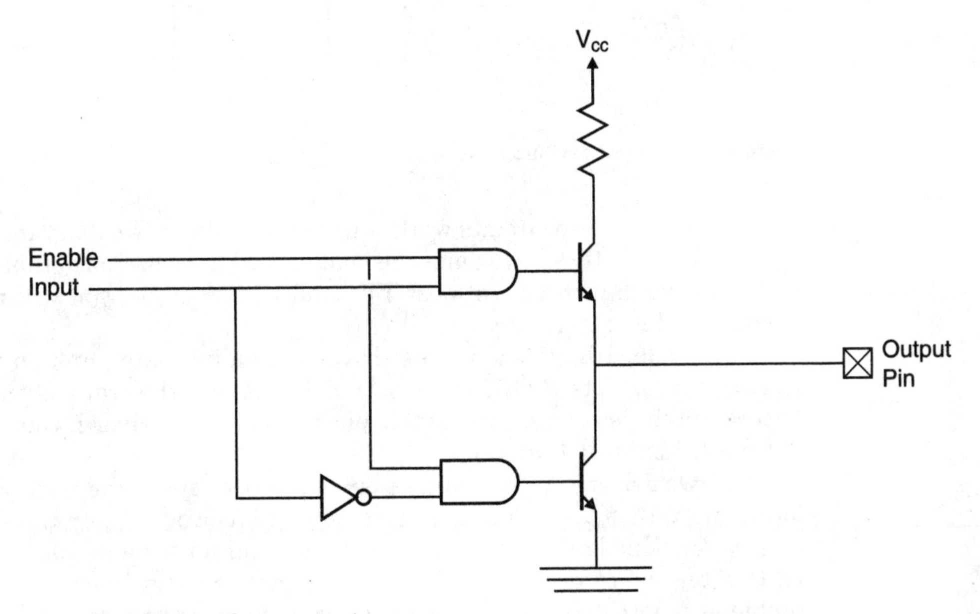

Figure 9.37 Tristate driver operation.

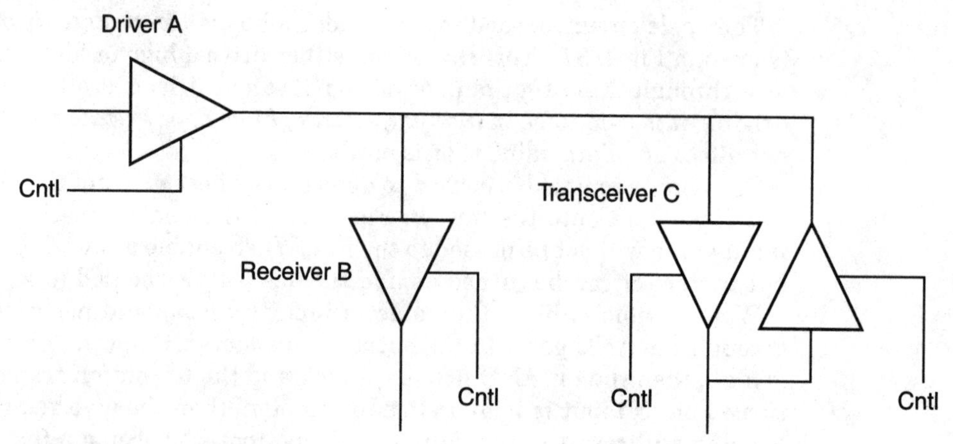

Figure 9.38 Electronic bus information.

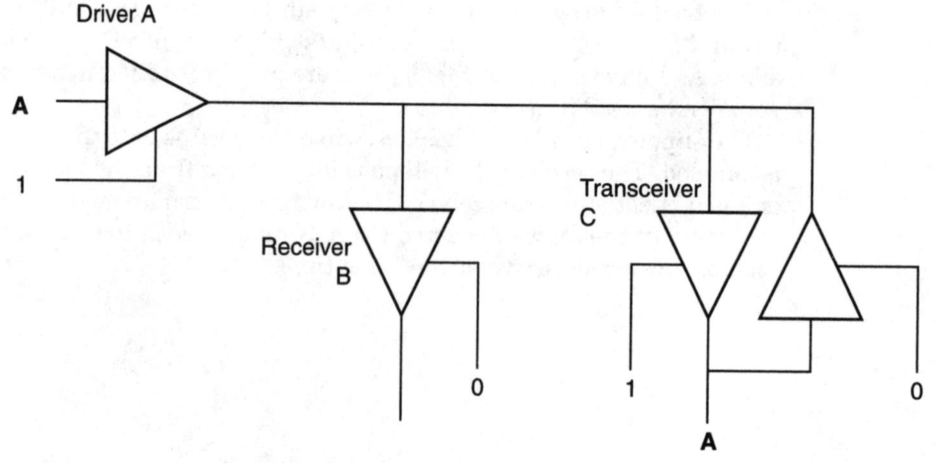

Figure 9.39 Bus write to transceiver.

To show how these circuits work, I have created the two diagrams, Fig. 9.39 and Fig. 9.40. In these two diagrams, note that I have assumed that Ctrl equal to 1 is enabled, whereas in most TTL and CMOS chips, you will find that a tristate enable is normally low.

In Fig. 9.39, I have shown the driver A sending data through the bus to transceiver C. Note that Ctrl for both driver A and the transceiver C receive tristate buffer is active. During this period of time, receiver B could be active, but not transceiver C driver.

If the two drivers in this example circuit were enabled and were sending different data, then the actual data level on the bus would be unknown (or *indeterminate*). This is known as *bus contention* and must be avoided at all costs. Bus contention can cause subtle problems that are very hard to find, or gross problems where driver chips burn out for no apparent reason.

In Fig. 9.40, I show the function of transceiver C changed to driver and receiver B enabled to receive the data from the transceiver C driver. As in the previous case, driver A cannot be enabled, to avoid the possibility of bus contention, but I should also point out that the transceiver C receiver cannot be enabled to avoid a possible race condition.

A *race condition* is a situation similar to bus contention where both components of a combined driver/receiver in a transceiver are enabled at the same time. In this case, when the data changes in the driver, it takes a certain amount of time (*gate delay*) to get to the bus. When the data gets to the bus, there is another gate delay caused by the receiver before this data is driven onto the net that the driver is on. Depending on the speed of the signals being sent and the characteristics of the tristate buffers, the value being driven onto the two buses can become rapidly changing (oscillating) and indeterminate.

It may seem like race conditions are something that can happen only in very fast bus data transfers, but they can happen any time there is a potential difference between two drivers in a feedback circuit. The easiest way to avoid this situation is to never have more than one driver active on the same bus (which is the case at the circuit end of the transceiver) at the same time.

There are, however, cases where active multiple drivers present on a bus are desired. The most popular one is the *dotted AND* bus, in which open-collector and open-drain drivers are used with a single, negatively active receiver. As is shown in Fig. 9.41. a pull-up resistor is used with a number of transistors, wired in the open-collector configuration on a bus with a single receiver. When any of these transistors is turned on, the bus will be pulled to ground and the input to the receiver will be a 0.

The reason why this circuit is called a "dotted AND" bus is that the only way the voltage level at the receiver's input is high is for all the drivers to be high. If any of the drivers are low, then the receiver's input will be low. The dotted

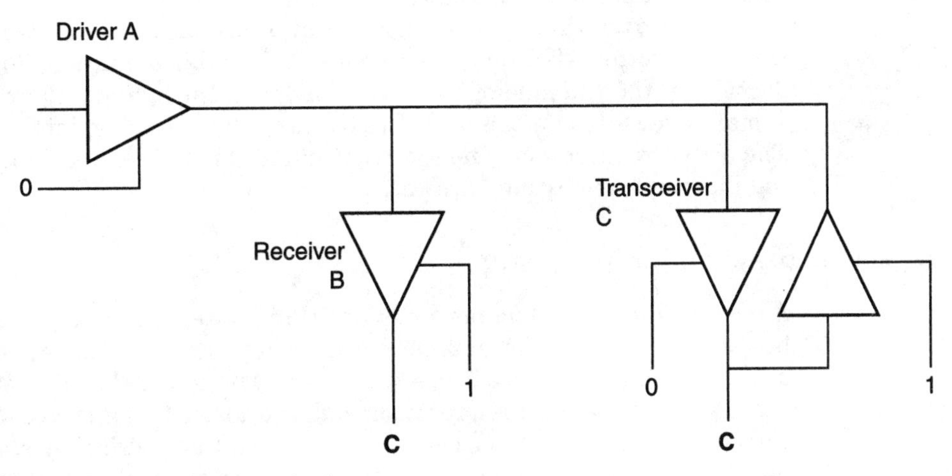

Figure 9.40 Bus write to transceiver.

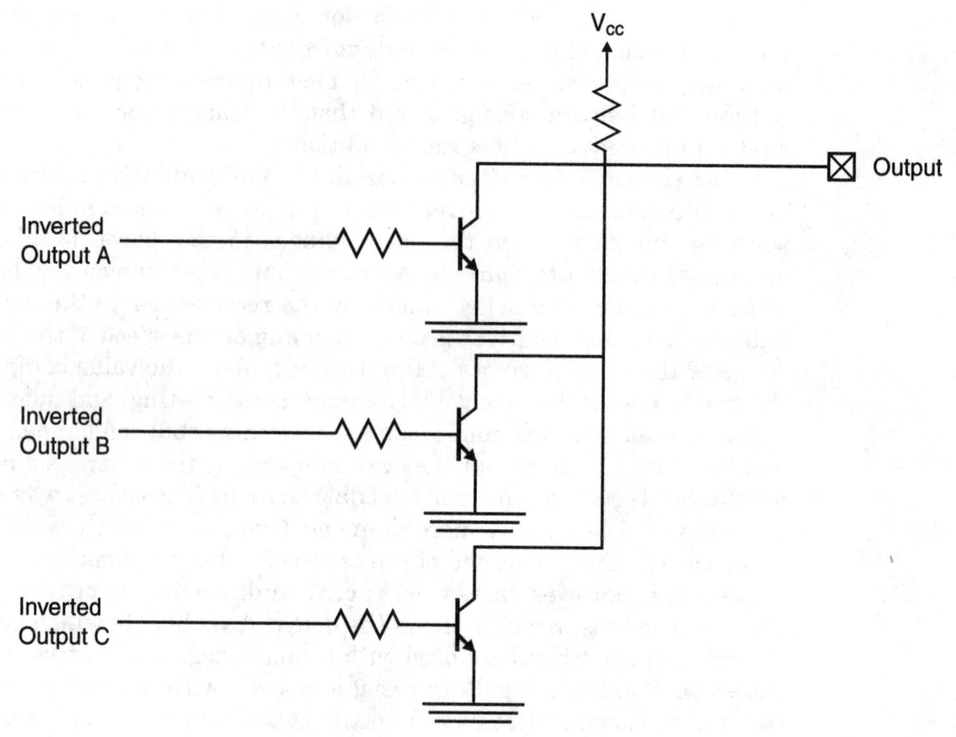

Figure 9.41 Dotted AND digital circuit.

AND bus is often used in microprocessor systems for interrupt requests—the processor can respond to only one interrupt request, but multiple devices can request the interrupt. Once the processor recognizes that there is an interrupt request, it can poll (read) each device that can request an interrupt to find out which one is making the request.

At the start of this section, I noted that a bus has only one signal on it at a time. The dotted AND bus may seem to violate this definition, but it actually doesn't. In the example of the processor interrupt request, there is only one signal or message on the dotted AND bus—the processor interrupt request. The actual request, once the interrupt handler is invoked, is found by polling the interrupt requesting hardware.

Project 15—Parallel Bus Operation

A basic computer system normally has three buses: an address bus, a control bus, and a data bus. The address bus is used to select which device and which address within the device is being accessed. The control bus is used to initiate the transfer between the processor and the device being accessed. Normally the control bus consists of negative active read and write lines. Finally, the data bus can be driven either by the computer system's processor or by the accessed device. Normally, the accessed device drives the data bus during reads while the processor drives the data bus during writes. A write/read memory cycle for a single memory device connected to a computer system processor is shown in Fig. 9.42.

In the Fig. 9.42 timing diagram, the computer system processor is shown to first drive the address of the device to be accessed. This operation allows the address decoders on the bus to select which device is to be active. Next the data to be stored into the device is put on the bus, and the _WR (write) control bit is toggled, causing the data to be latched into the accessed device. For most memory devices, the data is *latched,* or saved, on the rising edge of the _WR pulse.

After the data has been written to the device, the processor relinquishes control of the address and data bus (the control bus is held inactive) until it is time to access the bus again.

The second time the bus device is accessed, it is to read back the data that was stored in it. In this case, the device address is put on to the address bus, and the _RD (read) line is active. Notice that the data is active only when the _RD line is active. After the read has taken place, the address bus is returned to its original state (the processor has relinquished control of it).

Explaining and understanding the operation of the parallel buses used in a computer system can be difficult—fortunately, this can be easily simulated, as I show in this project, which provides two 4-bit memory devices that can be selected by a computer processor (which is you in this case). The schematic for the circuit is shown in Fig. 9.43.

Note that in Fig. 9.43 I combined the circuits for a bus into one thick line on the schematic. This is a common convention used in circuit diagrams to show

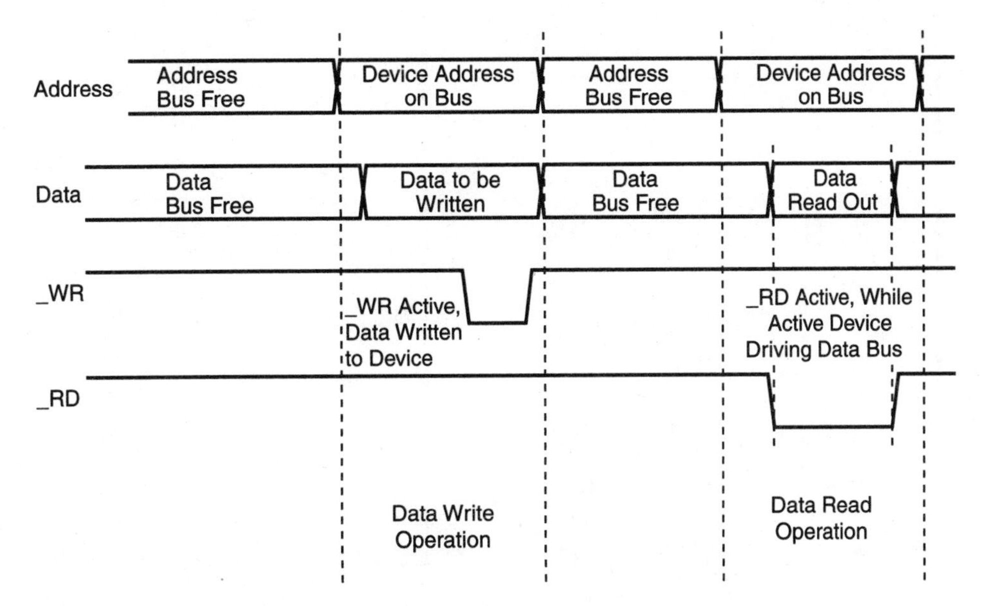

Figure 9.42 Project 15—simulated bus read and write.

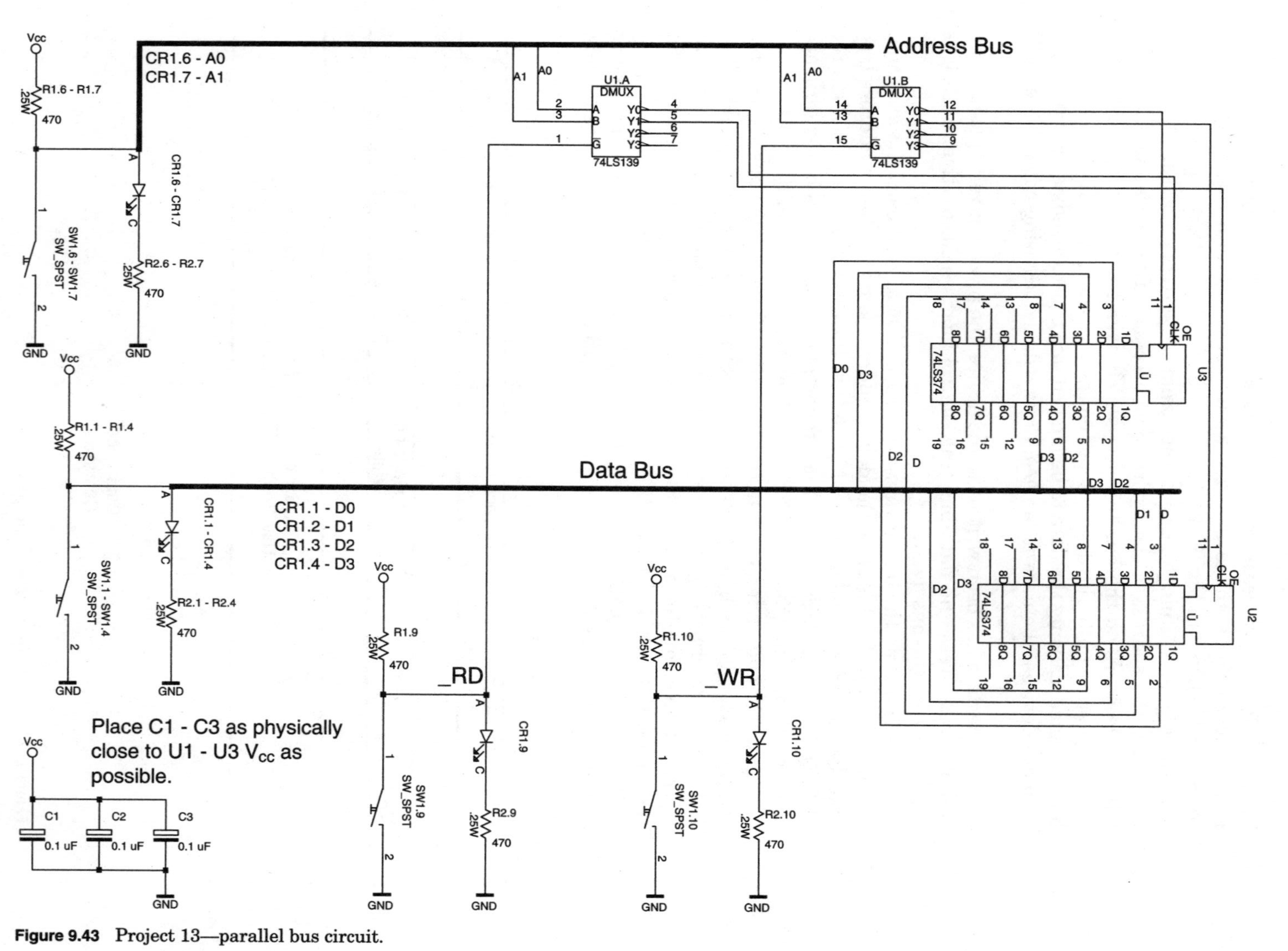

Figure 9.43 Project 13—parallel bus circuit.

how circuits are wired and is one that you should follow as well when you are drawing your own schematics. Using the thick bus line convention eliminates the need to draw multiple lines for each circuit in a bus.

The bill of materials for this project is:

Part	Description
U1	74LS139 dual 1-to-4 demultiplexer
U2, U3	74LS374 octal D flip-flop with tristate outputs
C1–C3	0.1-μF tantalum capacitors
Miscellaneous	+5-V power supply PCB, interface PCB, 65-row breadboard, wire

In this circuit, I made the data bus only 4 bits wide. The reason for this was to simplify the wiring and avoid needing additional switches from what is available in the Interface PCB. With 4 bits and two registers, I felt that the concept of bus operations with data transfer, addressing, and the _RD and _WR lines would be acceptably demonstrated.

This project can be built on a breadboard by using the layout I show in Fig. 9.44. In this diagram, because of space limitations, I have not shown how the second 74LS374 (U3) is wired onto the breadboard. The second 74LS374 is wired identically as the first 74LS374 except that the output enable (used to read the contents of the 74LS374) and the clock (used to write to the 74LS374) pins are driven from different outputs of the 74LS139.

With this wiring, the least significant 4 bits of the first 74LS374 can be accessed at address 0b000, while the least significant 4 bits of the second

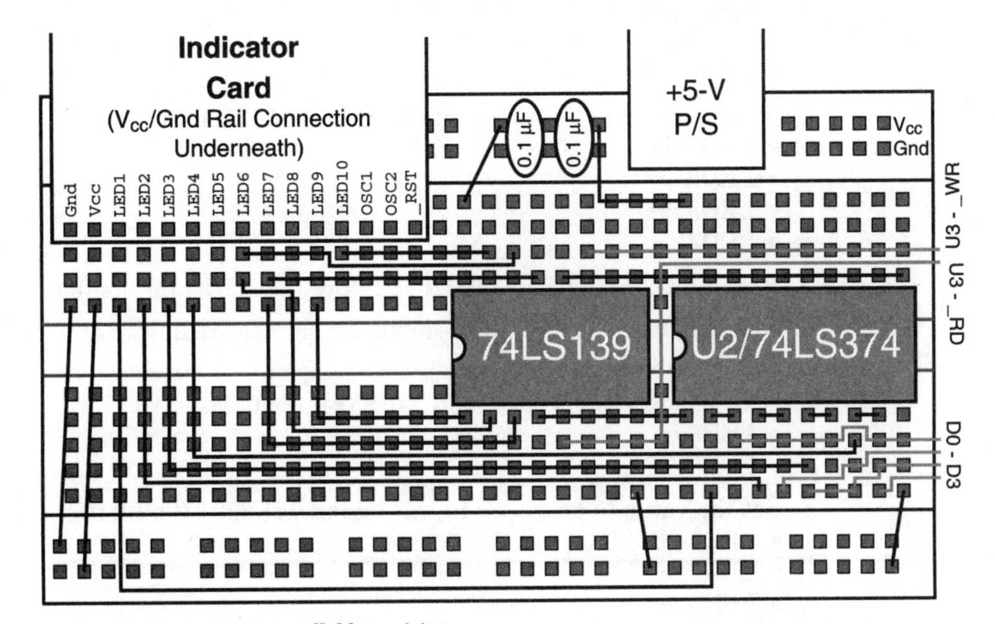

Figure 9.44 Project 15—parallel bus wiring.

74LS374 can be accessed at address 0b001. If any other addresses are specified on LEDs 6 or 7, then neither 74LS374 will be an address.

To write 4 bits to the first 74LS374 (at address 0b000), you should follow this process with the Interface PCB:

1. Set the address of the first 74LS374 (0b000). For this case, set:
   - LED6 off
   - LED7 off
2. Set the value to be written into the first 74LS374 on LED1 through LED4. If 0b01010 is to be written into the 74LS374, then set:
   - LED1 off
   - LED2 on
   - LED3 off
   - LED4 on
3. Press the _WR button (LED10) to store the data into the first 74LS374.
4. Release the address and data buses by setting LED1, LED2, LED3, LED4, LED6, and LED7 on.

Reading back the data in the 74LS374 is similar.

The reason why I used two memory chips (two 74LS374s in a single package) was to show how device addressing worked. To store 0b01001 in the first 74LS374 and 0b00110 into the second 74LS374 and read back the contents of the first 74LS374 without affecting the second 74LS374, you should follow these steps:

1. Set the address of the first 74LS374 (0b000):
   - LED6 off
   - LED7 off
2. Set the value to be stored into the first 74LS374 (0b01001):
   - LED1 on
   - LED2 off
   - LED3 off
   - LED4 on
3. Press the _WR button (LED10) to store 0b01001 into the first 74LS374.
4. Release the data and address bus by setting LED1, LED2, LED3, LED4, LED6, and LED7 on.
5. Set the address of the second 74LS374 (0b001):
   - LED6 on
   - LED7 off
6. Set the value to be stored in the second 74LS374 (0b00110):
   - LED1 off
   - LED2 on
   - LED3 on
   - LED4 off

7. Press the _WR button to store the data into the second 74LS374.
8. Release the Data and Address bus by setting LED1, LED2, LED3, LED4, LED6, and LED7 on.
9. Set the address of the first 74LS374 (0b000):
   - LED6 off
   - LED7 off
10. Press the _RD button (LED9) to drive the data stored in the first 74LS374 onto the data bus. LED1 through LED4 will indicate the value in the first 74LS374.
11. Release the address bus by setting LED6 and LED7 on.

That's really all there is to reading and writing to devices on a computer system's parallel bus. Steps 9 through 11 can be repeated for reading the contents of other devices (like the second 74LS374). Steps 1 through 4, or 5 through 8 can be repeated for writing different values to the devices.

Project 16—Bus Contention

The biggest danger in implementing a multiple-driver bus is the opportunity for two or more drivers to be active on the bus at the same time. If this happens, the actual data on the bus is unknown (*indeterminate* is the correct term), and some potential data is lost or corrupted. The term used for this condition is *bus contention*. It should be avoided at all costs within your system.

Along with data being lost during bus contention, there is the situation where you have multiple drivers working against each other. In this case, you have current flowing from one device to another, and potentially damaging currents can pass between them. In this case, the circuits can be burned out, or in situations where the drivers are capable of driving large currents, circuit traces and connectors can be damaged.

To be honest, the opportunity for hardware to be damaged when bus contention is taking place is quite small, but bus contention still must be avoided when an application is designed to avoid the loss, or change of data. In this project, I want to set up a little experiment to show how bus contention can affect the data and voltage levels driven onto a bus.

The Project 14 circuit is quite simple, as can be seen in the schematic Fig. 9.45. Two drivers of a 74LS125 tristate buffer are connected together with their inputs and output enables controlled by the switches on the interface PCB. The bus output is passed to an LED so you can monitor the behavior of the circuit.

The bill of materials for the project is:

Part	Description
U1	74LS125 quad tristate buffer
C1	0.1-μF tantalum capacitor
Miscellaneous	Breadboard, +5-V power supply, interface PCB, wire

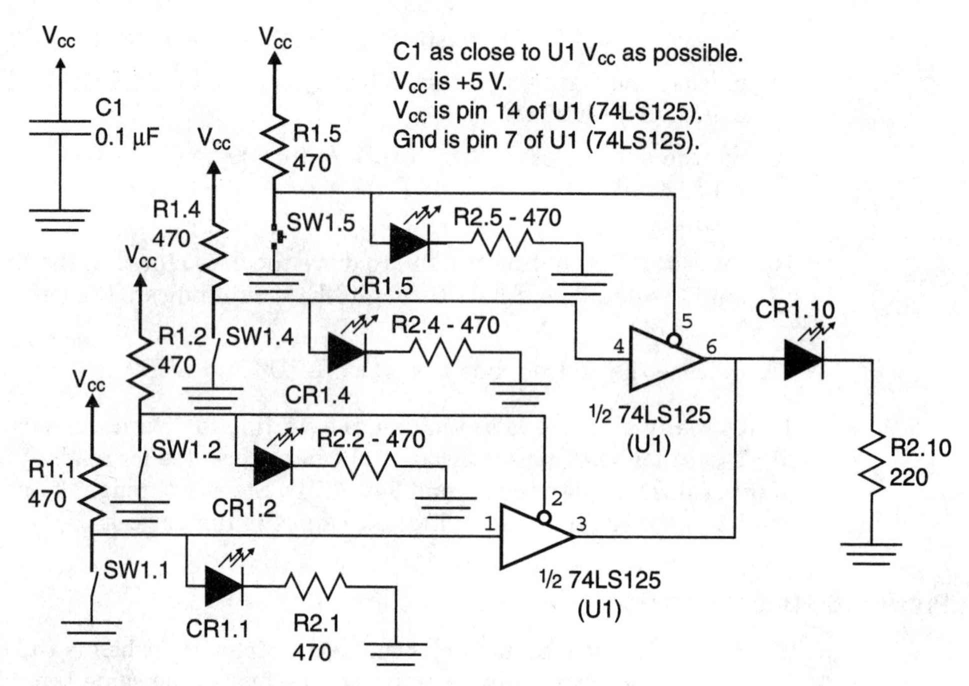

Figure 9.45 Schematic demonstrating bus contention.

Once you have the parts together, the circuit can be built on a breadboard as shown in Fig. 9.46.

For my prototype, I named the first tristate buffer A (pins 1, 2, and 3 of the 74LS125) and the second tristate buffer B (pins 4, 5, and 6). After applying power, I enabled the two outputs [pins 2 and 5—0 (LED off) input] and then varied the inputs and recorded the voltage results in the table below. I pulled on and off the connection to the LED to see what kind of interaction it had with the output voltage.

A B	Bus LED	LED voltage	No-LED voltage
0 0	Off	0.22	0.13
0 1	Off	0.61	0.54
1 1	On	3.51	3.67
1 0	Off	0.62	0.55

From the results, I can see that the low output always overrides the high output when the two drivers are driving different levels. At no time did the 74LS125 get noticeably warmer (something that always has to be considered). It should not be surprising that the bus LED did not turn on when the two drivers were at different levels—the 0.6 V (approximately) is not a large enough voltage drop for the LED to light.

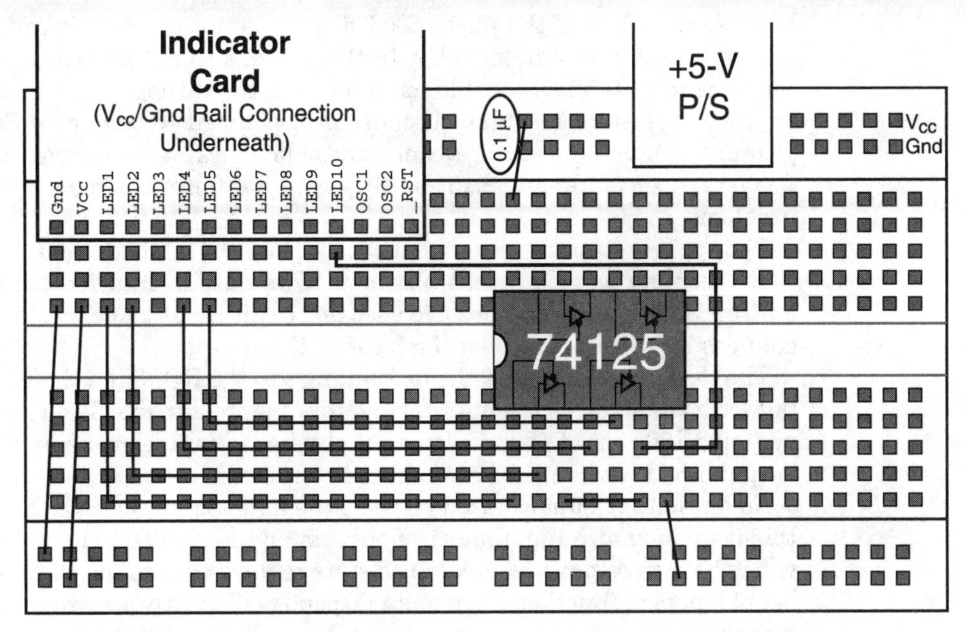

Figure 9.46 Project 16—bus contention project wiring.

What was surprising for me was the negligible difference the LED circuit made to the measured voltage level on the bus. I was expecting the LED circuit (with the two current-limiting resistors) to have a much greater effect on the actual voltage on the bus.

At these bus voltage levels, there are cases where you should consider that the voltage levels (and potential current source/sink of the opposing drivers) could cause other circuits to behave unpredictably. If this circuit were providing current for a bipolar transistor, you would probably find that the current supplied would be sufficient to turn the transistor on. Similarly, depending on the color of the LEDs, you may find that the voltage is high enough for the LED to light in a way that a person looking at the LED as an indicator would think that it was on.

If this circuit were driving a CMOS logic input, you might find that the output was at an intermediate state, with the CMOS transistors connected to the input acting as if they were an amplifier rather than a binary gate. In this case, the CMOS gate may randomly set the output voltage level, or pass it along without modification.

State Machines

One of the more interesting devices you can work with is known as the *state machine.* This class of circuit allows you to create a surprisingly complex application with a simple, single ROM and a couple of latches instead of a complex microcontroller. State machines are not widely used in modern applications because the costs of the parts needed to make up the circuit can very easily exceed that of a microcontroller. In the projects that I present in this book, I implement state machine–like circuits with digital logic, but I have avoided creating true state machines because programmed ROMs or an EPROM programmer would be needed. Despite these logistical issues, *state machines* are a good concept to know about as they give a different perspective on solving problems.

The typical state machine is shown in Fig. 9.47. This circuit consists of a ROM (usually EPROM) that has part of its output data fed back as a *state address.* Other address lines are used as circuit inputs, and the state machine changes its state address on the basis of these inputs.

The clock is used to pass the new address to the ROM and then pass the output from the ROM to the outputs and input state circuits. The two latches are operated 180° out of phase to prevent glitches from unexpectedly affecting any output circuits when the ROM changes state.

As few output bits are used as the state address as possible. The reason for this is to maximize the number of outputs and minimize the number of states, which have to be programmed. Each state requires two to the power of the number of inputs to function. Each state responds differently according to the inputs it receives.

A typical application for state machines is a traffic light. If a push button crossing light, as shown in Fig. 9.48, were implemented, a state machine circuit, wired as shown in Fig. 9.49, could be used.

In normal operation (which is known as *state 0*), the green light is on and the button is not pressed. If the button is pressed, then execution jumps to state 1, which turns on the yellow light for 5 sec (states 2, 3, 4, and 5), after which the red light is put on for 26 sec (states 6 to 31). If the button is

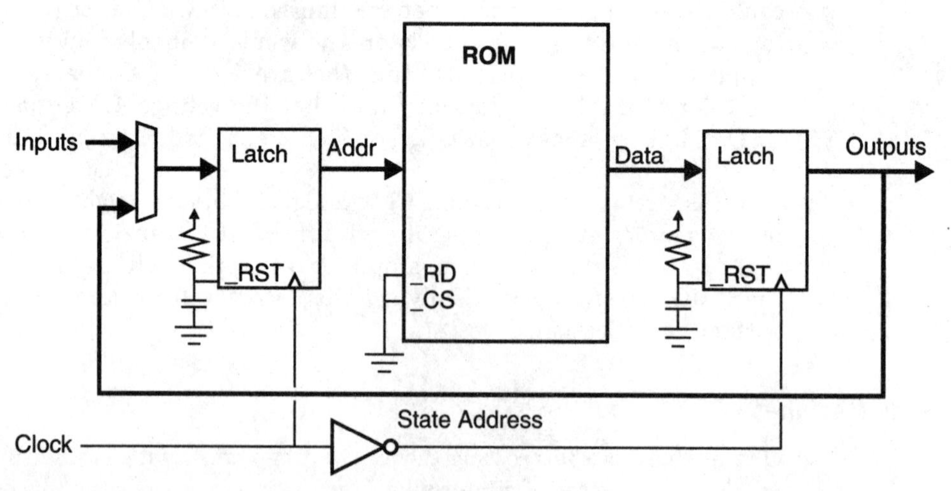

Figure 9.47 Basic state machine circuit.

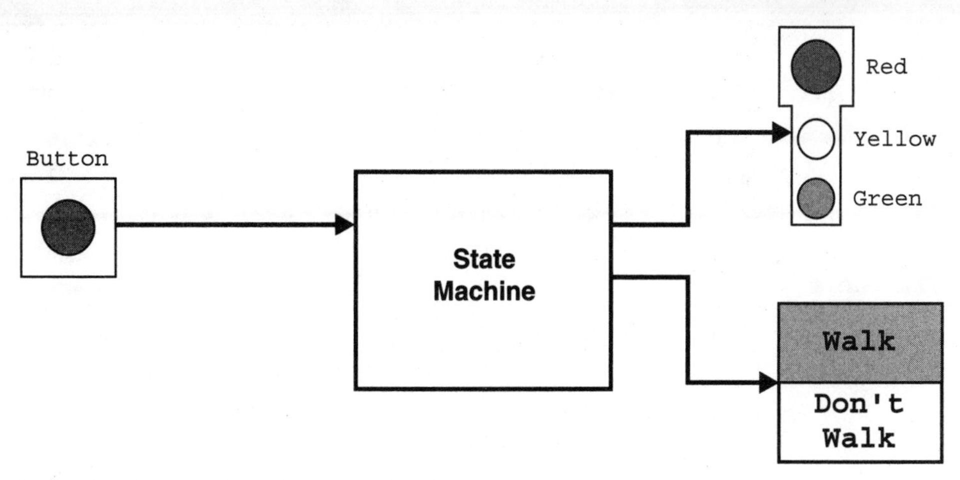

Figure 9.48 Traffic light state machine block diagram.

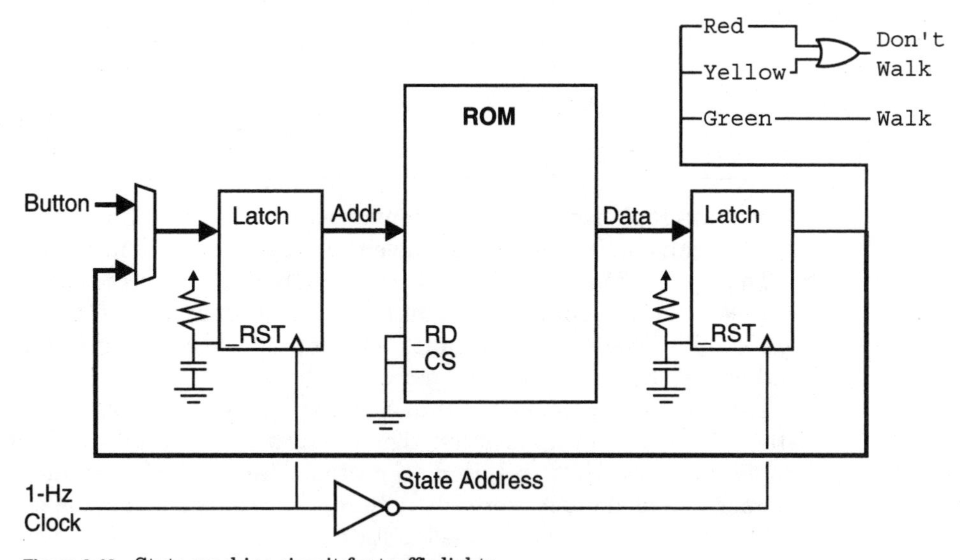

Figure 9.49 State machine circuit for traffic lights.

pressed during states 7 to 31, then execution jumps to state 6 to reset the timer.

To keep the circuit simple, I want to use an 8-bit data bus ROM with six inputs (five state, one button). This means that 2^6 (or 64) states are required in the ROM. These states are expressed in the table below. The reset on the input address latch is used to reset the state to 0 on the power-up. The button is assumed to be up if a 1 is returned and down if a 0 is returned.

State	Inputs		Outputs				Comments
	Button	State	Green	Yellow	Red	State	
0	1	0	1	0	0	0	Power-up
Button 0	0	0	0	1	0	1	
press 1	×	1	0	1	0	2	Reset
2	×	2	0	1	0	3	
3	×	3	0	1	0	4	No button
4	×	4	0	1	0	5	Yellow on
5	×	5	0	0	1	6	Red on
6	1	6	0	0	1	7	
6	0	6	0	0	1	5	
7	1	7	0	0	1	8	
7	0	7	0	0	1	5	
8	1	8	0	0	1	9	
8	0	8	0	0	1	5	
⋮							
31	1	31	1	0	0	0	Timer done
31	0	31	0	0	1	5	

This state table would then be converted into bits and burned into the ROM. An × means both input states have the same result on outputs.

This application is reasonable to code and build, but a problem arises with very complex state machines (ones that require tens of inputs and hundreds of different states). These state machines are normally hard coded into an ASIC chip rather than built from discrete parts like those I have shown for this application.

The reason for placing it within a chip is to give more outputs as well as more states in a custom application. The depth and the width of the data in real applications are better suited to custom chips, which can have noncustom memories added much more easily than where only commercial chips are used.

In the example above, I used a state machine with a 1-sec clock. Obviously in this situation there can be problems (such as the missed input if the button is pressed for less than 1 sec or it isn't released after it is pressed). This function makes state machines unattractive for rapidly changing inputs, and any kind of sophisticated real-time processing of inputs simply cannot be done economically with the state machine. When I say not economically, I am thinking in terms of the memory and properly programming the many states.

Project 17—Digital Dice

To finish this chapter, I wanted a project that would be pretty easy to implement, but give a kind of a "gee whiz" impression to you before going on to the actual projects. I thought creating a circuit that would simulate the action of a single die would be interesting and would give me a chance to demonstrate how state machines work. The final project presented here works quite well, but it was a surprising amount of work for me to get running.

I knew this application would be somewhat complex to get running, but I was surprised at the amount of work it actually took. This was largely a result of my not trying to simulate the action of the application before implementing it and not really thinking of how I had wired the LEDs. As I describe the application, I will discuss these problems and how I worked around them.

The digital dice circuit is moderately complex and uses the 4060 oscillator circuit that was built earlier in the chapter for the binary clock. This clock drives a pair of 74LS193 counters (shown in Fig. 9.50) that allow the clock to pass quickly when a start button is pressed, but slows down the passed clock after the button is released. The passed clock goes to another 74LS193, which counts down repeatedly from 10, passing the most significant 3 bits to a combinatorial circuit that displays the value as an LED "die" as shown in Fig. 9.51. The result is a random display that gives the appearance of a die.

The bill of materials for this project is:

Part	Description
U1	4060B CMOS oscillator/counter chip
U2, U3, U7	74LS193 counter
U4	74LS32 quad two-input OR gate
U5	74LS04 hex inverter
U6	74LS74 dual D flip-flop
U8	74LS02 quad two-input NOR gate
U9	74HC04 hex CMOS inverter
D1–D7	Red LEDs
Y1	32.768-kHz Crystal
R1	10-MΩ, ¼-W resistor
R2	10k, ¼-W resistor
R3–R9	470-Ω, ¼-W resistors
R10	470k, ¼-W resistor
R11	4.7M, ¼-W resistor
C1	100-pF capacitor
C2	15-pF capacitor
C3–C12	0.1-μF, 16-V tantalum capacitors
SW1	Momentary on SPST push button switch
Miscellaneous	Breadboard, +5-V power supply, wire

The clock circuitry, in its final form, appears to be quite simple; when the start button (SW1) is pressed, the 74LS74 D flip-flop is reset and the DISPCLK chip begins to run. As the start button is pressed, the reset clock is cleared and its input clock (the borrow from DISPCLK) is disabled. The counter output from the reset clock is used as the reload value of DISPCLK.

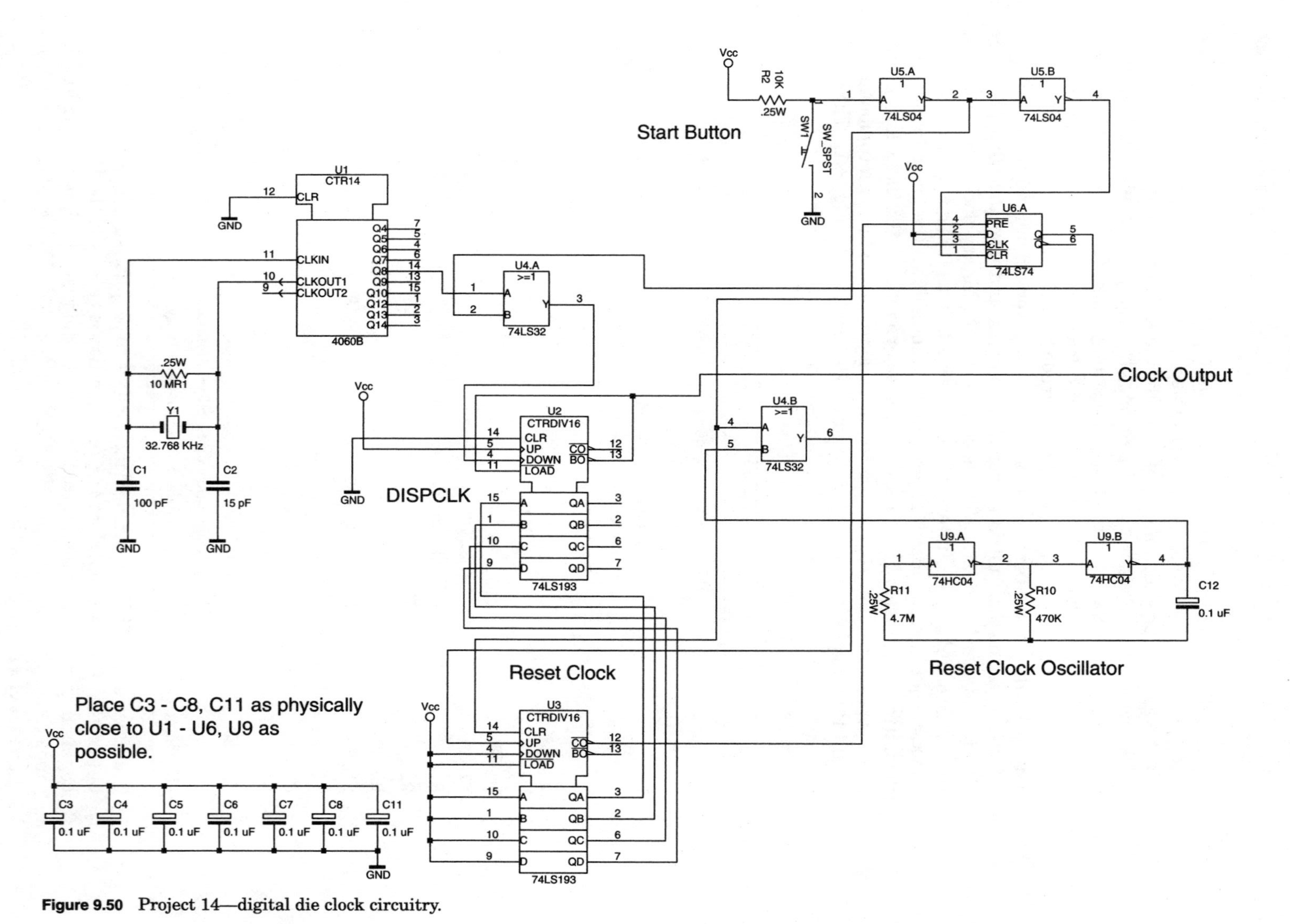

Figure 9.50 Project 14—digital die clock circuitry.

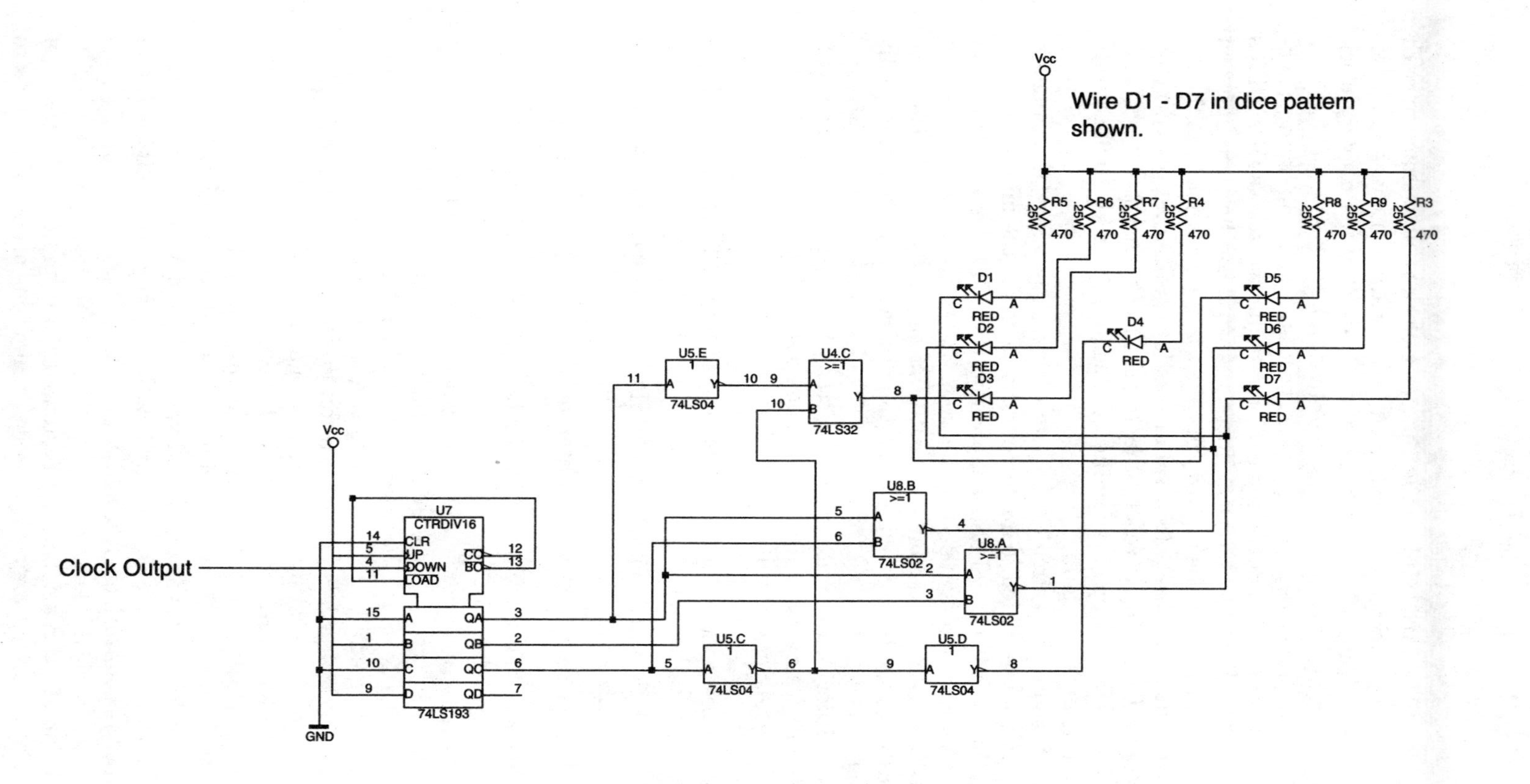

Figure 9.51 Project 14—digital die display circuitry.

DISPCLK runs backward, and, while the reset clock is held inactive with the clear pin active, DISPCLK is reloaded, after each borrow, with zero. During this state, DISPCLK pulses borrow quite quickly (once every clock cycle), and is passed along to the display counter (U7).

When the button is released, the reset clock begins to count up (using a separate clock output from the 4060B), and each new value is used as DISPCLK's reload value. As the reset clock's value increases, so does the time between pulses of the clock output. To the viewer, this gives the impression that the die is slowing down. Once the reset clock value overflows, or makes the carry bit active, the 74LS74 D flip-flop is reset, stopping the clock output.

Once the 74LS74 D flip-flop is reset, then the counters are all stopped until the start button is pressed again.

When I first designed this circuit, this operation did not seem that complex. The problem came when I tried to implement the circuit in the minimum number of chips. I found that I could not reliably stop DISPCLK because the carry output of the reset clock changes state every time a clock pulse was passed to it.

The big mistake with this circuit was that I did not really think it through. I tried to borrow from the binary clock's time set circuitry without really understanding what I was doing.

Starting over, I decided to understand exactly what I wanted to have happen in the circuit. Using the three paragraphs above, I was able to list the three states that the clock circuitry worked with:

State 1: Reset clock and DISPCLK running. Any time DISPCLK underflows (borrow is active), the latest reset clock value is used for reloading the 74LS193.

State 2: Reset clock overflowing while DISPCLK was running. When this happens, DISPCLK should be stopped.

State 3: Both reset clock and DISPCLK stopped.

Written out this way, the three operating states seem quite simple—the obvious question is why didn't I do this right from the beginning?

The best answer to this question is that I was being arrogant and thinking I could work out this circuit in my head. I was wrong, and it cost me several hours before I came to the conclusion that I should start over and work through the circuit again by understanding exactly what I want of it.

When I started over, I decided to use a D flip-flop to control the state operations. This flip-flop was reset by the start button and set when the reset clock overflowed. By implementing the circuit in this way, I was able to control the point at which the clock stopped to display the final die value.

Once this was done, the circuit was working properly; while the start button is pushed, the clock output was at the full speed of the 4060 output (128 Hz). When the start button is released, the reset clock increments until it overflows (at which point the counting operation stops).

While the reset clock is incrementing, each time DISPCLK underflows, it picks up the value of the reset clock and counts down from here. The result of this happening is an appearance that the die oscillations are slowing down to the point where it stops. Using the relaxation oscillator with the reset clock (which is a 10-Hz clock), you should see the die slow down over a 1.6-sec period once the start button is released.

In Fig. 9.50, you will see that I added a separate relaxation oscillator to run the reset clock. When I first built the circuit, I connected the reset clock to one of the unused 4060 oscillator outputs. When I did this, the circuit ran quite well, but I found that the values output were not all that random—the value 4 (display counter equal to 3) came up much more often than any other value.

The reason for this seemed to be that I used a clock that was in phase and in frequency for both the DISPCLK and reset clock. By adding the relaxation oscillator to the reset clock, I found that the results were much more random.

A good question about this relaxation oscillator would be why didn't I use two relaxation oscillators with a 74HC04, instead of a relaxation oscillator with an extra 4060. If I were going to build this circuit in production, I would probably do that—for this project, I decided to leave the 4060 in place and let you experiment with it to see the repeating 4 output (or other values) using different 4060 outputs.

Coming up with the output circuitry for the seven LEDs was actually quite challenging. Over the course of 2 hours, I tried to come up with the most efficient way of representing them in a way that a 3-bit counter (I use only the most significant 3 bits of the 74LS193) could drive them. Part of the problem was that I tried to come up with an exceedingly clever way of doing it (like using a 74LS47 seven-segment LED display driver) until I came up with the combinatorial circuits that can be seen in Fig. 9.51.

Before reading on and finding out how *I* did it, why don't you spend a few moments trying to come up with your own way of implementing seven LEDs for a die. I think you'll find that there is more thought required than you would expect at first blush.

When I started to think about how I wanted to represent the LEDs, I started off by numbering out each LED as:

```
1       5
2   4   6
3       7
```

This diode numbering has been used in the schematic shown in Fig. 9.51.

To make the job simpler and not have to find equations for seven different LEDs, I combined the LEDs wherever possible. The result was to group the corner and side LEDs, and I referred to them as:

1–7

2–6

3–5

Note that I work with opposite corners. When the values 2 and 3 were displayed, I wanted to use opposite corners; this complicated the circuitry somewhat, but makes the operation of the project more attractive.

With a reference system in place, I then decided to try and come up with a method of implementing these LEDs. Using the convention of 1 for on and 0 for off, I created the table:

Die value:	1	2	3	4	5	6
Counter value:	0	1	2	3	4	5
2–6	0	0	0	0	0	1
1–7	0	1	0	1	1	1
3–5	0	0	1	1	1	1
4	1	0	1	0	1	0

I put 2–6 at the top of the table because it is active only when a 6 is being shown. I put 4 at the bottom of the table because it is the only LED lit for 1. I also put in the counter value (which is zero based) to allow me to develop output equations without a lot of work converting the die value into the counter output value.

The output equations I came up with for the different LEDs are:

$$2\text{–}6 = C_0 \cdot C_2$$

$$1\text{–}7 = C_0 + C_2$$

$$3\text{–}5 = C_1 + C_2$$

$$4 = !\, C_0$$

You may be surprised to see that the output equation for 2–6 is not:

$$2\text{–}6 = C_0 \cdot !\, C_1 \cdot C_2$$

as this is the *actual* counter value for die value 6. Adding the $!\, C_1$ term is not required because the count value never goes above 5. If I had implemented the circuit with all three counter outputs, the number of gates required to drive the LEDs would have been doubled. This is a simple optimization that you should look for—make sure you understand the boundaries of the values input into combinatorial circuits.

These equations were implemented with a 74LS04 hex NOT gate chip and a 74LS32 quad two-input OR gate, and when I connected them up, I found that nothing looked right. Sometimes none of the LEDs were turned on, and other times the three across were turned on. The circuit didn't appear to be working correctly.

Fortunately, I realized pretty quickly what the problem was—I was using positive logic for the output (I was used to the interface PCB used with the other experiments), but I had wired the LEDs to light negative actively.

I should point out here something about LEDs and the convention they use for wiring. In the interface PCB, the LEDs are wired to light on a positive (high) output from whatever driver is on the circuit. I did this to make the initial operations of the logic circuits easier for you to see as well as make the interface PCB simpler and cheaper for you to build.

Normally, LEDs are wired as being negatively active. The simple reason for this is that most types of logic (TTL and CMOS included) are not designed with the ability to source relatively large amounts of current (such as is required for LEDs). Instead, the high current drivers are implemented to sink current. This means that to safely wire LEDs into your application, they should *always* be wired with their anodes connected to V_{cc} and their cathodes wired to logic outputs.

Realizing this mistake, I changed the table above to:

Die value:	1	2	3	4	5	6
Counter value:	0	1	2	3	4	5
2–6	1	1	1	1	1	0
1–7	1	0	1	0	0	0
3–5	1	1	0	0	0	0
4	0	1	0	1	0	1

and the output equations to:

$$2\text{–}6 = !\,C_0 + !\,C_2 = !\,(C_0 . C_2)$$

$$1\text{–}7 = !\,C_0 . !\,C_2 = !\,(C_0 + C_2)$$

$$3\text{–}5 = !\,C_1 . !\,C_2 = !\,(C_1 + C_2)$$

$$4 = C_0$$

In the new output equations, using NOR gates would seem to make the circuit much easier to wire. For the 2–6 output, I used the original equation (I didn't want to add a 74LS00 quad NAND chip for one gate) along with two available inverters. These are the equations that were used in the schematic in Fig. 9.51.

With the die output working properly, I then noticed the nonrandom output that I discussed above. The solution to this problem was to change the reset clock's input from a multiple of the clock used for DISPCLK to one that would be running at a completely different frequency.

Looking at the schematics, there are a number of places where you would feel that the application could be optimized. As I indicated above, a second relaxation oscillator could be added to run DISPCLK. Along with this, you will find that you could eliminate the need for the 74LS04 hex inverter chip in the application. By using NOR gates instead of OR gates for the clock control gates and using extra NOR gates as inverters, you could conceivably simplify this circuit by two chips.

One concept I haven't discussed yet is the use of multiple input gates as single input gates. As I have shown in Fig. 9.52, you can implement a NOT gate by tying together the inputs of an unused NOR gate. This trick also applies to NAND gates.

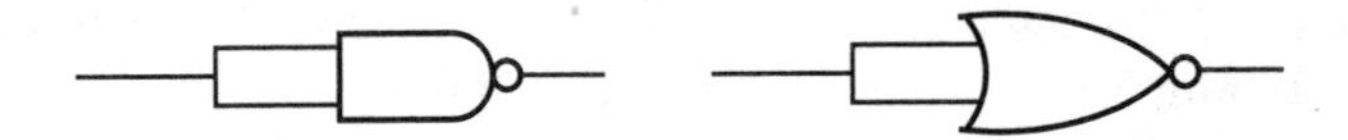

A NAND or a NOR with all inputs tied together is equivalent to a NOT:

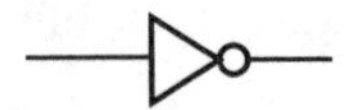

Figure 9.52 Using a NOR gate to simulate a NOT.

Two chips may not sound like a lot, but if you were in a manufacturing situation, these two chips could mean literally hundreds of thousands to millions of dollars in saved parts inventory and manufacturing costs. It could also mean a smaller final product or one that could be "swept" into a smaller-function PLD.

I did not optimize the circuit as I have described here because it would have made the final circuit just about incomprehensible and would lead to the question why did I put in NOR gates if I was just going to tie their inputs together. In this case, I had to balance showing a circuit that was reasonably easy to understand versus one that is very cryptic.

Chapter

10

Hardware Interfacing

So far in this book I have concentrated just on basic digital electronics and the common TTL chips that are used to produce applications that process data. While I have presented some information on button and switch input and LED status output, I really haven't gone into detail about interfacing digital circuits to different outside circuits. I find the most interesting aspect of working with electronics is how it interfaces to other devices and the real world. At the end of this chapter, I have included a tool that can be used to interface with different digital devices.

I also present some of the circuits needed to help implement and support digital electronic applications. In studying these circuits, you should get a better idea of how digital electronic applications are implemented.

When you look at the project presented in this chapter, you will find that it is very complex, with circuits that seem almost incomprehensible. In this application as well as others that you will see and work with, you will also find that there are cases where a very simple function seems to be implemented by very complex circuitry, as well as cases where very complex functions can be implemented with next to no gates.

This is very typical of digital electronics; the complexity of a circuit is dependent on its function, the available parts, and the designer. Each of these factors has about an equal impact on the complexity of the circuit and makes it hard to predict how an application will turn out. I will make a few comments about these three issues that you should keep in mind when you go through the projects.

The first is, when I present examples of how interfacing is done, I always try to err on the side of correctness. This is to say that when I design an interface to a device, I make sure that the correct interfacing waveform and signals are passed to it. There are several cases where I could have modified the interface protocols, but I have kept within the spirit of the datasheet of the affected parts. This is to ensure that I have created circuits that are as robust as possible. Improving on the interface is something that all designers will try to do if it makes the interface faster or better able to work with the interface functions already built into the application.

Second, as part of the effort to come up with the most "correct" use of different interfaces, I have also come up with specialized experiments and tests that I use to verify that the interface works as I expect. When I show these experiments in a project description, I recommend that you work through them along with the project to see how the device works or test out part of the application before you go on to the next subsystem.

Third, when you look at a manufacturer's book of published datasheets, you will find that there are a plethora of chips available, some of which provide exactly the function that you need for your application. What you will often find is that these specialized chips were built for a specific customer or they have been "obsoleted" because of relatively poor sales. The TTL standards have been around for over 30 years, so you shouldn't be too surprised if many of the documented chips are no longer available.

In this book, I have worked at specifying chips that are available through a wide range of distributors and retailers and are considered standard devices. This has made some of the projects in the book more complex, but for the most part, it has resulted in applications that you should be able to copy quite easily.

Last, you will find that as a designer you will think differently or approach problems differently from other designers. In the projects presented in this book, you should have realized that I have primarily focused on developing projects that are based on the NAND gate. In creating these projects, you will see common functions convoluted into something quite unrecognizable.

A good example of this was what I did with the XOR gate earlier in the book. Instead of using the straight circuit specified in most texts, I used the identities and laws in Appendix B to derive a number of boolean logic output expressions that could be used in different situations.

Even though I have tried to explain every instance in which I have done something (not so) clever, there will be situations where you do not understand why I have wired a circuit in a certain way. In these cases, I suggest that you try to follow through the text to see what was done and come up with your own way of implementing the function. If you come up with something that is smaller, faster (fewer gate delays), or cheaper, please let me know.

If you are familiar with my microcontroller books, you will probably look at the list of interfaces that I have given and you will see that I have taken out quite a few that you have used with the Microchip PICmicro microcontroller or 8051. In compiling the list of devices that you can interface to in this chapter, I have chosen only the ones that I felt could be interfaced to with a relatively simple interface. This is not to say that applications like Hitachi 44780–based LCDs cannot be controlled by TTL logic; instead I have just tried to keep the complexity of the applications to the level where you can easily see how they work.

Combining Input and Output

Often when working on applications, you will find some situations where peripheral devices will use more than one pin for I/O. Another case would be when you are connecting two devices, one input and one output, and would like to combine them somehow so you can reduce the number of I/O pins required. Fewer I/O pins means you can use less complex circuits and avoid complex application wiring. In this section, I will present you with two techniques for doing this and the rules governing their use. The ideas I am presenting here may at first appear problematic and possibly seem to cause problems with bus contention, but they really do work and can greatly simplify your application.

The most obvious ways of connecting two drivers together is to use tristate drivers, as discussed in the previous chapter. This type of connection, shown in Fig. 10.1, allows two devices with drivers and receivers to communicate. This method works well, but it is the most complex of the three methods that I will show in this section. The reason why it is complex is that the two circuits need to determine when they should be transmitting (driving the net) to avoid bus contention.

Open-collector drivers can be used in this application to avoid the possibility of bus contention, as is shown in Fig. 10.2. The pull-up ensures that the bus

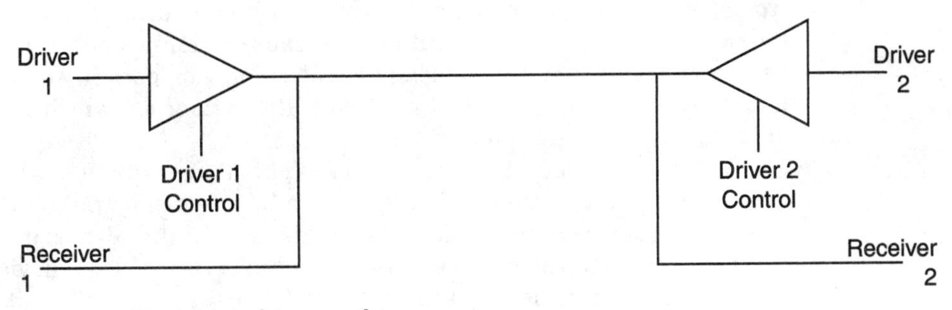

Figure 10.1 Two tristate drivers on the same net.

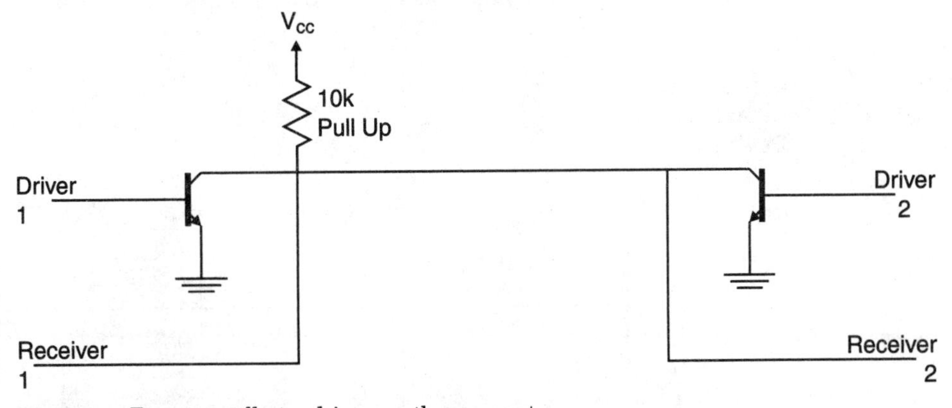

Figure 10.2 Two open collector drivers on the same net.

can float high. In this case, either driver can be active on the bus without there being a concern for bus contention. There is, however, the opportunity for "collisions" on the bus when both devices are driving at the same time. Normally an open-collector bus like this is used with intelligent devices that can determine if there was a collision and resolve it by waiting some length of time before resending the data.

When you interface the bidirectional digital logic I/O pin to a driver and a CMOS receiver (such as a memory with a separate output and input), you can use a resistor to avoid bus contention at any of the pins, as shown in Fig. 10.3.

In this situation, when the bidirectional I/O pin is driving an output, it will be driving the data-in pin register, regardless of the output of the data-out pin. If the bidirectional and data-out pins are driving different logic levels, the resistor will limit the current flowing between the bidirectional and the memory data-out pin. The value received on the data-in pin will be the bidirectional device's output.

When the bidirectional device is receiving data from the memory, the I/O pin will be put in *input* (or *high-impedance*) mode and the data-out pin will drive its value to not only the bidirectional device's I/O pin but the data-in pin as well. In this situation, the data-in pin should not be latching any data in. To avoid this, in most cases where this circuit is combining input and output, the two input and output pins are on the same device and the data mode is controlled by the bidirectional device to prevent invalid data from being input into the device. This is an important point because it defines how this trick should be used. A common use for this method of connection data in and data out pins is used in Serial Peripheral Interface (SPI) memories, which have separate data input and output pins.

The second trick is to have button input along with an external CMOS device receiver. As is shown in Fig. 10.3, a button can be put on the same net as an input device and the bidirectional device pin that drives it.

When the button is open or closed, the bidirectional logic device can drive data to the input device, and the 100k and 10k resistors will limit the current flow between V_{cc} and ground. If the bidirectional logic device is going to read, the button high (switch open) or low (switch closed) will be driven on the bus

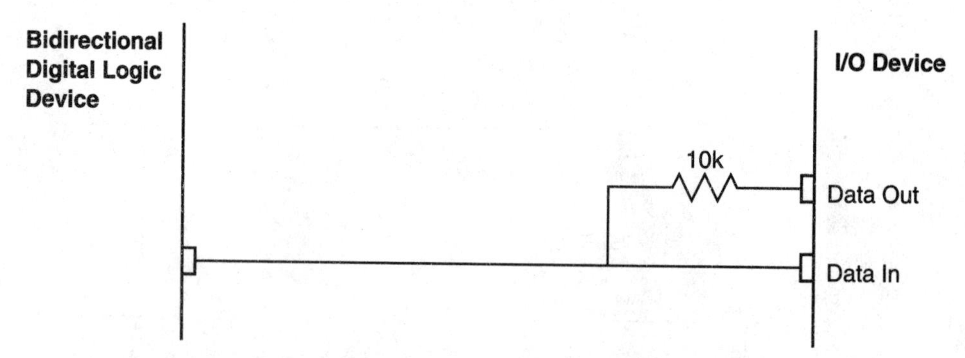

Figure 10.3 Combining CMOS I/O on one pin.

at low currents when the pin is in input mode. If the button switch is open, the 100k resistor acts like a pull-up and a 1 is returned. When the button switch is closed, there will be approximately a 0.5 V across the 10k resistor, which will be read as a 0.

The button with the two resistors pulling up and down is like a low-current driver and the voltage produced is easily overpowered by active drivers. As in the previous method, the external input device cannot receive data except when the bidirectional device is driving the circuit. A separate clock or enable should be used to ensure that input data is received when the bidirectional device is driving the line.

Two points about this method: (1) The second method can be extrapolated to work with a switch matrix keyboard; this can become very complex, but it will work. (2) A resistor-capacitor network for debouncing the button cannot be used with this circuit because the resistor-capacitor network will slow down the response of the bidirectional device driving the data input pin and will cause problems with the correct value being passed between the chips. When a button is shared with an input device, as shown in Fig. 10.4, software button debouncing will have to be done by an intelligent device, such as a microcontroller.

For the method of including multiple drivers on the same net shown in Figs. 10.3 and 10.4, I specified CMOS drivers because CMOS does not require actual current flow as does TTL. You *may* be able to use TTL drivers with these circuits, but you may find they are unreliable. To avoid problems with invalid currents being available to CMOS receivers, just use the last two circuits with CMOS digital logic.

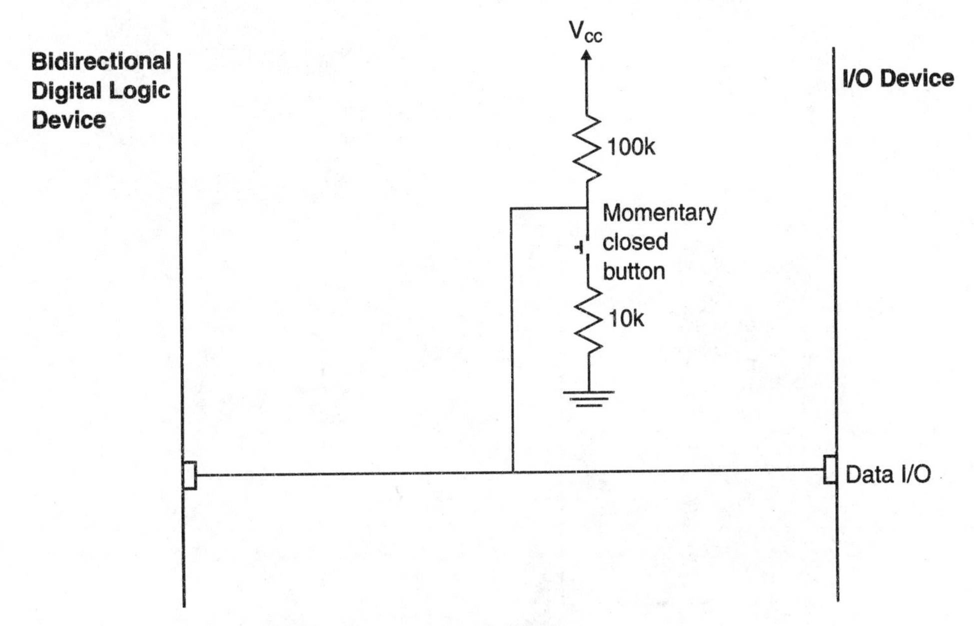

Figure 10.4 Combining CMOS button input with I/O pin.

LEDs

The most common form of output for a digital electronics device is the *light-emitting diode* (LED). As an output device, it is cheap and easy to wire to a digital device. Generally, LEDs require anywhere from 5 to 20 mA of current to light (which is often within the sink/source specification for most digital electronic chips' output pins). But what should be remembered is that LEDs are diodes, which means current flows in one direction only. In practice, the typical circuit that I use to control an LED from an I/O pin is that shown in Fig. 10.5. With this circuit, the LED will light when the output pin is set to 0 (ground potential). When the pin is set to input, or outputs a 1, the LED will be turned off.

In this book, I have used the output pin to drive the current to the LED directly so that a 1 will be on and a 0 will be off. This convention is a lot more intuitive to people new to electronics than relying on the reverse. The reason for using the reverse as a matter of course is that most different types of logic devices are designed to current-sink larger amounts of current than they source.

I assume that all LEDs have a voltage drop of 2.0 V across them. As I indicated earlier in the book, a normal silicon diode has a voltage drop of typically 0.7 V. You should realize that an LED is not a "typical" diode and should not be used for any purpose except as a user display. The LED may not have the speed or current-carrying characteristics of traditional diodes.

The 220-Ω resistor is used for current limiting and will prevent excessive current that can damage the microcontroller, LED, and the power supply. Assuming that there is a 2.0-V drop across the LED, the current flowing through the LED can be calculated:

$$\begin{aligned}\text{Current} &= V/R \\ &= (5\text{ V} - 2\text{ V})/220\ \Omega \\ &= 13.6\text{ mA}\end{aligned}$$

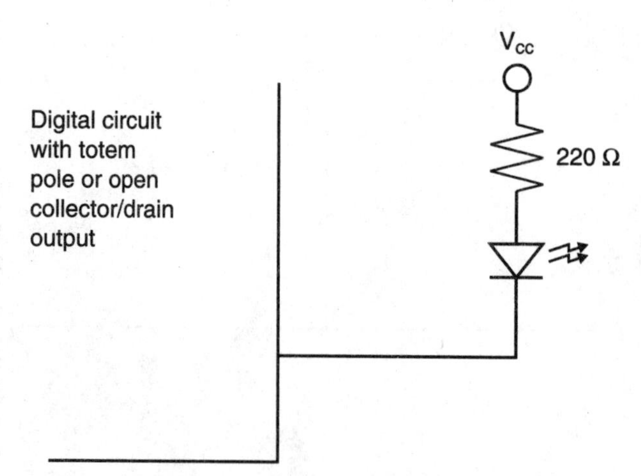

Figure 10.5 LED connection to a digital driver.

This current is approximate, as the applied voltage and voltage drop across the LED may change. The 220-Ω resistor gives a good range of currents in these situations without causing the LED to dim or not light at all.

Some CMOS (and TTL) devices have current-limiting capabilities built in and can drive LEDs without any current-limiting resistors. In these cases, you may want to dispense with the current-limiting resistor—but I would recommend that you don't. The current-limiting resistor will limit the amount of total current used by the application and may allow you to use battery power instead of having to provide a household mains power supply.

Probably the easiest way to output numeric (both decimal and hex) data is via *seven-segment LED displays.* These displays were very popular in the '70s (if you're old enough, your first digital watch probably had seven-Segment LED displays), but they have been largely replaced by *liquid crystal displays* (LCDs).

Seven-segment LED displays (Fig. 10.6) are still useful devices that can be added to a circuit without a lot of effort. By turning on specific LED (each of which lights up a segment in the display), the display can be used to output decimal numbers.

Each one of the LEDs in the display is given an identifier and a single pin of the LED is brought out of the package. The other LED pins are connected

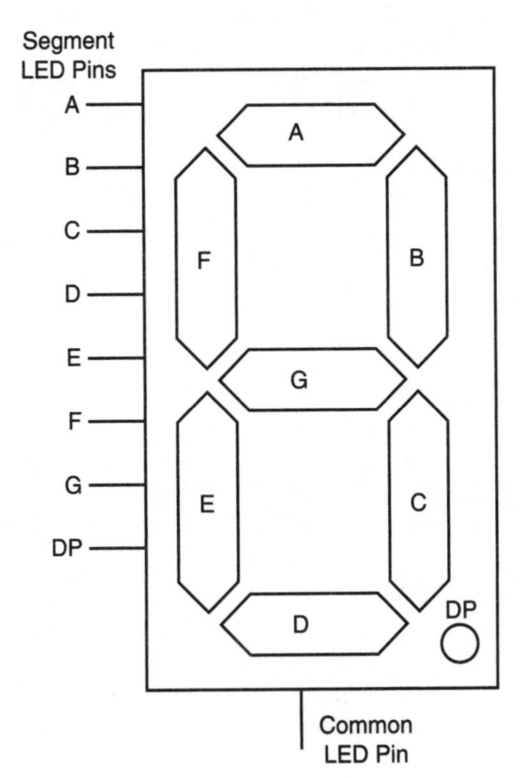

Figure 10.6 Organization of a seven-segment LED display.

together and wired to a common pin. This common LED pin is used to identify the type of seven-segment display (as either *common cathode* or *common anode*).

Wiring one display to a digital device is quite easy. It is typically wired as seven [or eight, if the decimal point (DP) is used] LEDs wired to individual pins.

You can develop your own circuit for decoding an incoming binary number and display it on a seven-segment LED. Many courses in basic electronics will have you do that to try and understand how to optimize a boolean logic circuit. Developing the logic functions for a seven-segment LED display comes under the heading of being too much work for the return that you will get. As I will show later in this chapter, I will use an LED driver chip to provide the decoding and seven-segment LED display driving.

A common method of wiring multiple seven-segment LED displays together is to wire them all in parallel and then control the current flow through the common pin. Because the current is generally too high for a single TTL or CMOS pin, a transistor is used to pass the current to the common power signal. This transistor selects which display is active. To avoid flicker, I generally run the application's switching so that each digit is turned on/off at least 50 times per second. The more digits you have, the faster you have to cycle the interrupt handler (i.e., 8 seven-segment displays must cycle at least 400 digits per second, which is twice as fast as four displays).

In Fig. 10.7, four common-cathode seven-segment displays are shown connected to a microcontroller. In this circuit, the controller will shift between the displays, showing each digit in a very short "time slice." This is usually done by using a clock or in a microcontroller in a timer interrupt handler that is initiated by timer.

You may feel that assigning a specific bit to select each display LED is somewhat wasteful (at least I do). I have used high-current TTL demultiplexer (i.e., 74S138) outputs as the cathode path to ground (instead of discrete transis-

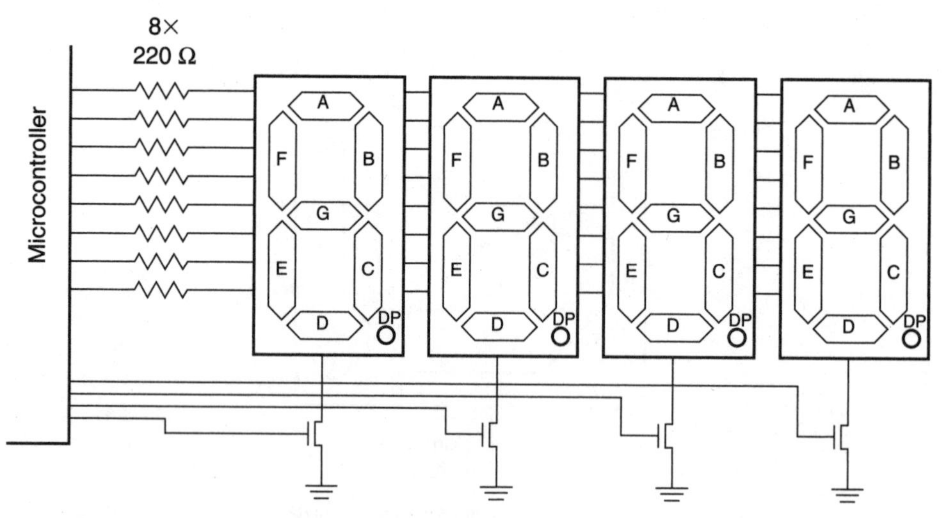

Figure 10.7 Wiring 4 seven-segment LED displays.

tors). When the output is selected from the demultiplexer, it goes low, allowing current to flow through the LEDs of that display (and turning it on). This actually simplifies the wiring of the final application as well. The only issue is to make sure the demultiplexor output can sink the maximum of 35 to 140 mA of current that will be sunk through the common cathode connection.

In addition to seven-segment displays, there are 14- and 16-segment LED displays available that can be used to display alphanumeric characters (A to Z and 0 to 9). By following the same rules as used to wire up a seven-segment display, you shouldn't have any problems with wiring the display to digital electronic chips. The problem will be in finding a decoder/display device that can be used to display the characters. If you are going to work with a 14- or 16-segment LED display, then I recommend that you use a microcontroller, as it will be cheaper and will allow you to develop your own character sets.

Asynchronous Serial Communications

Asynchronous, or non-return-to-zero (NRZ), long-distance communications came about as a result of the Baudot teletypewriter. This device mechanically (and, later, electronically) sent a string of electrical signals (which we would call bits) to a receiving printer.

This data packet format is still used today for the asynchronous transmission protocols used in most computer systems. With the invention of the teletype, data could be sent and retrieved automatically without having an operator sitting by the teletype all night (unless an urgent message was expected—normally, the nightly messages could be read in the morning).

Before going on, there is one point some people get unreasonably angry about and that's the definition and usage of the terms *data rate* and *baud rate.*

The baud rate is the maximum number of possible data bit transitions per second. This includes the start, parity, and stop bits at the ends of the data packet shown in Fig. 10.8, as well as the 5 data bits in the middle. I use the term *packet* because we are including more than just data (there is also some additional information in there), so *character* or *byte* (if there were 8 bits of data) is not an appropriate term.

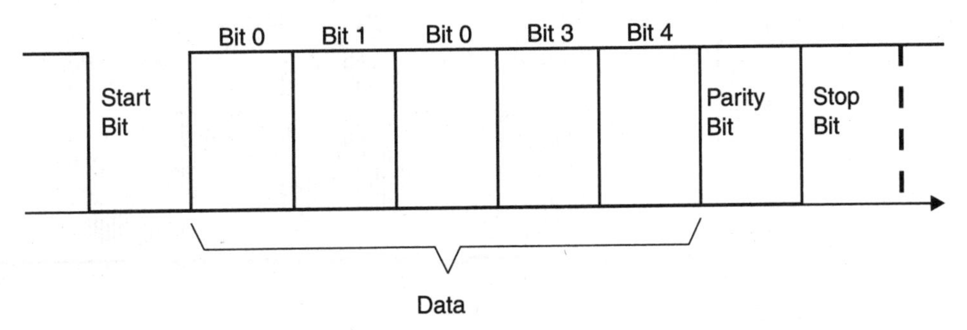

Figure 10.8 Baudot asynchronous serial data.

This means that for every 5 data bits transmitted, 8 bits in total are transmitted (which means that nearly 40 percent of the data transmission bandwidth is lost in teletype asynchronous serial communications).

The *data rate* is the number of data bits that are transmitted per second. For this example, if you were transmitting at 110 baud (which is a common teletype data speed), the actual data rate is 68.75 bits per second (or, assuming 5 bits per character, 13.75 characters per second).

I tend to use the term *data rate* to describe the *baud rate.* This means that when I say *data rate,* I am specifying the number of bits of *all* types that can be transmitted in a given period of time (usually 1 second). I realize that this is not absolutely correct, but it makes sense to me to use it in this form and I will be consistent throughout the book (and I will not use the term *baud rate*).

With only 5 data bits, the baudot code could transmit only up to 32 distinct characters. To handle a complete character set, a special 5-digit code was used to notify the receiving teletype that the next 5-bit character would be an extended character. With the alphabet and most common punctuation characters in the primary 32, this second data packet wasn't required very often.

In the data packet diagram in Fig. 10.8, there are 3 control bits. The start bit is used to synchronize the receiver to the incoming data. In most UARTs (universal asynchronous receiver/transmitters), there is an overspeed clock shown in Fig. 10.9 (running at 16 times the incoming bit speed) that samples the incoming data and determines whether the data is valid.

When waiting for a character, the receiver hardware polls the line repeatedly at 1/16 bit period intervals until a 0 (space) is detected. The receiver then waits half a cycle before polling the line again to see if a "glitch" was detected

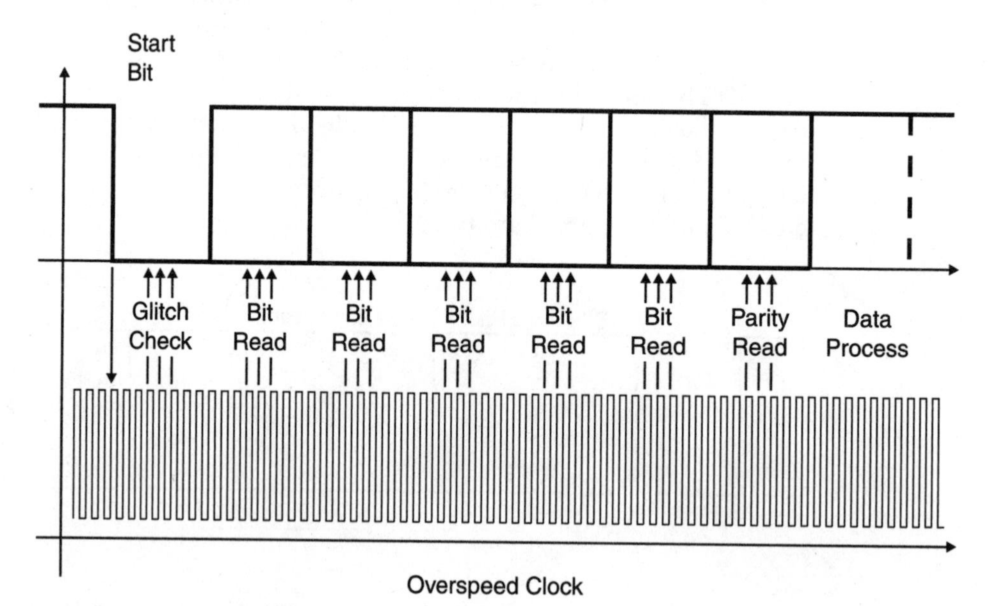

Figure 10.9 Reading an asynchronous data packet.

and not a start bit. Note that polling will take place in the middle of each bit to avoid problems with bit transitions (or if the transmitter's clock is slightly different from the receivers, to minimize the chance of misreading a bit).

Once the start bit is validated, the receiver hardware polls the incoming data once every bit period multiple times (again to ensure that glitches are not read as incorrect data).

The stop bit was originally provided to give both the receiver and the transmitter some time before the next packet is transferred (in early computers, the serial data stream was created and processed by the computers and not custom hardware as in modern computers).

The *parity* bit is a crude method of error detection that was first brought in with teletypes. The purpose of the parity bit is to indicate whether the data was received correctly. An *odd* parity meant that if all the data bits and parity bits set to a "mark" (1) were counted, then the result would be an odd number. *Even* parity is checking all the data and parity bits and seeing if the number of mark bits is an even number. Along with even and odd parity, there are *mark, space,* and *no* parity. *Mark parity* means that the parity bit is always set to a 1, *space parity* means always having a 0 for the parity bit, and *no parity* means eliminating the parity bit altogether.

The most common form of asynchronous serial data packet is 8-N-1, which means 8 data bits, no parity bit, and 1 stop bit. This reflects the capabilities of modern computers to handle the maximum amount of data with the minimum amount of overhead and with a very high degree of confidence that the data will be correct.

I stated that parity bits are a crude form of error detection. I said that because they can only detect 1 bit error (i.e., if two bits are in error, the parity check will not detect the problem). If you are working in a high-induced-noise environment, you may want to consider using a data protocol that can detect (and, ideally, correct) multiple bit errors.

In the early days of computing (the 1950s), while data could be transmitted at high speed, it couldn't be read and processed continuously. So, a set of "handshaking" lines and protocols were developed for what became known as *RS-232 serial communications.*

Within the original RS-232 implementation, the typical packet contained 7 bits, (which is the number of bits each ASCII character contains). This simplified the transmission of human-readable text, but made sending object code and data (which were arranged as bytes) more complex because each byte would have to be split up into two "nybbles" (which are 4 bits long). Further complicating this is the need for the first 32 characters of the ASCII character set to be reserved as "special" characters (carriage return, backspace, etc.). This meant the data nybbles would have to be converted (or shifted up) into valid characters (this is why, if you ever see binary data transmitted from a modem or embedded in an e-mail message, data is either sent as hex codes or the letters A to Q). With this protocol, to send a single byte of data, two bytes (with the overhead bits resulting in 20 bits in total) would have to be sent (and, surprisingly enough, to send data would take twice as long as sending a text file of the same length).

As I pointed out above, modern asynchronous serial data transmission normally sends 8 bits at a time, which will avoid this problem and allow transmission of full bytes without breaking them up or converting them.

DTE stands for *data terminal equipment* and designates the connector used for computers (the PC uses this type of connection). *DCE,* or data communications equipment, applies to modems and terminals that transfer the data.

Understanding which RS-232 model a piece of equipment fits under is critical to successfully connecting two devices by RS-232. With a pretty good understanding of the serial data, the actual voltage signals can be examined.

As I've mentioned above, when RS-232 was first developed into a standard, computers and the electronics that drive them were still very primitive and unreliable. Because of that, we've got a couple of legacies to deal with.

The first is the voltage levels of the data. A mark (1) is actually −12 V and a space (0) is +12 V. Figure 10.10 shows what this looks like.

From Fig. 10.10, you should see that the hardware interface is not simply a TTL or CMOS level buffer. Later in this section, I will introduce you to some methods of generating and detecting these interface voltages. Voltages in the switching region (±3 V) may or may not be read as a 0 or 1, depending on the receiver hardware. You should always make sure the voltages going into a RS-232 circuit are in the valid regions to ensure all receivers will properly receive the data.

Of more concern are the handshaking signals. These six additional lines (which are at the same logic levels as the transmit/receive lines shown in Fig. 10.10) are used to interface between devices and control the flow of information between computers.

The request-to-send (RTS) and clear-to-send (CTS) lines are used to control data flow between the computer (DCE) and the modem (DTE device). When a PC is ready to send data, it asserts (outputs a mark) on RTS. If the DTE device is capable of receiving data, it will assert the CTS line. If the PC is unable to

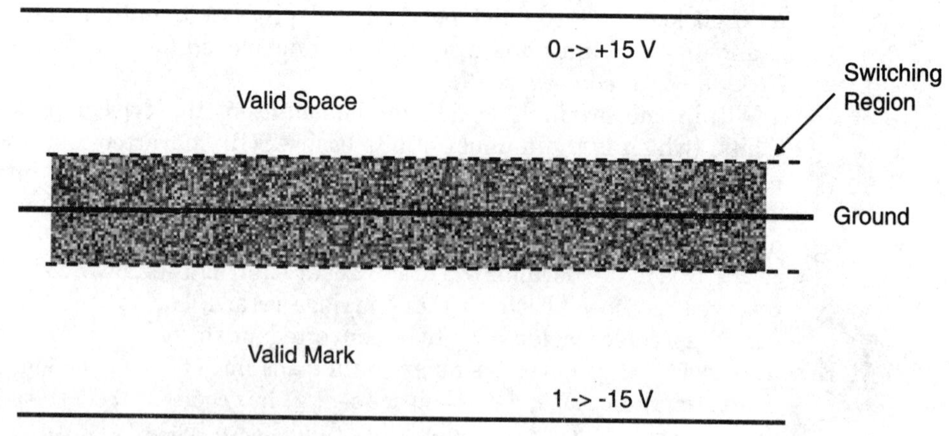

Figure 10.10 RS-232 voltage levels.

receive data (i.e., the buffer is full or it is processing what it already has), it will deassert the RTS line to notify the DTE device that it cannot receive any additional information.

The data transmitter ready (DTR) and data set ready (DSR) lines are used to establish communications. When the PC is ready to communicate with the DTE device, it asserts DTR. If the DTE device is available and ready to accept data, it will assert DSR to notify the computer that the link is up and ready for data transmission. If there is a hardware error in the link, then the DTE device will deassert the DSR line to notify the computer of the problem.

There are two more handshaking lines that are available in the RS-232 standard that you should be aware of, even though chances are you will never connect anything to them. The first is the data carrier detect (DCD), which is asserted when the modem has connected with another device and they have established a communication link. The ring indicator (RI) is used to indicate to a PC whether the phone on the other end of the line is ringing or is busy. This line is very rarely used in modern applications. The ring indicator function is normally provided within the AT modem command set protocol.

There is a common ground connection between the DCE and DTE devices. This connection is critical for the RS-232 level converters to determine the actual incoming voltages. The ground pin should never be connected to a chassis or shield ground (to avoid large current flows or be shifted and prevent accurate reading of incoming voltage signals). Incorrect grounding of an application can cause the computer, or the device it is interfacing, to reset or the power supplies to blow a fuse or burn out. The latter consequences are unlikely, but I have seen it happen in a few cases.

To avoid these problems, make sure that chassis and signal grounds are separate or connected by a high-value resistor (hundreds of kilohms).

Before going too much further, I should expose you to an ugly truth: the handshaking lines are almost never used in RS-232 communications. The handshaking protocols were added to the RS-232 standard when computers were very slow and unreliable. In this environment, data transmission had to be stopped periodically to allow the receiving equipment to catch up.

Today, this is much less of a concern and normally three-wire RS-232 connections are implemented as in Fig. 10.11. I normally accomplish this by shorting the DTR/DSR and RTS/CTS lines together at the external device end. The DCD and RI lines are left unconnected.

With the handshaking lines shorted together, data can be sent and received without the need for software to handle the different handshaking protocols.

I want to make a few points about three-wire RS-232. The first is that it cannot be implemented blindly; in about 20 percent of the RS-232 applications that I have worked on over the years I have had to implement some subset of the total seven-wire protocol lines (transmit, receive, ground, and four handshaking lines). Interestingly enough, I have never had to implement the full

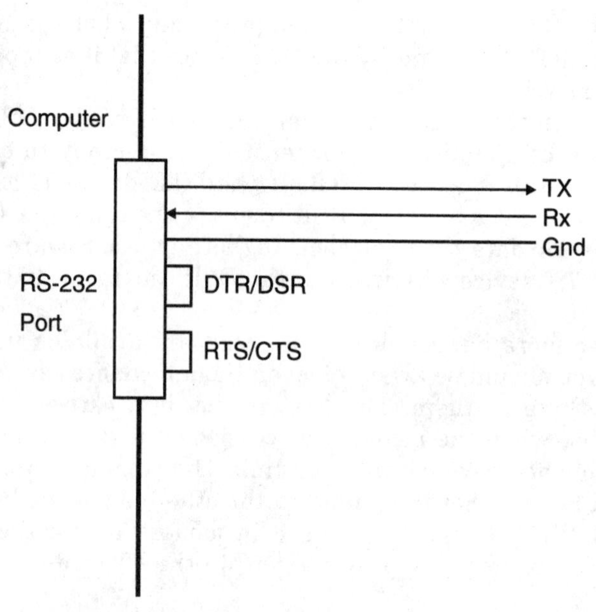

Figure 10.11 Typical RS-232 wiring.

hardware protocol. This still means that four out of five times, if you wire the connection as shown in Fig. 10.11, the application will work.

The second point is that, with the three-wire RS-232 protocol, there may be applications where you don't want to implement the hardware handshaking (the DTR, DSR, RTS, and CTS lines); you may want to implement software handshaking. There are two primary standards in place. The most common standard for doing this is the XON/XOFF protocol, in which the receiver sends an XOFF (DC3 or character 0x013) when it can't accept any more data. When it is able to receive data, it sends an XON (DC1 or character 0x011) to notify the transmitter that it can receive more data.

The third aspect of the RS-232 I want to discuss is the speeds at which data is transferred. When you first see the speeds (such as 300, 2400, and 9600 bits per second), they seem rather arbitrary. The original serial data speeds were chosen for teletypes because they gave the mechanical device enough time to print the current character and reset before the next one came in. Over time, these speeds have become standards and as faster devices have become available, they've just been doubled (i.e., 9600 bps is 300 bps doubled 5 times).

To produce these data rates, most USARTs use a clock 16 times the data rate. Most even-megahertz operating clocks are divided by integers to get the nominal RS-232 speeds. This might seem like it won't work out well, but because of RS-232's strange relationship with the number 13, the situation isn't as bad as it may seem. If you invert (to get the period of a bit) the data speeds and convert the units to microseconds, you will discover that the periods are almost exactly divisible by 13. This means that you can use an even-megahertz oscillator in the hardware to communicate over RS-232 at standard frequencies.

For example, if you had a circuit running with a 5-MHz instruction clock and you wanted to communicate with a PC at 9600 bps, you would determine the number of cycles to delay by:

1. Finding the bit period in microseconds. For 9600 bps, this is 104 μsec.
2. Dividing this bit period by 13 to get a multiple number. For 104 μsec, this is 8.

Now, if the external device is running at 20 MHz (which means a 200-nsec cycle time), you can figure out the number of cycles as multiples of 8 × 13 in the number of cycles in 1 μsec. To get the total number of cycles for the 104-μsec bit period in the circuit clocked at 5 MHz, you simply evaluate:

$$\text{S cycles}/\mu\text{sec} \times 13 \times 5\ \mu\text{sec/bit} = 1300\ \text{cycles/bit}$$

The device you are most likely to interface to is the PC and its serial ports consisting of basically the same hardware and BIOS interfaces that were first introduced with the first PC in 1981. Since that time, a 9-pin connector has been specified for the port (in the PC/AT) and there has been one significant hardware upgrade, introduced when the PS/2 was announced. For the most part, the serial port has changed the least of any component in the PC for the past 17 plus years.

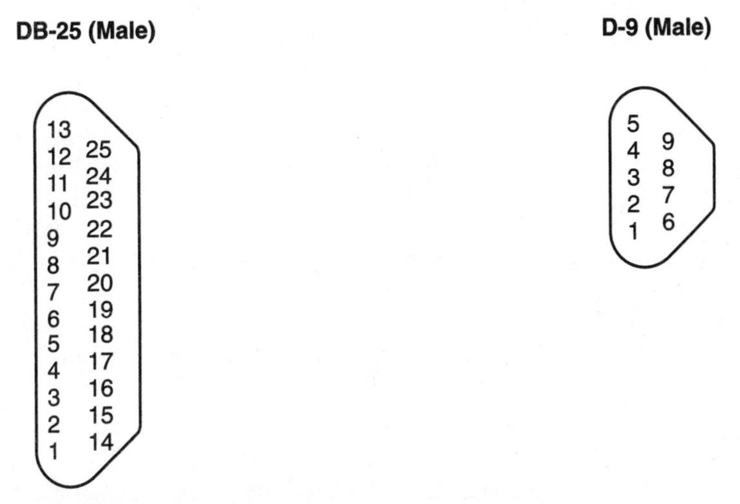

Figure 10.12 IBM PC DB-25 and D-9 RS-232 connectors.

Either a male 25-pin or male 9-pin connector is available on the back of the PC for each serial port. These connectors are shown in Fig. 10.12 on the previous page. These connectors are wired as:

Pin name	25 pin	9 pin	I/O direction
TxD	2	3	Output (O)
RxD	3	2	Input (I)
Gnd	7	5	
RTS	4	7	O
CTS	5	8	I
DTR	20	4	O
DSR	6	6	I
RI	22	9	I
DCD	8	1	I

The 9-pin standard was originally developed for the PC/AT because the serial port was put on the same adapter card as the printer port, and there wasn't enough room for the serial port *and* parallel port to both use 25-pin D-shell connectors. Actually, I prefer the smaller-form-factor connector.

RS-232 serial communications have a reputation for being difficult to use, and I have to disagree. Implementing RS-232 isn't very hard when a few rules are followed and you have a good idea of what's possible.

When implementing an RS-232 interface, you can make your life easier by doing a few simple things. The first is the connection. Whenever I do an application, I standardize on using a 9-pin D shell with the DTE interface (the one that comes out of the PC) and use standard "straight-through" cables. The external devices are wired as DCE. In doing this, I always know what my pin-out is at the end of the cable when I'm about to hook up an external device to a PC.

By making the external device always DCE and using a standard pin-out, I don't have to fool around with null modems or making my own cables.

When I am creating the external device, I also loop back the DTR/DSR and CTS/RTS data pairs inside the external device, rather than at the PC or in the cable. This allows me to use a standard PC and cable without having to do any wiring on my own or any modifications. It actually looks a lot more professional as well.

Tying DTR/DSR and CTS/RTS also means that I can take advantage of built-in terminal emulators. Virtually all operating systems have a built in "dumb" terminal emulator that can be used for debugging the external device without requiring the PC code to run. Getting the external device working before debugging the PC application code should simplify the work that you have to do.

As I went through the RS-232 electrical standard earlier in this section, you were probably concerned about interfacing standard, modern technology (i.e., TTL and CMOS) devices to other RS-232 devices. This is actually a legitimate concern because without proper voltage level conversion, you will not be able to read from or write to external TTL or CMOS devices. Fortunately, this conversion isn't all that difficult, or rather I should say there are methods that can make it quite easy.

If you look at the original IBM PC RS-232 port specification, you'll see that 1488/1489 RS-232 level converter circuits were used for the RS-232 serial port Interfaces. The pin-out and wiring for these devices in a PC are shown in Fig. 10.13.

I should make a few comments about the 1488/1489 components. The first comment applies to transmitting data; each transceiver (except for 1) is actually a NAND gate (with the inputs being *X*A and *X*B outputting on *XY*). When I wire in a 1488, I make sure that the second input to a driver is always pulled high (as I've done with 2B in Fig. 10.13.

The second comment has to do with the 1489 receiver. The *X*C input is a flow control for the gates (normally RS-232 comes in the *X*A pin and is driven as TTL out of *XY*). This pin is normally left floating (unconnected).

These chips are still available and work very well (up to the 115,200 bps maximum PC RS-232 data rate). I'd never use them in my own projects because the 1488 (transmitter) requires ±12-V sources in order to produce valid RS-232 signal voltages.

I want to present to you three methods that you can choose from for converting RS-232 signal levels to TTL/CMOS (and back again) when you are creating digital electronic RS-232 interfaces. These three methods do not require ±12 V and in fact just require the +5-V supply that is used for logic power.

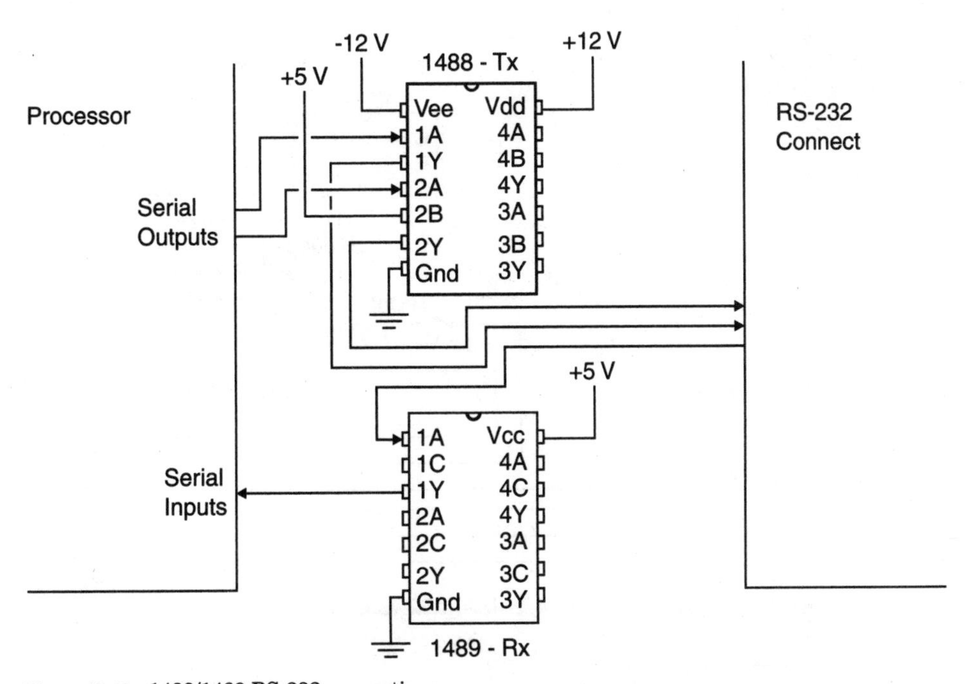

Figure 10.13 1488/1489 RS-232 connections.

The first method is to use an RS-232 converter that has a built-in charge-pump to create the ±12 V required for the RS-232 signal levels. Probably the most well-known chip that is used for this function is the MAXIM MAX232 (see Fig. 10.14). This chip is ideal for implementing three-wire RS-232 interfaces (or add a simple DTR/DSR or RTS/CTS handshaking interface). Ground for the incoming signal is connected to the processor ground (which is not the case's ground).

In addition to the MAX232, MAXIM and some other chip vendors have a number of other RS-232 charge-pump-equipped devices that will allow you to handle more RS-232 lines (to include the handshaking lines). Some charge pump devices are available that do not require the external capacitors that the MAX232 chip does; they will simplify the layout of your circuit (although these chips do cost quite a bit more).

The next method of translating RS-232 and TTL/CMOS voltage levels is to use the transmitter's negative voltage, as shown in Fig. 10.15. This circuit relies on the RS-232 communications running only in half-duplex mode (i.e., only one device can transmit at a given time). When the external device wants to transmit to the PC, it sends the data either as a mark (leaving the voltage being returned to the PC as a negative value) or as a space (turning on the transistor and enabling the positive voltage output to the PC's receivers). If you go back to the RS-232 voltage specification drawing (Fig. 10.10), you'll see that +5 V is within the valid voltage range for RS-232 spaces.

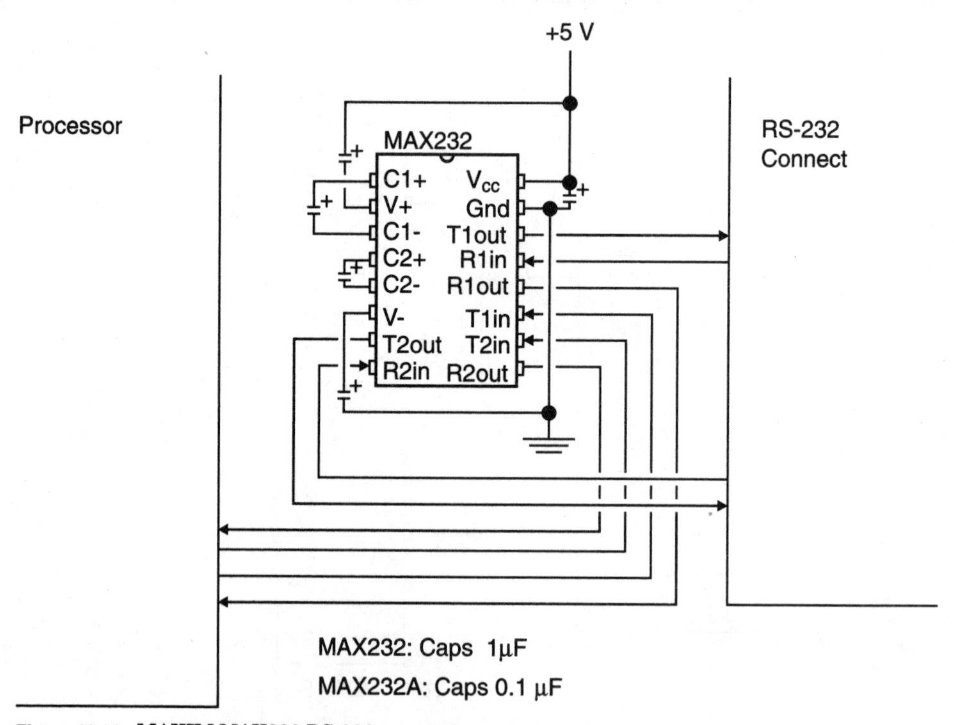

Figure 10.14 MAXIM MAX232 RS-232 connections.

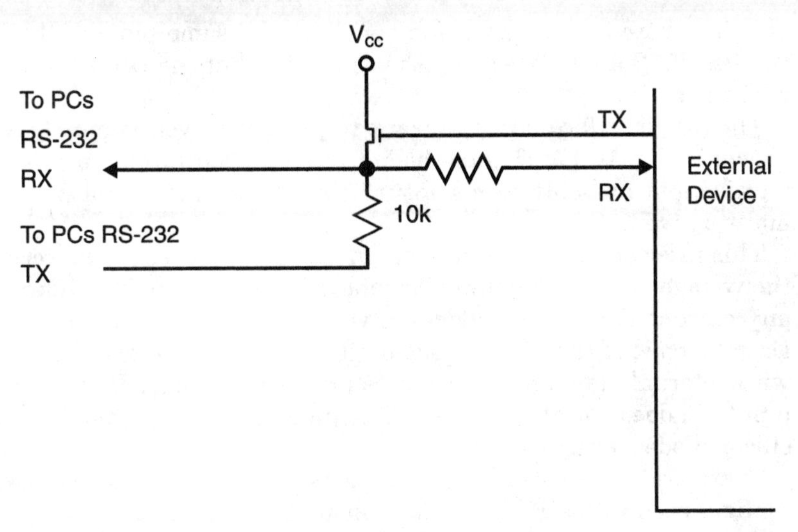

Figure 10.15 RS-232 to external device connection.

This method works very well (consuming just about no power) and is obviously a very cheap way to implement a three-wire RS-232 bidirectional interface.

There are a few comments I want to make about this circuit. In the diagram above, I show an N-channel MOSFET device simply because it does not require a base-current-limiting transistor the same way a bipolar transistor would.

When the external device transmits a byte to the external device through this circuit, it will receive the packet it's sent because this circuit connects the external device's receiving pin (more or less) directly to its transmitting pin. The software running in the external device (as well as the external device) will have to handle this.

You also have to make absolutely sure you are transmitting only in half-duplex mode. If both the internal device and the external device attempt to transmit at the same time, then both messages will be garbled. Instead, the transmission protocol that you use should wait for requested responses, rather than send them asynchronously (or, you could have one device—either the PC or external devices—wait for a request for data from the other).

Another issue is that data out of the external device will have to be inverted to get the correct transmission voltage levels (i.e., a 0 will output a 1) to make sure the transistor turns on at the right time (i.e., a positive voltage for a space). Unfortunately, this means that the built-in serial port for many microcontrollers cannot be used because it cannot invert the data output as required by the circuit. An inverter could be put between the serial port and the RS-232 conversion circuit to avoid this problem. However, there is a chip, the Dallas Semiconductor DS275, that basically incorporates the circuit above (with a built-in inverter) into the single package shown in Fig. 10.16. I should point out that there are two part numbers with the same pin-out, the DS1275 and the DS275. Both work exactly the same way, but the DS275 is a later version of the part.

The last interface circuit I want to present to you is simply a resistor, as shown in Fig. 10.17. This method of receiving data from an RS-232 Device to a logic input probably seems absurdly simple and could not work. But it does, and very well.

This interface works by relying on clamping diodes in the receiver holding the voltage at the maximum allowable for the receiver. The 10k resistor limits any current flows and provides a voltage drop as shown in Fig. 10.18. The actual circuit of the 10k current-limiting resistor and external device I/O pin with internal clamping diodes appears in Fig. 10.19. If the external device input pin does not have internal clamping diodes, then you can add your own silicon diodes for this function.

There are a few rules for this implementation. Some people like to use this configuration with a 100k+ resistor instead of the 10k shown above. Personally, I don't like to use anything higher than 10k because of the possibility of induced noise with a CMOS input (this is less likely with TTL) causing an incorrect data

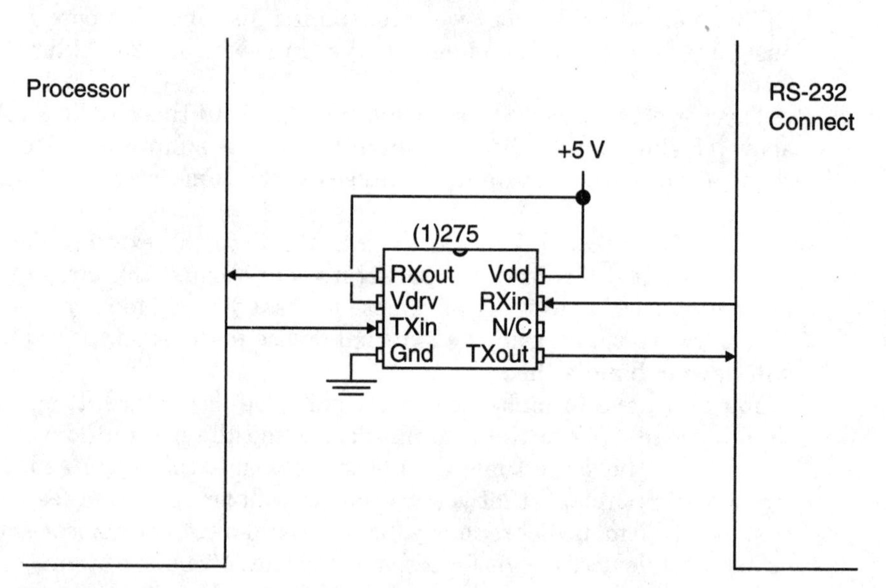

Figure 10.16 Dal Semi (1)275 RS-232 interface.

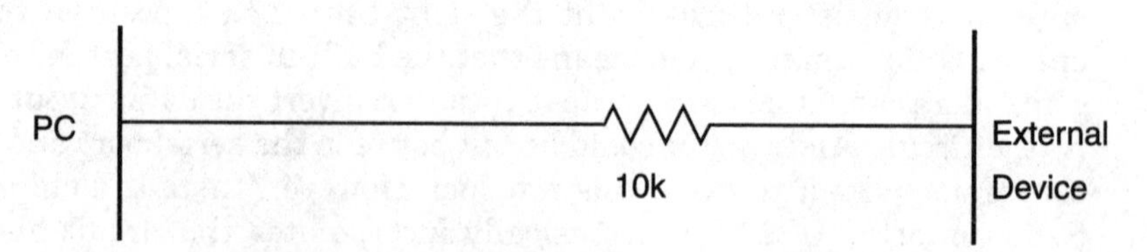

Figure 10.17 Simple RS-232 to TTL/CMOS voltage conversion.

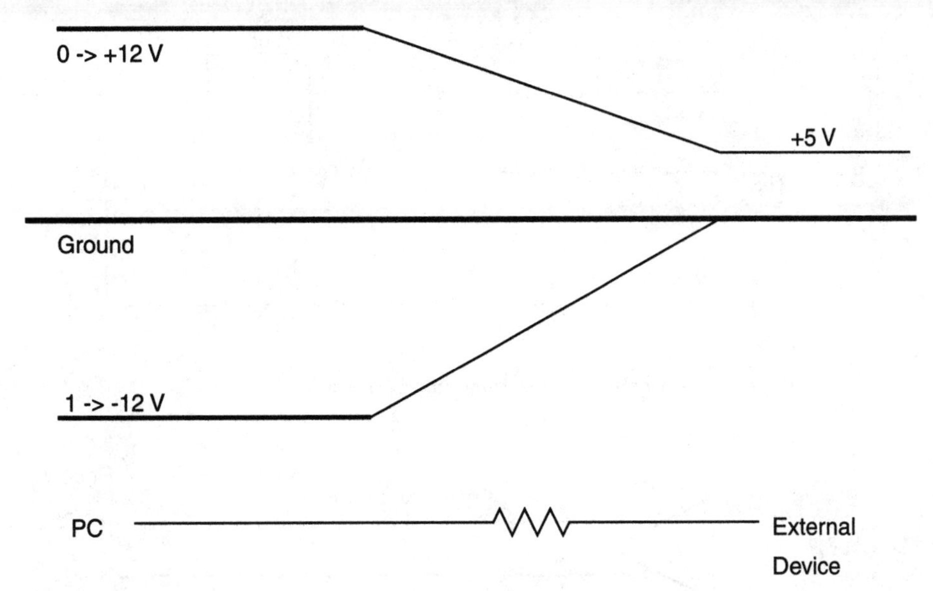

Figure 10.18 RS-232/resistor voltage conversion.

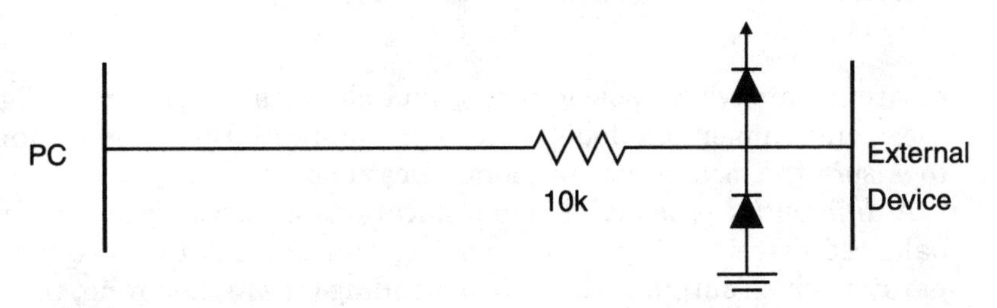

Figure 10.19 Simple RS-232 to TTL/CMOS voltage conversion with clamping diodes.

to be read. A 10k resistor will maximize the current required to change the state of the I/O Pin.

With the availability of many CMOS devices requiring minimal amounts of current to operate, you might be wondering about different options for powering your circuit. One of the most innovative that I have come across is using the PC's RS-232 ports itself as powering devices that can be attached by the circuit shown in Fig. 10.20.

When the DTR and RTS lines are driving a space, a positive voltage (relative to ground) is available. This voltage can be regulated and the output used to power the devices attached to the serial port (up to about 5 mA). For extra current, the TX line can also be added into the circuit, with a break being sent from the PC to output a positive voltage.

So far in this section I have discussed single-ended asynchronous serial communications methods such as RS-232 and direct NRZ device interfaces. These interfaces work well in home and office environments, but can be unreliable in

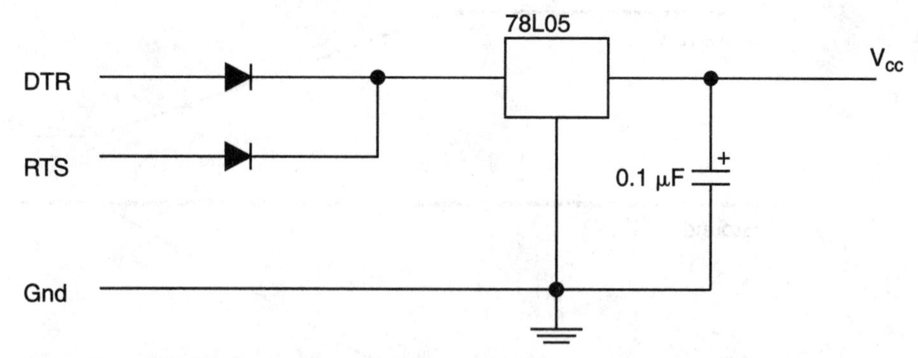

Figure 10.20 "Stealing" power from the PC's serial port.

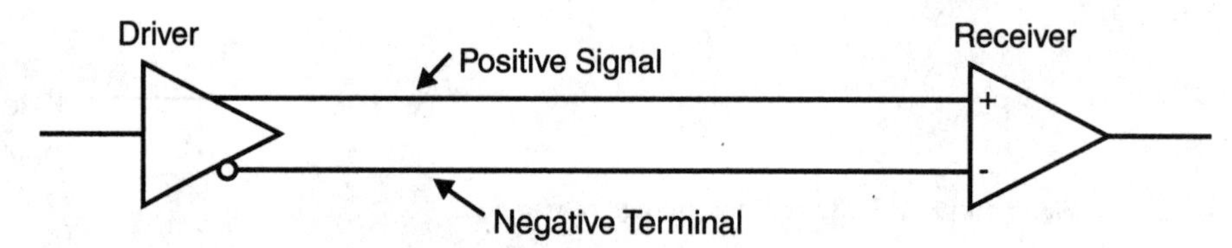

Figure 10.21 Differential pair serial data transmission.

environments where power surges and electrical noise can be significant. In these environments a double-ended or differential pair connection is optimal to ensure the most accurate communications.

A *differential pair* serial communications electrical standard consists of a balanced driver with positive and negative outputs that are fed into a comparator, which outputs a 1 or a 0, depending on whether or not the positive line is at a higher voltage than the negative. Figure 10.21 shows the normal symbols used to describe a differential pair connection.

There are several advantages to this data connection method. The most obvious one is that the differential pair doubles the voltage swing sent to the receiver, which increases its noise immunity. This is shown in Fig. 10.22; when the positive signal goes high, the negative voltage goes low. The change in the two receiver inputs is 10 V, rather than the 5 V of a single line, assuming the voltage swing is 5 V for the positive and negative terminals of the receiver. This effective doubling of the signal voltage reduces the impact electrical interface has on the transmitted signal.

Another benefit of differential pair wiring is that if one connection breaks, the circuit will operate (although at reduced noise rejection efficiency). This feature makes differential pairs very attractive in cars, aircraft, and spacecraft where loss of a connection could be catastrophic.

To minimize AC transmission line effects, the two wires should be twisted around each other. Twisted pair wiring can either be bought commercially or made by simply twisting two wires together; twisted wires have a characteristic impedance of anywhere from 30 Ω upward.

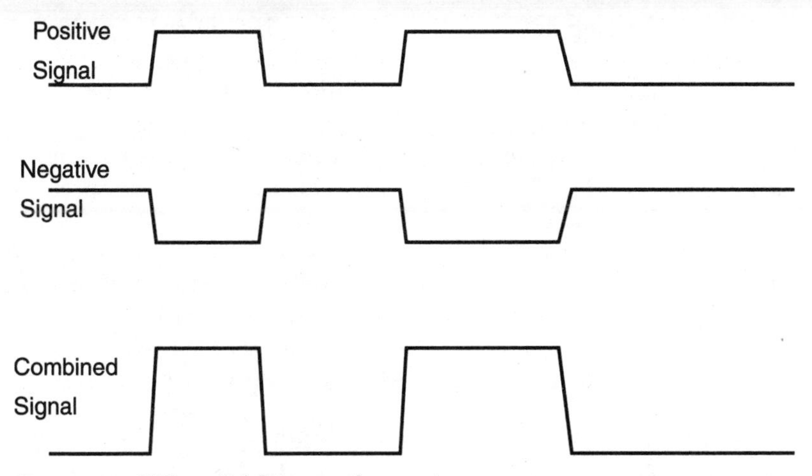

Figure 10.22 Differential data waveform.

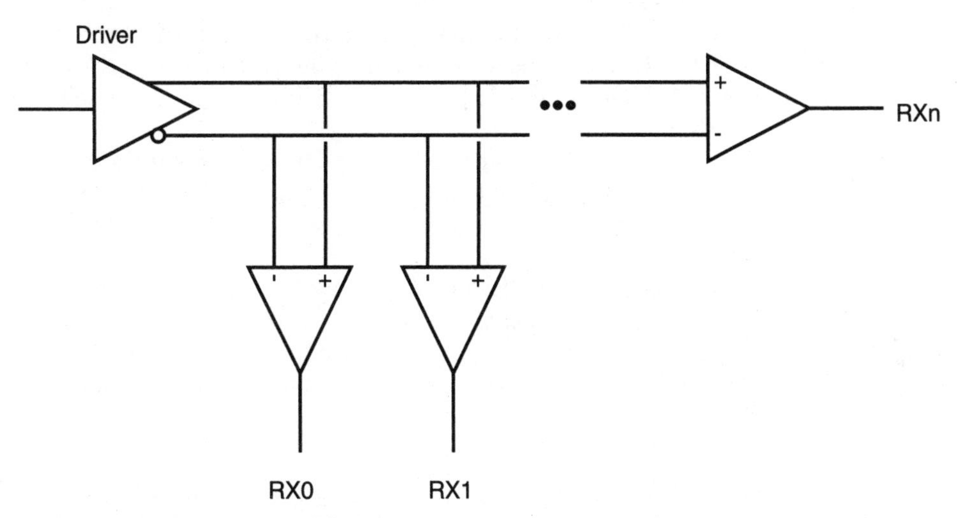

Figure 10.23 Multiple-receiver RS-422.

A common standard for differential pair communications is RS-422. This standard, which uses many commercially available chips, provides:

1. Multiple receiver operation
2. Maximum data rate of 10 Mbit/sec
3. Maximum cable length of 4000 meters (with a 100-kHz signal)

Multiple-receiver operation, shown in Fig. 10.23, allows signals to be broadcast to multiple devices. The maximum distance and speed changes with the number of receivers of the differential pair and its length. The specified 4000 m at 100 kHz or 40 m at 10 MHz illustrates this balancing between line length and data rate. For long data lengths, a few-hundred-ohm terminating resistor may be required between the positive terminal and negative terminal at the end of the lines to minimize reflections coming from the receiver and prevent them from affecting other receivers.

RS-422 is not as widely used standard as you may expect; instead, RS-485 is much more popular. RS-485 is very similar to RS-422, except it allows multiple drivers on the same network. The common chip is the 75176, which has the ability to drive and receive on the lines as shown in Fig. 10.24. In the right 75176 of Fig. 10.24, the RX and TX and two enables are tied together. This results in a two-wire differential I/O device. Normally, the 75176s are left in RX mode (pin 2 reset) unless they are driving a signal onto the bus. When the unused 75176s on the lines are all in receive mode, any one can take over the lines and transmit data.

As with RS-422, multiple 75176s (up to 32) can be on the RS-485 lines with the capability of driving or receiving. When all the devices are receiving, a high, or 1, is output from the 75176. This means the behavior of the 75176 in the RS-485 (because these are multiple drivers) is similar to that of a dotted AND bus; when one driver pulls down the line, all receivers are pulled low. For the RS-485 network to be high, all unused drivers must be off or all active drivers must be transmitting a 1. This feature of RS-485 is taken advantage of in small system networks like CAN.

The only issue to be on the lookout for when creating RS-485/RS-422 connections is to keep the cable polarities correct (positive to positive and negative to negative). Reversing the connectors will result in lost signals and misread transmission values.

Another common method of serially transmitting data asynchronously is to use the Manchester encoding format (Fig. 10.25). In this type of data transfer, each bit is synchronized to a start bit and the following data bits are read with the space dependent on the value of the bit. In this type of data transmission,

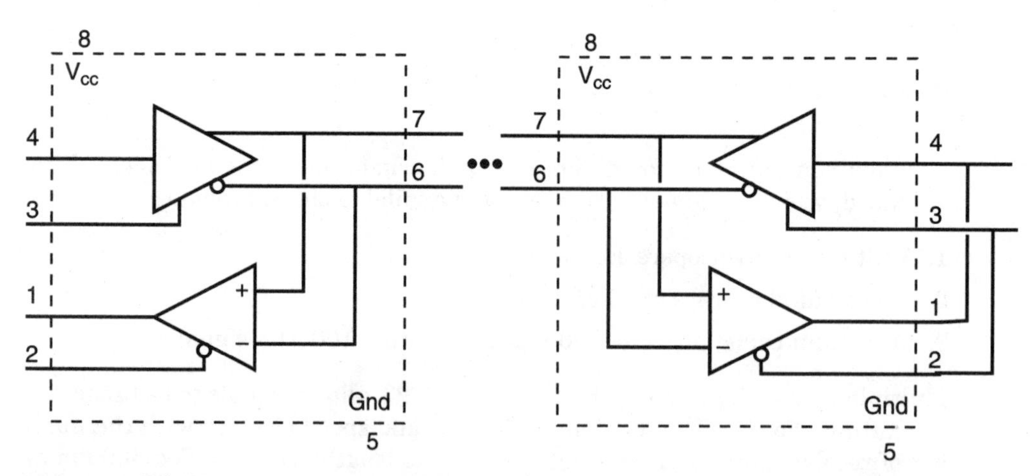

Figure 10.24 RS-485 connection using a 75176.

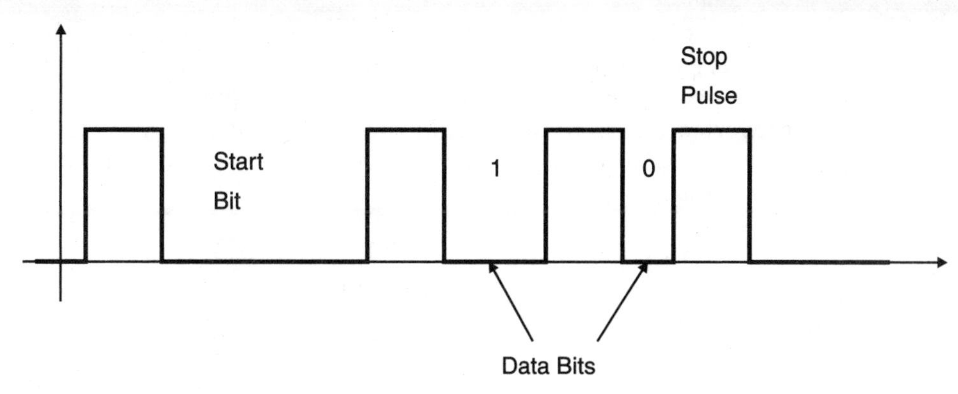

Figure 10.25 Manchester-encoded serial data.

the data size is known, so the stop pulse is recognized and the space afterward is not treated as incoming data.

Manchester encoding is unique in that the start bit of a packet is quantitatively different from a 1 or a 0. This allows a receiver to determine whether the data packet being received is actually at the start of the packet or somewhere in the middle (and should be ignored until a start bit is encountered).

Manchester encoding is well suited for situations where data can be easily interrupted or the receiver becomes active in the middle of a data packet. Because of this, it is the primary method of data transmission for infrared control (such as used in your TV's remote control).

Phototransistors and Optoisolators

Earlier in the book, I introduced light-emitting diodes (LEDs) as simple diodes that emit light when current passes through them. The change in voltage potential as the electrons flow over the PN junction results in a release of equivalent energy in the form of photons. This operation tends to be well understood because it can be observed visually (pass current through an LED and watch it produce light).

When I wrote this description, I should have asked, "What happens if *photons* are injected into a PN junction?" The obvious (and correct) answer is that current would be produced by the photons entering the PN junction. There are diodes, known as *photodiodes,* designed for accepting light to produce current.

Taking this as true, as a follow-up, you might ask why expensive solar cells are used when inexpensive LEDs could be used to generate electricity. The answer to this question comes down to the relatively poor efficiency of LEDs in converting electrical energy into light. For most LEDs, the fraction of light energy released versus the amount of energy available within the LED is less than 5 percent (although there are some LEDs that can convert light at much higher efficiencies). Current solar cell technology is running at better than 15 percent and can be built in very large sizes (as I write this, the first of the four solar panel "wings" of the International Space Station have just been erected, spanning 270 ft).

To give you an idea of how much power can be generated by a photodiode, consider the example of a phototransistor out in the sunlight (sunlight provides about a kilowatt of power over every square meter). If the photodiode converted 5 percent of the light that reaches it into electrical energy and the diode has a surface area of 1 mm^2, then the possible power output could be calculated as:

$$\text{LED power output} = \frac{\text{sun power}}{\text{unit area}} \times \text{surface area} \times \text{efficiency}$$

$$= 1\ \text{kW/m}^2 \times 10^{-6}\ \text{m}^2 \times 5\%$$

$$= 10^3 \times 10^{-6} \times 0.05\ \text{W}$$

$$= 50\ \mu\text{W}$$

To get 1 W of power (200 mA at 5 V) out of an array of LEDs, 20,000 of them would be required! To make matters worse, I have assumed best-case conditions for the power available from the sun (the full kilowatt is available only at the equator at noon) and a diode that is 1 mm^2 in size is actually very large. Obviously, this is not a practical way to generate power.

Despite this, photodiodes are used for controlling circuits. The small amount of power output by the photodiode can be amplified to give a circuit an indication of whether the light to the diode is blocked. Developing effective amplifiers for the photodiode's signal is difficult and can be costly. This often makes it impractical to use photodiodes for light level tests or on/off tests.

Instead of the photodiode, a *light-dependent resistor* (LDR) can be used for light-level sensing. This device consists of a resistor that changes value according to the amount of light striking it. LDRs are quite inexpensive and easy to use when wired into voltage divider circuits where an analog-to-digital converter is available to convert the voltage level into a binary value that can be responded to.

The drawback of LDRs is their poor frequency response. Where the power output of photodiodes changes almost immediately after a change in light, LDRs can take many microseconds (or even milliseconds) to respond to changing light levels.

The characteristics of photodiodes and LDRs result in the need for a device that is relatively inexpensive and can provide easily measurable changes for quickly changing light levels. The solution to this problem is quite elegant. Instead of providing a base current to a transistor from an LDR or photodiode, the base current can be provided by light falling on a transistor and used to produce a base-emitter current that would turn on the transistor. The resulting device (called, not surprisingly, a *phototransistor*) has the symbol shown in Fig. 10.26. A photoresistor can directly control circuits by controlling current passing through its collector/emitter.

Going back to the current calculation for the photodiode, you should realize that there is a very small amount of current being produced to turn the tran-

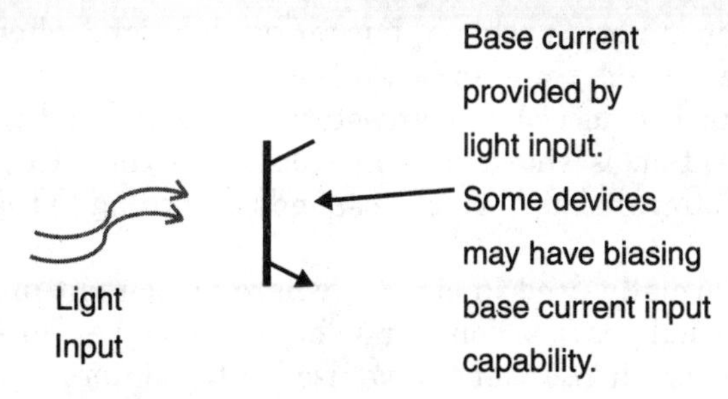

Figure 10.26 Circuit diagram of a phototransistor.

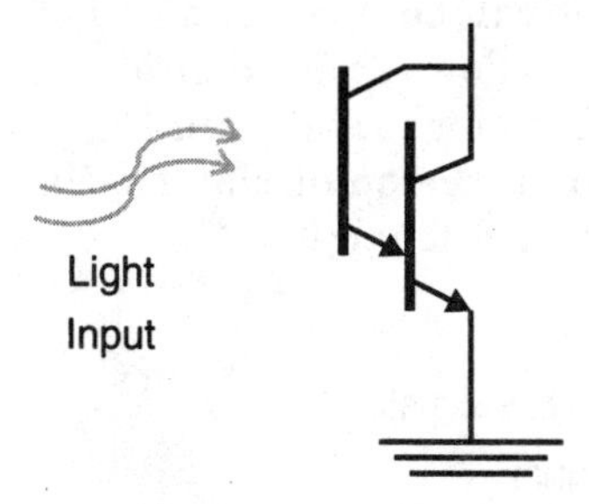

Figure 10.27 Phototransistor wired in the Darlington array configuration.

sistor on. For a typical bipolar transistor with a beta of 150 (like a 2N3906), the total current controllable by the device would be about 7.5 mA (and again, this is under best-case conditions). A current of 7.5 mA is quite a bit better than the 50 μA of the original photodiode, but still not enough to control most circuits.

There are some phototransistors that have a third pin for providing a biasing base current. This bias base current will allow different light levels to affect the operation of the phototransistor differently.

However, the current available from the phototransistor is enough to drive another transistor wired to it in the Darlington array configuration (see Fig. 10.27). The second transistor in the Darlington array is the actual driver transistor for the circuit. This method of current "amplification" is used for many different applications and is often built into "standard" device packages to provide high-current control of different devices. Transistors built in the Darlington array configuration are often used for providing control for DC motors, relays, and other magnetic devices that require more current than is typically available in a digital electronics circuit.

In a Darlington array, phototransistors can easily control 20 mA or more of current in real-world applications. Phototransistors built in a Darlington array configuration are quite inexpensive and available in different models. I should point out that, whereas in regular transistors you can expect a frequency response of many tens (if not hundreds) of megahertz, with phototransistors you can typically expect a response time of about 4 μsec (250 kHz). For most applications, this is not a problem, but for applications where a rapidly changing light source is used, this can be an issue.

I'm pointing this out because phototransistors are often used as part of an optoisolator (the operation is shown in Fig. 10.28). In this circuit, an LED shines light onto a phototransistor, which responds according to the amount of light it receives.

Optoisolators are normally used in circuits where one circuit must be separate from another. Normally this is done for safety reasons (i.e., an external circuit is powered by mains wiring), but I have used optoisolators as a method of controlling different voltage levels in a circuit. Using an optoisolator, a 12-V power supply can be controlled by a +5-V power supply without having to calculate how the transistors in the circuit will be turned on and off.

An example of using an optoisolator to allow a +5-V digital electronic circuit to control the operation of a +12-V DC motor is shown in Fig. 10.29. Besides providing an easy way to control a motor, the optoisolator isolates any noise generated by the motor from the control electronics.

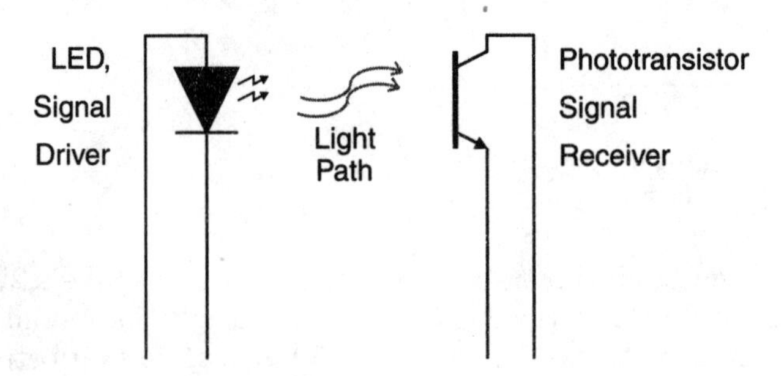

Figure 10.28 Elements of an optoisolator.

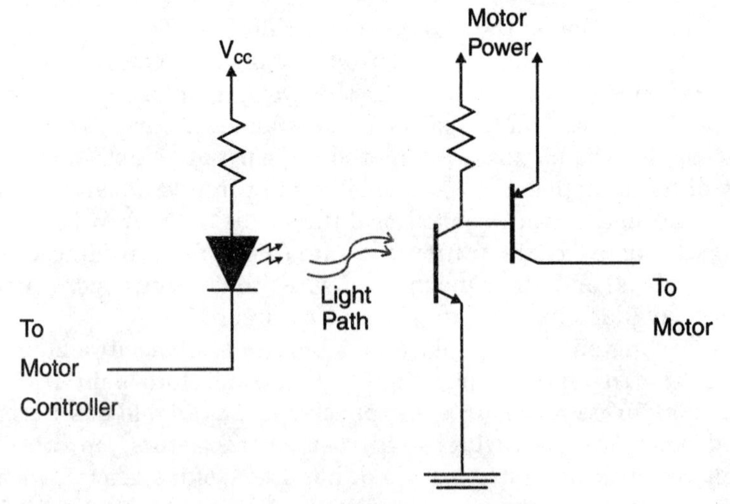

Figure 10.29 Optoisolator as part of a motor controller.

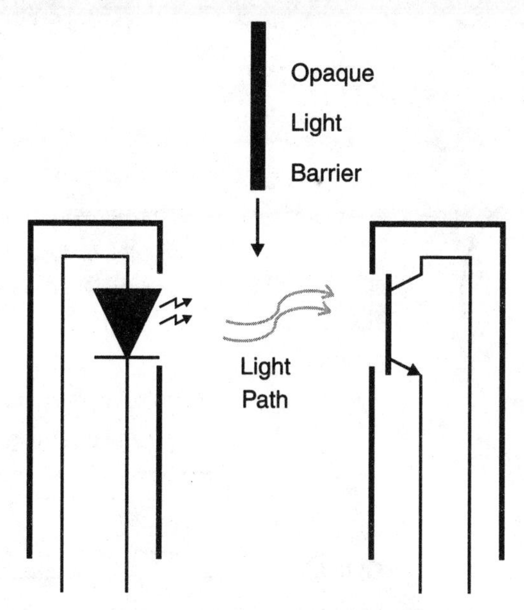

Figure 10.30 Operation of an optical interrupter.

Another very common application for optoisolators is as an optical interrupter, in which the light beam between the LED and the phototransistor can be broken by some opaque barrier to indicate an input condition. The operation of the optical interrupter is shown in Fig. 10.30.

A very common use for optical interrupters is in personal computer "mice." To determine whether or not the mouse has been moved and in what direction it has been moved, a slotted wheel's position and movement is sensed by two optical interrupters as shown in Fig. 10.31. The two optical interrupters are placed in such a way that they may or may not be on at the same time. Depending on which way the wheel turns, one of the optical interrupters will be turned on and off before the other. An intelligent device within the mouse times the transitions between the ons and offs and tracks which optical interrupter goes on first. With this information, the movement of the mouse can be detected along with the direction and speed that it is going at.

Switch Matrix Keypads

If you have an application that requires a large number of buttons for user input, you could use the button debouncing techniques presented earlier in the book, along with some method of prioritizing the different possible inputs and combining the inputs into a simple binary number. While this method is possible, it is very expensive and quite complex to implement.

Most PC keyboards and numeric keypads do not use this method; instead the keys and buttons are arranged in rows and columns and they can be drawn

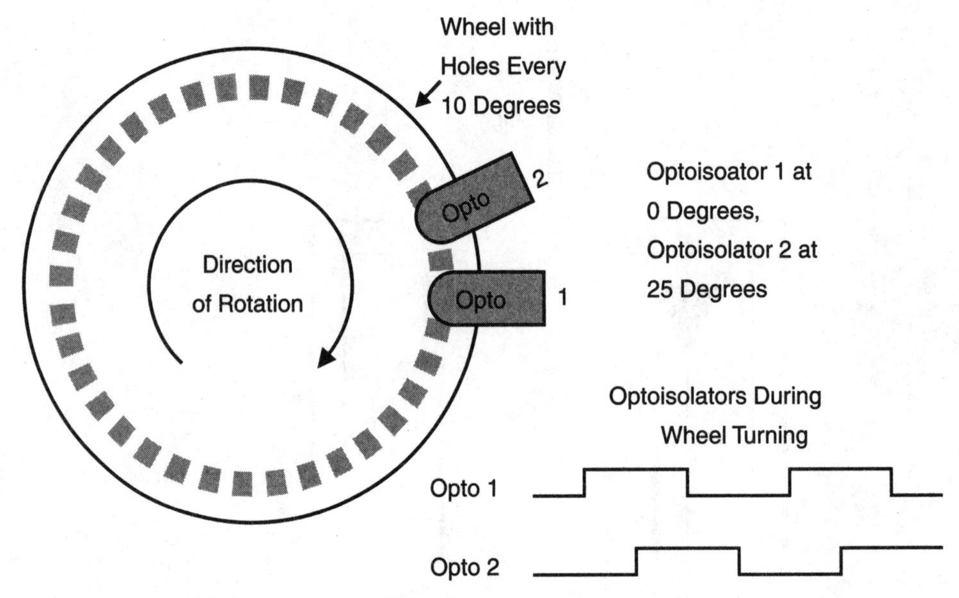

Figure 10.31 An optical interrupter to detect rotary movement.

out in such a way that they look like a matrix. A momentary-on switch is placed at the intersection of each row and column as shown in Fig. 10.32. This *switch matrix* provides the ability to scan a large number of button inputs with a relatively small number of lines. Your PC's 104/105-key keyboard usually has a 22 by 7 matrix connection to a microcontroller, which scans through the keys and reports any key presses according to the algorithms that I present later in this section.

In this section, I will describe how *keypads* are scanned rather than *keyboards.* I define a keypad as a switch matrix array that has 32 or fewer buttons. This probably seems like an arbitrary decision point, but I chose it because there is a commercially available chip that I will describe in this section that makes the reading of these smaller switch matrix arrays much easier.

While switch matrix keypads are an extension to buttons with many of the same concerns and issues to watch for, they are several orders of magnitude of complexity more difficult to work with in simple digital logic systems. This extra complexity is why I use software pseudocode to explain the operation of reading a switch matrix keypad in this section instead of trying to show how the keypad is read by logic gates.

The diagram shown in Fig. 10.32 may not look like the simple button, and a similarity with the simple buttons I have presented so far will not be very obvious. The relationship will become more obvious when I add pull-ups on the

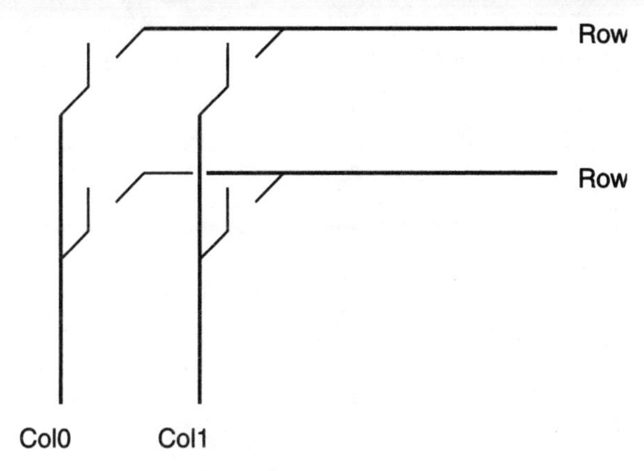

Figure 10.32 2 × 2 switch matrix.

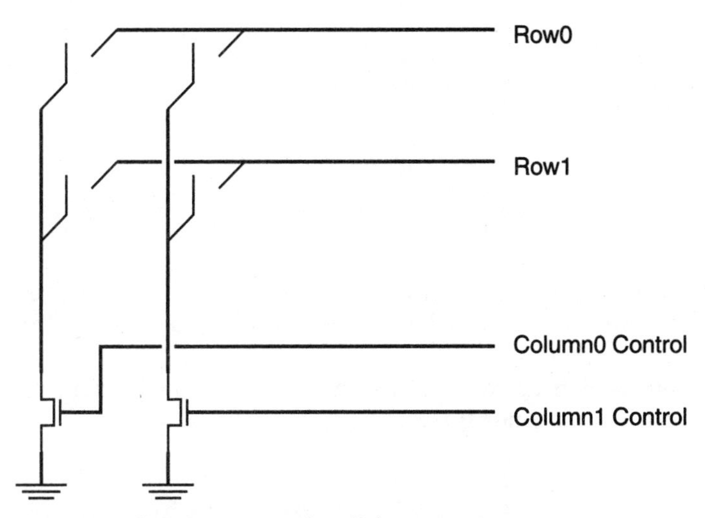

Figure 10.33 Switch matrix with pull-down transistors.

rows and switchable ground connections on the columns, as I show in Fig. 10.33. In this case, by connecting one of the columns to ground, if a switch is closed, the pull-down on the row will connect the line to ground. When the row is polled by an I/O pin, a 0, or low voltage, will be returned instead of a 1 (which is what will be returned if the switch in the row that is connected to the ground is open).

In this case, the keyboard can be scanned for any closed switches (buttons pressed) by using the code:

```
int KeyScan()                    // Scan the Keyboard and Return when a
{                                // key is pressed

int i = 0;
int key = -1;

 while (key == -1) {

   for (i = 0; (i < 4) & ((KeyPort & 0x00F) == 0x0F0); i++);

   switch (KeyPort & 0x00F) {// Find Key that is Pressed
     case 0x00E:                 // Row 0
       key = i;
       break;
     case 0x00D:                 // Row1
     case 0x00C:
       key = 0x04 + i;
       break;
     case 0x00B:                 // Row2
     case 0x00A:
     case 0x009:
     case 0x008:
       key = 0x08 + i;
       break;
     else                        // Row3
       key = 0x0C + i;
       break;
     }//end switch
   }// end while
  return key;
} // End KeyScan
```

The KeyScan function will return only when a key has been pressed. This routine will not allow keys to be debounced or other code to execute while it is executing.

These issues can be resolved by calling the KeyScan routine once every 5 msec or so (using a timer interrupt handler):

```
IntKeyScan( )                    // 5 msec Interval Keyboard Scan
{

int i =0;
int key = -1

 for (i = 0; (i <4) & ((KeyPort & 0x00F) == 0x00F)); i++);

 if (KeyPort & 0x00F) != 0x00F) { // Key Pressed
   switch (KeyPort & 0x00F) {// Find Key that is Pressed
     case 0x00E:                 // Row 0
       key = i;
       break;
     case 0x00D:                 // Row1
     case 0x00C:
       key = 0x04 + i;
       break;
     case 0x00B:                 // Row2
     case 0x00A:
     case 0x009:
     case 0x008:
```

```
        key = 0x08 + i;
        break;
      else                      // Row3
        key = 0x0C+i;
        break;
    } //end switch
    if (key == KeySave) {
      keycount = keycount + 1;// Increment Count
      if (keycount == 4)
        keyvalid = key;         // Debounced Key
    } else
      keycount = 0;             // No match - Start Again
    KeySave = key;              // Save Current key for next 5 msec
  }                             // Interval
} // End IntKeyScan
```

This routine will set the keyvalid variable to the row/column combination of the key button (which is known as a *scan code*) when the same value comes up four times in a row. This time scan is the debounce routine for the keypad. If the value doesn't change for four intervals (20 msec in total), the key is determined to be debounced.

There are two things to notice about this code. The first is that in both routines I handle the row with the highest priority. If multiple buttons are pressed, then the one with the highest bit number will be the one that is returned to the user. The code does not keep track of which key was pressed first.

The second point is that this code can have an autorepeat function added to it very easily. To do this, a secondary counter has to be first cleared and then incremented each time the keycount variable is 4 or greater.

The code and methodology for handling switch matrix keypad scans I've outlined here probably seems pretty simple. Depending on your familiarity with programming and different microprocessors and microcontrollers, you will probably realize that implementing these functions could be done even more simply in assembly language programming. You should also realize that this code would be quite difficult to implement using just logic chips.

To avoid the complexities of trying to develop TTL logic that will carry out the functions described in the pseudocode presented above, I suggest that you use the 74C922 keypad decoder chip (Fig. 10.34). This chip can be used to debounce and encode up to 32 buttons (although 16 is the normal maximum) and carries out button debouncing internally in addition to keeping track of two currently held down keys when new keys are pressed.

The 74C922 is quite easy to wire to a 4×4 (16-button) switch matrix keypad, as shown in Fig. 10.35. The two capacitors are used to create a relaxation oscillator within the chip that is used to scan through the buttons as well as provide a debounce delay count for the application. The two capacitor values are calculated as:

$$C_{osc} = \text{scan rate}/10{,}000$$

$$C_{kbd} = 10\ C_{osc}$$

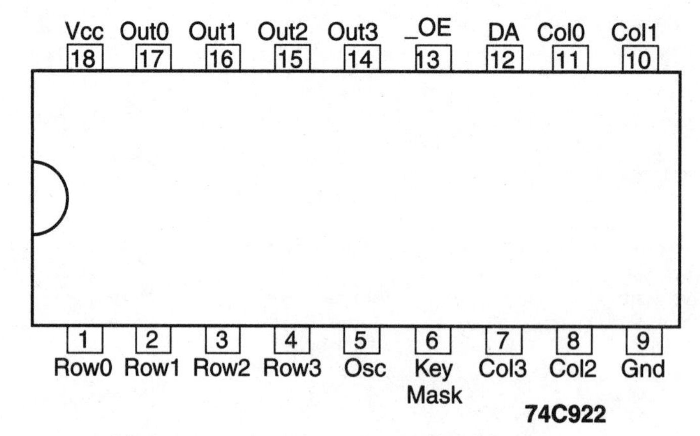

Figure 10.34 Sixteen-key switch matrix keypad controller.

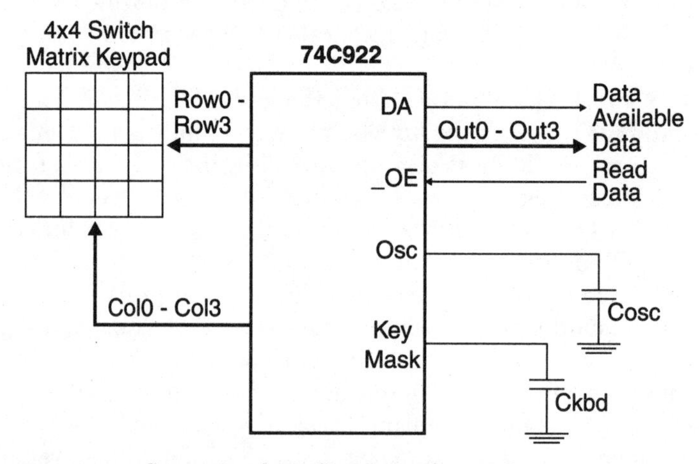

Figure 10.35 Connecting the 74C922 keypad controller.

As I have indicated, I like a debounce interval of 20 msec. Plugging this into the formulas above, I get values of 2 μF and 20 μF. When I build my own applications, I tend to have a lot of 10-μF (for power filtering) and 1-μF (for MAX232 RS-232 level converters) electrolytic capacitors on hand. I have not found any problems with using these components, and I would recommend that you use them as well to avoid having to stock multiple capacitor values for different applications.

By doubling up row sensors of the 74C922, you can add a number of additional keys to the application.

Analog Signal I/O

So far in this book, when I have described methods for interfacing digital devices to analog devices, I have been concerned only with passing steady-state data values. When I say *steady state,* I am referring to relatively slowly changing inputs that can be read by using simple ADC technologies. For this and the following sections, I am going to describe a number of methods that can be used to pass analog information *dynamically* to digital devices and back again.

Before going too far in the description of the methodologies used, I want to point out that the actual signals being worked with are not that complex—for the most part, I will be simply working with single frequencies and different voltage levels. You will not be creating a "SoundBlaster" like a synthesizer/audio digitizer; instead I will discuss basic analog I/O interfacing and present you with the basic circuits that can be used with it.

When I discuss analog signal I/O, I will be focusing on audio-frequency signals for the simple reason that you can easily hear the input and output. In this section, I want to discuss the different aspects of audio signals and present you with the background for the circuits that are needed to work with them. Much of this information will be a review, and I do not go very deeply into the technical aspects of the science of acoustics, but I will go into the basic concepts and how they affect interfacing digital electronics to analog signals.

Most radio and audio signals repeat for some period of time during their operation. A plucked violin string will output a sound (*signal*) that is characterized by how far the string moves and at what rate the string moves (*oscillates*) back and forth. As shown in Fig. 10.36, the distance the string moves back and forth is proportional to amplitude of the sound produced by the instrument. The *amplitude* of the sound is what we refer to as *loudness.* The larger the amplitude of the sound wave, the louder it will seem.

Amplitude is measured as the peak-to-peak height of the wave, or the distance from the bottom of the wave to the top. For sound, this measurement is impractical, but the amplitude of electrical signals is used to characterize the signal. When working with analog electrical signals, you can very easily measure amplitude using a digital multimeter or an oscilloscope.

Also shown in Fig. 10.36 is the *frequency* of the signal. In the case of sound, which is produced by the movement of some object that compresses the

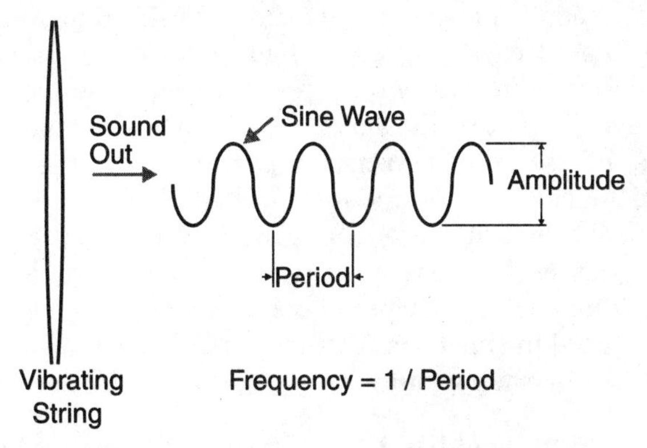

Figure 10.36 Audio signal produced by a violin string.

surrounding air, the frequency of the sound is equivalent to the movement of the object. The faster the object moves back and forth, the higher the frequency and the higher the pitch of the sound.

While the amplitude of an electrical signal can be very easily measured by a digital multimeter or an oscilloscope, measuring the amplitude of a sound wave is not so simple. In this case, the sound has to be converted into an electrical signal that then can be measured. The amplitude of a sound wave (also known as *sound pressure*) is usually in the units of decibels (dB). Decibels are specified by the formula:

$$\text{dB} = 20 \log (\text{signal/reference})$$

and are most often used to describe the level of noises relative to the smallest possible audible sound (which is 0 dB). For example, someone talking next to you is producing "noise" (which may or may not be in quotations, depending on the person) at 60 dB. Rearranging the formula, you can find that this speech is 1000 times louder than the softest sound you can hear.

To see the logarithmic relationship of the decibel scale, consider a "MegaDeath" rock concert that runs at about 115 dB. In this case, the signal is over 550,000 times louder than the softest possible sound, or over 550 times louder than normal conversation.

The decibel scale and unit of measurement is often used with electrical signals as well where the amplification or attenuation (reduction) of a signal is measured relative to the input signal (the *reference* in the dB formula above). I will explain this in more detail later in this section.

One of the most difficult concepts for people to understand about analog signals and sound is that it is made up of many other signals. The basic signal is the sine wave, and multiple applications of this signal can be used to produce any analog signal that you can imagine. This is quite a powerful statement and one that you should think about.

The sine wave is a drawing of the vertical position of a point on a disk when the disk is rolled for some period of time. You are probably familiar with the sine wave, but if you aren't, I suggest that you ask a math teacher or an expert in electronics to recommend some texts explaining basic trigonometry.

In the meantime, there are only a few things that you need to know about sine waves for this description. The wave representation of sound in Fig. 10.36 is actually a sine wave. To plot a sine wave, the sin or cos buttons on your scientific calculator can be used. The two critical parameters of the sine wave are its amplitude (peak-to-peak measurement described above) and its period.

Sin is the normal abbreviation used for sine, and cos is the abbreviation used for cosine. The cosine wave is identical to the sine wave, but it appears 90° (or $\pi/2$ radians, which will be discussed below) ahead of the sine wave. For the information presented in this book, I will work only with a sine wave function, which is described by the formula:

$$\text{Wave height} = A \sin (2\,\pi t/\text{period})$$

A true sine wave has a positive and negative value.

The period of the sine wave can be measured in degrees, or for electronics, in radians. Since one full cycle of a wheel covers 2 times π times the radius of the wheel, the period of a sine wave is measured in terms of radians, which is multiplied by two times π to get the current position of the wheel.

When you learned about sines and cosines, you may have been taught with the period described in degrees (with one full turn of the wheel being 360°). The use of radians is a convention that is used within electronics, and I recommend that when you are working in electronics, you set your calculator to "rad." To convert from degrees to radians, you can use the formula:

$$\text{Radians} = 2\,\pi \times \text{degrees}/360$$

Using this formula, 45° will be π divided by 4 or 0.7854.

To convert time into radians for use with a sine wave, you should use the formula:

$$\text{Time in radians} = 2\,\pi t/\text{period}$$

Adding a sine wave's peak-to-peak amplitude to the sine wave formula is very simple—it is just a constant multiplied by ½. The extreme values for a sine wave are $+1$ and -1, which gives you a maximum peak-to-peak value of ½. Normally in mathematics texts, you will see the amplitude of a sine wave multiplied by 1, but in electronics, this will result in twice the expected amplitude of the wave.

With this basic understanding of what a sine wave is, we can now look at how many different sine waves are combined to produce different signals. When I produce a clock signal, I always strive to make it as close to a perfect *square wave* as possible. A perfect square wave has equal high and low times or 50 percent duty cycle, and vertical rises and falls when transitioning from low to high and back to low again. It is probably surprising, but the square wave is an excellent example of how a number of different sine waves are combined in order to produce a perfect square wave.

In the case of the square wave, numerous sine waves of differing periods are added upon each other to produce the actual signal. In Fig. 10.37, I show how two different sine waves are added together to create a signal that approximates a square wave. This method can be used to produce many other different and arbitrary waveforms.

Note that the frequency of each sine wave that makes up the square wave is a multiple of the square wave frequency. Each different-frequency sine wave is known as a *harmonic*. The sine wave at 3 times the frequency of the square wave is known as the *third harmonic*; the sine wave at 5 times the frequency of the square wave is known as the *fifth harmonic,* and so on.

Also note that the amplitude of each sine wave changes with the harmonic in a way that is predictable. In fact, the square wave function could be described as:

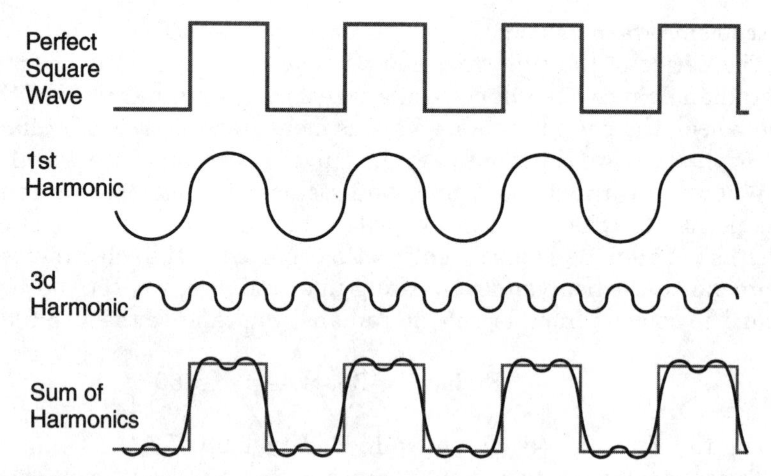

Figure 10.37 Square wave produced by sine waves.

$$\text{Squarewave}\ (t) = A\ (\sin \frac{2\pi t}{P} + \frac{1}{3} \sin \frac{6\pi t}{P} + \frac{1}{5} \sin \frac{10\pi t}{P} + \ldots)$$

or as a sum of terms using the formula:

$$\text{Squarewave}\ (t) = \sum_{i=0}^{A\ \infty} \sin \frac{2\ (2i + 1)\ \pi t/P}{2i + 1}$$

In this format, the function is known as a *series*.

I found series fascinating when I learned about them in university. By using series, you can simplify many mathematical processes or produce arbitrary waveforms like the square wave demonstrated here. In fact, many analog signal *function generators* internally produce a series of sine waves and mix them together to produce an arbitrary signal out.

In Fig. 10.37, I used the example of how two sine waves can be mixed together to form an approximation of a square wave. In commercial function generators, four or five sine waves are mixed together to form the actual function. I am pointing this out to indicate that just a few sine waves are required to produce a very good example of an arbitrary waveform function.

The obvious question from this introduction is whether or not actual square waves are produced in TTL logic by using sine waves. There is a seemingly simple test for this, and that is to filter out the harmonics of the signal and see if you are left with a sine wave. I originally wanted to include an experiment for you to try and filter out the upper harmonics, but this turned out to be much more difficult than I expected, so I am going to just explain the theory (along with what the problems were that I encountered).

A *filter* is simply a device that outputs less of a signal than what is put into it. Your sunglasses are a filter because they only let a fraction of the total light through to your eyes. For digital electronics, you will be most concerned with a *low-pass filter,* which filters out high frequencies as shown in Fig. 10.38.

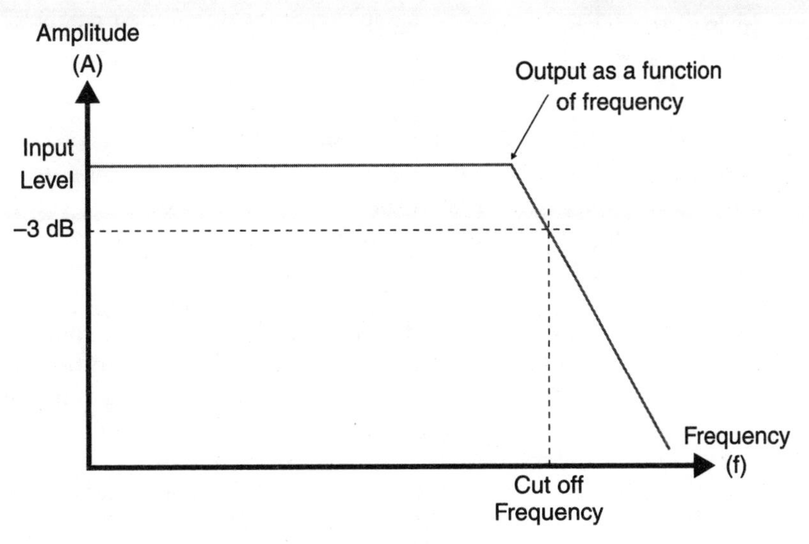

Figure 10.38 Operation of a low-pass filter.

Input frequencies are output from a low-pass filter as long as they are less than the *cutoff frequency.* At the cutoff frequency, the low-pass filter has attenuated the input signal by 3 dB (the amplitude of the output signal is 0.708 of its input value). As the frequency increases, the amount of attenuation increases.

To demonstrate the different high-frequency components of a square wave, I wanted to pass the output of the 1-kHz oscillator through a low-pass operational amplifier (best known as an *op amp*) circuit.

An op amp is a differential amplifier whose output is a combination of two compared inputs. The amplifier output ideally has infinite gain (although in reality, it is on the order of 10,000 times) with infinite impedance on the two inputs and zero impedance on the outputs. In Fig. 10.39, a typical op amp amplifier circuit is shown.

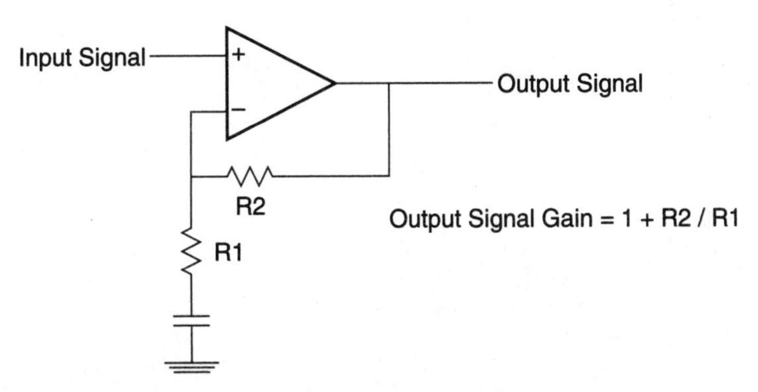

Figure 10.39 Operational amplifier signal amplification.

In Fig. 10.39, the amplitude of the "Input Signal" will be changed by using the formula:

$$\text{Gain} = 1 + \text{R2/R1}$$

In this circuit, the output signal's amplitude will always be greater than the input signal's. As I will show in the next section, the power voltage applied to the op amp limits the actual magnitude of the output signal. If the amplified value is greater than the applied voltages, then the output signal will be "clipped' to this maximum voltage.

A common low-pass filter circuit for the op amp is the *two-pole Butterworth low-pass filter circuit* shown in Fig. 10.40. I like this circuit because the values for the two resistors (R) and two capacitors (C) are the same and the circuit can be used to amplify the resulting output signal.

When I passed the 1-kHz oscillator signal from the interface PCB to a Butterworth filter set to a 1.591-kHz cutoff frequency (using 1k resistors and 0.1-μF capacitors), I got an input and output waveform similar to the one shown in Fig. 10.41. The output signal may *appear* to approximate a sine wave, but note that its high point is closer to one end of the high square wave (although it is not at the end as you might expect).

The reason why the output is not a true sine wave is that the input is not a true square wave. There is a small RC component that can be seen in the signal. The resulting waveform requires a much more complex series than the simple series shown above for the basic square wave.

With some work in ensuring that the input signal has absolutely "square" edges and using a filter with more poles, you should be able to see a first-order harmonic sine wave much more clearly than you would see in this example. If you look at *The Art of Electronics,* second edition, there is an example circuit used to produce a sine wave with a six-pole low-pass filter and circuitry designed to get as perfect a square wave input as possible for controlling a telescope drive.

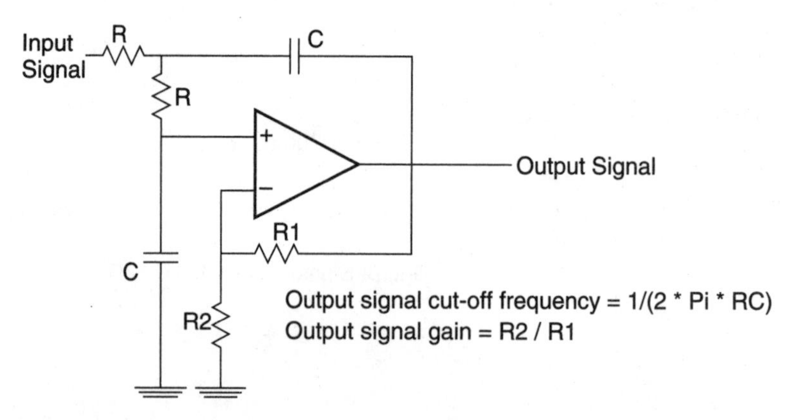

Figure 10.40 Two-pole Butterworth low-pass filter.

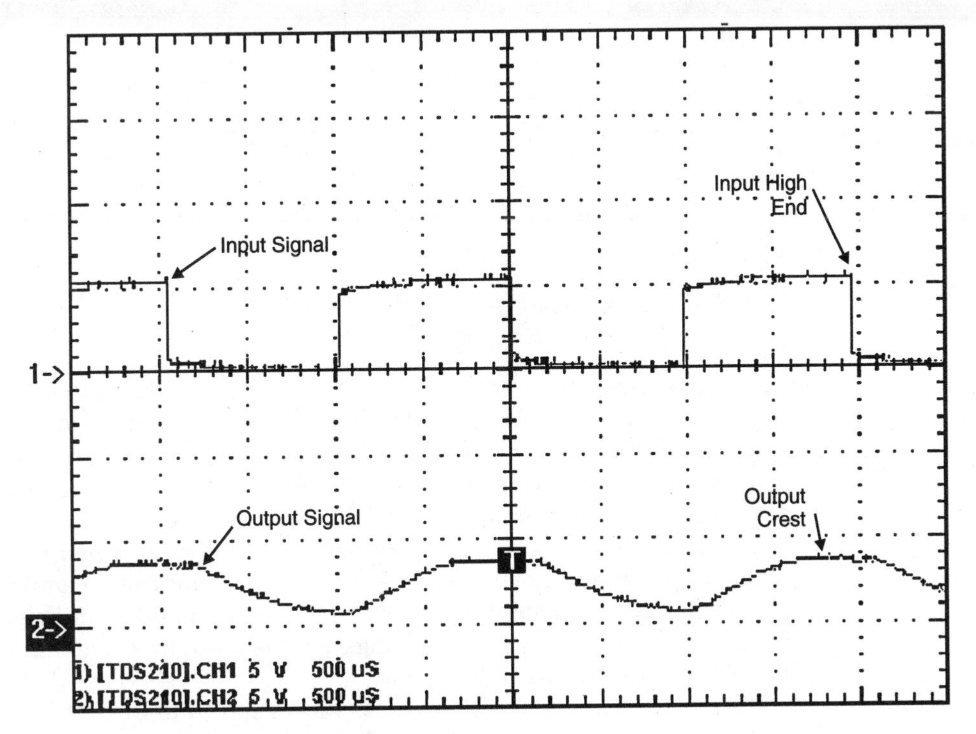

Figure 10.41 Two-pole Butterworth filter operation.

There are many different types of filters. Along with the *low-pass filter,* there are the *high-pass filter,* which allows only signals *above* the cutoff frequency and the *band-pass filter,* which allows only frequencies within a specific range to be passed. You will find that the low-pass filter will be the most useful one when interfacing analog signals to digital devices. The reason for this is that you will be interested in sampling one frequency and anything above it will be noise, which can interfere with the circuit.

The use of operational amplifiers, as I have shown in this section, probably seems quite simple, but they can be very difficult to work with. This is especially true when they are used with complex (*multipole*) filters. To their credit, they are generally quite inexpensive, and there are a number of circuits in books and on the Internet that you can use to implement different filters and functions.

Now that I have shown you how to deconstruct a square wave into its constituent parts and obtain just a sine wave, you are probably wondering how to reconstruct the signal back into a signal that can be used by digital logic. It will seem ironic (and overly complex), but I will put the filtered circuit into a high-gain amplifier and clip the voltage to the level that is required. This process is shown in Fig. 10.42.

In Fig. 10.42, the microphone picks up noise and passes it to a low-pass filter. This filter outputs the lowest expected frequency of the input signal. To do

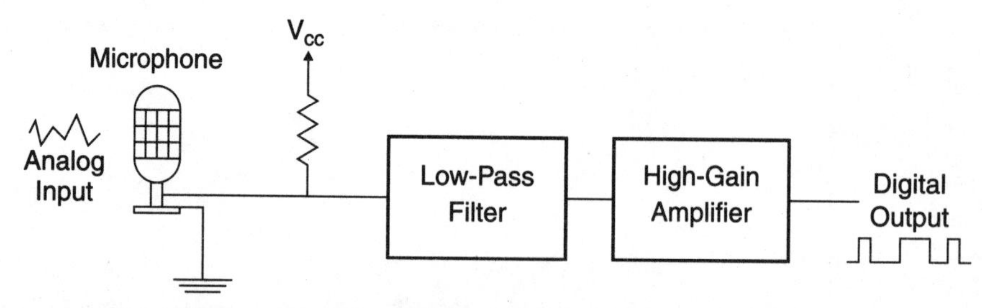

Figure 10.42 Typical microphone to digital electronics circuit.

this, I generally set the cutoff frequency at $1^1/_2$ times the base signal that I am expecting. When I do this, the third and higher harmonics are effectively filtered out of the input. In the example square wave filter I showed above, note that the low-pass filter has a cutoff frequency of 1500 Hz in order for me to observe the first harmonic of a 1000-Hz square wave.

The next stage performs a high order gain (100 times or more) which results in the output signal having a larger magnitude than the op amp's power supply. When this happens, the output signal "maxes out" or is "clipped" at the power supply extremes. The resulting signal is a square wave that can be input into standard TTL and CMOS logic.

The standard audio input device, as I have shown in Fig. 10.42, is the *microphone*. The microphone is a device that produces a voltage when a diaphragm is displaced by sound pressure. This voltage is proportional to the movement of the diaphragm, which is proportional to the sounds in the air. The original microphones used carbon powder to generate the electrical signal, but modern devices use a piezoelectric element instead. The connection of a microphone into a circuit is very simple and should follow the circuit shown in Fig. 10.42, with the pull-up as the microphone changes its resistance as the diaphragm is moved back and forth.

For audio output, you should not be surprised to see a *speaker* being used for the function. A speaker is a device that converts either electrical or current signals into mechanical motion (often using a piezoelectric device) that in turn moves a diaphragm, which interacts with the air to produce a sound. I will lump together piezoelectric devices and wire-wound speakers (the type you have in your stereo or TV) in this book and just call them speakers.

To wire a speaker to a digital output, I use the circuit shown in Fig. 10.43. In this circuit, when the output changes state, the speaker will deform, "click," and return to its original state (because of the capacitor charging and the speaker not having any current across it). This probably seems surprising, as the speaker does not stay in the deformed state until the output changes it again.

The reason for this can be seen in the oscilloscope picture in Fig. 10.43. When the speaker deforms, large kickback voltages are produced that last a relatively long time. These voltages can cause upsets in certain chips, so the capacitor is added to the circuit to filter out large voltages. This doesn't mean

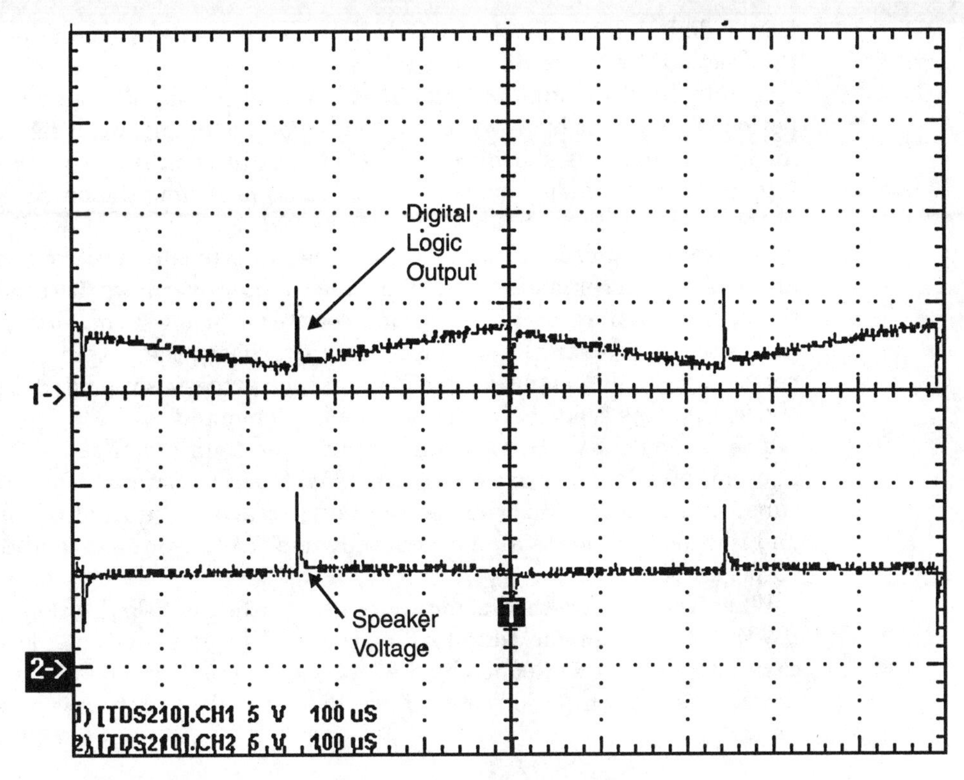

Figure 10.43 Circuit to drive a speaker from a digital logic device.

that the driver isn't "backdriven," but the voltage applied to the output is much less and not as potentially damaging.

There is a lot of material in this section. I hope that you understand that my goal in showing you all this material is to help you understand what is involved in designing simple circuits that allow digital circuits to interface with analog signals. In these few pages, I have done nothing more than just scratch the surface of this topic, but hopefully I've given you some ideas on how to work with analog circuits. You should also understand that you should not be looking at performing complex tasks on analog signals with simple digital circuits. In the following sections, I will present more information that will show you how to better interface digital electronic devices in analog-world situations.

Pulse-Width-Modulation I/O

Most other digital devices do not handle analog voltages very well. This is especially true for situations where high-current voltages are involved. The best way to handle analog voltages is to use a string of varying wide pulses to indicate the actual voltage level. This string of pulses is known as a *pulse-width-modulated* (PWM) analog signal and can be used to pass analog data from a digital device, control DC devices, or even output an analog voltage. In this section, I want to discuss PWM signals.

A pulse-width-modulated signal is a repeating signal that is on for a set period of time that is proportional to the voltage being output. A pulse-width-modulated signal is shown in Fig. 10.44. I call the *on time* the *pulse width* in Fig. 10.44 and the *duty cycle* is the percentage of time the on time is relative to the pulse-width-modulated signal's *period.*

To output a pulse-width-modulated signal, typically two counters (*timers*) are used with a common clock; when one counter overflows, it resets itself and the second counter. Until the second counter overflows, the output of the circuit is set to 1. When the second counter overflows, the output of the circuit is reset until the first counter overflows and the process is repeated. Figure 10.45 shows how this type of circuit could be implemented.

The example PWM generator circuit uses counters that can be reloaded upon an overflow positive pulse. The PWM period counter (the first counter) runs continuously, and when it overflows (reaches the final count), it resets and reloads the count value for not only itself but also the second counter (the on-period counter).

When the PWM period counter resets, it sets the S-R flip flop, driving the PWM output high for the start of the PWM signal output. The on-period counter is reset and reloaded by the PWM period counter and runs until it overflows. When the on-period counter overflows, the PWM output is halted and the

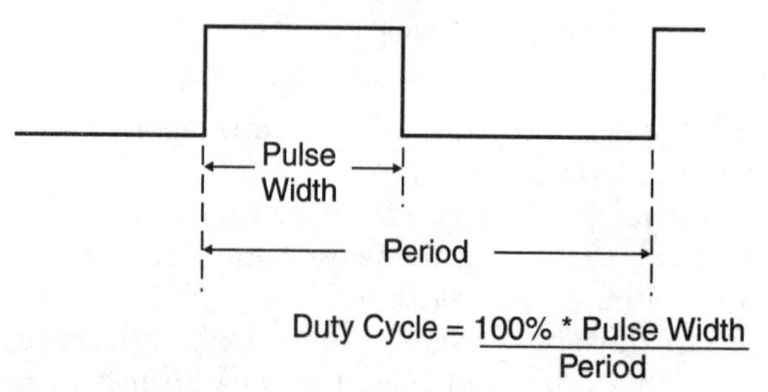

Figure 10.44 Pulse-width-modulated signal waveform.

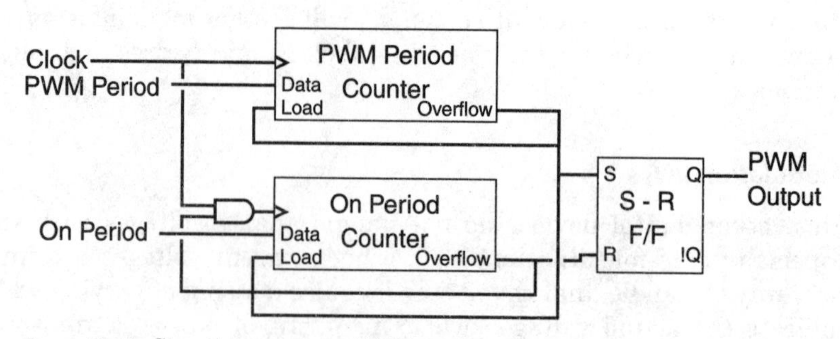

Figure 10.45 Circuit to generator PWM outputs.

counter also stops running until the PWM period counter reloads it, which resets the overflow output and allows the counter to drive the on-period counter once more.

The hardware circuit I've shown here should look quite simple, and it actually is. An actual implementation will vary, depending on whether the counters can be reset by the overflow (most TTL chips can) and whether a full on and a full off are required.

This circuit is really designed to have at least one clock cycle on and off. If this is not acceptable, then the circuit will have to be modified to allow for the on period to be either zero or equal to the value of the PWM period counter. The on period value greater than the PWM period counter is a situation that should be avoided.

Passing analog data back and forth between digital devices in any format is not going to be accurate because of the errors in digitizing the value and restoring it. This is especially true for pulse-width-modulated signals that can have very large errors due to the sender and receiver not being properly synchronized and the receiver not starting to poll at the correct time interval. In fact, the measured value could have an error of upward of 10 percent of the actual value. This loss of data accuracy means that the analog signals should not be used for data transfers; instead the analog values should be digitized by an ADC and the values then sent to the other device for processing.

In using a PWM for driving an analog device, it is important to make sure the frequency is faster than what a human can perceive. As noted in the section LEDs above, this frequency is 30 Hz or more. But for motors and other devices that may have an audible whine, the pulse-width-modulated signal should have a frequency of 20 kHz or more to ensure that the signal does not bother users (although it may cause problems with their dogs).

The problem with the higher frequencies is that the granularity of the pulse-width-modulated signal decreases. This is due to the inability of the digital device to change the output in relatively small time increments from on to off relative to the size of the pulse-width-modulated signal's period.

Complex Analog Output

From time to time, you will have to create circuits that drive out specific analog voltages at different times. While there are commercially available digital-to-analog converters (DACs), these chips are usually quite expensive and difficult to interface by using just simple digital logic. The commercial DACs usually can convert a 6- to 8-bit digital number into an analog voltage, but for most applications this is overkill because you need just one or two different voltage levels that you can control. In this section, I will introduce to you what I call a *multiple-output voltage divider*. This circuit is very easy to interface and can be used in a variety of different situations quite cheaply.

Earlier in the book, I introduced you to a single-voltage-output voltage divider like the one in Fig. 10.46. In this circuit, the output voltage is determined by the formula:

$$V_{out} = V_{in}\left(\frac{R_s}{R_s + R_n}\right)$$

where R_n is the resistance between the tap and V_{cc} and R_s is the resistance between the tap and ground.

A *variable-resistance* voltage divider can be implemented as part of a *resistor ladder* as shown in Fig. 10.47. In this circuit, the voltage output is dependent on which is the highest transistor that is turned on (and is pulling the circuit down). This circuit is designed to output a set number of specific analog voltages, including V_{in} and Gnd. These different voltages can be used in applications where specific and not general voltage levels are required for the application.

To calculate the output voltage of the circuit, the same formula is used as for a regular voltage divider, but I modify it to indicate the dependence of the output on which resistor is pulled to ground:

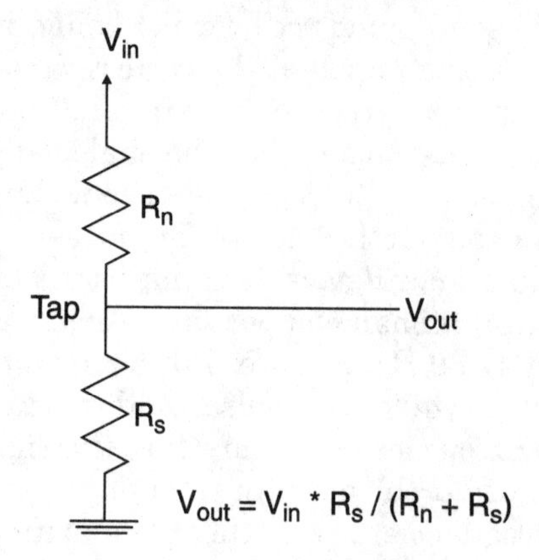

Figure 10.46 Voltage divider circuit.

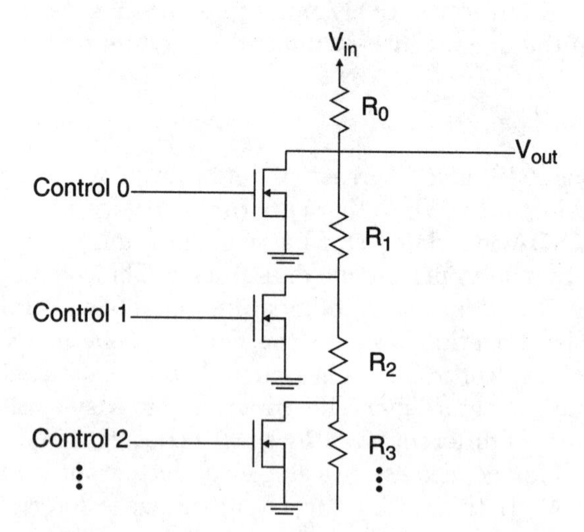

Figure 10.47 Voltage ladder circuit.

$$V_{out} = V_{in}\left(\frac{R_1 + \ldots + R_n}{R_0 + R_1 + \ldots + R_n}\right)$$

This formula takes into account the situation where Control 0 is active and V_{out} is pulled to ground, but does not properly take into account the situation where all the control transistors are turned off. In this situation, the output voltage is equal to the input voltage of the circuit.

From this formula, V_{out} for the case where Control 2 is on is:

$$V_{out} = V_{in}\left(\frac{R_1 + R_2}{R_0 + R_1 + R_2}\right)$$

As you may have already realized, this circuit outputs the correct voltage only if there is no load on it. If this circuit is used to drive some kind of current load, the actual voltage output will change (usually in an unexpected way). To prevent this a *voltage follower,* or buffer, has to be used to provide drive current to an external device. A simple op amp can be used for this purpose when it is wired as a unity-gain amplifier as shown in Fig. 10.48.

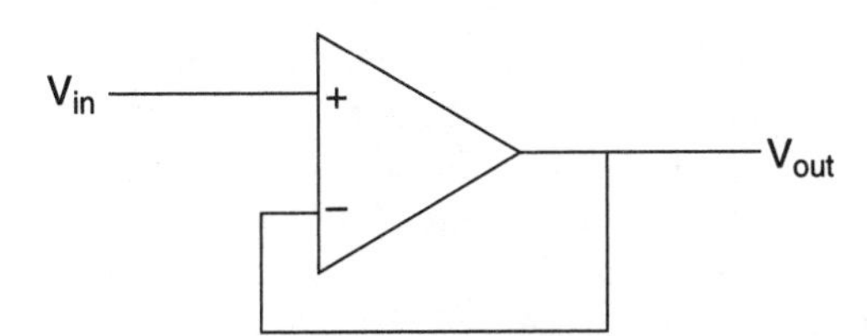

Figure 10.48 Voltage buffer unity-gain amplifier.

Calculating the values for the different resistors in the circuit can be surprisingly difficult. When you are calculating values for your own applications, you should first select the value of R_0 such that it will have the correct characteristic resistance for the circuit. If you use the op-amp buffer, this is not critical, but for other applications, where the analog voltage is going to drive an analog device, matching the impedance is quite critical. For example, coaxial cables used with TV signal transmission have a characteristic impedance of 75 Ωs.

When you are calculating these resistances, you will find that you will get values that are not standard resistor values. You can put standard resistors in parallel or series to get the actual resistances.

Once this is done, I suggest you build your circuit on a breadboard and test out the different voltage levels. Chances are everything will be correct, but you may find you have to tweak some resistances because of unexpected interactions with the different circuits on the board.

You should have noticed in Fig. 10.47 that I specified that N-channel MOSFETs should be used for the switches instead of NPN bipolar transistors.

The reason why I did this was to avoid the 0.7-V drop across the bipolar transistor's base-emitter PN junction. You can build multiple-output voltage dividers with NPN bipolar transistors, but you may find that the transistors interact with resistor circuits. The N-channel MOSFETs offer a minimum of interaction with the resistor circuit and allow the resistance values to be calculated quite accurately.

I must point out that this is not a precision device. While you can get acceptably good results for most applications, this circuit should not be considered adequate for providing reference voltages. This is especially true if you are using an op-amp buffer to drive the output to current loads. The op amp should be as low-noise a device as possible to make sure the output is stable and not too noisy.

Relays and Solenoids

Some real-life devices that you may have to control by digital electronics are electromagnetic devices like relays, solenoids, and motors. These devices cannot be driven directly by a TTL or CMOS drivers because of the current required and the noise that electromagnetic devices generate. This means that special interfaces must be used to control electromagnetic devices.

The simplest method of controlling these devices is to just switch power on and off to the coil in the device. The circuit shown in Fig. 10.49 works with relays (as is shown), solenoids (which are coils that draw an iron bar into them when they are energized), and DC motors (which will turn in only one direction).

In this circuit, the digital logic turns on the Darlington transistor pair, causing current to pass through the relay coils, closing the contacts. To open the relay, the output is turned off (or a 0 is output). The shunt diode across the coil is used as a kickback suppressor. When the current is turned off, the magnetic flux in the coil will induce a large back electromagnetic force, or EMF (voltage), which has to be absorbed by the circuit or there may be a voltage spike that can

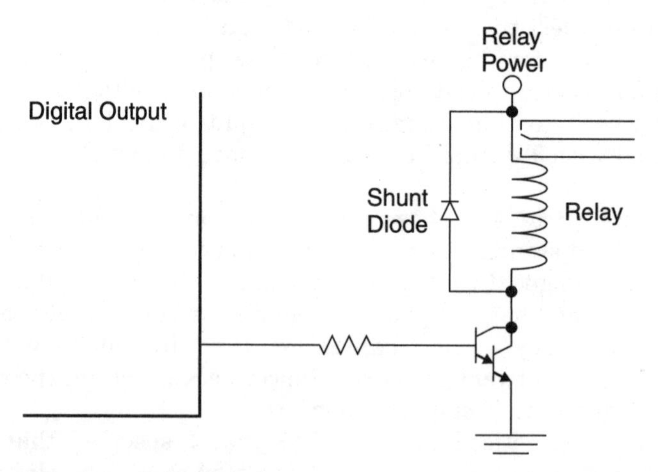

Figure 10.49 Relay control.

damage the relay power supply and the controlling digital electronics. This diode must *never* be forgotten in a circuit that controls an electromagnetic device. The kickback voltage is usually on the order of several hundred volts for a few nanoseconds. This voltage causes the diode to break down and allows current to flow, attenuating the induced voltage.

Instead of designing discrete circuits to carry out this function, I like to use integrated chips for the task. One of the most useful devices is the ULN2003A (see Fig. 10.50) or the ULN2803 series of chips, which have Darlington transistor pairs and shunt diodes built in for multiple drivers.

DC and Stepper Motors

Motors can be controlled by exactly the same hardware as shown in the previous section, but as I noted above, they will run in only one direction. A network of switches (transistors) can be used to control turning a motor in either direction; this is known as an H bridge and is shown in Fig. 10.51.

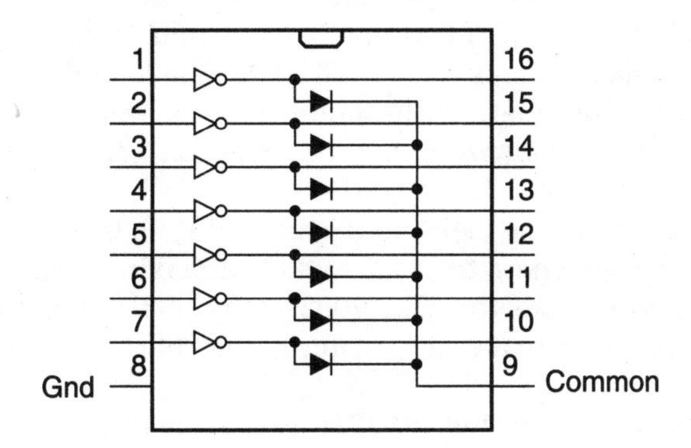

Figure 10.50 ULN2003A driver array.

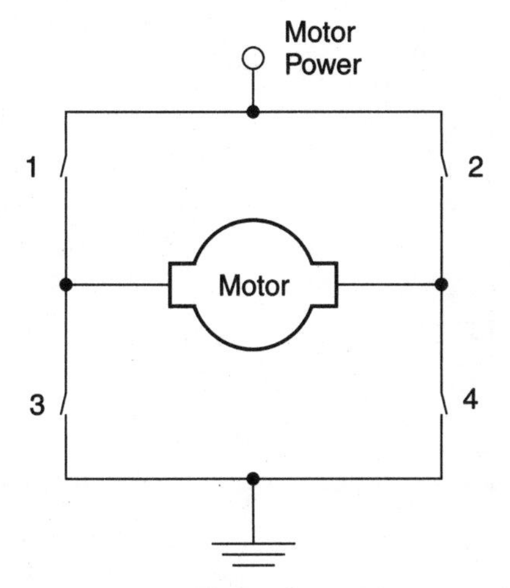

Figure 10.51 H bridge motor driver.

In this circuit, if all the switches are open, no current will flow and the motor won't turn. If switches 1 and 4 are closed, the motor will turn in one direction. If switches 2 and 3 are closed, the motor will turn in the other direction. Both switches on one side of the bridge should *never* be closed at the same time, as the motor power supply will burn out or a fuse will blow because there is a short circuit directly between motor power and ground.

Controlling a motor's speed is normally done by pulsed control signals in the form of the pulse-width-modulated (PWM) signal discussed previously in this chapter (see Fig. 10.44). This will control the average power delivered to the motors. The higher the ratio of the pulse width to the period, the more power delivered to the motor.

The frequency of the PWM signal should be greater than 20 kHz to avoid the PWM from producing an audible signal in the motors as the field is turned on and off. This can be very annoying and damaging in some motors.

In the same way the ULN2003A was used to simplify the wiring of a relay control, the 293D (Fig. 10.52) or 298 chip can be used for controlling a motor.

The 293D chip can control two motors (one on each side), connected to the buffer outputs (pins 3, 6, 11, and 14). Pins 2, 7, 10, and 15 are used to control the voltage level (the switches in Fig. 10.51) of the buffer outputs. Pins 1 and 9 are used to control whether the buffers are enabled. These can be PWM inputs, which makes control of the motor speed very easy to implement.

V_s is the +5 V used to power the logic in the chip and V_{ss} is the power supplied to the motors and can be anywhere from 4.5 to 36 V. A maximum of 500 mA can be supplied to the motors. Like the ULN2003A, the 293D contains integral shunt diodes. This means that to attach a motor to the 293D, no external shunt diodes are required.

In the example circuit shown in Fig. 10.52, you'll notice that I've included an optional "snubber" resistor and capacitor. These two components, wired across the brush contacts of the motor, will help reduce electromagnetic emissions

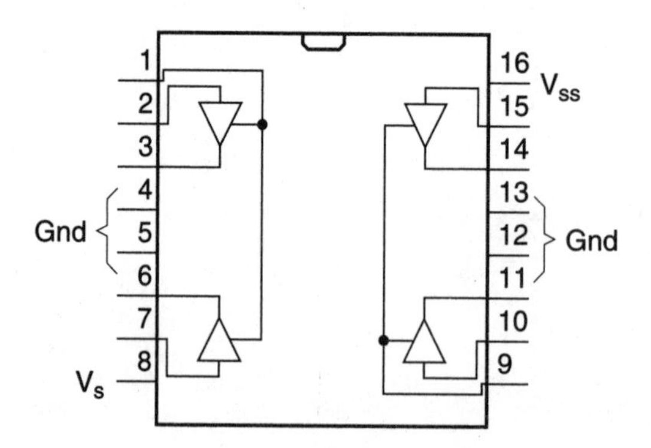

Figure 10.52 293D H bridge motor driver.

and noise spikes from the motor. In the motor control circuits that I have built, I have never found them to be necessary. But if you find erratic operation from the microcontroller when the motors are running, you may want to put in the 0.1-μF capacitor and 5-Ω (2-W) resistor snubber across the motor's brushes as shown in Fig. 10.53.

There is an issue with using the 293D and 298 motor controller chips, and that is that they are bipolar devices with a 0.7-V drop across each driver (for 1.4 to 1.5 V for a dual driver circuit as shown in Fig. 10.53). This drop, with the significant amount of current required for a motor, results in a fairly significant amount of power dissipation within the driver. The 293D is limited to 1 amp total output and the 298 is limited to 3 amps. For these circuits to work best, a significant amount of heat sinking is required.

To minimize the problem of heating and power loss, I have recently been looking at using power MOSFETs to control motors. The IRF510 and IRF9510 N-channel and P-channel MOSFETs are commonly used for relatively small (less than an amp of drive current) motors.

Stepper motors are much simpler to develop control functions for than regular DC motors. This is because the motor is turned one step at a time or can turn at a specific rate (specified by the speed in which the steps are executed). In terms of the hardware interface, stepper motors are a bit more complex to wire and require more current (meaning that they are less efficient), but these are offset by the advantages in control.

A *bipolar* stepper motor consists of a permanent magnet on the motor's shaft that has its position specified by a pair of coils (Fig. 10.54). To move the magnet and the shafts, the coils are energized in different patterns to attract the magnet. For the example above, the following sequence would be used to turn the magnet (and the shaft) clockwise.

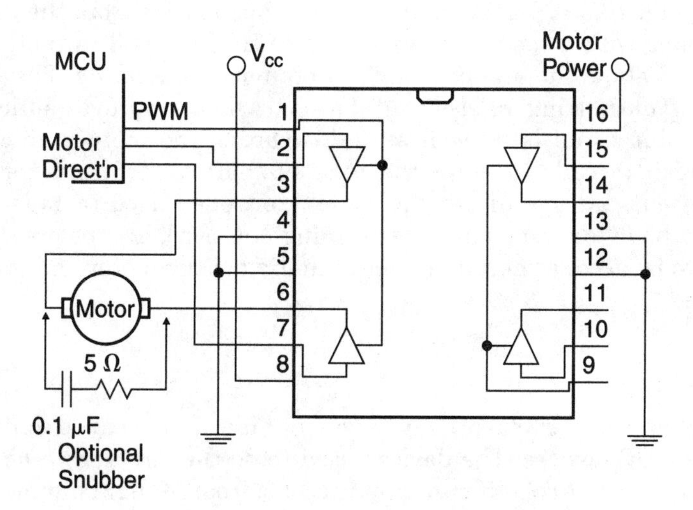

Figure 10.53 Wiring a motor to the 293D.

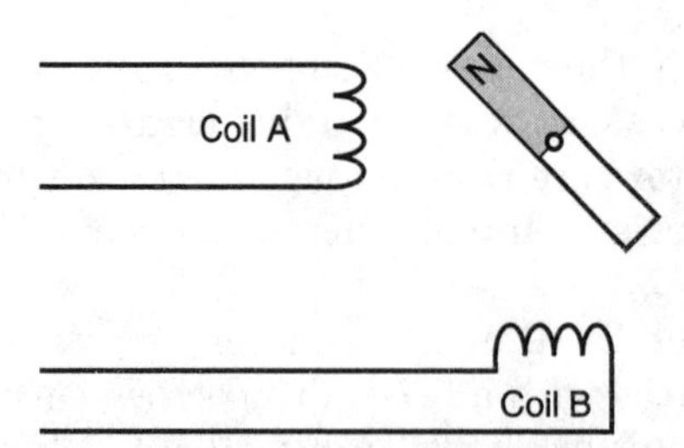

Figure 10.54 Stepper motor.

Step	Angle, degrees	Coil A	Coil B
1	0	S	
2	90		N
3	180	N	
4	270		S
5	360/0	S	

In this sequence, coil A attracts the north pole of the magnet to put the magnet in an initial position. Then coil B attracts the south pole, turning the magnet 90°. This continues on to turn the motor 90° for each step.

The output shaft of a stepper motor is often geared down so that each step causes a very small angular deflection (a couple of degrees at most rather than the 90° in the example above). This provides more torque output from the motor and greater positional control of the output shaft.

A stepper motor can be controlled by something like a 293D (each side driving one coil). But there are also stepper motor controller chips, like the UC1517 (Fig. 10.55). In this chip, a step pulse is sent from the digital logic control along with a specified direction. The INH pin will turn off the output drivers and allow the stepper shaft to be moved manually. The UC1517 is capable of outputting *bilevel* coil levels (which improves efficiency and reduces induced noise) as well as *half-stepping* the motor (which involves energizing both coils to move the magnet/shaft by 45° and not just 90°). These options are specific to the motor/controller used (a bipolar stepper motor can have four to eight wires coming out of it), and before deciding on features to be used, a thorough understanding of the motor and its operation is required.

AC Current Control

It will probably be very surprising to you, but it is quite easy to control AC currents using DC devices. The devices used to do this are known as triacs and are used for a variety of different applications, from AC light dimmers to motor controllers.

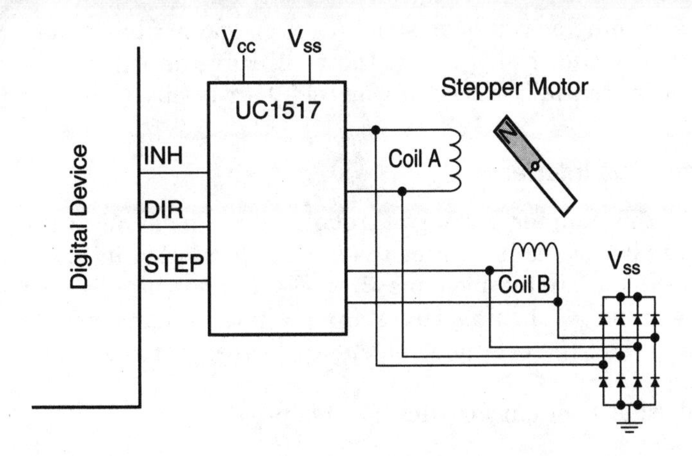

Figure 10.55 UC1517 control of a stepper motor.

Triacs are interesting devices and come under the heading of *thyristors,* which are used to switch AC signals on and off. Triacs do not rectify the AC voltage because they consist of two *silicon controlled rectifiers* (SCRs) that allow the AC current to pass without any clipping. A typical circuit for triacs is shown in Fig. 10.56.

Triacs do not allow AC current to pass unless their gates are biased relative to the two AC contacts. When controlling triacs using digital logic circuits, I will pull their gates to logic ground. The current required to close a triac can range from 10 mA to several hundred milliamperes. For household applications, a 10- to 25-mA-controlled triac can be used, which is well within the current output capabilities of TTL chips.

When triacs are used in light dimmer applications, the point at which the AC wave crosses the zero voltage point is identified within the circuit and then the triac is opened for a specific fraction of the wave. This process is similar to controlling a DC motor with a PWM signal.

A word of warning and caution; AC can be tricky, with large energies occurring when you least expect them. Even with household wiring, there is are opportunities to get seriously burned or electrocuted. If you are not sure of

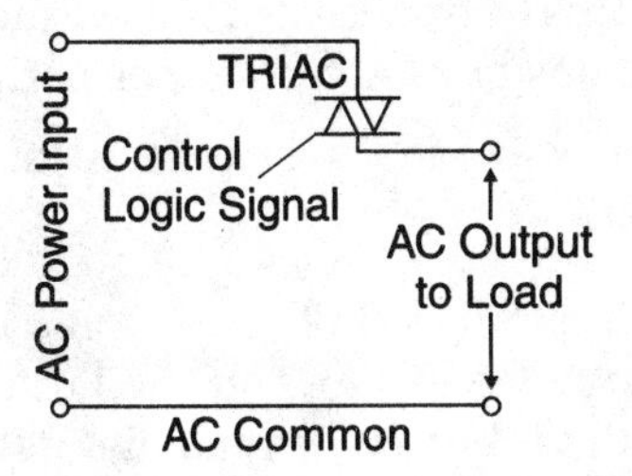

Figure 10.56 Typical triac AC control circuit.

what you are doing or you have some concerns, do not build the circuit and if you have built it, don't plug it into the wall! Have an expert check over your circuit before potentially harming yourself, your family, or your home.

Project 17—Hexadecimal Bus Interface

To finish off this chapter, I will go through a complex application that demonstrates a number of the interfaces that I have presented in this and the earlier chapters. This circuit took me 3 weeks of working part time to get it functioning properly. The final circuit is a peripheral that you can add to a computer system (like the example one shown in the next chapter), and it works quite well.

The goals that I set out for building this project were:

1. Provide a simple programmable peripheral for use with bus-based computer systems
2. Work with the 74C922 to provide a hex keypad with four additional buttons
3. Use LED displays for data and addresses and pass data between the displays as the keypad buttons are pressed
4. Interface to a bus with only one TTL Input
5. Demonstrate a complex wire-wrapped board for the book

To operate the hexadecimal bus interface, I decided on using four buttons to select different operating modes for the tool. These buttons are:

1. _User/peripheral (_U/P)—When in user mode, the board will be able to drive the buses _RD and _WR lines, allowing it to be used as a simulated processor for accessing peripherals. In this mode, the value entered into the address is driven onto the bus and data is either read or written, depending on the state of the _R/D button. In peripheral mode, the card will compare addresses on the bus and either drive the 8-bit data value in response to a read (_RD active) or will save the 8-bit driven value and display it on the LEDs. The data and address values can be updated at any time in this mode.
2. _Address/data (_A/D)—Determines to which display, address or data, the keystroke will be passed to.
3. _Read/write (_R/D)—As noted above, this button is used in user mode to determine whether data is driven onto the bus or read from other peripherals.
4. Enter—Used to initiate a bus read or write in user mode. After the data is sent, the address counter is incremented.

I am happy to say that I succeeded in meeting all of these goals and the circuit operated as I had specified. The circuit was built on the large wire-wrap board shown in Fig. 10.57. I built a small 20-button keypad and used common-cathode dual seven-segment displays for the current address and data values. Three 5-mm LEDs were used to display the state of the control buttons.

Figure 10.57 Hexadecimal bus interface.

I don't recommend building this circuit because of its complexity and power consumption. The circuit has several thousand wires, which will take a week or more to wire-wrap (I don't think you can fit this circuit on any kind of single breadboard). There were times while building this circuit when I thought that I was about to go blind. As I will discuss below, a number of the components did not work as I expected (or desired), which meant I had to rebuild sections of the circuit repeatedly, which increased the complexity of the final circuit and made tracing through the various subsystems much more difficult.

I am amazed that the board I chose ended up being the right size for the project. When I first counted out the number of chips that I expected to have to use, I came up with 33. The total was a couple more, and I was amazed at how accurate I was, considering the sections that had to be redesigned.

Along with this complexity, the circuit requires about 800 mA. This is much more than the 7805 +5-V power supply PCB that comes with this book can comfortably source. I ended up using a 5-amp bench supply, and I noticed something about this circuit that I haven't experienced in circuits in 20 years or so—the board and the chips got warm. I've been working with CMOS-based circuits for so long, I had forgotten what is was like to work on a board that varied in temperature from being the same as my body to about 120°F (50°C).

In Fig. 10.57, you can see a rather large (doubled, actually) heat sink on my prototype. As I added the subsystems to the project, I discovered the current requirements (largely due to the LEDs) going up, and I had a problem with the 7805 that I used going into "thermal shutdown." As I will discuss later in this section, I was unable to get sufficient current out of the 7805 to run the circuit correctly. I left the 7805 with its heat sink on the board, but powered the application from a bench supply.

The biggest mistake that I made when I built this project is that I built it backward. Looking at Fig. 10.57, you will see that I have placed the keypad on the left-hand side of the board. Unfortunately, I am right handed. I noticed the problem when I had finished wiring in the keypad. As this took me about 3 hours of work and I was starting to realize that this was a huge project, I decided to leave it and build the project as is. While the hexadecimal bus interface works fine, I found that my prototype is somewhat clumsy to work with—especially reaching across it to press a button.

If you are thinking of taking the circuit and designing a PCB for it, I would like to discourage you from doing this as well. Using simple TTL (or CMOS, to reduce the power consumption) chips for this type of project is simply not the optimal solution. When I build this circuit again (it actually is a very useful circuit, which makes finding a better solution desirable), I will use either a microcontroller (like the Microchip PICmicro MCU, which has an 8-bit bus interface built in) or a PLD. By doing this I will reduce the chip count (and power requirements) significantly and also be able to debug the application much more easily.

You can build this circuit with CMOS logic to minimize the current required, but you will still be left with the daunting task of wiring it. Modern technology can really simplify this application and avoid many of the problems I had in building it.

As I indicated above, I wanted to demonstrate this application as a complex wire-wrapped project. I used machined-socket wire-wrap pins, which although somewhat more expensive, are more reliable and take multiple wraps and unwraps better than cheaper sockets make out of stamped metal. On the back side of the wire-wrap PCB, I placed a 0.1-μF capacitor as shown in Fig. 10.58. The capacitor is soldered to the socket and, in turn, the socket is soldered to the PCB for mechanical stability.

To build the 20-pin keypad, I first soldered in PCB-mount push button switches, then wired them together as a switch-matrix keypad. I used stripped, 20-gauge, solid-core wire for the column connections and 28-gauge wire-wrap wire for the row connections. The result of 3 hours' work is not pretty (as can be seen in Fig. 10.58), but it does work very well. To label the switches, I used an adhesive-tape labeler with the smallest font available.

When I designed the circuit, I broke it up into four subsystems, as I have shown in Fig. 10.59. When I first started building the application, I started with the keypad, then went on to the data display, address display/compare, and interface control. This was a reasonable way of going about the work, but unlike other projects, such as the clock, the different subsystems in this project are very tightly coupled together, with many wires and quite a bit of crit-

Figure 10.58 Back side of hexadecimal interface showing keypad wiring and back side decoupling capacitor.

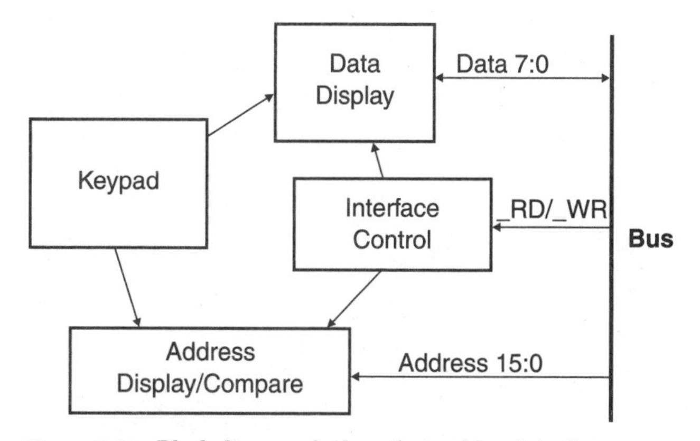

Figure 10.59 Block diagram for hexadecimal bus interface.

ical signal timings between the different circuit elements. This was the cause of the extended period of time it took to get the project running.

When I first designed this application, I did not use any kind of circuit simulation. Looking back at a number of the problems that I had to work through, I could have saved myself some time if I had. I simply did not expect to have major problems with the application. The problems that I did have were certainly not major, but simulating the subsystems in different operating modes could have minimized them.

In this section, I will treat the four different subsystems as four different projects. Each will have its own schematic and each will have its own bill of material. Each different circuit can be used as the basis for another project, or swept up into a PLD to reduce the amount of work you have to go through. Note that in the schematics, I keep track of the components used and the reference designators indicate that the four subsystems are actually part of one large application.

Project 17a—Keypad

The first subassembly that I built was the keypad. I had never worked with the 74C922 before and I thought this would be a good opportunity. Looking at the chip's datasheet, you can see that, while it is designed for a 16-button keypad, it can be expanded quite easily to a 32-button keypad. As I've shown in Fig. 10.60, the top four buttons are used for the four control buttons I presented at the start of this section.

Other than this modification, the use of the 74C922 is very straightforward, and to avoid any potential timing problems, I decided to simply use three of these four buttons to toggle D flip-flops. Looking at Fig. 10.60, you will see a number of other circuits [namely, the multiple NOT gates on data available (DA) as well as the single-shots and the 74LS74 D flip-flop] that indicate that I had some problems with getting the keypad working properly.

The bill of materials for the keypad is:

Part	Description
U1	74C922 keypad controller
U2	4049 CMOS inverter
U3, U7	74LS74 dual D flip-flop
U4	74LS123 dual single-shot
U5	74LS00 quad two-input NAND gate
U6	74LS139 dual 2-to-4 decoder
R1, R3	15k, ¼-W resistor
R5	100k, ¼-W resistor
R6–R8	470-Ω, ¼-W resistor
C1	10-μF, 35-V electrolytic capacitor
C2	1.0-μF, 16-V electrolytic capacitor
C3, C5	100-pF capacitor (any type)
C6–C12	0.1-μF, 16-V tantalum capacitors
D1–D3	5-mm red LEDs
D4	1N914 silicon diode
SW1–SW20	Momentary-on PCB-mount switches

When I first wired this circuit, it seemed to work correctly, although when I connected the button indicator LEDs, I discovered my first problem. When one of these buttons was pressed, the indicator LED would toggle correctly, but

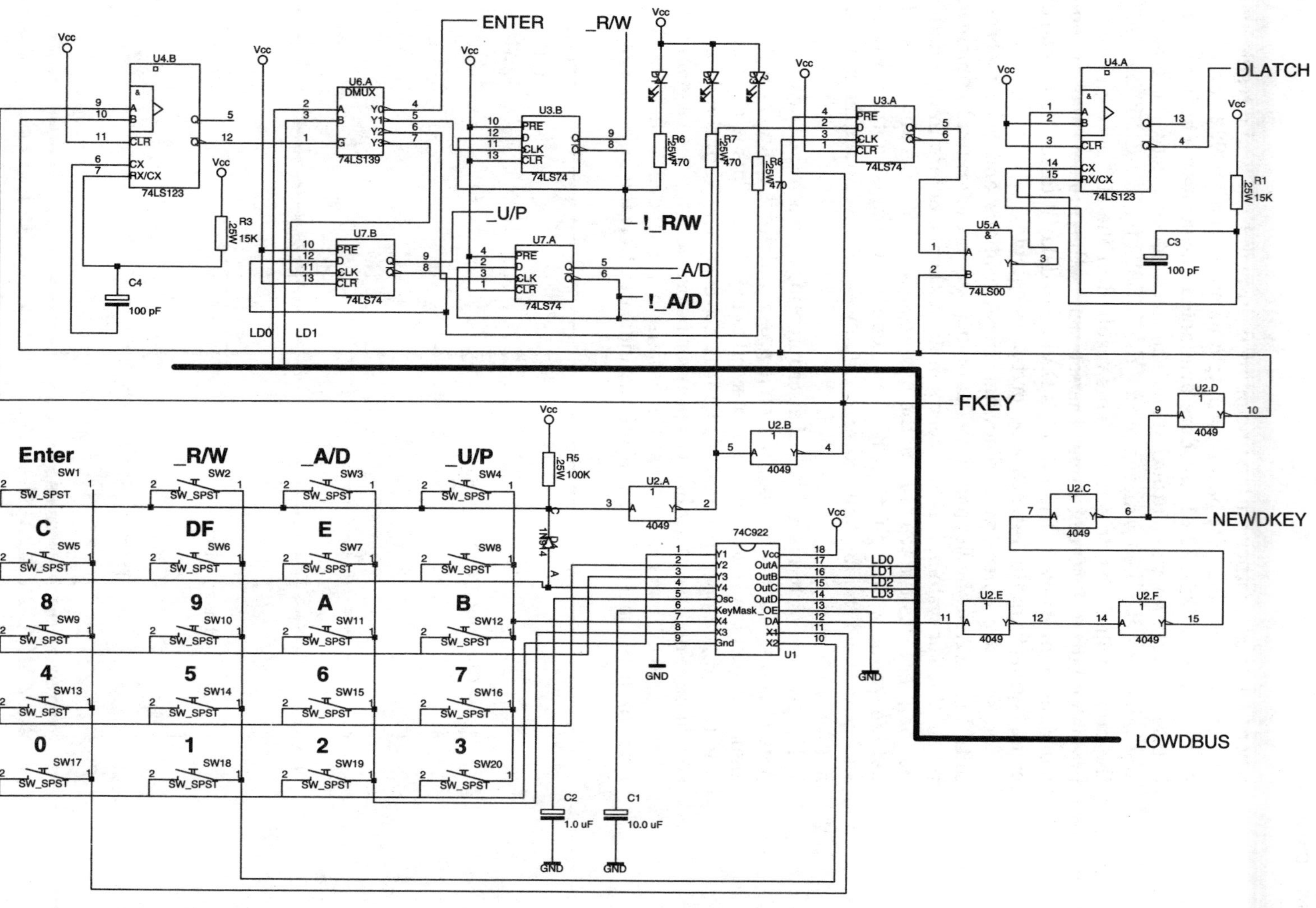

Figure 10.60 Project 17a—keypad circuitry.

when I pressed another button (including 0 through F), it would toggle again. The problem was found to be a slow output on one of the 74C922's output pins. As I've shown in Fig. 10.61, the data-available line was active before a correct button row/column was output. When the clock became active, the *previous* key value was still being driven out by the 74C922 for about 150 nsec before the proper value was driven out. This resulted in the 100-nsec glitch on the unrelated button LED clock line. The solution to this problem was to add the four CMOS inverters on the 74C922's DA line to delay it until the keypad data being output from the 74C922 was correct and stable.

The second problem was somewhat more insidious and wasn't observed until I connected the keypad to the data display. In the original implementation of this subsystem, I used a very simple method of determining whether a button press was the top four or the hex values—if both DA and the top-four buffered value were high, then it was a top-four key press. Otherwise, if DA was active, and the top-four buffered value was inactive, then the hex values had been pressed.

The problem with this method of discriminating between the two types of key presses is that when a top-four button is pressed, there is an interval when the signals match the conditions for a hex value key press. This is a result of the way the 74C922 operates, and, while it is not documented in the datasheet, I should have realized that this was possible before building the original circuit.

In Fig. 10.62, you can see that the buffered top-four FKEY line is active for more than 100 msec before the 74C922 indicates that the key press is active.

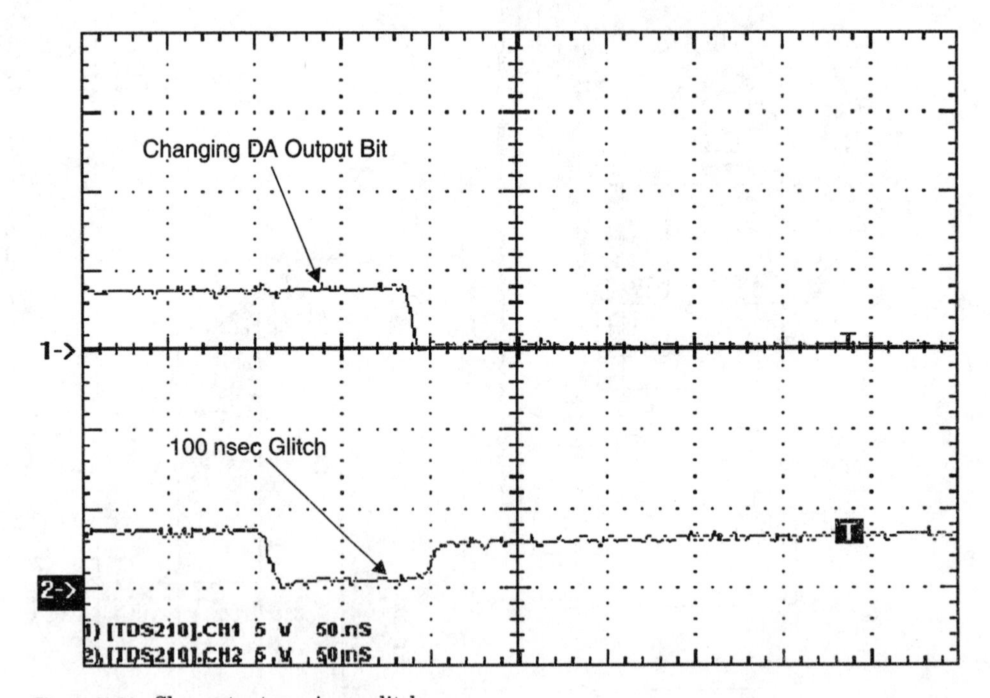

Figure 10.61 Slow output causing a glitch.

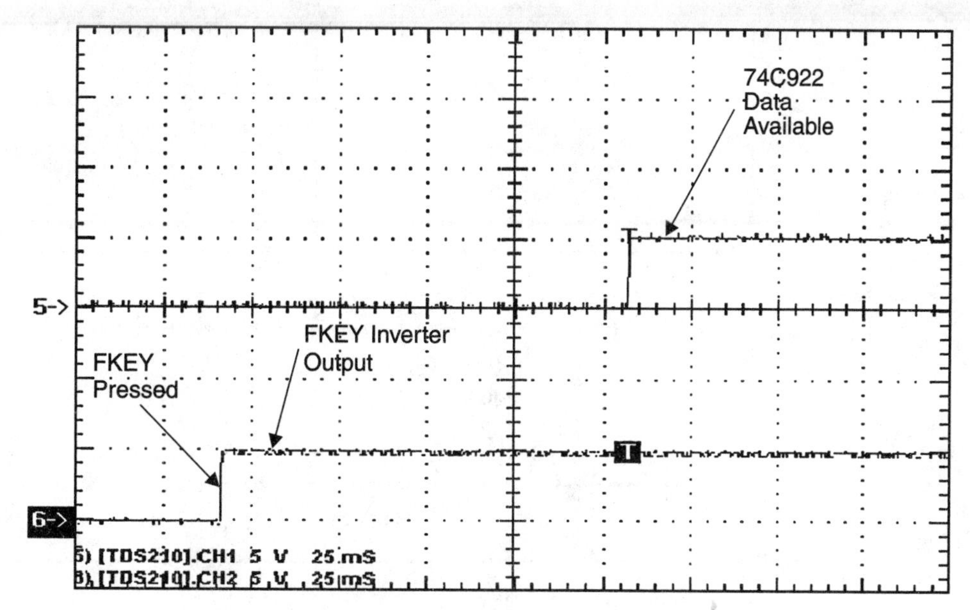

Figure 10.62 DA and inverted FKEY press.

This delay is the debouncing function of the 74C922. Similarly, if you look at the waveform when the button is released (seen in Fig. 10.63), you will see that there is a similar delay between the top-four FKEY line being inactive to DA becoming inactive. When the top-four button is released, the condition indicating a hex value key press, like the one shown in Fig. 10.64, exists and the hardware responds to it appropriately.

Looking back at the subsystem schematic (Fig. 10.60), you will see the D flip-flop that I used to avoid the problem with the offset FKEY line. This flip-flop is set when the FKEY line becomes loaded with a 1. When the D flip-flop is loaded with a 1, the NAND gate connected to its negated output will always output a 1.

When the DA line becomes active, the FKEY status is loaded into the flip-flop. If FKEY is inactive, then the negative output value will be high, allowing the NAND to output a low voltage and trigger the 74LS123 single-shot. The purpose of the two 74LS123 single-shot is to provide a data toggle function close to the time the key is pressed, rather than have data latched in when the key is released (which is not intuitive for users).

Using the 15k resistors and 100-pF capacitors in the 74LS123 gives an output pulse 250 nsec in duration. This pulse is long enough for virtually any type of logic to execute without any problems.

This circuit can be used in a variety of situations where multiple debounced button inputs are required. Despite the problems that I had with developing it, the resulting circuit works very well and can be used for a variety of different applications to give quite a "professional" feel to the application.

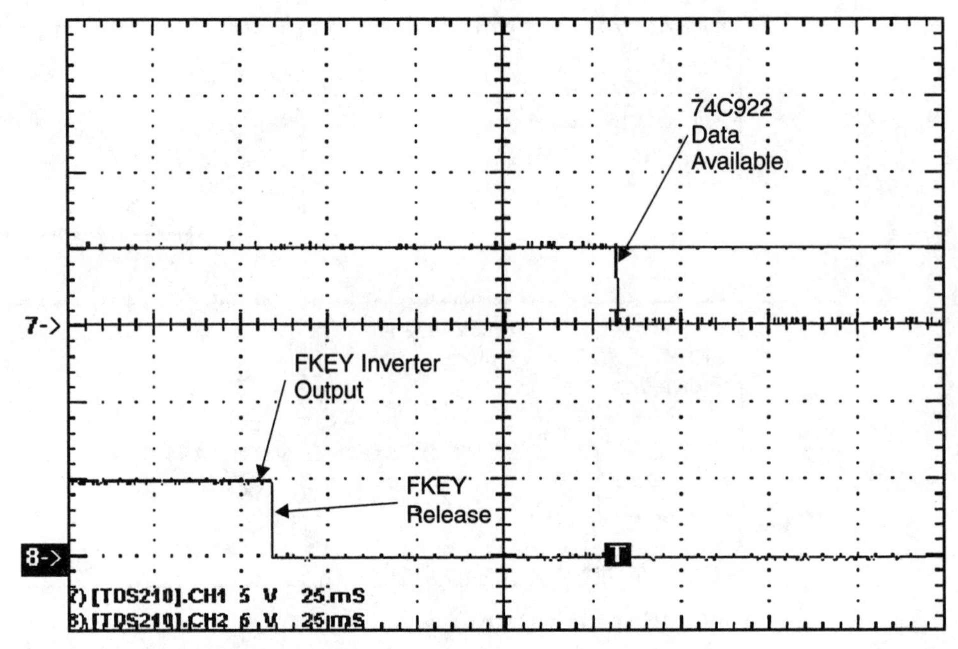

Figure 10.63 DA and inverted FKEY release waveforms.

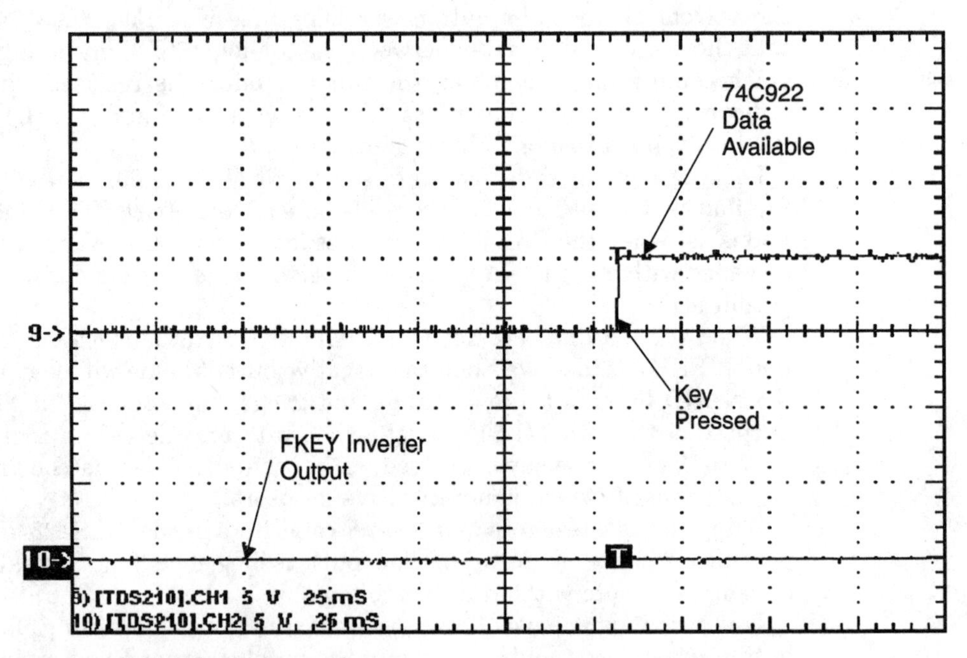

Figure 10.64 Digit key press waveforms.

Project 17b—Data display

The next subsystem that I added to the project was the data display. This circuit consists of two 7-segment LED common-cathode displays driven by 9368 hex LED drivers and a 74LS574 used to save and pass along data. When a bus access takes place, the 74LS574 passes data 8 bits at a time. When a keypad button is pressed, then the low 4 bits in the 74LS574 are passed to the high 4 bits of the 74LS574. At the same time as the low 4 bits are being passed to the high 4 bits, the low 4 bits are loaded with the 4-bit hex value. The final circuit is shown in Fig. 10.65.

The bill of materials for the data display subsystem is:

Part	Description
U8, U9	DM9368 hexadecimal seven-segment LED display drivers
U10	74LS574 8-bit D flip-flop
U11, U12	74LS244 8-bit tristate driver
U13	74LS245 bidirectional 8-bit tristate driver
CR1–CR2	Common-cathode, seven-segment LED displays
C13–C18	0.1-μF, 16-V tantalum capacitors
J1	18-pin DIP socket

I had very few problems with this circuit—actually it seemed to provide a good test base for the other subsystems, as I indicate in this section. The transfer of data from one part of the chip to another is a result of using edge-triggered D flip-flops. When the chip's clock line goes high, the value at the input is saved—no data is passed through to show up at a different part of the chip.

The only special feature of this circuit and how it is connected to the keypad subsystem is that during the keypad update, I enable the NEWDKEY tristate buffers before and after the data transfer takes place. By doing this, I make sure that all the voltage levels have time to settle at their correct values before the data is passed.

Chances are, if I were to use the same line as the data clock for the data output enable it would work. But, I would not count on this being 100% reliable, and it might work differently for different chips. It is always a good design practice to create a window that is larger than the time needed to pass the data.

Project 17c—Address display/compare

The next subsystem I created was the address display/compare circuit. This circuit displays and saves keypad input (selected by the _A/D button) and compares the address data to the current bus data. If the address on the bus matches the comparison value in the address buffers, it allows a read or a write of the display data circuitry. This function may seem to be similar to the circuitry presented for

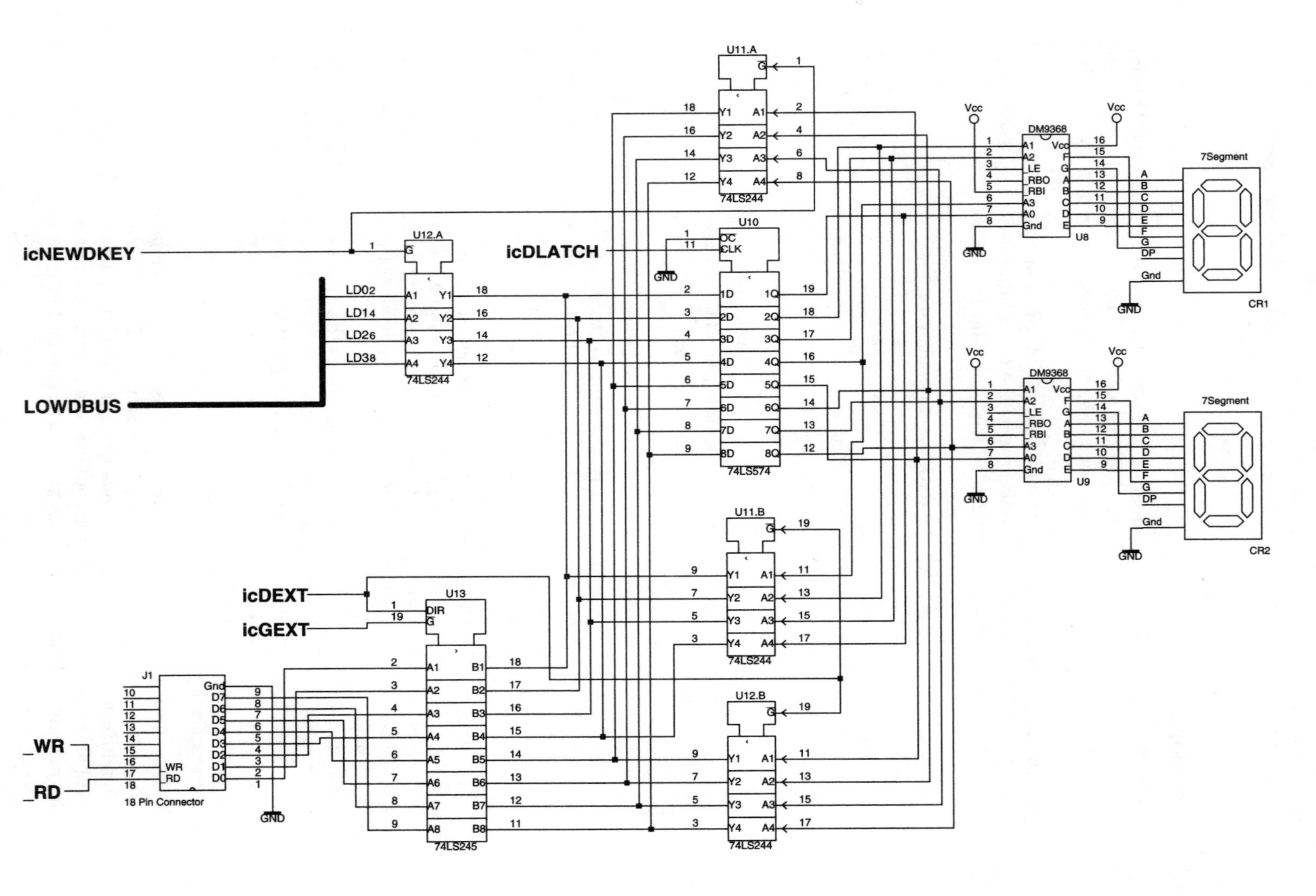

Figure 10.65 Project 17b—data display circuitry.

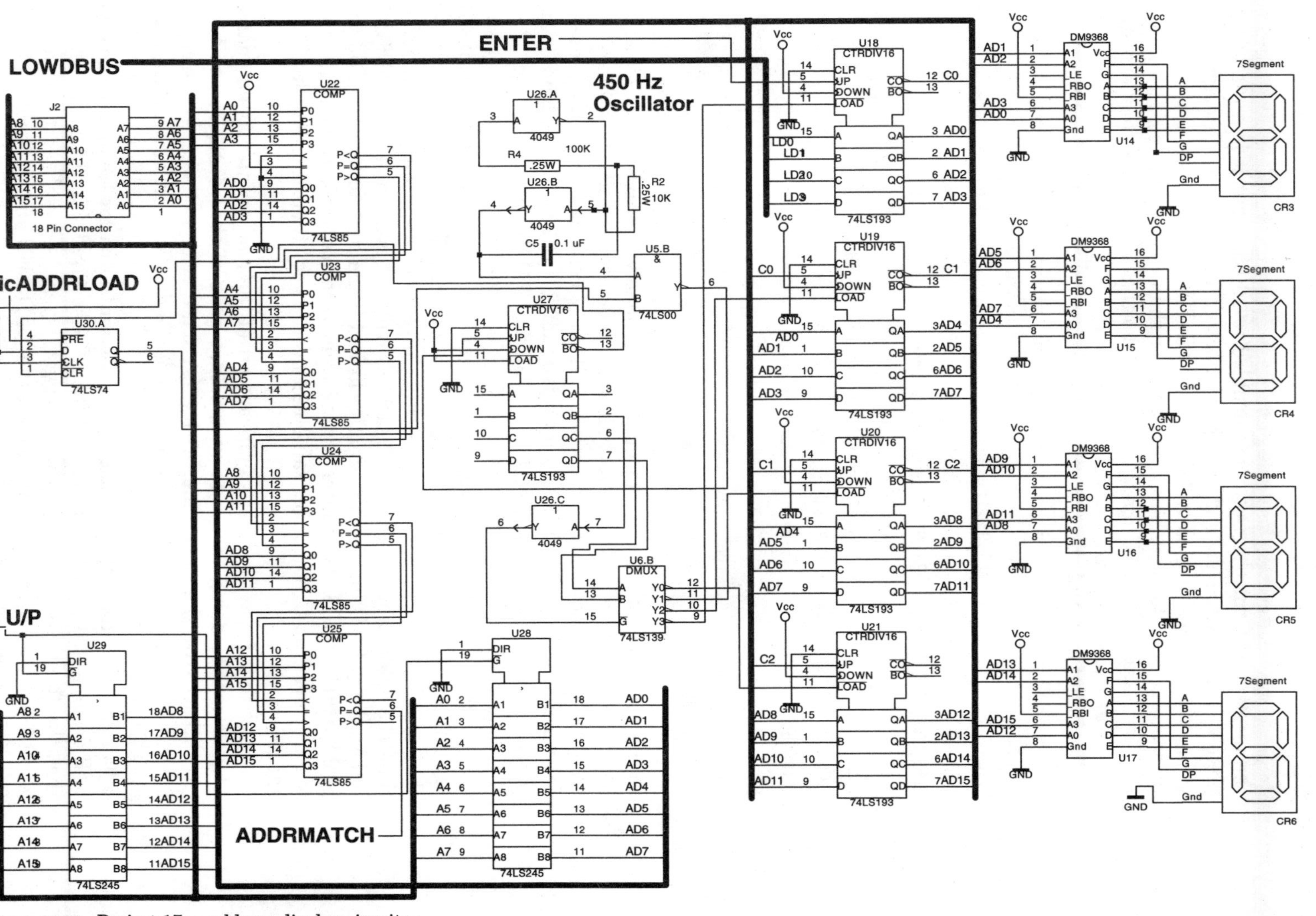

Figure 10.66 Project 17c—address display circuitry.

the display data subsystem (except doubled to 16 bits), but it is actually completely unique, with some different problems that I had to work through.

The circuitry for this subsystem is shown in Fig. 10.66. Part of the reason for its complexity is its use of 4-bit chips, rather than 8- (or even 16-) bit counters and comparators. I used the 74LS193s and 74LS85s because they are standard, easily found chips.

The bill of materials for this subsystem is:

Part	Description
U14–U17	DM9368 hexadecimal seven-segment LED display drivers
U18–U21	74LS193 4-bit counters
U22–U25, U27	74LS85 4-bit magnitude comparators
U26	4049 CMOS inverter
U28–U29	73LS245 bidirectional 8-bit tristate driver
U30	74LS74 dual D flip-flop
CR3–CR6	Common-cathode, seven-segment LED displays
R2	10k, ¼-W resistor
R4	100k, ¼-W resistor
C5, C19–C35	0.1-μF, 16-V electrolytic capacitor

The operation of the keypad update is probably confusing—I'm sure you don't understand why I have included the relaxation oscillator, the counter, and the 3- to 8-decoder. The reason for all these chips is to copy the contents from a downstream 74LS193 to an upstream 74LS193 without having any problems with the data flowing through all four 74LS193 chips.

Unlike the 74 series edge-triggered flip-flops used in the data display subsystem, the 74LS193 features a *transparent* latch, which means that while the _LOAD pin is active (low), the input data is output. This value is saved when the _LOAD line goes high. If all four chips have their _LOAD input pins active, then the value input at the first 74LS193 will be saved in *all* the 74LS193s unless the _LOAD line is individually clocked (from high to low) to avoid this problem. The 74LS193 counter counts from zero to 15 (0b01111) and stops when it overflows. When the value has overflowed, then the data has been shifted in.

Notice that I use the least significant bit of the load counter as an enable. The reason for doing this is to ensure that none of the 74LS193s can have _LOAD active at the same time. With this circuit, after 4 bits have been copied from one 74LS193 to the one above it, there is a dead spot in which none of the _LOAD pins are active.

The fact that the 74LS193 uses a transparent latch for loading data is not well documented in the datasheets. When I first wired this circuit, I assumed that the flip-flops in this chip were edge triggered—I realized they weren't when I powered up the circuit, turned the _A/D LED off, and saw that each key press resulted in the same number being displayed. Going back to the datasheets, I almost had to work through the logic diagram of the 74193 to understand what was happening.

Before working through the problem above, I had a much more upsetting problem, and that was, after I had built this circuit, I discovered that the keyboard problems that I thought I had fixed had returned. When I was pressing value keys, the function key LEDs would turn on and off. I originally thought I had a problem with a wire dropped onto the back of the card and something was shorting out.

Fortunately, I followed my own advice for debugging problems and looked at the power supply before getting out a magnifying glass to inspect the back side of the card. When I put my digital multimeter on the V_{cc} and Gnd pins of a chip, I discovered that the applied power was at 4.44 V. Wondering if I had a loading problem, I put an oscilloscope onto a couple of the card's signal lines and saw the waveform shown in Fig. 10.67.

At this point, I was still a using standard 7805 voltage regulator circuit (the same as the one on the +5-V power supply PCB that comes with the book), and I noticed that the 7805 was getting quite warm. I added two large heat sinks and a higher unregulated supply to the card, and it started working normally again. When I looked at the signals on the board, they looked like Fig. 10.68.

The lesson in this is to make sure that V_{cc} never goes below 4.75 V (which is 5 percent under a nominal 5.0 V). Most circuitry will work at 4.5 V, but if you go below 4.5 V, you will have problems. Staying at 4.75 V or above (to 5.25 V) will avoid any potential low-voltage problems in your applications.

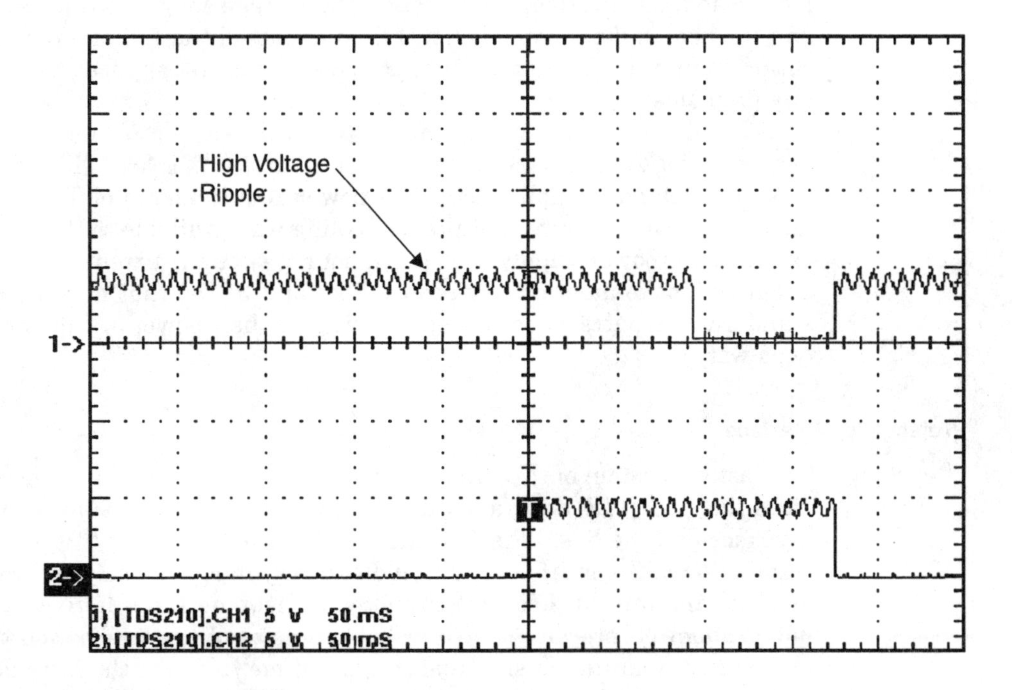

Figure 10.67 Operation at 4.44 V.

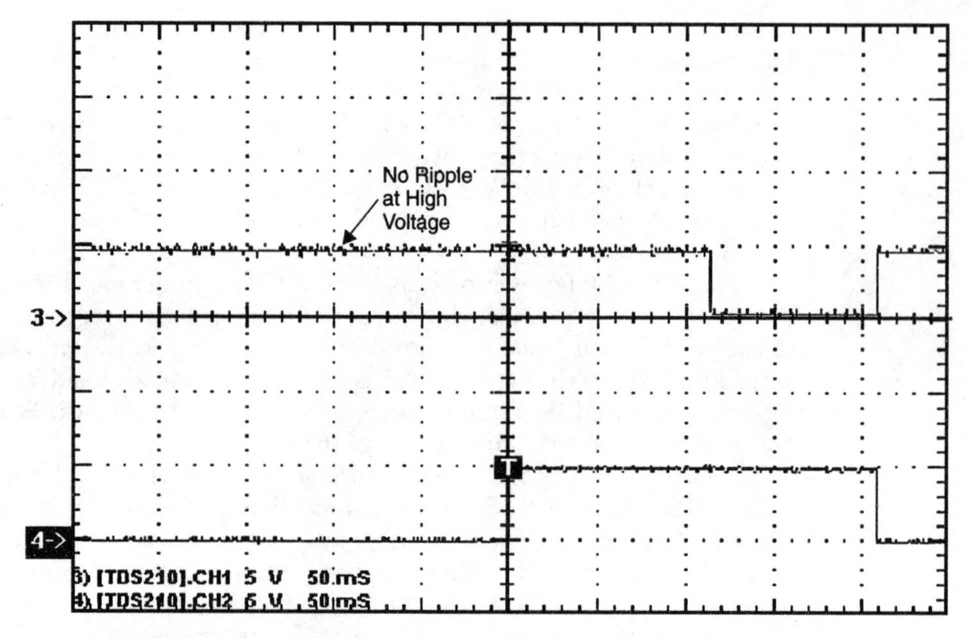

Figure 10.68 Operation at 4.87 V.

I ran with the large heat sinks until I put in the comparators and 74LS245 for the address interface, and I found the voltage sagging again below 4.75 V and the heat sinks extremely hot. I then bypassed the 7805 and ran the application from a bench supply. Once this was done, the application ran without any problems.

As I indicated at the start of this section, this was a problem I hadn't had to deal with for 20 years. Back then, very complex TTL logic (like this circuit) was used for most applications; the power requirements of TTL would be a major concern to ensure that proper cooling was available to the circuit board as well as a power supply that could source enough current. Modern CMOS circuitry (not to mention microcontrollers and PLDs) that have now replaced standard TTL parts run at a small fraction of their power and do not get nearly as warm.

Project 17d—Interface

The last subsystem of this circuit is the interface control. This subsystem provides the _RD/_WR, data, and address interface between the hexadecimal bus interface and the bus. When I designed the subsystems for this application, I wanted to make sure that they would run very independently and without the need for the interface control circuitry. In keeping the different subsystems independent of interface control, I kept the size of the application that I had to debug at a minimum size, and it allowed me to debug the interface control at my leisure, making sure that the circuit worked properly.

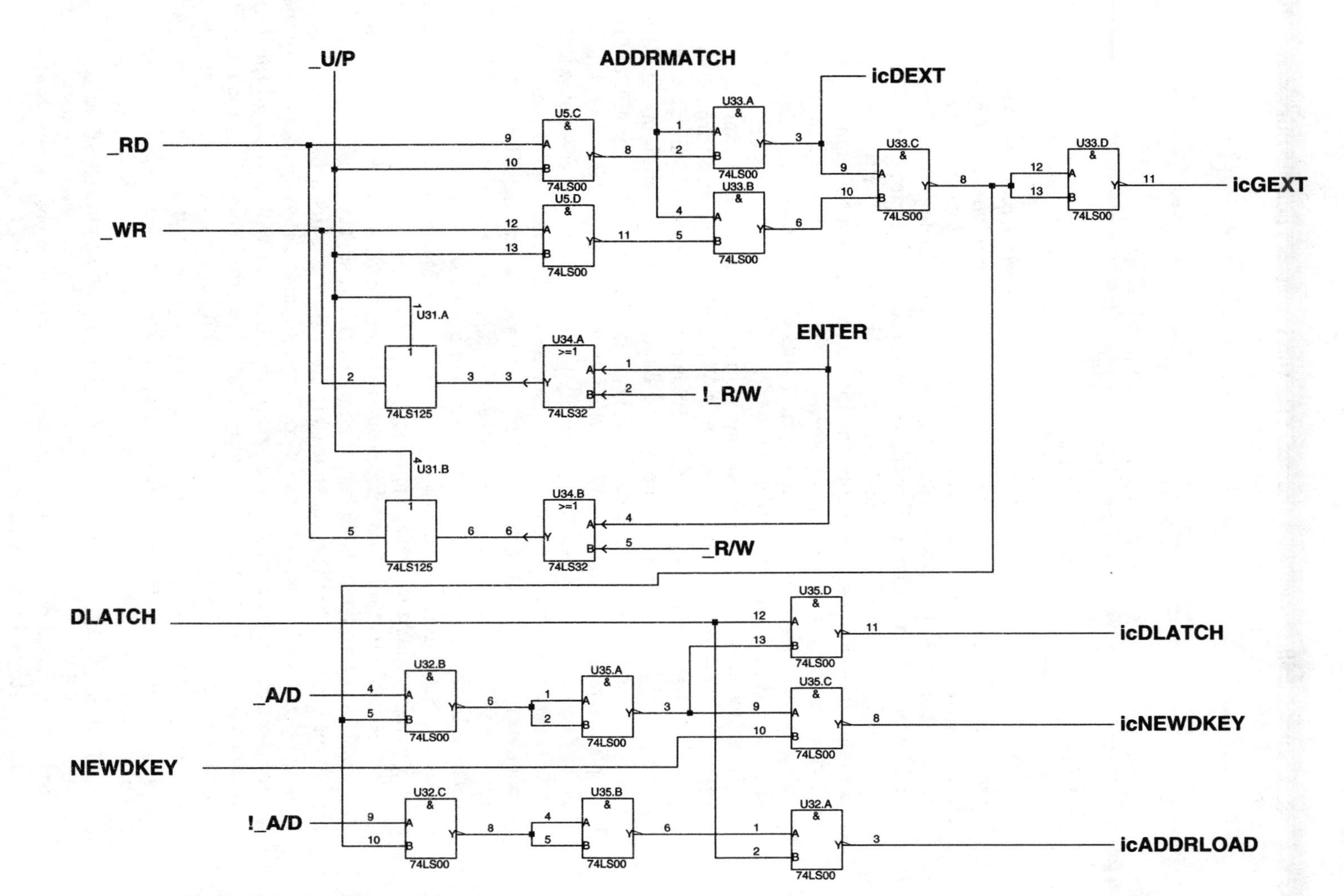

Figure 10.69 Project 17d—interface control.

Figure 10.69 shows the schematic for the interface control subsystem. The bill of materials for the circuit is listed below:

Parts	Description
U31	74LS125 quad tristate driver
U32–U33, U35	74LS00 quad two-input NAND gate
U34	74LS32 quad two-input OR gate
C36–C40	0.1-μF, 16-V tantalum capacitors

While I said I would create this circuit at my leisure, it really worked without any problems right from the start. I felt I had the operation of the circuitry very well understood by the time I added it to the circuit. The two biggest issues I had with adding these circuits was making sure that I had enough space for the chips and I could find the different lines to tap into to add the interface control function.

When I designed the application's subsystems originally, I tried to keep the signals that would be modified by the interface control subsystem easy to find. To help facilitate this, I used a different color wire for the signals that I was going to wire into interface control, and I always wired them last, to make sure that I could easily unwrap them.

Interface control provides two important functions to the hexadecimal bus interface. The first is driving the _RD/_WR signals correctly throughout the project in user mode. This was done by using 74LS123 single-shots with 250-nsec delays (the same as the ones used for the keypad subsystem), and I followed the same rules that I used for writing to the data display and address display/compare subsystems. When the key value button was pressed, a window of valid data (address and optionally the 8-bit data) was created, with the actual transfer taking place in a 250-nsec period. In this mode, I mask off the two inputs by using 74LS00 NAND gates, to avoid the issue of trying to determine whether or not the active _RD/_WR signals were caused by the PCB itself or by the external circuitry. When I connected the hexadecimal bus interface to a 6264 static random-access memory (SRAM) chip, I was able to read and write to it without any problems.

The second function of interface control is to provide an interface to an intelligent bus controller in peripheral mode. In this case, the _RD/_WR lines are used for input (rather than output as in the user mode described in the previous paragraph). To the user, the hexadecimal bus interface works exactly the same way as it does in user mode except that the user will notice that the data will be updated periodically. The only area where this could be a problem is if, while the user is updating the data, the intelligent peripheral overwrites it. I tested this mode using the computer processor presented in the next chapter.

Figure 10.70 shows all the decoupling capacitors used in this application. These capacitors are wired as shown in Fig. 10.58, with each of them soldered to the back side of a wire-wrap socket. It is typical to draw them this way to show the common function and avoid making the other schematics for the application cluttered. The appropriate capacitors have been added to the various subsystems' bills of materials.

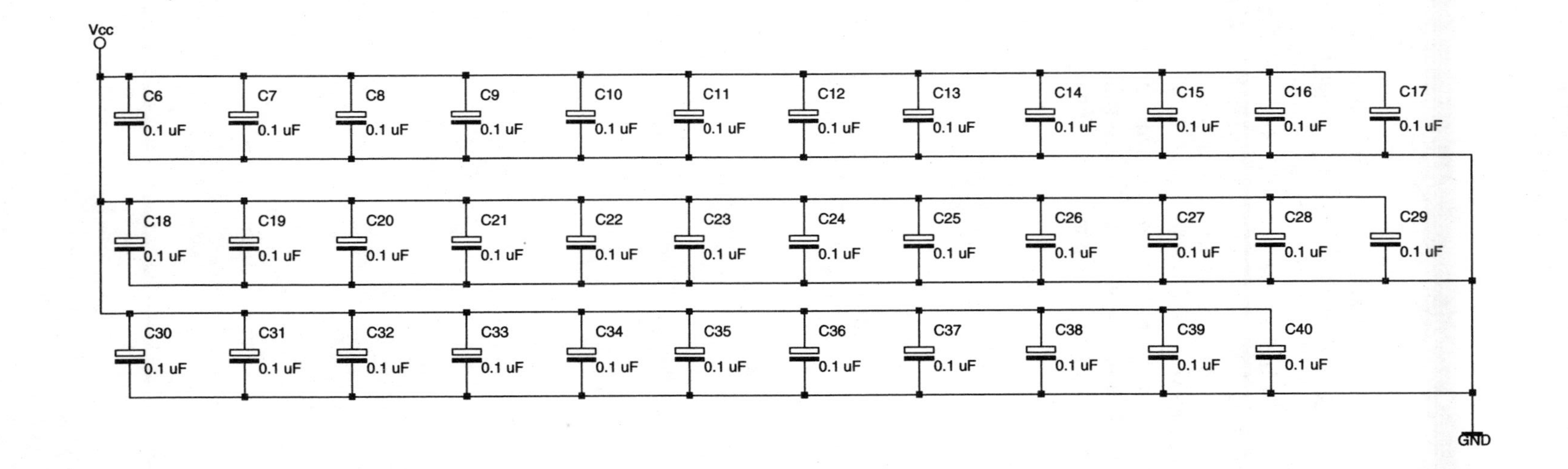

Place + terminal as close to U1 through U35's V_{cc} pins.

Figure 10.70 Project 17—capacitor farm.

One thing I have left out in this project description is how I interface from the hexadecimal bus interface to another circuit. As can be seen in Fig. 10.57, I used 18-pin DIP ribbon-cable connections. The ribbon cables that I used were bought at a surplus store in Toronto. This is not to say that you could not build these cables yourself, but it will be somewhat expensive, and if you don't have the right tools (a special kind of press), it will be a difficult process.

This is probably the easiest method of connecting the hexadecimal interface to another circuit—other methods could involve multiple wires, card edge connectors, or even building the circuit into the application, but these all involve quite a bit more work. In the next chapter, I will use the interface PCB for a similar function, and I recommend that you use this version of the hexadecimal bus interface, as it is much cheaper and easier to work with in the prototype environment.

Chapter

11

Computers as Digital Devices

To round out this book, I thought I would look at the most famous digital device of all and the one that is probably recognized as the most important innovation of the twentieth century—the computer. In this chapter, I will review the different parts of a computer and explain the digital devices behind them. Once I have reviewed the different parts of the computer system, I will show how they all work together in a design for a simple computer system that you can build for yourself, totally out of logic chips. The computer system will not use a microprocessor nor will it use any VLSI chips except for memory.

It will probably surprise you, but there really isn't one defining person or technology that is responsible for the computer. When the first computers were being postulated in the late 1930s and early 1940s, along with the basic concepts of boolean arithmetic needed for implementing a computer system, the actual circuits for creating the system were unknown.

In the development of this book, I have worked through much of the background technology needed to create a computer system. With this knowledge explained, I can now move on and show how a circuit that takes a series of instructions can be used to carry out "intelligent" tasks such as controlling processes, calculating results, or playing a game with you.

The computer system that I will present here probably has 8 times more memory and is an order of magnitude faster than the first computer systems built in the mid-1940s. The early computers used vacuum tubes and often had a mean time between failure (MTBF) of several minutes and a mean time to repair (MTTR) measured in days. The circuit that you will build with standard TTL chips will have an MTBF on the order of years and a MTTR that can be measured in single minutes. In terms of power, the 1940s computers required a hydroelectric substation, while the computer presented here could conceivably be powered from alkaline radio batteries. The biggest contrast of them all is that the first computers were built on the same amount of area as several basketball courts while the computer system that I will present here will take up about the same area as a sheet of paper.

As impressive as this comparison is, it is nothing compared to what can be accomplished when integrated circuit technology can be used. Microcomputer systems that are many times more powerful than the computer described here and can be built on a silicon die a tenth of the area of your thumbnail are commonplace and inexpensive. These chips are built into everything from a toaster to a jet fighter.

Processor Architectures

In this section, I want to introduce you to many of the concepts regarding the design and implementation of simple computer processors. I will give you some background in the different processor types and explain the advantages/disadvantages of the different types and their features. This information is not meant to provide you with a complete understanding of computer processor architecture design, but should help explain some concepts behind the computer systems that you work with and the one that will be presented at the end of the book.

Currently, many processors are called *RISC* (reduced instruction set computer—pronounced *risk*) because there is a perception that RISC is faster than *CISC* (or complex instruction set computer). This can be confusing because there are many processors available that are identified as RISC-like, but are in fact CISC processors. And, in some applications, CISC processors will execute code faster than RISC processors or execute applications that RISC processors cannot.

What is the real difference between RISC and CISC? In CISC processors, there tends to be a large number of instructions, each carrying out a different permutation of the same operation (accessing data directly, through index registers, etc.) with instructions perceived to be useful by the processor's designer.

In a RISC system, the instructions are at a bare minimum to allow users to design their own operations, rather than use what the processor designer has given them. Below, I show how a stack "push and pop" would be done by a RISC system in two instructions. The two simple constituent instructions can be used for different operations (or "compound instructions" such as this).

This ability of being able to write to all the registers in the processor as if they were the same is known as the *orthogonality* or *symmetry* of the processor. This feature allows some operations to be unexpectedly powerful and flexible. This can be seen by comparing conditional jumping in the two different computer architectures. In a CISC system, a conditional jump is usually based on status register bits. In a RISC system, a conditional jump may be based on a bit anywhere in memory. Having the ability to branch on the basis of any bit in a system can greatly simplify the operation of flags in an application and how they control executing code.

For a RISC system to be successful, more than just reducing the number of things that are done in an instruction has to be carried out in the processor architecture design. The flexibility of the different instructions, and thus how the processor executes instructions, can be increased to the point to where a very small instruction set—one that is able to execute in very few instruction cycles—can be used to provide extremely complex functions in a surprisingly efficient manner.

Many years ago, the United States government asked Harvard and Princeton Universities to come up with a computer architecture to be used in computing tables of naval artillery shell distances for varying elevations and environmental conditions.

Princeton's response was for a computer that had common memory for storing the control program as well as variables and other data structures. It was best known by the chief scientist's name, John Von Neumann. Figure 11.1 is a block diagram of the Von Neumann architecture.

The memory interface unit is responsible for arbitrating access to the memory space between reading instructions (on the basis of the current program counter) and passing data back and forth with the processor and its internal registers.

It may at first seem that the memory interface unit is a bottleneck between the processor and the variable/RAM Space—especially with the requirement for fetching instructions at the same time. In many Princeton-architected processors, this is not the case because the time required to execute an instruction is normally used to fetch the next instruction (this is known as prefetching). Other processors (most notably the Pentium processor in your PC) have separate program and data "cache" memory that can be accessed directly while other memory accesses are taking place.

In contrast, Harvard's design used separate memory banks for program storage, the processor stack, and variable random-access memory (RAM). Fig. 11.2 is the block diagram for a "typical" Harvard processor.

The Princeton architecture won the competition because it was better suited to the technology of the time. Using one memory was preferable because of the unreliability of then current electronics (this was before transistors were in general use); a single memory and associated interface would have fewer things that could fail.

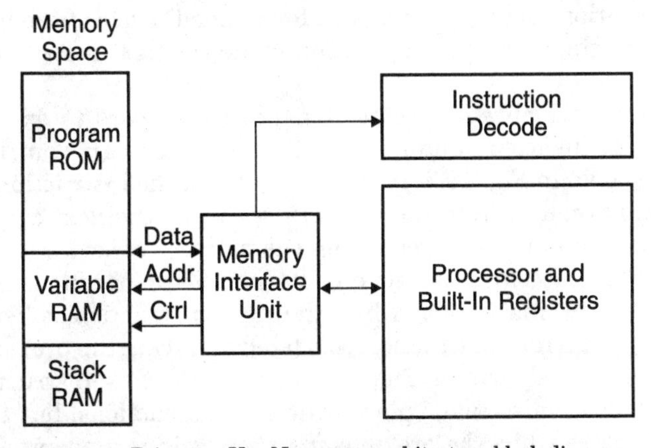

Figure 11.1 Princeton Von Neumann architecture block diagram.

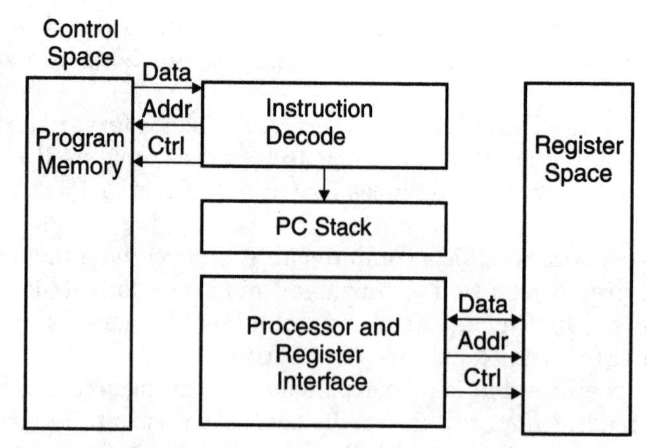

Figure 11.2 Harvard architecture block diagram.

The Von Neumann architecture's largest advantage is that it simplifies computer design because only one memory is accessed. For PCs, workstations, and servers, its biggest asset is that the contents of RAM can be used for both variable (data) storage as well as program (instruction) storage. This means that applications can be loaded from another source (e.g., a hard disk) and stored anywhere in memory for use by the application.

The Harvard architecture was largely ignored until the late 1970s, when microcontroller designers realized that the architecture had advantages for the devices they were currently designing. Most modern small processors use the Harvard architecture because it allows the read-only memory of the application, which is stored in nonvolatile memory, to be kept separately from application variables, which are kept in volatile static RAM (SRAM) memory.

The Harvard architecture tends to execute instructions in fewer instruction cycles than the Von Neumann architecture. This is because a much greater amount of instruction parallelism is possible in the Harvard architecture. *Parallelism* means that instruction fetches can take place during previous instruction execution and not either wait for a "dead" cycle of the instruction's execution or stop the processor's operation while the next instruction is being fetched.

For example, if a Princeton-architected processor were to execute a read byte and store in the accumulator instruction, it would carry out the instruction sequence shown in Fig. 11.3. In the first cycle of the instruction execution, the instruction is read in from the memory space. In the next cycle, the data to be put in the accumulator is read from the memory space.

The Harvard architecture, because of its increased parallelism, would be able to carry out the instruction while the next instruction is being fetched from memory (the current instruction was fetched during the previous instruction's execution). As shown in Fig. 11.4, executing this instruction in the Harvard architecture also takes place over two instructions, but the instruction read takes place while the previous instruction is carried out. This allows

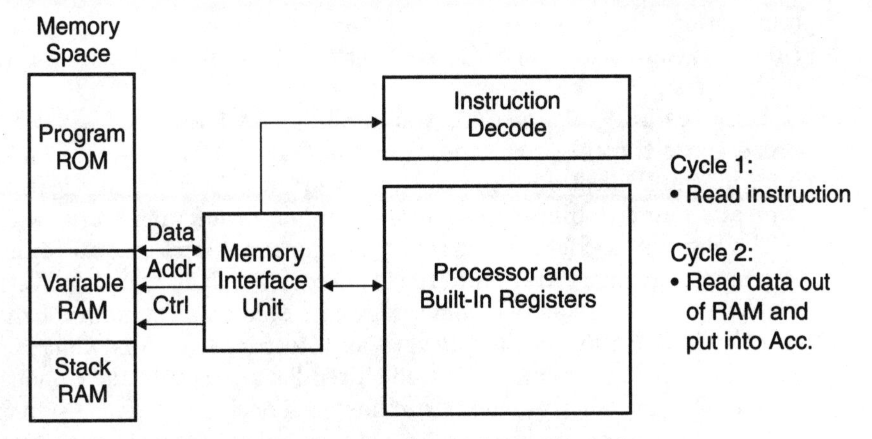

Figure 11.3 "Move accumulator register" in the Princeton architecture.

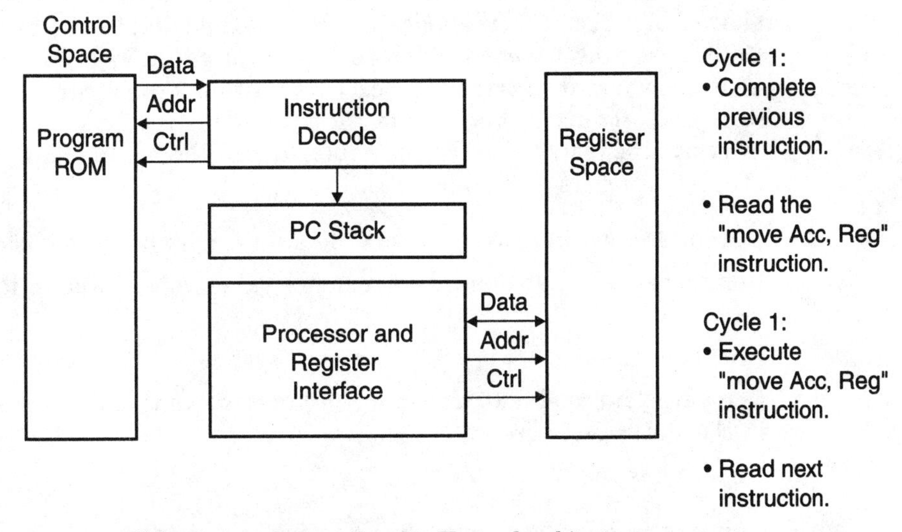

Figure 11.4 "Move accumulator register" in Harvard architecture.

the instruction to execute in only one instruction cycle (while the next instruction is being read in).

This method of execution (parallelism), like RISC instructions, also helps instructions take the same number of cycles for easier timing of loops and critical code. This point, while seemingly made in passing, is probably the most important aspect that I would consider in choosing a processor architecture for a timing-sensitive application.

I should caution you that simple performance comparisons may not be representative of all the processors using these two types of architectures. The comparison that matters is the actual application, and different architectures and devices will offer unique features that may make it easier to do a different application. In some cases, certain applications will not only be more efficiently executed by a specific architecture but can *only* be done by a specific architecture.

After reading this section, you probably feel that a Harvard-architected processor is the only way to go. But the Harvard architecture lacks the flexibility of the Princeton in the software required for some applications that are typically found in high-end systems such as servers and workstations.

The Harvard architecture is really best for processors that do not process large amounts of memory from different sources (which is what the Von Neumann architecture is best at) and have to access this small amount of memory very quickly. This feature of the Harvard architecture (which is what is used in the PICmicro's processor) makes it well suited for microcontroller applications.

Once the processor's architecture has been decided on, the design of the architecture goes to the engineers who are responsible for implementing the design in silicon. Most of these details are left "under the covers" and do not affect how the application designer interfaces with the application. There is one detail that can have a big effect on how applications execute and that is whether the processor is a *hard-coded* or *microcoded* device. I will go through the two different types to show how instructions are executed within a computer processor.

Each processor instruction is in fact a series of instructions that are executed to carry out the instruction. For example, to load the accumulator in a processor, the following steps could be taken:

1. Output address in instruction to the data memory address bus drivers.
2. Configure internal bus for data memory value to be stored in the accumulator.
3. Enable bus read.
4. Compare data read in to zero or any other important conditions and set bits in the status register.
5. Disable bus read.

Each instruction for a processor has a series of steps that must be executed in order to carry out the instruction's function. To execute these steps, the processor is designed to either fetch these series of instructions from a memory or execute a set of logic functions unique to the instruction.

A microcoded processor is really a processor within a processor. In a microcoded processor, a state machine executes each different instruction as the address to a subroutine of instructions. When an instruction is loaded into the instruction holding register, certain bits of the instruction are used to point to the start of the instruction routine (or microcode), and the μcode instruction decode and processor logic executes the microcode instructions until an instruction end is encountered. This is shown in Fig. 11.5.

I should point out that having the instruction holding register wider than the program memory is not a mistake. In some processors, the program memory is only 8 bits wide, although the full instruction may be some multiple of

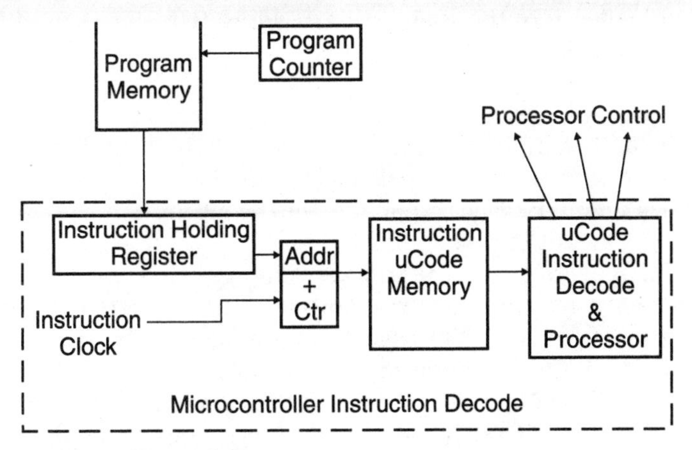

Figure 11.5 Microcoded processor.

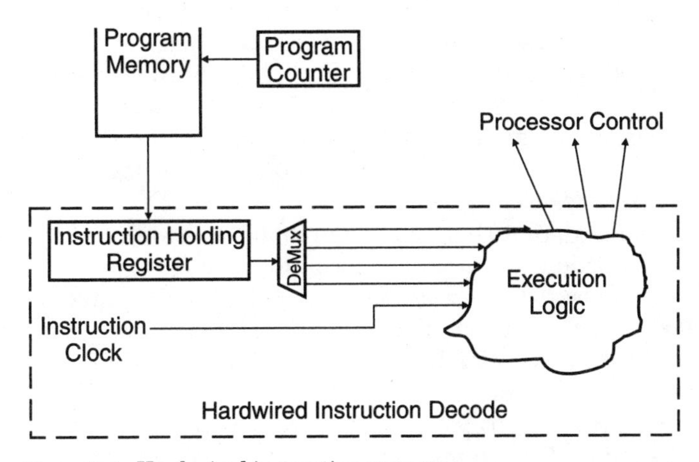

Figure 11.6 Hard-wired instruction processor.

this (for example, in the 8051 most instructions are 16 bits wide). In this case, multiple program memory reads take place to load the instruction holding register before the instruction can be executed.

The width of the program memory and the speed at which the instruction holding register can be loaded is a factor in the speed of execution of the processor. In Harvard-architected processors, like the Microchip microcontroller, the program memory is the width of the instruction word and the instruction holding register can be loaded in one cycle. In most Princeton-architected processors, which have an 8-bit data bus, the instruction holding register is loaded through multiple data reads.

A *hard-wired* processor uses the bit pattern of the instruction to access specific logic gates (possibly unique to the instruction), which are operated as a combinatorial circuit to carry out the instruction. Figure 11.6 shows how the instruction loaded into the Instruction Holding Register is used to initiate a specific portion of the "Execution Logic", which carries out all the functions of the instruction.

Each of the two methods offers advantages over the other. A microcoded process is usually simpler than a hard-wired one to design and can be implemented faster with less chance of having problems at specific conditions. If problems are found, revised "steppings" of the silicon can be made with a relatively small amount of design effort.

A great example of the quick and easy changes that microcoded processors allow occurred a number of years ago when IBM wanted to have a microprocessor that could run 370 assembly language instructions. Before IBM began to design its own microprocessor, it looked around at existing designs and noticed that the Motorola 68000 had the same hardware organization (or architecture) as the 370 (although the instructions were completely different). IBM ended up paying Motorola to rewrite the microcode for the 68000 and came up with a new microprocessor that was able to run 370 instructions, but at a small fraction of the cost of developing a new device.

A hard-wired processor is usually a lot more complex because the same functions have to be repeated over and over again in hardware—how many times do you think that a register read or write function has to be repeated for each type of instruction? This means that the processor design will probably be harder to debug and be less flexible than a microcoded design, but instructions will execute in fewer clock cycles.

This brings up a point that you are probably not aware of. In most processors, each instruction executes in a set number of clock cycles. This set number of cycles is known as the *instruction cycle* and is measured as a multiple of *clock cycles.* If each instruction cycle in a processor takes four clock cycles, then a device running at 4 MHz is executing the instructions at a rate of 1 million instructions per second.

Using a hard-coded over microcoded processor can result in some significant performance gains. For example, the original 8051 was designed to execute one instruction in 12 cycles. This large number of cycles requires a 12-MHz clock to execute code at a rate of 1 million instructions per second (MIPS), whereas the PICmicro has to have a clock that runs at 4 MHz to get the same performance.

Instructions and Software

When a processor executes *program* or *code* statements, it is reading a set of bits from program memory and using the pattern of bits to carry out specific functions. These bits are known as an *instruction,* and each pattern carries out a different function in the processor. A collection of instructions is known as a *program.* When the program is executing, the instructions are fetched from a program memory by using a *program counter* address. After an instruction is executed, the program counter is incremented to point to the next instruction in program memory.

There are four types of instructions:

1. Data movement
2. Data processing
3. Execution change
4. Processor control

The data movement instructions move data or constants to and from processor registers, variable memory, and program memory (which, in some processors, are the same thing) and peripheral I/O ports. There can be many types of data movement instructions, depending on the processor architecture, number of internal addressing modes, and the organization of the I/O ports.

The basic register that is used in most small processors is the accumulator. This register stores temporary values as well as the results of a bitwise logical and arithmetic operation. The status register is a complementary register that stores the comparison results of the operation that updated the accumulator. The comparisons consist of whether the result stored in the accumulator is equal to zero or an add operation had a result that was greater than 255 (for an 8-bit processor).

In Harvard architecture processors, variable memory and program memory are separate, and the I/O registers are usually part of the variable memory space (although I/O registers can be in a their own separate space). All Harvard-architected processors have only five addressing modes:

1. Immediate value contents put into the accumulator.
2. Variable/register contents put into the accumulator.
3. Indexed address variable/register put into the accumulator.
4. Accumulator contents put into a variable/register.
5. Accumulator contents put into indexed address variable/register.

I say *only* five addressing modes because, as you investigate other processor architectures, you will find that many devices can have more than a dozen different ways of accessing data within the memory spaces. The five methods above are a good base reference for a processor and can provide virtually any function that is required of an application. The only missing function that I feel is necessary in a processor is the ability to "push and pop" data on a stack.

A stack is a set of RAM registers that can be accessed not by a specific address, but by the first element in the stack. The operation of a stack is shown in Fig. 11.7. When data is stored on the stack, the save operation is called a *push*. When data is taken off the stack and read, the operation is called a *pop*.

Since the stack is accessed from only one point and data is stored on the stack like papers in a pile, the last piece of data that is put onto the stack is the first one that will be read back out. This method of operation is described in the stack memory type name: *LIFO,* which stands for *last in, first out*. Stacks are useful for implementing subroutine calls and returns.

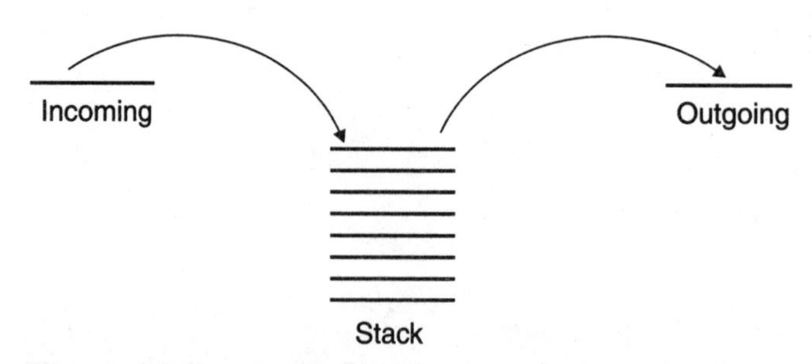

Figure 11.7 Stack data flow.

The Princeton processor architecture has all its I/O registers, RAM, and program memory in one addressing space. This means that while the same five addressing modes listed above are available, you may find opportunities that weren't available to you in the Harvard-architected processor.

Data processing instructions consist of basic arithmetic and bitwise operations. A typical processor will have the following data processing instructions:

1. Addition
2. Subtraction
3. Incrementing
4. Decrementing
5. Bitwise AND
6. Bitwise OR
7. Bitwise XOR
8. Bitwise negation

These instructions work the number of bits that is the data word size (for most small processors, this is 8 bits). Many processors are capable of carrying out multiplication, division, and comparisons on data types of varying sizes (and not just the word size). For most microcontrollers, the word size is 8 bits and advanced data processing operations are not available.

Execution change instructions include gotos (or jumps), calls, interrupts, branches, and skips. For branches and gotos, the new address is specified as part of the instruction. Branches and gotos are very similar, except branches are used for "short jumps," which cannot access the entire program memory and are used because they take up less memory and execute in fewer instruction cycles. These execution changes are known as *nonconditional* because they are always executed when encountered by the processor. Skips are instructions that will skip over the following instruction. Skips are normally conditional and based on a specific status condition.

An example of using status bits is the way a 16-bit variable increment is implemented in an 8-bit processor. If the processor's zero flag is not set after the low 8-bit increment, then the following instruction, which increments the upper 8-bit increment, is skipped. But, if the result of the lower 8-bit increment is equal to zero, then the upper 8-bit increment is executed:

```
Increment      LowEightBits
SkipIfZero
 Increment     HighEightBits
```

The skip is used in many small processors to provide conditional execution. In other processors, there are conditional branches or gotos, which means that the skip instruction is not always needed.

Other execution change instructions include the call and interrupt, which causes execution to jump to a routine and return back to the instruction after the call/interrupt instruction. A *call* is similar to a branch or goto and has the address of the routine to jump to included in the instruction. The address jumped to is known as a *routine* and includes a return instruction at its end, which returns execution to the previous address.

There are two types of interrupts. *Hardware interrupts* can be thought of as calls to routines that are initiated by a "hardware event." *Software interrupts* are instructions that make calls to interrupt handler routines. Software interrupts are not often used in smaller microcontrollers, but they are used to advantage in the IBM PC (and devices based on the PC's processor architecture).

In most processors, the program counter cannot be accessed directly, to jump or call arbitrary addresses in program memory. In some processors, the program counter cannot be updated by the contents of a variable. In these cases the program counter can be directly accessed and updated. Care has to be taken in updating the processor's program counter to make sure the correct address is calculated before it is updated.

Processor control instructions are specific and control the operation of the processor. A common process or control instructions is "sleep," which puts the processor (and microcontroller) into a low-power mode. Another processor control instruction is the interrupt mask, which stops hardware interrupt requests from being processed.

Buses and Device Addressing

Earlier in the book, I presented the concept of *buses* and gave a simple example of an addressed bus that could be used for accessing 4 data bits in a register. While the example probably seemed very simple, it was actually a very good example of how a computer processor accesses data. The peripheral and memory functions of a computer system are separate from the processor and are accessed by using an addressed bus.

An addressed computer bus has three different types of buses built into it:

Data bus

Address bus

Control bus

The *data bus* is used to transfer data between the processor, peripherals, and memory. In modern systems, the data bus is usually some multiple of 8, bits (where 8 bits is a *byte*). Your PC probably has a 64-bit data bus, which allows two 32-bit words of data to be transferred at any time. Most small processors have only an 8-bit bus for data.

The *address bus* contains the address of the device and a register or memory location within that device. The address is normally sent before the data in a processor transfer, to allow the computer system to determine where the data is going to be transferred from before the data transfer takes place.

The *control bus* is output by the computer processor to indicate to the memory and peripheral devices that a data transfer is going to take place. This bus can be as simple as two lines, one to indicate that a read is about to take place and one to indicate a write is about to take place. As I will discuss below, the control bus can become very complex as the data transfer speed increases or the functionality of the peripheral increases.

In Fig. 11.8, I show an example processor read and write with the address latch enable (ALE) becoming active with the transfer address to notify the memory and peripherals that a data transfer is about to take place. Next, the read and write control bits (_RD and _WR, respectively) become active to indicate to the devices what type of transfer is to take place and to coordinate it with the processor.

It may be surprising to see that in Fig. 11.8 I have shown the _WR line to be active much longer than the _RD line. The reason for this is to reflect the operations of different devices. In static random access memory (SRAM) devices, before a read or a write can take place, the pathways within the chip must be opened for the data transfer. If the write data is being driven while this is taking place, then the write operation, which may take some time, can take place as soon as possible. Another reason for having the low write pulse is to allow the write I/O device to set up while the data is active, with the data latched at the rising edge of _WR. When reading data from an SRAM device, the reverse

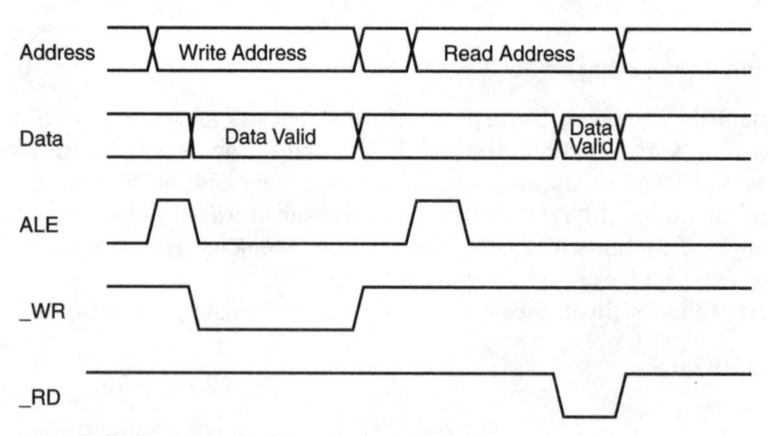

Figure 11.8 Memory and I/O bus timing waveforms.

isn't necessarily true; once the data pathways are set up, the read value can be driven out very quickly.

The important point to remember about reading and writing to peripherals and memory is that, while the processor data transfer is active, the data is correct. It is a good idea to always keep the data valid to the limits of the regions where _RD and _WR are active to make sure that the processor has passed/accepted the data correctly. Not having valid data within these regions could result in incorrect data being transferred.

I should point out that most processors make sure that the data is valid outside the _WR window and is read inside the _RD window to make sure that no data is corrupted because of normal gate delays within the chips.

An important aspect of computer processor design is the placement and wiring of peripheral and memory devices. When you look at the datasheet for a computer processor, you will see a diagram like the memory map shown in Fig. 11.9.

The figure shows a 12-address bit processor that has an 8-bit data bus. In this implementation, the processor accesses I/O devices in the first 256 bytes of memory, its variable memory in the second 256 bytes of memory, and its read-only memory (ROM) in the final 3584 bytes of memory. In the memory map, the start and stop addresses for each of these regions is spelled out and shown.

The problem with this diagram is that it doesn't explain how the different devices are wired to the processor. In Fig. 11.10, I have shown how the processor could be wired to the three busses and, using an "address decoder," how each device is accessed. In Fig. 11.10, I have assumed that the I/O devices and memories have positive active chip enables and each device has at least the

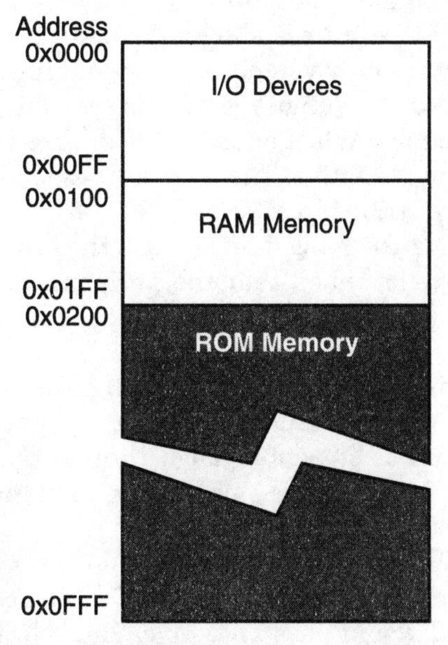

Figure 11.9 Memory map for processor.

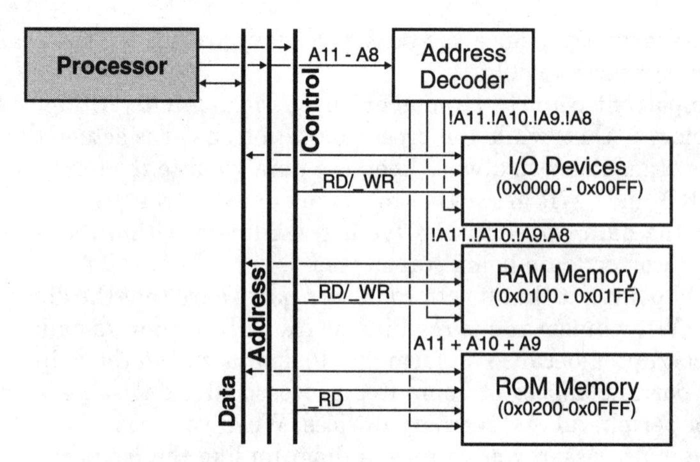

Figure 11.10 Memory and I/O bus device topology.

first 8 bits passed to it. The first two devices (I/O devices and RAM memory) are both read/write devices, so both the _RD and _WR lines from the processor are passed to them. The ROM memory cannot be written to, so only the processor's _RD line is passed to it.

This diagram may be referred to the bus *topology* because it shows the interrelations between the different devices and the buses. For more complex systems, where there are multiple buses and peripheral devices that can autonomously perform data transfers, the topology becomes much more complex, and choosing the correct one becomes critical to the efficient operation of the computer system.

In this example, the upper four address lines are continually monitored by the address decoder, and it continually outputs a chip select for whatever device is currently addressed by the processor. The devices that themselves transfer data to or from the data bus only when the _RD/_WR lines are active.

The name *chip select* is used because, in most computer systems, single chips provide the memory and I/O functions. When one of these devices is addressed, the chip select pin on the part is made active. The chip will respond to any _RD or _WR operations that take place with chip select active.

In this example, I have not used the ALE line because the address decoder provides this function, and I have assumed that the peripheral and memory devices can respond to the changing address conditions in time for the data transfers to take place.

Elsewhere in the book, I have used the 74LS138 and 74LS139 decoder chips for the function provided by the address decoder in Fig. 11.10. For most applications, chips like these can be used, but note that if they were used in this application, the ROM memory address range would have to be made up of a combination of the different bits.

If I were to create this application on my own, I would probably use the 74LS138, but combine it with some logic, to just select the I/O devices and RAM memory. For the ROM memory, I would pass an active chip select to it when neither of the other two devices were selected.

The bus example shown in this section and used by the example processor is very simple and not very challenging to add peripherals to. If you compare what I've shown in this section to your PC's PCI (or even ISA) parallel buses, you will probably see some points of similarity, but there will be quite a few additional control lines that are not explained here.

The bus examples given in this book are the basic information that you will require to implement input/output to a computer processor. In a modern high-speed system, there are additional control lines built into the buses for high-speed data transfers, data transfers initiated by peripheral equipment and not the processor [direct memory access (DMA) or bus multimastering], and interrupt requests.

These functions are outside the scope of this book, and, while the need for many of these different functions is explained in this book, I do not go into detail on how they can be implemented or used. For this information you will have to look at more complete or advanced books on computer processor design and implementation.

Peripheral Functions

All computer systems have I/O functions. These functions can range from I/O pins of remarkable simplicity, literally just a pull-up and a transistor, to full Ethernet interfaces or video on-screen-display functions. After you understand a computer system's processor architecture and address buses, the next area to work through is the peripherals, understanding how I/O pins work as well as how more complex functions are carried out.

When I said an I/O pin could be as simple as a transistor and a pull-up resistor, I wasn't being facetious. The Intel 8051 uses an I/O pin that is this simple, as shown in Fig. 11.11. This pin design is somewhat austere and is designed so that the pin can be used as an input when the output is set high so another driver on the pin can change the pin's state to high or low easily against the high-impedance pull-up.

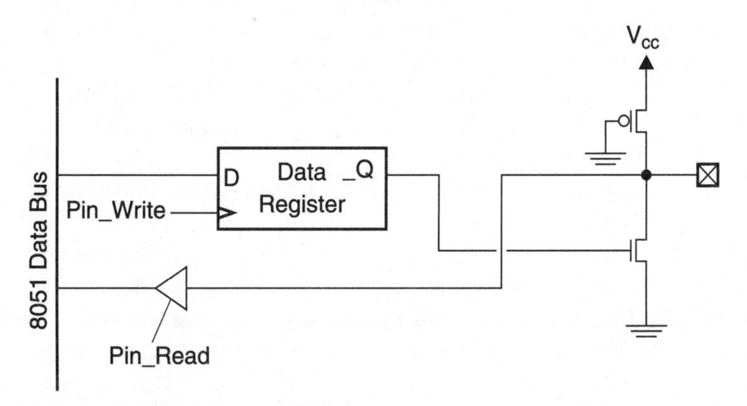

Figure 11.11 8051 parallel I/O pins.

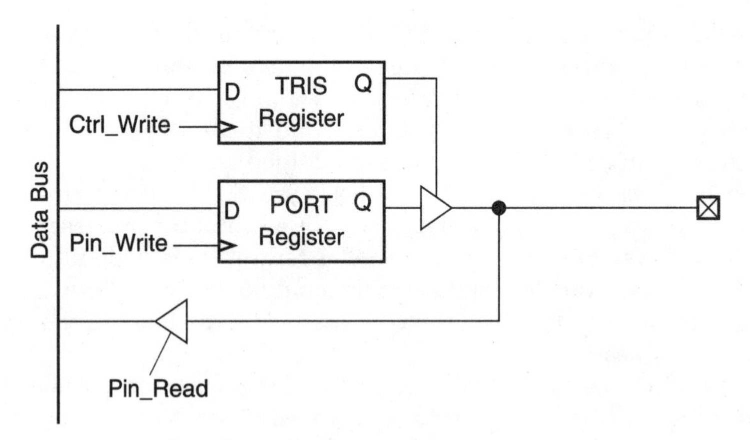

Figure 11.12 Standard I/O pin block diagram.

A more typical I/O pin is shown in Fig. 11.12. This configuration is used in Microchip PICmicro microcontroller. This output pin provides "tristatable" output from the control register. This pin can be used for digital input as well (with the output driver turned off). This is a common I/O pin design and uses the TRIS control to select the pin mode as input or output.

A computer system may also have more advanced peripheral I/O functions such as serial I/O. These peripheral functions are designed to simplify the operation or interfacing to other devices. I call this a *peripheral function* because it provides a service that is not central to the execution of the computer system, but is needed to provide a full set of features to the system itself.

Software Development Tools

Chances are you are familiar with many of the tools that are used for developing application software for computer systems that I will discuss in this section. You may be surprised at this statement because you have never done any programming before. Chances are you have helped create a Web page; in doing this you are actually creating a computer program and you will have used tools that are similar to the ones that I will discuss in this section.

An editor is an application program that runs on a PC or workstation to allow a human-readable file to be created or changed. The editor can also be used for looking at files (which is known as *browsing*). A standard Microsoft editor, like WordPad (Fig. 11.13), is a Word-based editor. The cursor, which is the vertical bar where characters will be placed, is moved onto the window either by using the arrow keys or by using the mouse and setting its position by a left click. When I'm editing a file, I very rarely use the mouse, instead I use the arrow keys, home, end, page up, and page down almost exclusively.

The following table lists the standard Microsoft editor operations:

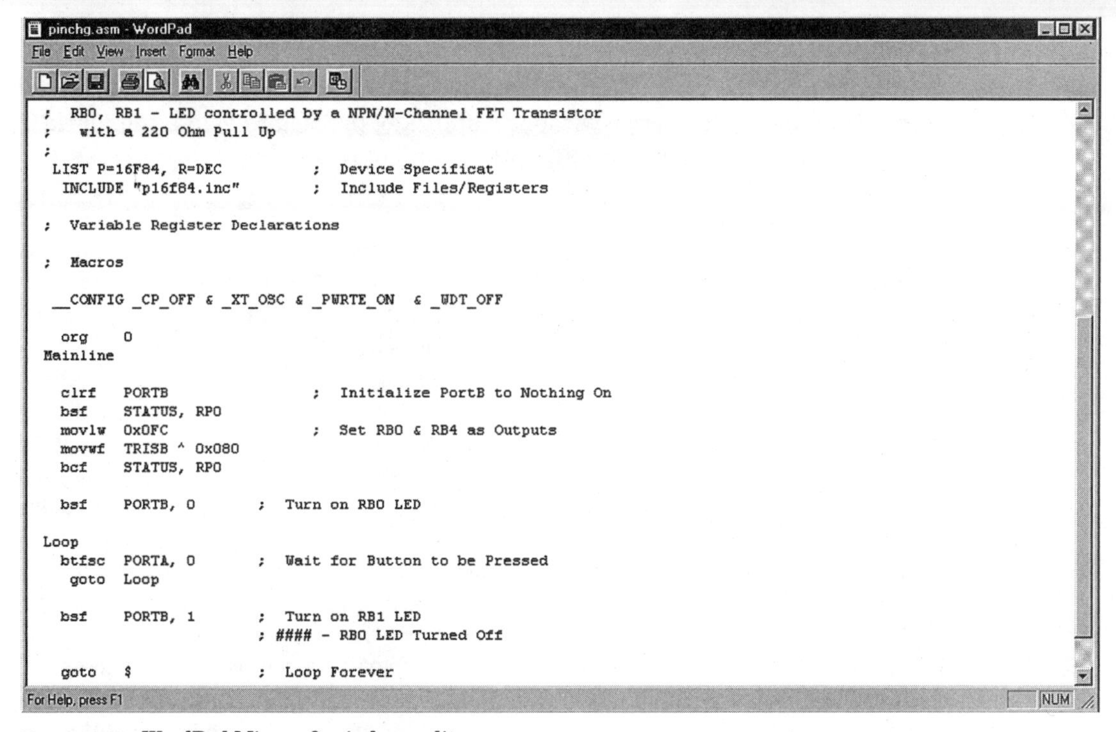

Figure 11.13 WordPad Microsoft windows editor.

Keystrokes	Operation
Up arrow	Move cursor up one line
Down arrow	Move cursor down one line
Left arrow	Move cursor left on character
Right arrow	Move cursor right on arrow
Page up	Move viewed window up
Page down	Move viewed window down
Ctrl—left arrow	Jump to start of word
Ctrl—right arrow	Jump to start of next word
Ctrl—page up	Move cursor to top of viewed window
Ctrl—page down	Move cursor to bottom viewed window
Home	Move cursor to start of line
End	Move cursor to end of line
Ctrl—home	Jump to start of file
Ctrl—end	Jump to end of file
Shift—left arrow	Increase the marked block by one character to the left
Shift—right arrow	Increase the marked block by one character to the right
Shift—up arrow	Increase the marked block by one line up
Shift—down arrow	Increase the marked block by one line down
Ctrl shift—left arrow	Increase the marked block by one word to the left
Ctrl shift—right arrow	Increase the marked block by one word to the right

To delete and move text, I use the cut and paste functions. To select text to cut and paste, I mark the text first, pressing a shift key while moving the cursor to highlight the text to be relocated. If you have marked text incorrectly, then simply move the cursor without a shift key pressed to delete the marked text. Pressing your mouse's left button and moving the mouse across the desired text will also mark it. Left-clicking on another part of the screen will move the cursor there and eliminate the highlighting for the text.

Next, the keystroke Ctrl—X is used; this removes the marked text from the file and places it into the windows clipboard. Ctrl—C copies the marked text into the clipboard and doesn't delete it. To put the text at a specific location within a file after the current cursor location, Ctrl—V is used.

Note that I do not use delete or insert. Deleting marked text destroys it completely whereas Ctrl—C saves it in the clipboard so it can be restored if you made a mistake. The insert key toggles between data insert and replace mode for Microsoft keystroke–compatible editors. Normally when an editor boots up, it is in insert mode which means all keystrokes are placed. This is the preferable mode to be in.

In a line-based editor, a CR/LF character combination terminates each line displayed on the screen. In a Microsoft-compatible editor, the CR/LF is used to separate paragraphs and lines are broken up on the display in order to show all the data on them. If you were to look at a paragraph produced by a Microsoft-compatible editor on a line editor, you would find that the paragraph would be one line and most of it would not be displayed because it was past the right edge of the editor's text window. The advantage of a line-based editor is the ease in which blocks of data can be moved without the relative positions being changed.

Once the application code has been created, you are ready to convert it into machine language that can be executed by the computer system. To convert application source code that you created on the editor into machine language, you must use a *compiler* or an *assembler* with a linker to create a file that the computer system can work with.

When you first see the results of a compiler converting high-level source code into assembly language, you will probably regard the process of changing high-level language source code into assembly code as more like magic than a series of mathematical operations executed by a computer. As I will show in this section, compilers work through a series of reasonably simple rules to convert high-level language statements into assembly language. There is no magic involved, even though the results sometimes are pretty amazing.

Modern compilers also look for opportunities to simplify assembly code, further resulting in smaller and more efficient applications. If you are a beginner to assembly language programming, you should not be surprised to discover that modern compilers can produce more efficient assembly code than *you* can.

Once you have compiled or assembled your application and linked it into a format that can be used by the computer system, you should simulate it to see if there are any gross errors. This is an important step before running the code in a system, where you will be able to observe what is happening only at a very gross level.

A simulator consists of a software model of the processor, which can be controlled and through which you can pass test I/O signals back and forth to the software model processor. It is important to remember that the processor and I/O in a simulator are software models and some situations that can come up in simulator hardware cannot be properly simulated.

I tend to think of a simulator as a collection of "black boxes" that are controlled by the simulator host software, as shown in Fig. 11.14. I drew the simulator this way because it allows different boxes to be swapped in and out to make up different part numbers without significant effort to create simulators for different functions. There are probably more actual simulator modules used in commercial products, but for the purposes of this discussion, the five presented in Fig. 11.14 are adequate.

To help you to debug your applications, the I/O pins have the capability of being driven with external inputs. This is the stimulus driver box in Fig. 11.14. It can either be directly used in changing I/O register bits or used to run a stimulus file. A *stimulus file* is a file that is created with I/O information that can be processed by the simulator processor without much intervention by the user.

The next step up from a simulator is an *emulator*, which is a hand-made device connected to your development PC or workstation and allows you to monitor application execution in hardware. The emulator block diagram looks like Fig. 11.15.

An advantage of an emulator over a simulator is that actual pin I/O signals can be observed, both from the processor's perspective and from the circuit's. In addition, the emulator often has the same hardware as the actual device, so there are no missing peripheral interface functions. The best method of providing an emulator is to use a bond-out chip, which is an actual processor with specific access provided to the processor so that it can be started or stopped and provide emulator-unique instructions.

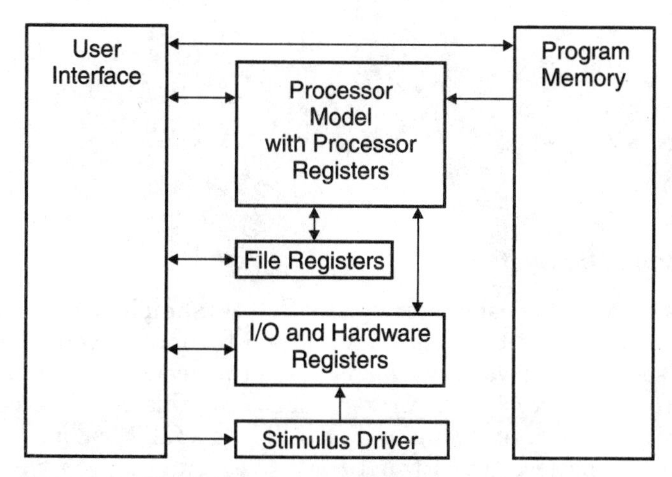

Figure 11.14 Simulator software architecture.

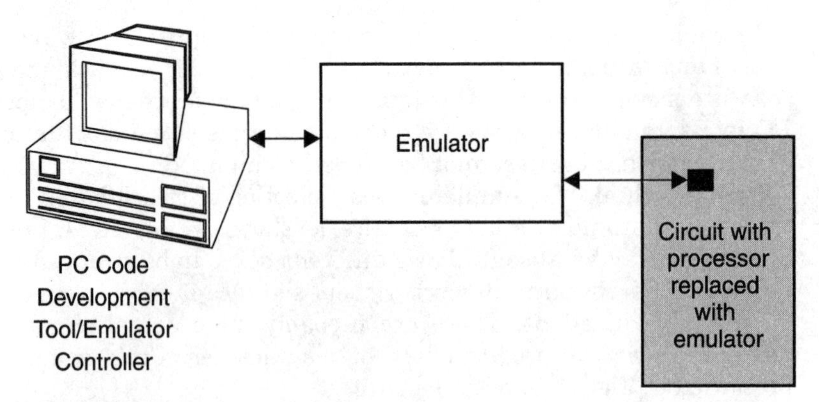

Figure 11.15 Emulator circuit operation.

There are two disadvantages to working with an emulator. The first one is cost. Emulators generally cost $2000 or more and require separate pods for each device being emulated. This cost is not significant for many companies, but for small companies and individuals, it can be. To help offset the cost, there are a number of simplified emulators. These devices often provide many of the same features as a full emulator, but are restricted in different ways.

When I'm developing software, I always find that I'm the happiest (most productive, debug fastest) using an integrated development environment (IDE). This extends to PC programming, where I always liked the original Borland Turbo Pascal and the modern Microsoft Visual Basic and Visual C++ development tools (part of Microsoft's Visual Development Studio integrated development environment).

An integrated development environment integrates all the software development tools I've described in the previous sections:

Editor

Assembler

Compiler

Linker

Simulator

Emulator

Project 18—TTL Chip Computer System

I've always wanted to design a computer that was implemented in TTL technology and could be built in just a few hours. I thought it would be a fascinating way of seeing what was involved in doing the design and debugging of the circuits involved. Part of my interest in designing this circuit was to see how minimal a computer design could be and still be useful. I had a lot of fun working through the project, and I think that if you build it, you will learn a lot about digital electronics, computers, interfacing, and design tradeoffs.

To start off the project, I created a set of requirements for it:

1. Build the entire circuit on a reasonable size breadboard.
2. Program/control the operation of the computer by using a circuit like the hexadecimal bus interface, but without its complexity.
3. Do not use any programmable parts.
4. Power the circuit by the +5-V power supply provided with the book.
5. Clock and control the operation of the computer by using the Interface PCB provided with the book.
6. Except for using a 6264 8k-by-8 SRAM, the entire design would be designed with LS TTL logic.
7. All the chips that were used should be easily found or bought.

The actual implementation of the circuit that I came up with is shown in Fig. 11.16. As you can see, I almost kept everything to a single breadboard, although I had to add two breadboards to the end of the circuit to be able to create the application.

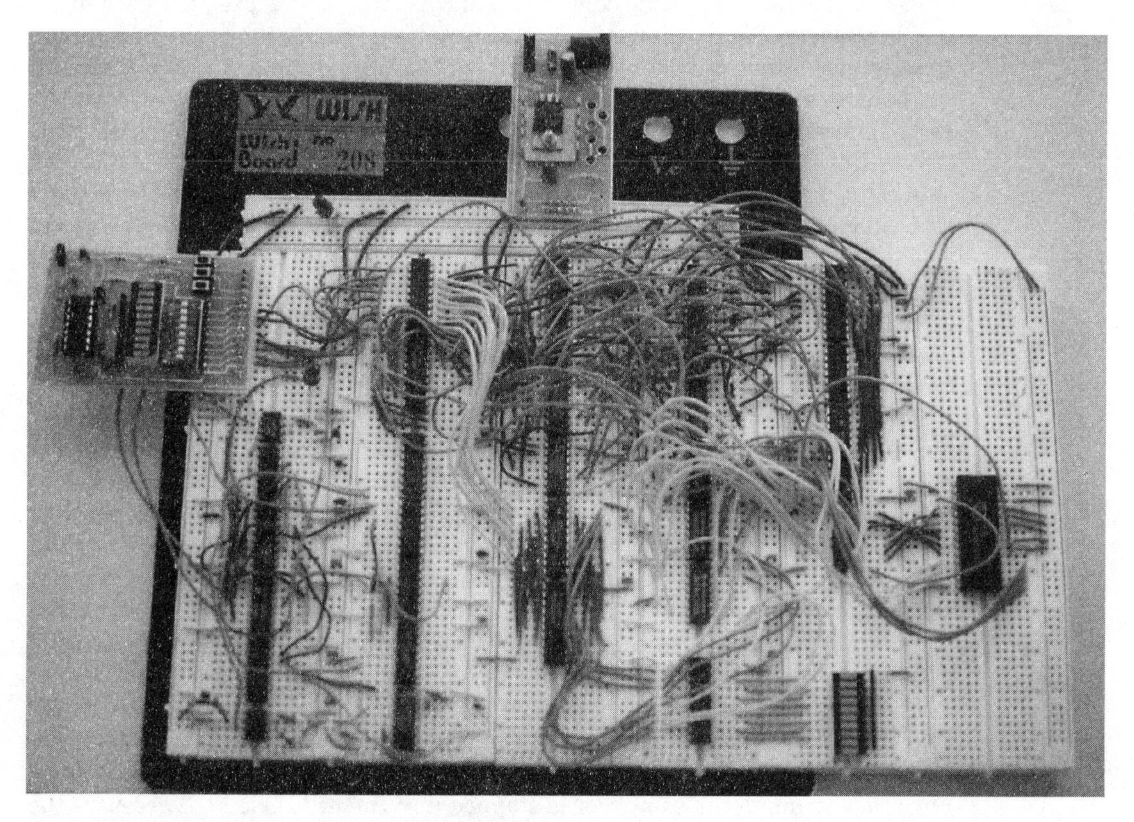

Figure 11.16 TTL computer built on breadboard.

You will find that it will take you five or more hours to wire this circuit, and I suggest you load up on precut wires. I also suggest that you work through the application using the instructions I've listed below and take several breaks. It would not be unreasonable to wire this circuit over three evenings, breaking it up as I have suggested in the book.

This is a large project (similar in size and scope to the hexadecimal bus interface). It will not work if you try to rush through building it or if you are tired. As I discuss how the circuit is built, notice that I will specify tests to work through to make sure everything is wired correctly. Failing to do these tests and not progressing until they have passed will cause problems for you in getting the full application working.

Despite these seemingly dire warnings, the circuit isn't that hard to build, and it is just amazing to watch working.

To meet the requirements I've laid out, I worked through several different computer architectures and arithmetic logic unit (ALU) designs until I came up with the one presented here. The actual circuit could have been made quite a bit simpler, except for the two requirements of using the interface PCB and for not using programmable parts.

This circuit could have used the hexadecimal bus interface, which would have reduced quite a bit the complexity of this circuit. The problem with using the hexadecimal bus interface is the time it requires to build it and the requirement that it be built by wire wrapping or some other permanent method of wiring. In the actual circuit, I use two LED bargraph displays (one on the interface PCB to monitor data and one to monitor address), which will provide all the needed functions.

Ideally, I would have used the seven-segment LED displays (and the DM9368 drivers) of the hexadecimal display to make the operation easier to read. If you have a large breadboard, or are going to build the circuit on a wire-wrapped breadboard, you may consider making this modification to the circuit (as well as using a hex keypad for data entry). The Interface PCB does provide the needed functions, but as you will see, it can be quite cumbersome.

If I were to use programmable parts, I could have greatly simplified the control functions provided in the computer. Instead of using a counter and a decoder for providing timing signals, I could have used an EPROM loaded with microcode for the project. I didn't use an EPROM for two reasons. The first is programming—inexpensive EPROM programmers can be difficult to find, and building one, while not that hard, requires software and other resources that I didn't want to provide in the book. The second reason is that I wanted to avoid doing a microcoded processor, instead working through the logic needed to control the processor functions.

The block diagram I came up with for the computer is shown in Fig. 11.17. The computer is an 8-bit processor, which means it processes data 8 bits at a time with an 8-bit address bus, giving it a maximum of 256 memory locations. For programming, the computer is stopped and the interface PCB accesses the data bus and controls the computer's program counter.

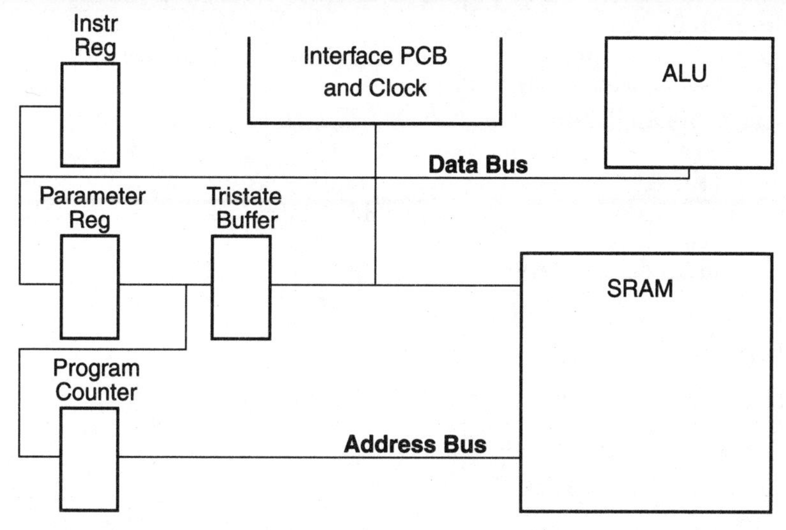

Figure 11.17 TTL computer block diagram.

The interface PCB will be responsible for reading and writing to the SRAM to program it. After the application has completed, the interface PCB can be used to read back the contents of the SRAM to see the results of the operation. To monitor the address, I have added an LED bargraph display.

You may be wondering why I have specified an 8k-by-8 SRAM for the computer processor if I am going to be using only 256-by-8 of it. I specified the larger SRAM because I had a number on hand and the 6264 part is very easy to find and buy.

I decided upon a 16-bit instruction word, with the first byte being the instruction and the second byte being a parameter to the instruction. This results in the 256-byte address space having only enough space for a maximum of 128 instructions. The actual address space available to the application is less because the 256-byte address space is also used for variables and, if you wanted to expand the processor, for I/O functions as well.

Since I use the same address space for program memory, variables, and possibly I/O, the computer architecture is of the Princeton type. As the early computers did, I used the Princeton architecture because it resulted in a simpler hardware design.

To execute each instruction, both bytes have to be read in and saved. Depending on whether or not the address pointed to by the second byte or the second byte itself is used will affect how the ALU executes the instruction.

You probably realize that there are a few deficiencies in this design. These are primarily the lack of a program counter stack and the lack of indexed addressing. Both these features can be implemented with self-modifying code. While not the ideal way of doing things, adding the circuitry for these functions would have made the circuit much more complex.

To get the basic block diagram shown in Fig. 11.17, I worked through several iterations of what I felt would result in the simplest amount of wiring. While the address and data buses go to various locations around the circuit (which means a lot of wiring), the actual application is quite simple, and if you keep the interconnected chips close together, you shouldn't have any problems with wiring the application together.

With the basic block diagram and the instruction decided on, I next decided to design the ALU. With a minimum circuitry requirement, I wanted to figure out a way to create a processor that could manipulate data in the following ways:

1. Load an accumulator
2. Add to an accumulator
3. Subtract from an accumulator
4. Bitwise AND accumulator
5. Bitwise OR accumulator
6. Bitwise NOT accumulator
7. Bitwise XOR accumulator
8. Shift accumulator contents left and right

Along with providing these functions, I also wanted to be able to have conditional execution change. This meant that a zero and a carry flag were needed.

To implement all these functions in a minimum amount of space, I decided to go back to one of the first concepts that I have presented in this book, and that was use the concept that TTL logic is based on the NAND gate. So, to create this computer's ALU, I decided to put in the ability to NAND the accumulator with another value in addition to providing a simple adder. The resulting block diagram for the ALU is shown in Fig. 11.18.

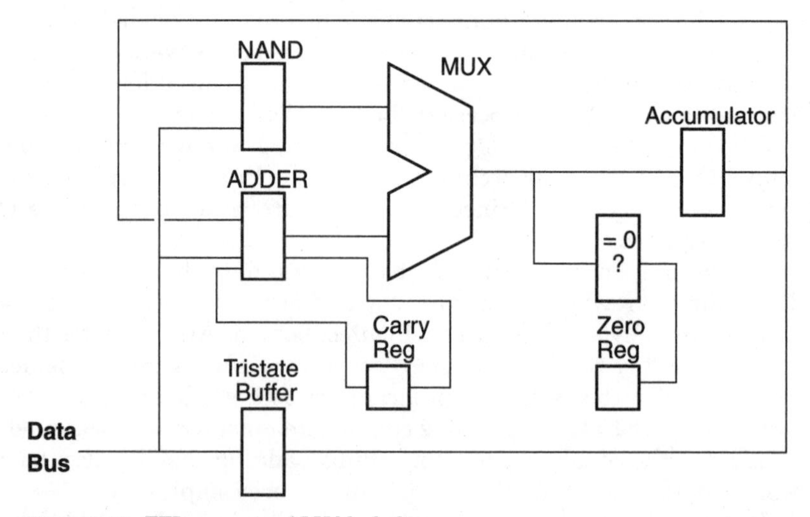

Figure 11.18 TTL computer ALU block diagram.

With the computer block diagram set up and an ALU chosen for it, I then went to work on designing the instruction operation. The first task was to decide on the instructions that were to be provided. With the very simple nature of this computer system in mind, I decided on four instructions: NAND, ADD, GOTO, and SAVE. The operation of the different instructions is explained in the table below.

Instruction	Operation	Flags modified/used
`NAND #Immediate`	`ACC = !(ACC.Immediate)`	Zero
`NAND Variable`	`ACC = !(ACC.[Variable])`	Zero
`ADD #Immediate`	`ACC = ACC + Immediate`	Carry, zero
`ADD Variable`	`ACC = ACC + [Variable]`	Carry, zero
`SAVE Variable`	`[Variable] = ACC`	None
`GOTO Address`	`PC = Address`	None
`GOTOC Address`	`if (Carry == 1) then PC = Address`	Carry
`GOTOZ Address`	`if (Zero == 1) then PC = Address`	Zero

The two addressing modes used in this computer system are the *Explicit,* in which an 8-bit value is explicitly used for NAND and ADD, and the `Variable`, which is the address in the SRAM where the parameter can be found—I used the square brackets ([and]) around `Variable` to indicate that the contents at `Variable` are accessed, not the numeric value of `Variable`. The `Address` in the `GOTO` instructions is an explicit value as well. These addressing modes will make more sense as I demonstrate some software below.

As I will show later in this section, these eight instructions can be combined to provide all the instructions used in a traditional processor. These instructions may seem arbitrary, but remember that the whole computer requires less than 40 chips and, not counting the SRAM, is implemented in less than 1000 transistors.

I decided that I would go with an eight-clock-cycle instruction cycle. This would allow me to set up the reads and writes needed for the instructions without changing the cycle and to keep as many signals as possible the same through the eight effective instructions.

Working through the instruction cycle waveforms took up the most time in designing the computer. When I was doing this, I wanted to establish a simple basic timing diagram that could be used to help me work through the control circuitry design. When I finished this work, I came up with the standard template I had drawn on a piece of graph paper, shown in Fig. 11.19.

In Fig. 11.19, you can see that I have broken each of the eight instruction cycles into separate zones in which the instruction execution is carried out. You might feel this was an inefficient way of implementing the instruction operation, but it allowed me to use simpler instruction decode circuitry than I would have needed if I were to have a different number of clock cycles for different instructions. The different instruction clock cycle zones are explained in more detail in the following table.

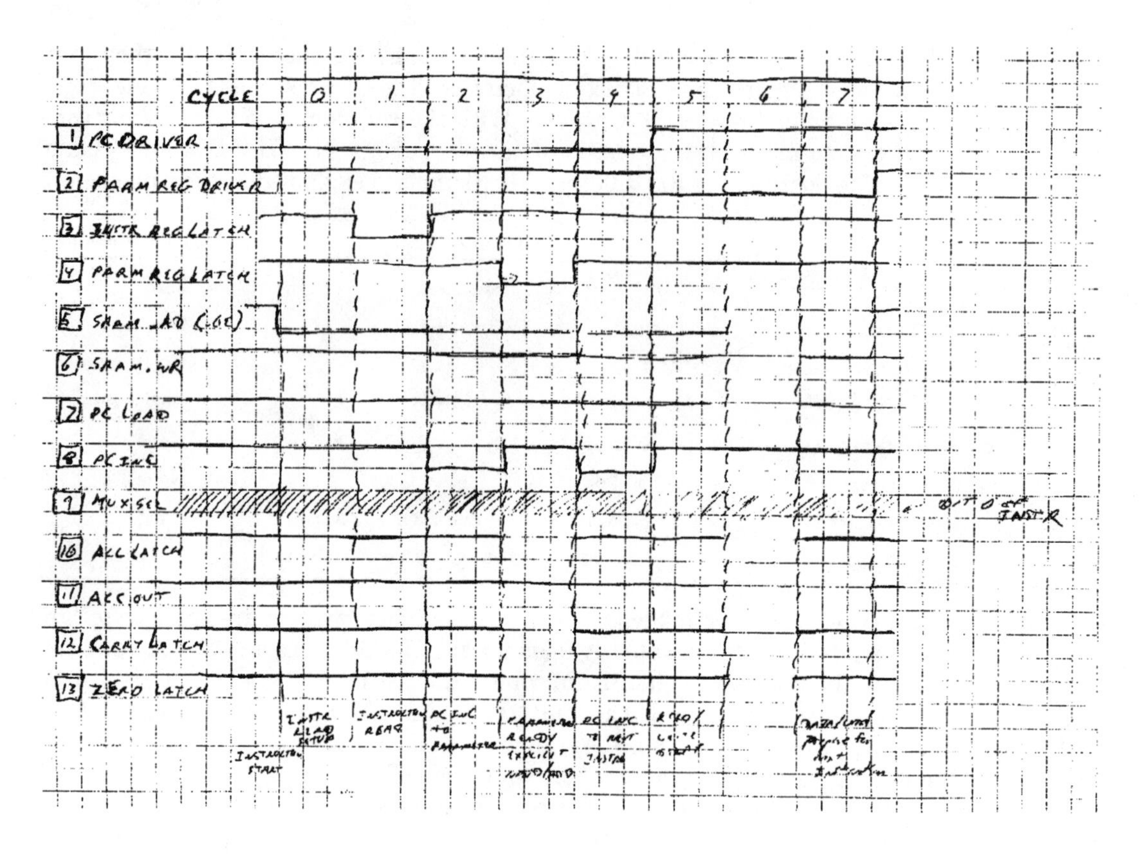

Figure 11.19 TTL computer timing blank sheet.

Instruction cycle zone	Description
0	Instruction start. Drive program counter onto the bus.
1	Read first byte of instruction into instruction register.
2	Increment program counter to point to the second instruction byte.
3	If explicit parameter for NAND/ADD, execute instruction. Also read in contents of SRAM pointed to by the second instruction byte.
4	Increment the program counter to the first byte after the current instruction.
5	Turn off PC driver and drive the address bus with the second instruction byte.
6	If a `Variable` instruction (NAND/ADD/SAVE) then execute the operation. If `GOTO`, update the program counter.
7	Leave drivers as is to make sure there is no possibility of a driver disabled while a write is taking place.

This sequence of operations and the resulting waveforms were designed to minimize the possibility that there could be a timing problem between a driver and a receiver. Note that in the SRAM timing waveforms, I have assumed that the SRAM will output whatever address it is given, even if it changes during output (_OE active). This is not a very contentious assumption, as most memory chips (not just SRAM) will work in this way for just this type of application.

When I was happy with the template and how the computer processor was going to work, I then photocopied it and used it to plan out how the different instructions would execute. Figure 11.20 shows one of these sheets that has been filled in for the NAND/ADD instructions with an explicit parameter.

In this instruction, the NAND/ADD operations execute while the parameter register is being loaded. In the `Variable` version of this instruction, the NAND/ADD operations execute in cycle 6 of the instruction, using the parameter register as the address driven into the SRAM.

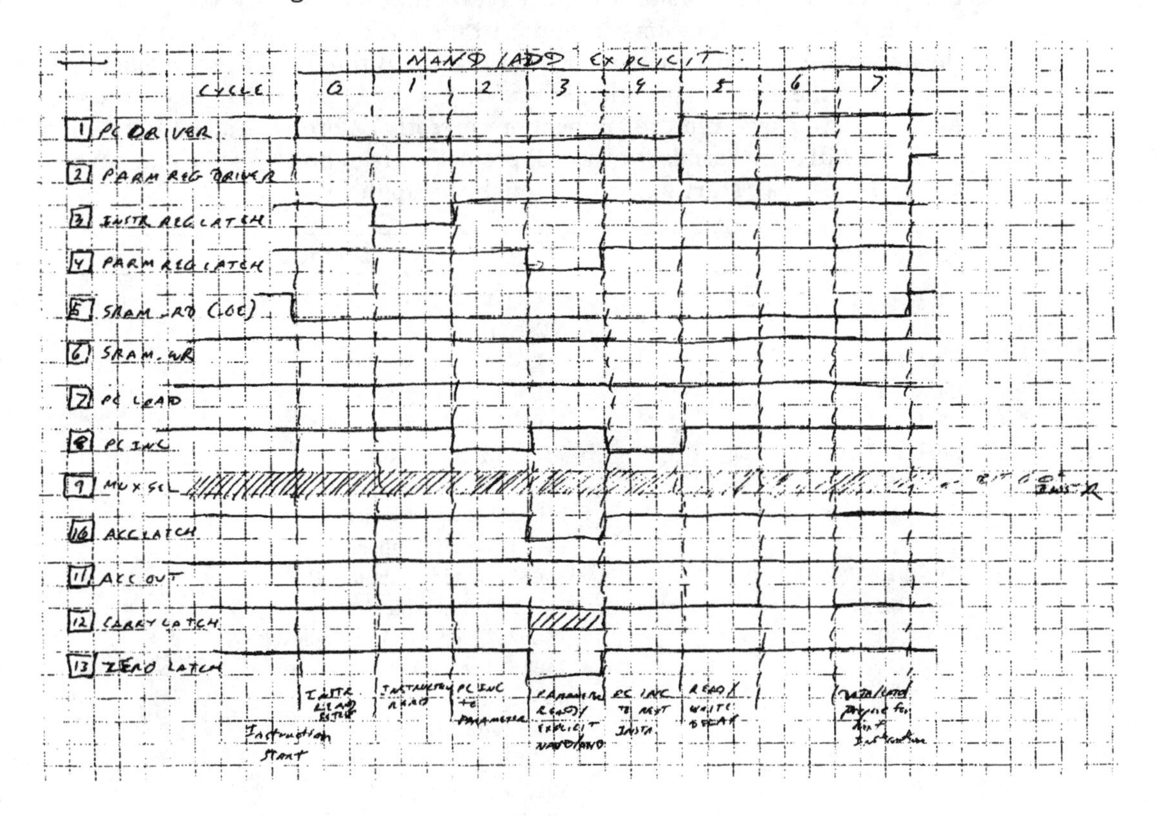

Figure 11.20 TTL computer NAND/AND timing.

This type of tool, while seemingly very crude, is extremely effective in showing how the circuit works and looking for any situations where there are access conflicts (bus contention) or if data does not have a path to its destination. The copies of the two diagrams do not reflect the number of times I drew out the processor architecture or the number of different sample timing waveforms that I drew to finally get the sample blank that I used for the application.

If you are ever faced with designing a complex digital logic circuit, then I suggest you follow the path I have used here: first create a high-level block diagram, next work out the timing for the circuit, and finally specify the different logic functions. By following these steps, you should end up with a working circuit first off, and even if you don't, the problems should be relatively minor.

I used the same process for the hexadecimal bus interface circuit, and, while I wasn't 100 percent right (I had to go back and fix three parts of the circuit), the process was a lot easier and the circuits that do work are very solid.

With the timing diagrams in hand, I then sat down and started designing the application. In this write up, I will walk you through each of the major subsystems of the circuit and provide you with a set of tests that you can use to make sure you've wired the application correctly before moving on to the next step. The bill of materials for the full project is given below so you can have all the parts ready to start working through the application.

Part	Description
U1, U14, U27–U28, U30	74LS245 8-bit tristate buffer
U2	NE555 timer chip
U3, U19	74LS74 dual D flip-flop
U4, U9, U33	74LS04 hex inverter
U5, U10, U15–U16	74LS00 quad two-input NAND
U6, U31, U34–U35, U37	74LS08 quad two-input AND
U7, U25–U26	74LS193 4-bit counter
U8	74LS138 3-to-8 decoder
U11	74LS123 dual single-shot
U12–U13, U32, U36, U38	74LS32 quad two-input OR
U17–U18	74LS283 4-bit adder
U20–U21	74LS257 four 2-to-1 multiplexor
U22–U24	74LS574 octal D flip-flop
U29	6264 8k-by-8 SRAM
D1	5-mm red LED
D2–D9	10-LED bargraph display
R1, R4	10k, ¼-W resistor
R2	470-Ω, ¼-W resistor
R3	1-MΩ, ¼-W resistor
R5	4.7k, ¼-W resistor
R5–R13	470-Ω, 8/9 resistor SIP
C1–C3, C6–C43	0.1-μF, 16-V tantalum capacitors
C4–C5	100-pF capacitor (any type)
Miscellaneous	+5-V power supply, interface PCB, breadboard, wire

In many ways, this application is typical of the early projects presented in this book, except that it is very complex. This is probably the limit to which I would expect a breadboard circuit could be built. If you look at the photograph of the circuit that I created (Fig. 11.16), I had to add two 6-in breadboards to the right of the largest board I could find to get the circuit wired together.

Along with making sure you have enough space for the circuit, also make sure you have enough wire. For this project, I went through six plastic boxes of wire. Before you start to build this project, I highly recommend that you have enough breadboard area and wires. It will be quite a bit more than you think you will need.

The first circuit I put together was the interface PCB/clock circuit that you can see in Fig. 11.21. In this circuit, I used the 2-MHz clock built into the interface PCB to run the circuit. The eight switches are used for programming the computer and the three remaining buttons are used for reading and writing program memory as well as changing the computer from program mode (PGM-MODE on Fig. 11.21) to run mode.

When the board first powers up, it is in PGMMODE and the Interface PCB can be used to load data into the computer using the procedures outlined below.

When the circuit is built, there are a number of tests that you can do to check out its operation:

1. On power up, the LED (D1) should light, indicating the circuit is in PGM-MODE. Pressing _RST on the interface board should toggle this LED, in addition to sending a pulse to 193CLEAR.
2. When the LED is on, by pressing LED9 (_RD), the DATADIR line should go low and a 250-nsec pulse should be output at PCINC. PCINC can be checked with a logic probe, and the actual length of time is not important.
3. When the LED is on, by pressing LED10 (_WR), the DATADIR line should go high and a 250-nsec pulse should be output at PCINC as well as a 500-nsec pulse output at _SRAMWR.
4. Turn the LED off by pressing _RST and check to see that the U7 counter is operating and that U8's outputs are cycling.

Next, you should assemble the ALU according to the schematic shown in Fig. 11.22. When I built this circuit, I started at the top of the board and worked my way to the right. I recommend that you create some kind of checklist for this wiring as you will end up with a real "rat's nest," and it is very easy to get confused here.

To Test this circuit, I recommend that you set up a 555 timer as a monostable output with a momentary-on button and use it to latch in the ACCLATCH data. Put some kind of data on the interface PCB and press the _WR (LED10) button to set U1 to drive the interface PCB data onto the data bus. You should disconnect the wire from U23's 1Q output (pin 19) and pull it high or low, depending on whether or not you want to test the NAND (high) or ADD functions. Next, press the ACCLATCH 555 pulse button to latch data through the ALU into the accu-

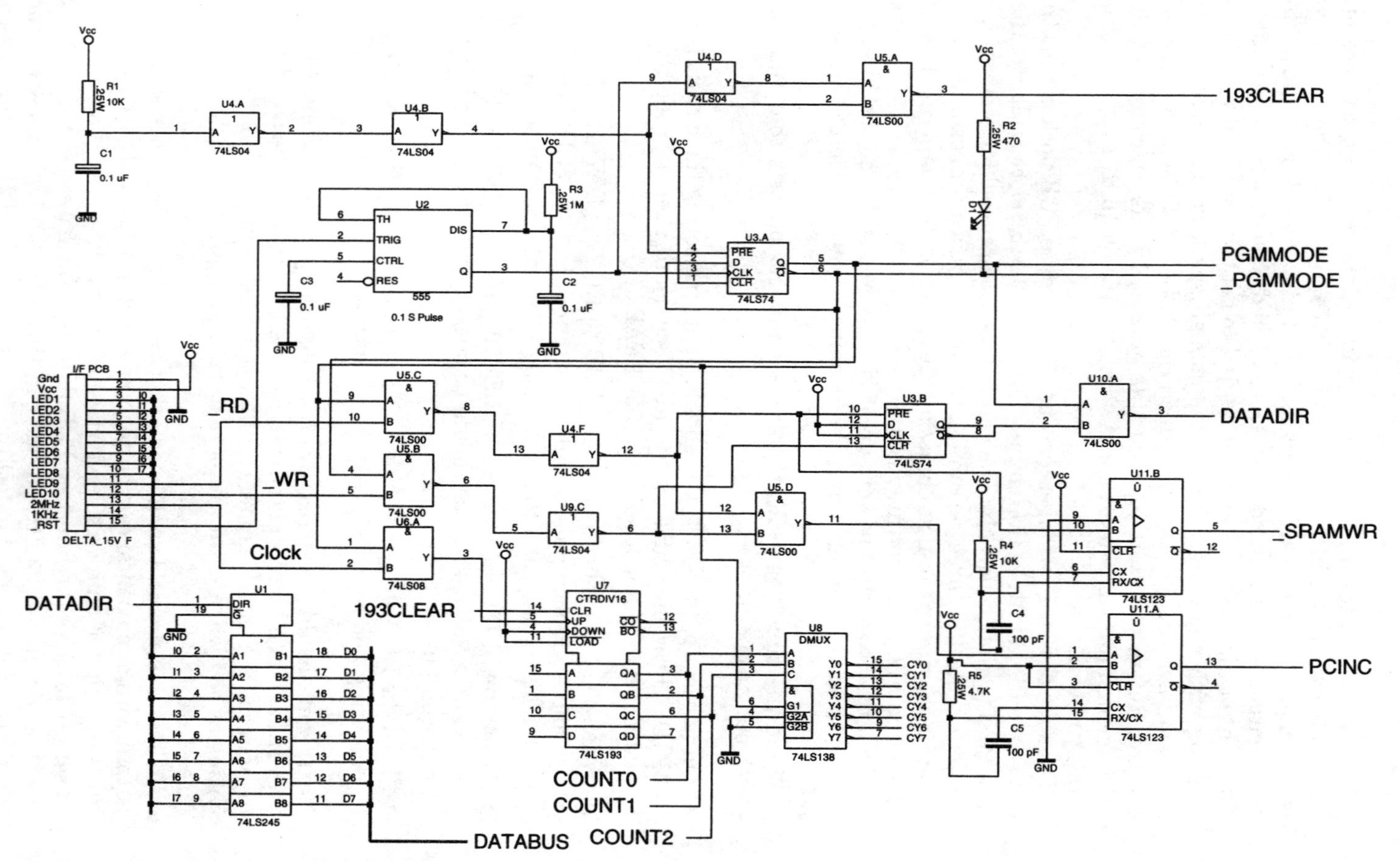

Figure 11.21 Project 18—interface PCB/clock circuitry.

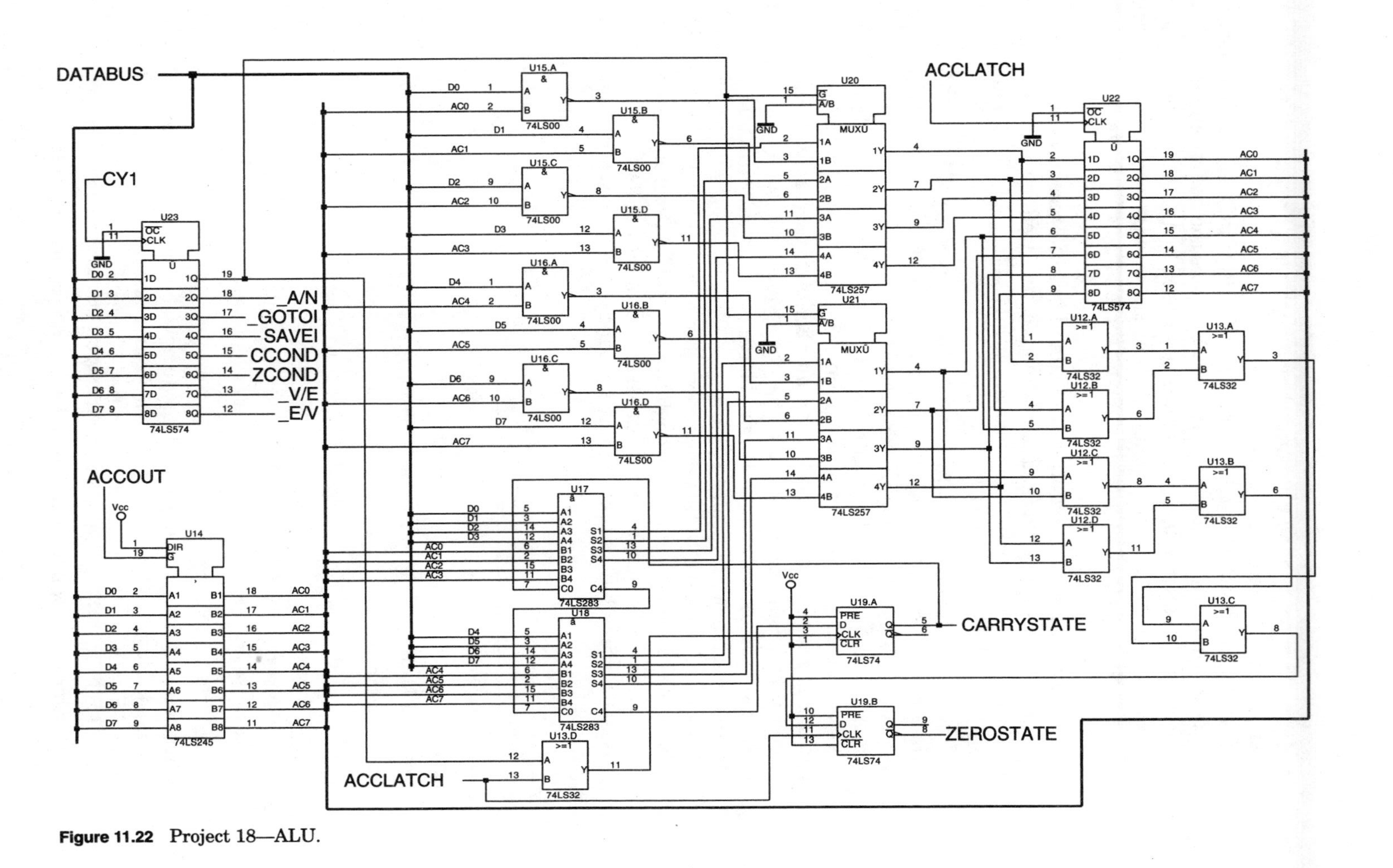

Figure 11.22 Project 18—ALU.

mulator. This may sound confusing, but it really is quite easy to test the ALU functions in this manner.

Now, you can work through the program counter (PC) and memory circuitry that is shown in Fig. 11.23. This circuitry provides the addressing information for the computer, as well as immediate data used by the ALU. This is one section where, with a little bit of planning, you can wire the various chips together very simply and efficiently, which means there is very little chance for wiring errors.

Testing this circuitry requires the same kind of procedure as that for the ALU, although what I did (and it worked out very well for me) was to connect the miscellaneous control logic (shown in Figs. 11.24 and 11.25) and test out each function in PGMMODE.

When I was happy with how the reading and writing of the SRAM worked, I then added the remaining logic from Figs. 11.24 and 11.25 and looked for opportunities to check the different functions. To be honest, I found that this section worked the first time, and I really didn't have any problems with it.

Note that for this section of the circuit, I did not work at "optimizing" the gates that I used. I know there are a number of unused gates throughout the circuit that I could have used and thereby reduced the total chip count, but I refrained from using them, as it would have complicated wiring the circuit and made it much more difficult to follow.

When you've gotten to this point, you should have a working computer!

Now, we can start working through the fun stuff of programming and seeing how it works. I have not created an assembler program for this computer, but it is not terribly hard to program it by hand, using the instruction specifications:

Instruction	Bit pattern
`NAND #Explicit`	`0b001000101`
`NAND Variable`	`0b010000101`
`ADD #Explicit`	`0b001000100`
`ADD #Variable`	`0b010000100`
`SAVE Variable`	`0b010001110`
`GOTO Address`	`0b001000010`
`GOTOC Address`	`0b001010010`
`GOTOZ Address`	`0b001100010`

These bit patterns can be related back to how the instruction register's bits are specified. Note that for many of these instructions, various bits are "don't cares." For the pattern above, I specified values for the don't care bits to make sure there isn't any problem with specifying the correct instructions.

To enter in an application, the following procedure is used:

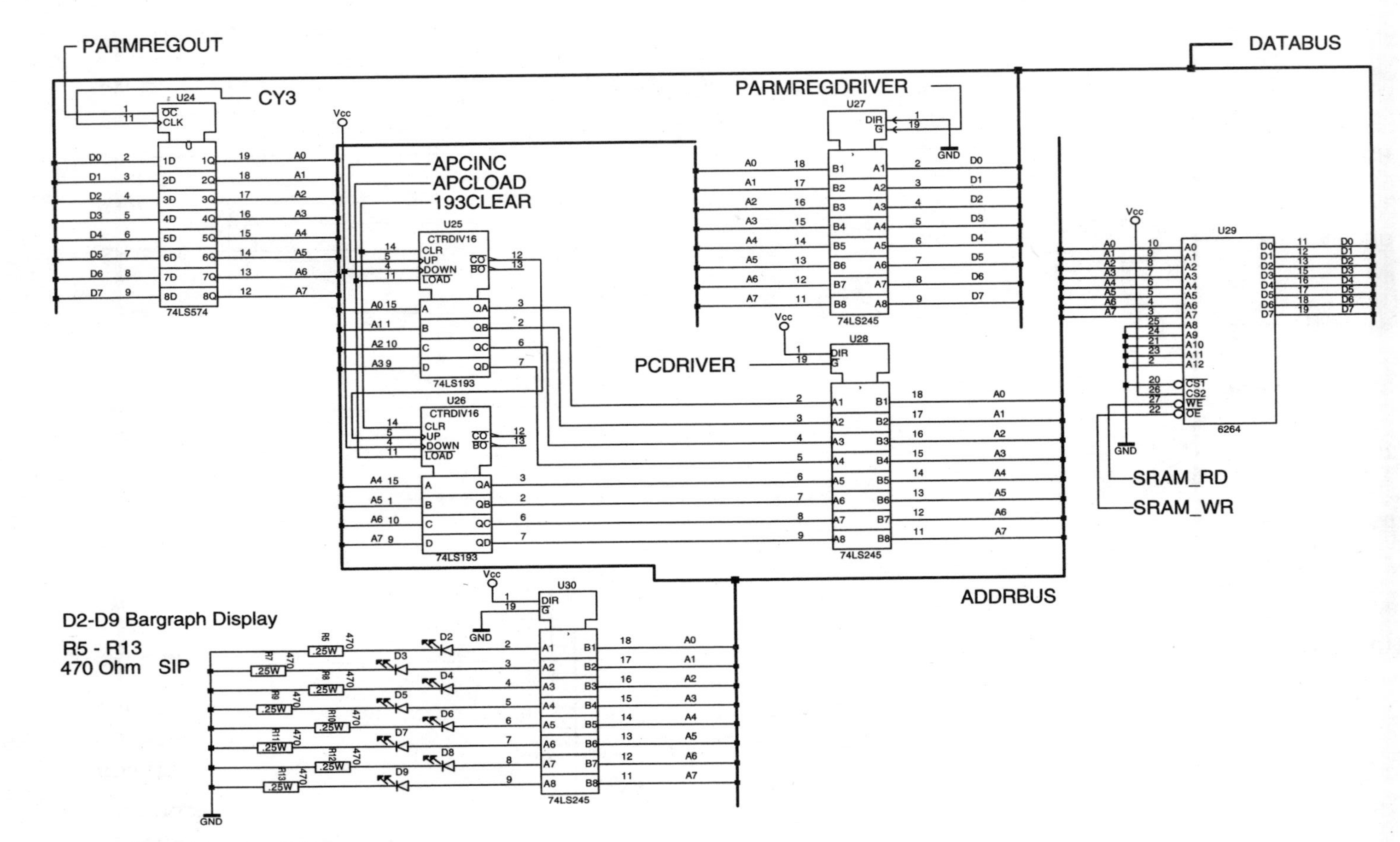

Figure 11.23 Project 18—PC and memory.

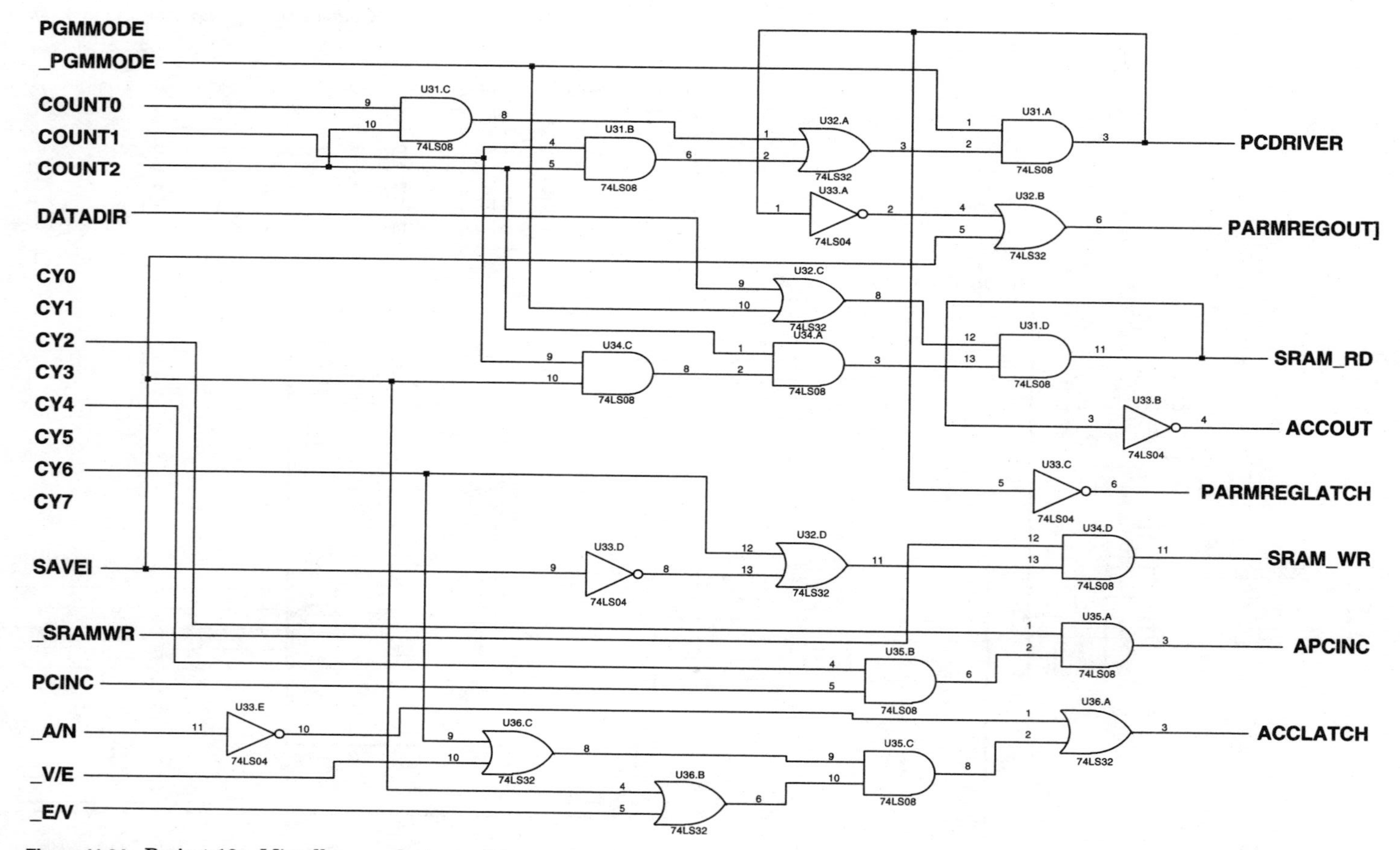

Figure 11.24 Project 18—Miscellaneous logic, part 1.

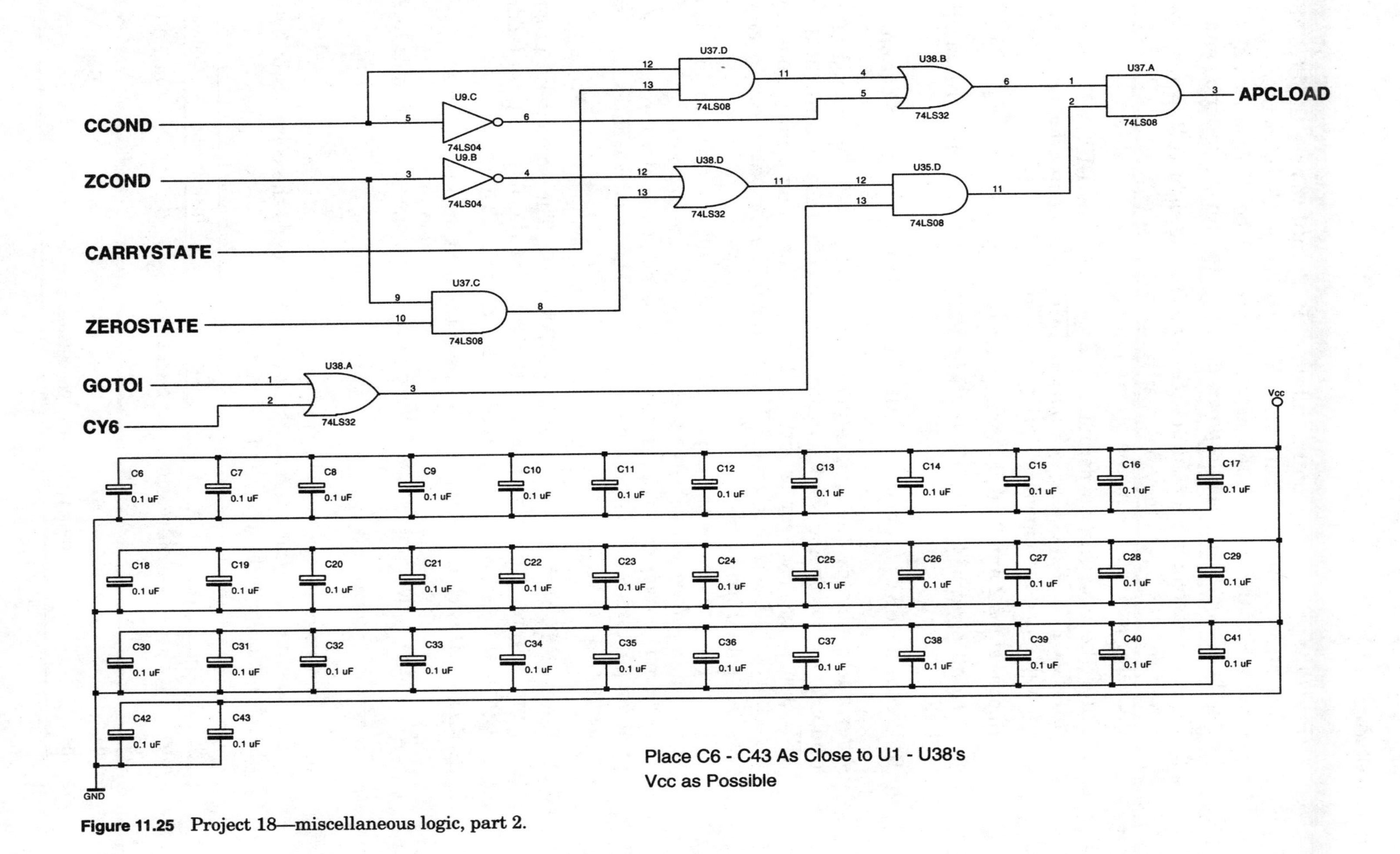

Figure 11.25 Project 18—miscellaneous logic, part 2.

1. Press the _RST button on the Interface PCB until you see the PGMMODE LED turn off and back on again. This will reset the computer's program counter—all the LEDs on the address display will be off. Another way of forcing the computer to reset for programming is to turn the power on and off.
2. Enter the instruction for the address by using the eight DIP switches on the interface PCB. Press the _WR button (LED10 on the interface PCB) to save the value and increment the program counter.
3. Enter the instruction's parameter using the eight DIP switches on the interface PCB. Press the _WR button (LED10 on the interface PCB) to save the value and increment the program counter.
4. Repeat steps 2 and 3 until the program is entered.
5. Turn all eight of the interface PCB's DIP switches off to turn on all the LEDs.
6. Press the _RST button on the interface PCB to execute the program.

After a few minutes, the application code should have been entered into the computer's memory. You can tell the application is finished because the address LED is set to the last address with the least significant bit being on, but darker than the other LEDs that are on. This is due to the computer reading an endless loop goto instruction.

If you want to review the application, then I will show how to do this as I explain how software applications are written for the computer.

As there is no way of looking at the computer's accumulator through the programming interface, when the application is finished, you can check the results by using this procedure:

1. Press _RST to turn on the PGMMODE LED. The computer's program counter is reset at this time and the value of the first address (address 0x000) in the SRAM is displayed at the interface PCB.
2. Press _RD each time you want to check the next memory address.

To create applications, you will have to set up a page with headings like:

```
Address Instruction Parameter Instruction Comments
```

before you start doing any kind of programming.

For the first application, how about one that reads the contents of byte in SRAM, increments it, and puts it back:

Address	Instruction	Parameter	Instruction	Comments
0x00	0b001000010	0b000000010	GOTO 2	Start ELOOP
0x02	0b001000101	0b000000000	NAND #0	Make start
0x04	0b001000101	0b011111111	NAND #FF	Endless loop
0x06	0b010001100	0b000000001	SAVE 1	

Address	Instruction	Parameter	Instruction	Comments
0x08	0b001000101	0b000000000	NAND #0	Read in A
0x0A	0b010000101	0b000010100	NAND A	
0x0C	0b001000101	0b011111111	NAND #FF	
0x0E	0b001000100	0b000000001	ADD #1	Increment
0x10	0b010001100	0b000010100	SAVE A	Save
0x12	0b001000010	0b000001000	GOTO 8	Loop again
0x14—address of A				

Note that at the end of the program, I specified the address of A. It will be located at address 0x012 and no code can execute over it.

Also note that the first instruction jumps to the second instruction and the second through fourth are changing the value of the parameter of this instruction to turn it into an endless loop. The reason for this is to stop execution of an application from changing a value after you have stopped it once. In this first program, this code will prevent A from being updated after the first time it has been written.

I like to use the basic application format:

Address	Instruction	Parameter	Instruction	Comments
0x00	0b001000010	0b000000010	GOTO 2	Start ELOOP
0x02	0b001000101	0b000000000	NAND #0	Make start
0x04	0b001000101	0b011111111	NAND #FF	Endless loop
⋮				
Application goes here				
0x??	0b001000010	0b0????????	GOTO ??	Endless loop
Variables go here				

For all my applications, after the first four instructions, I add my application, then the endless loop ("GOTO ??" instruction), followed by application variables.

To try out this application, follow the steps:

1. Program in the 18 bytes of the application using the programming procedures detailed above.
2. Press the interface PCB's _RST button to start the application executing. The PGMMODE LED should turn off.
3. Press the Interface PCB's _RST button again to go back into programming mode.
4. Now, press _RD (LED9 on the interface PCB) to review the application. You should see that address 1 has been changed by the code, and as you step along to address 0x012, you will see that A has been given a random value.

5. To see this happen again, toggle the Interface PCB's _RST button (the PGMMODE LED should turn off and on).
6. Next, press _RD (interface PCB LED9 button) again to increment to address 1.
7. Change the value for address 1 to 2 and press _WR to save it in memory.
8. Turn off all the LED switches on the interface PCB.
9. Press _RST on the interface PCB to execute the code.
10. Repeat steps 2 through 4 to look at the new value in A.

This application essentially generates a random number. Note that it executes a large fraction of the total instructions available to it.

The second example application loads A with zero and repeats until it is equal to 16:

Address	Instruction	Parameter	Instruction	Comments
0x00	0b001000010	0b000000010	`GOTO 2`	Start ELOOP
0x02	0b001000101	0b000000000	`NAND #0`	Make start
0x04	0b001000101	0b011111111	`NAND #FF`	Endless loop
0x06	0b010001100	0b000000001	`SAVE 1`	
0x08	0b010001100	0b000011100	`SAVE A`	Initialize A
0x0A	0b001000100	0b000000000	`ADD #0`	Clear carry flag
0x0C	0b001000101	0b000000000	`NAND #0`	Read in A
0x0E	0b010000101	0b000011110	`NAND A`	
0x10	0b001000101	0b011111111	`NAND #FF`	Endless loop
0x12	0b001000100	0b000000001	`ADD #1`	Increment
0x14	0b010001100	0b000011110	`SAVE A`	Save
0x16	0b001000101	0b000010000	`NAND #10`	Equal to 16?
0x18	0b001000101	0b011111111	`NAND #FF`	
0x1A	0b001100010	0b000001100	`GOTOZ 0C`	Not 16; loop again
0x1C	0b001000010	0b000011100	`GOTO 1C`	Endless loop
0x1E—address of A				

Note that in this code I cleared the carry flag before starting the loop to make sure that for the first operation an extra 1 isn't inadvertently added to A. Other than this, the code is quite straightforward and should be fairly easy to understand.

As you work through this application, you will realize that this is a *very* tedious way of interfacing to a computer.

At the start of the section, I mentioned that the NAND function of this computer would allow you to create all the basic programming instructions. In the table below, I have listed various functions and how they can be built out of the eight instructions available to you. Note that for the different NAND/ADD instructions, I have just placed *value* for the instruction parameter without regard to it being an explicit value or a variable address. TEMPACC is a variable used for temporary accumulator storage. Variables are labels with a colon (:) at their end.

Instruction	TTL computer instructions	Comments
Nop	GOTO SKIP	
	SKIP:	
ACC = value	NAND #0	ACC = ! (ACC.0)
		= ! (0)
		= 0x0FF
	NAND value	ACC = ! (0x0FF.value)
		= !value
	NAND #FF	ACC = ! (0x0FF.!value)
		= ! (!value)
		= value
ACC = ! ACC	NAND #FF	ACC = ! (ACC.0x0FF)
		= ! ACC
ACC = -ACC	NAND #FF	ACC = ! ACC
	ADD #1	ACC = ! ACC + 1
		= -ACC
ACC = 0	NAND #0	ACC = 0x0FF
	NAND #FF	ACC = ! (0x0FF . 0x0FF)
		= ! (0x0FF)
		= 0
ACC = ACC & value	NAND value	ACC = ! (ACC . value)
	NAND #FF	ACC = ! (! (ACC . value))
		= ACC . value
ACC = ACC	value	NAND #FF	ACC = ! (ACC . 0x0FF)
		= ! ACC
	SAVE TEMPACC	
	NAND #0	ACC = 0x0FF
	NAND value	ACC = ! value
	NAND TEMPACC	ACC = ! (! value . TEMPACC)
		= ! (! value . ! ACC)
		= value + ACC
ACC = ACC ^ value	SAVE TEMPACC	
	NAND #0	ACC = 0x0FF

Instruction	TTL computer instructions	Comments
	NAND value	ACC = ! value
	NAND TEMPACC	ACC = ! (ACC . ! value)
	SAVE TEMPXOR	
	NAND #0	ACC = 0x0FF
	NAND TEMPACC	ACC = ! ACC
	NAND value	ACC = ! (! ACC . value)
	NAND TEMPXOR	ACC =
		! ((! ACC . value) . ! (ACC . ! value))
ACC = ACC - value	SAVE TEMPACC	
	ADD #0	Carry flag = 0
	NAND #0	ACC = 0x0FF
	NAND value	ACC = ! value
	ADD #1	ACC = -value
	ADD TEMPACC	ACC = ACC - value
if (ACC > value)	SAVE TEMPACC	
goto LABEL	ADD #0	Carry flag = 0
	NAND #0	ACC = 0x0FF
	NAND value	ACC = ! value
	ADD #1	ACC = -value
	ADD TEMPACC	ACC = ACC - value
	GOTOC LABEL	
ACC = ACC << 1	SAVE TEMPACC	Rotate through carry
	ADD TEMPACC	ACC = ACC * 2
		Carry flag = ACC Bit7
	GOTOC SKIP	If carry set, then jump
		over next instruction
	ADD #FF	ACC = (ACC * 2) - 1
	SKIP:	
	ADD #1	ACC = ACC << 1
Loop n times	NAND #0	
	NAND n	ACC = ! n
	SAVE COUNT	
	LOOP:	
	:	Code inside loop
	NAND #0	
	NAND COUNT	ACC = n
	ADD #FF	ACC = n - 1

Instruction	TTL computer instructions	Comments
	NAND #FF	ACC = ! (n - 1)
	SAVE COUNT	
	GOTOC LOOP	

I have created three or four applications of my own on this computer, and it really is quite an interesting little system for learning how a computer works and how to program one at a very low level. It is unfortunate that there is no indexed addressing or subroutine calls in this computer, but they could be demonstrated by using self-modifying code.

This would also be an interesting processor to develop a simulator for, and it could be very useful if it were converted into VHDL (Verilog Hardware Definition Language) for use with a PLD. In this case, a more complex user interface could be developed.

If I were to work through this application again, there are two things that I would do differently. The first is to use a keypad, like the hexadecimal bus interface's with the seven-segment LED displays. This would make programming and review a lot easier and quite a bit faster. The second is to find some way of interfacing to I/O devices. The obvious way of doing this is to provide some kind of address selection between the SRAM and external devices, but I feel that a lot could be gained by making a second I/O space for the architecture.

Chapter

12

Creating Your Own Applications

Before going through the steps that I use to develop digital electronics applications, there are a few words of advice I want to give you. First of all, document everything. Get into the habit of carrying a notebook around with you so that if you get an idea on how to do something, you can record it. Human memory is a pretty fallible storage device and I've spent many hours trying to remember that great idea I had the day before at lunch.

Second, start small; I get a lot of e-mail from people who want to create a substantial project like the Lego MindStorms and ask me where they should start. My reply is to get a book and figure out how to work with boolean algebra and follow it up with creating simple circuits that you can use to test out your ideas. Once you have this background, you can work through the task of developing your "killer" application.

A large project for any beginner will start off strong and then become bogged down as the project seems to drag on and on. I know of some people who rise to the occasion of developing large applications and become experts through them, but these people are few and far between.

Next, don't settle on the first method you come up with. There's always more than one way to do something. Take, for example, developing the timing for a flashing LED; using digital electronics, the following methods could be used for developing the timer.

1. Slow clock
2. Fast clock with delay by a counter
3. 555 timer

Which method is best in your application? Spend a few minutes thinking of options that can make the application much easier later on.

It is unfortunate, but I usually discover better methods for carrying out something *after* I finish an application. If I figure out a better way, I'll usually keep it in my notebook for the next time rather than go back and recreate the application. I may go back and change the application, but that is only if the change offers a substantial improvement to what was already there (i.e., the circuit costs less to build or uses fewer parts). The basic rule here is "If it ain't broke, don't fix it."

Lastly, steal other people's ideas and methods. This does not mean literally stealing their circuit diagrams and using them unchanged, but instead understanding how their code, interfaces, and applications work and recreating them as they work best for you. This will not only help you avoid getting "blocked," but chances are you will learn from others and be able to use their ideas in ways that aren't readily apparent.

An excellent resource for this is Rudolf Van Graf's *Encyclopedia of Electronic Circuits* (more information about books is available in "Resources"). These seven books and others in the series provide a large selection of circuits from virtually every facet of electronics. Other resources include the various monthly electronics magazines (also listed in the appendixes), as these will have at least one digital electronics application or PICmicro application in them per month.

The Internet also has a plethora of circuits at various Web sites. I just want to caution you not to believe everything that you read on the Internet, as the information can be inaccurate (as any information can be). I've also found a number of Web pages that will offer a few "free" circuits as an inducement for you to buy design information from the page's owners or use them as consultants.

Requirements Definition

The most important thing you can do before starting to design a digital electronics application is to create a set of requirements. These requirements will help you to understand what is required to do the project and allow you to check off whether or not you are meeting the original design requirements. When you first start developing digital electronics applications, you should be very rigorous in defining the requirements of your application, as this will allow you to keep track of what you are doing and not get overwhelmed with the task in front of you.

The first thing that I like to do is to create a simple, one-sentence statement specifying what the project is supposed to do. Some sample statements based on the examples and projects in this book, are:

> I want to create a digital clock that uses a 32.768-kHz watch crystal for precise timing measurements.
>
> I want to demonstrate how a simple state machine can be created to control traffic lights.
>
> This project will convert an NRZ serial data stream to 8-bit parallel data at 1200 and 9600 bits per second.

Many of these application description statements can be found in the first paragraph of the application write-ups presented in this book. I always feel that you must be able to describe what you want to do, simply and concisely. This statement becomes a guide for you to follow when creating your other requirements definitions.

Next, the physical requirements for the application must be stated. These requirements specify the external requirements that will drive the circuit and software requirements that follow. A good list to have is:

1. What type of logic technology should be used in the project?
2. What is the clock speed of the project?
3. How much power does the project consume?
4. What is the power source?
5. What type of circuit carrier/board is to be used?
6. Who will be using the project?
7. What are the user interfaces?
8. What are the safety concerns?
9. How much should the project cost?
10. How long will the project take to develop?

For many of these questions, the answer will be "none" or "no restrictions." Even though some answers may be vague, you are still narrowing down what you want to do.

You may find that some of the answers will affect other questions' answers. For example, if you want to power the application by a battery, you will probably want to use CMOS TTL-compatible logic chips for their low power drain. But, if there are definite speed requirements, you may find that the CMOS TTL gate delay is too long to safely implement the application. In this case, you will probably have to use bipolar TTL chips, which have different (higher) power requirements than CMOS TTL and lower costs.

Developing the requirement plan is an iterative and recursive process, with each change affecting other answers. This process should not be undertaken in just a few minutes, but careful thought must be taken to ensure that the requirements are reasonable and achievable.

Along with the requirement plan, I find it useful to come up with a set of timing drawings (similar to the ones I showed in the TTL computer project). They do not have to be fancy or even drawn to any kind of scale. The important thing to remember with these drawings is to make sure you are clear on which signals operate at which times.

As you work through your project development, you may find that suppliers cannot reliably provide you with the parts that are required. If this is the case, you may find you have to go back and change your original defining statement. This is really not a serious problem. As you work through the requirements, if you work objectively with your requirements and the assumptions they're based on, you will find that you can get a solid set of requirements that will lead you to a successful application.

Developing a Qualification Plan

Part of the requirements that you define for your application should be an application "qualification plan." This plan consists of a list of tests that the application must pass before you can consider it ready for use. In the list of qualification tests that I present in this section, there will be some items that may seem obvious, but part of the purpose of the qualification plan is to give you a checklist for when you build other units or instruct others to build the application.

A typical qualification plan would consist of:

- Test for the power supply to supply 4.75 to 5.25 V of regulated power at the required current load.
- "Reset" driven at different voltage levels.
- The built in oscillator running when both oscillator pins are probed.
- User interface output functions are at the correct initial state on power-up.
- User interface output functions respond correctly to user inputs.
- Application outputs respond to user inputs as expected.
- Application outputs respond to timer or application input events as expected.
- Application test can detect all functional failures.
- Reliability calculations have been performed to ensure the application will not fail before its expected life is finished.
- Environmental qualification.
- FCC, BABT, and other emissions and regulatory testing
- User documentation is complete and correct.
- Manufacturer documentation is complete and correct.

The first three points of this list (checking power, reset, and clocking) may seem to be obvious and unnecessary, but I would argue that they are needed to ensure that the application will work reliably. If the application isn't executing, then they should be the first things that are looked at.

The power range that I specified in the list above probably seems quite restricted, especially if you want to have your application battery powered. If you are going to work with battery power, then you will have to make sure the parts selected will work over a wide voltage range, with the ability to reset or shut off if the power goes too low [this is known as a *brown out reset* (BOR)]. Most bipolar TTL active integrated circuits work reliably only within the ±5 percent window that I have specified in the requirements list above.

Note that for many different kinds of inputs and application circuits, a simple digital multimeter or logic probe will not be sufficient. Ideally an oscilloscope or logic analyzer should be used to look at the actual I/O and confirm that it is correct. If you do not have access to these tools, then you should work with subassembly experimental circuits in which you can test the functions before creating the larger circuit.

When qualifying applications, make sure that the electronics will work with a range of voltages and timings. I picked ±5 percent because this range should be appropriate for all integrated circuits that you will come in contact with. When you test for timings, if you have an oscillator built into the circuit, then you should vary its timings by up to 10 percent (both faster and slower) to make sure that poor-quality oscillators will not affect the overall performance of the application. These margins are known as *guardbands* and are designed to help you catch marginal or "flaky" circuit designs.

With digital logic, I recommend that all sequential devices should be *edge triggered* instead of *level triggered,* as the former type is much less sensitive to timing issues.

Application testing is something that is near and dear to my heart (mostly because I have spent almost all of my professional life ensuring that products work correctly). In the list of qualification items above, I noted that the test should be for all functional aspects of an application. If you've worked around electronic circuits for any length of time, you will know that this is almost always impossible because there are so many different ways in which a circuit can fail.

The important point about testing is to ensure the application will respond correctly to the expected inputs. There will be cases when unexpected inputs will be received (such as when the application connection is jostled or the circuit is zapped by static electricity), and these are hard to plan for. You should have a good idea of how the application works under normal operating conditions and be able to test these conditions.

It probably seems surprising that I have included the need for a reliability calculation here—especially since you are probably planning that your first applications are going to turn on an LED by a button's input. I will not go into reliability calculations in this book, but if you are going to develop an application for a commercial product, then you will probably want to do the calculations and even test out the application in an accelerated life test to make sure your calculations are correct.

Reliability qualification is not something that many companies worry about, and, with the ruggedness of modern components, it is not perceived as being a critical factor to a product's success. Personally, I believe that testing the reliability of a product is the most important indicator of the quality of the design and the quality of the components used. Admittedly, reliability testing can add significant costs to a product, but it can keep you from unhappy customers and lawsuits if your products fail prematurely.

Environmental qualification is simply ensuring that the product will run reliably in the location where it is expected to operate. Obviously for most of the applications presented here, this is not a concern because the circuits will be run on a bench or in a home or office environment. Part of the environmental testing is ensuring the application does not radiate excessive noise. For products to be sold commercially they will have to be FCC Class B or BABT (British Authority on Broadcasts and Transmissions) certified.

Many applications work in extremes of voltage, electrical noise, vibration, temperature, or humidity. Along with this, the applications may be sealed into a package and end up being heated by the circuits used within them. If any environmental extremes are expected, then the product should be tested and qualified for these environments, and the reliability calculations done to make sure the product will operate for its required lifetime.

For many of the environments circuits are used in, testing to industry or governmental specifications will be required. For example, if your application was going to be placed in an aircraft, then you should make sure that all specifications are met for equipment operating in this environment. Many of these specifications (especially in the aerospace arena) will seem difficult and expensive to meet, but remember that in many cases these specifications were a result of other, earlier products failing—sometimes harming people.

Last, make sure you have properly documented your application. I consider this aspect of project development to be part of the qualification plan because documentation for users as well as manufacturers is critical to having your product integrated. User interfaces should be designed to make the product easy and intuitive to use (more on this below). Making sure that the extra hardware required by the user is specified, as well as how to connect and power the application, is always necessary if anybody other than yourself is going to use the application or build it.

Documentation should be written with an electronic tool that is widely available to any potential users. If there is no single format that you feel comfortable with (for example, your product can be used with PCs, Macintoshes, or Unix workstations), then you should document how to use the product in straight ASCII text files.

If straight ASCII text files don't seem appropriate because formatted text is required with graphics, then I would suggest either Adobe .pdf format or using HTML (Hypertext Markup Language) for your documentation. Both data formats are widely available on different computer systems along with development tools that can be found fairly cheaply and easily.

Documentation format and content should be discussed with the target audience to ensure that it is written appropriately and will be usable. Providing documentation electronically is usually preferable because it can be easily distributed and updated without having to send out multiple text copies.

All these requirements can be summarized in an ISO 9000 registration. ISO 9000 (the generic term) is a certification required by the European Union and is used to document development and manufacturing processes. Even if you are a one-person company, ISO 9000 certification can be achieved for a modest cost and will add enforced rigor to your development process.

ISO 9000 is also required by many companies. For example, when I asked for quotes on the +5-V power supply PCB and interface PCB that are included in this book, only ISO 9000–certified companies were asked to respond.

User Interfacing

For the most part, digital logic does not lend itself to providing very good user interfacing. You are restricted to simple buttons and LEDs. Despite these limitations, you can provide some user interface capabilities to your application that will help make it easier to use. Often just a few labeled LEDs will change an application that seems to be impossible to use to one that is actually quite simple and intuitive to use.

By following a few simple rules, you can come up with a user interface for your application that will make operation of the application intuitive and easy to use effectively by anybody.

The first thing you should do is to understand what kind of feedback is appropriate for the application. For example, an oxygen sensor in an automobile does not require a human user interface, but an LED interface may be appropriate for service center maintenance. A burglar alarm circuit should probably have a light and a siren. A programmable microwave oven should have a keypad and LCD character display for setting the oven strength and cooking time.

When you are deciding what is the best way to interface to the user, remember that the user probably won't have a manual handy. If an operating mode has to be selected or data entered, the user will have to be guided through the process. This could be done in a manual, but manuals get lost. Ideally, you should have the device prompt the user with the operating instructions.

In many order circuits, you will see that LEDs are used for this function, as shown in Fig. 12.1. The biggest concern with using LEDs in a panel like the one in Fig. 12.1 is that the user will have trouble figuring out how to get to another mode or what to do next in the current mode. Unless a printed manual is included with the application, you will find users who have trouble figuring out how to work the application. Along with these problems, there are also the issues of labeling buttons and making sure their labels are appropriate. To avoid this, the panel should include the instructions necessary to use the complete project.

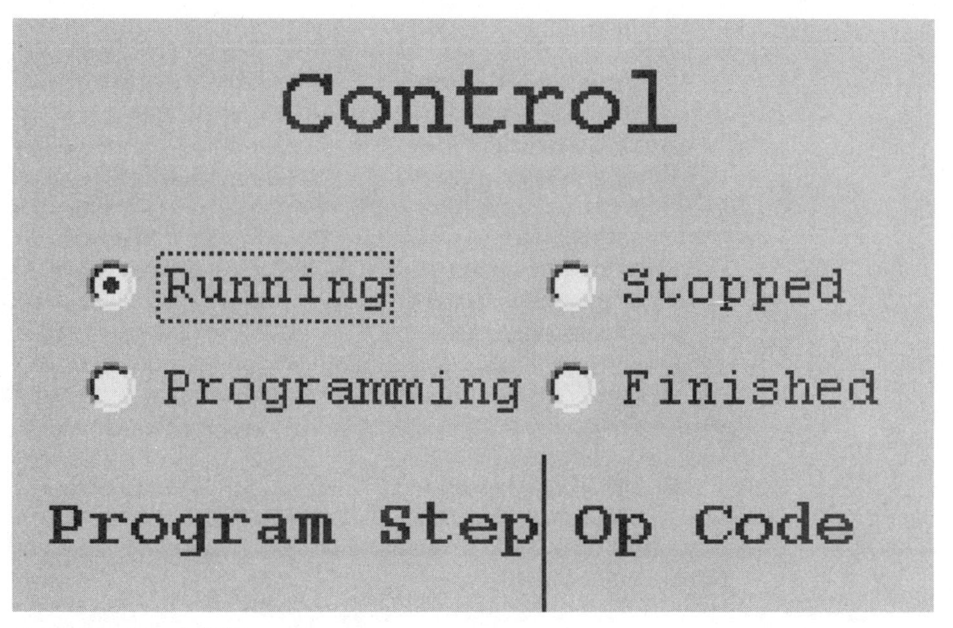

Figure 12.1 Sample LED panel.

When the circuit is connected to another device (or any two devices are wired together), there should be some kind of connection check (or "ping") to ensure that the link is active. For digital logic, I provide a pull-up on one of the devices with an LED. On the other device, I tie the mating pin to ground. This circuit will turn on the LED when the two boards are plugged together.

All inputs must be debounced either by hardware (if possible) or by software. The techniques I have shown for button debouncing are quite effective and should be appropriate for most cases. In the circumstances where they aren't (such as when a switch matrix keyboard is used), then an intelligent device (such as a microcontroller) will be required to control the application.

Chapter

13

Debugging Projects

Before ending this book, I want to discuss the techniques I use for efficiently characterizing and identifying problems so that they can be fixed. In this chapter, I will discuss the techniques I teach for product failure analysis, with the end goal being an understanding of exactly what is the problem and what is the best way to fix it. The process outlined here will help ensure that not only will the problem be identified in the most efficient manner, but that its properties will be recorded and the evidence saved so that this problem will not recur in the future.

You will find that following the steps and procedures I present in this chapter will allow you to fix problems faster than if you just look for the first problem and try to "nail it." This will seem like a paradox because I will be pushing for you to characterize and understand exactly what is happening and what the symptoms are telling you. When you understand exactly what is happening, you can design a fix that will address these problems and will result in the problem being resolved once and for all. By following the procedures outlined in this chapter, you will find that your success rate of finding and eliminating problems will be close to 100 percent—a much higher rate than you can expect from simply replacing the parts that you think are the problem after a few seconds' consideration.

At work, I try to discourage people from using the term *debug,* as it seems to denote a quick fix. Instead, I prefer that the term *failure analysis* be used. It's interesting to see how changing a simple word to a "high-falutin" term can change the perception of an operation that was once "quick and dirty" into a structured process that not only fixes the problem with a high degree of confidence but helps the application designer and manufacturer prevent it from happening again in the future.

So, instead of blindly going ahead and simulating the application in a three-step process, what I recommend for fixing a problem is:

1. Characterizing the problem
2. Hypothesizing what the cause of the problem is
3. Testing your hypothesis out

Note that I am using the term *application* in this chapter, rather than the more traditional *project* that I use throughout the book. The process outlined in this chapter can be used not only for hardware circuits, but also for software and virtually every instance in your life where you have to solve a problem. It may be hard to think through this process when your roof is leaking and rain is pouring in, but with some practice you'll find that you can look at every problem in this way and you will end up resolving it in the most efficient manner.

One attitude I feel that is critical for failure analysis is being suspicious of anything that doesn't "feel right." I have worked with a number of people over the years who have let problems escape out the door simply because they didn't react to something that did not work as expected once or twice during application development. If you have a suspicion that something isn't right, then you should go out and fix it before the problem escapes and you end up with a reputation for shipping shoddy products.

Characterizing Problems

Even if your circuit just seems to be sitting there and doesn't seem to be working at all, you can still get information out of it to help you figure out what the problem is. On the other end of the spectrum, if the application works "mostly okay" you should still be able to isolate where the errors are by understanding what it is actually doing. This process of looking at a failing application and trying to detail what its environment is along with its response is known as *characterization* and is the critical aspect of performing failure analysis. In this section I'll introduce you to the concept of error characterization and discuss some attitudes you should adopt when collecting the characterization data.

When I have any application that isn't running properly, the first thing that I do is to carry out a scan around the circuit to look at all the input and output voltage levels. I normally use a logic probe and I record all of my results on a sheet of paper. Digital logic levels can be checked with a logic probe, but not voltage levels. When you first work with your applications, you may want to check the signal pins with both a logic probe and a voltmeter until you get a good idea of what is happening and what you expect to see at the inputs and outputs of each chip.

When checking a circuit's clock or oscillator, check all pins with a logic probe or an oscilloscope probe to make sure the capacitance of the probe doesn't inadvertently set the oscillator running. I have seen several clock circuits where a missing capacitor was compensated for when a probe was connected to it.

Next, try to follow the signal flow through the circuit and try to record it at each stage.

Remember; you are not looking for where the data or signals go wrong. In the characterization step you are collecting data on what is happening in the circuit, not making conclusions on what is right and wrong. This is important because while the data or signal may be invalid at a certain point, the cause is not necessarily the chip that it is passing through. The root cause of the problem could be another input to the chip, which is causing the current signal's output to be incorrectly changed. The purpose of the characterization step is to get all the pertinent data regarding the failure so a hypothesis on what is the root cause can be made with a high degree of accuracy.

I call the state of mind that you need for characterizing the failure data "totally naïve." You are not expecting any results and you are not going to base any decisions on the observations.

This is actually very difficult for humans to do, and I know that I will often want to break off and look at what I think is the problem. I have only met one individual in my life whom I would consider "totally naïve" when looking for problems, and he is always the person who is able to find the true root cause of a problem.

When you have looked through the circuit and recorded your observations as to the behavior of the application, compare your notes with what the expected values are. This is an important point because I can't tell you how often I've thought I've seen a problem only to discover I miscounted and was looking at the wrong pin. This step is not fixing the problem; it is just making sure that you have not made any mistakes that will send you in the wrong direction in trying to figure out what the problem actually is.

Once I was giving a course on failure analysis to some technicians at work and presented the concept of being totally naïve when debugging circuits. The class and I were amused when one of the technicians shared with us the perspective he used when debugging defective boards. The technician said that he worked like that all the time; only he did it from the perspective of checking over his boss's work and hoping to find something he could hold over him. The more obvious a problem, the better because it made him feel better about what he would find. Thinking about it, I have never met anyone who was so happy to be looking for problems.

The important aspect of the characterization process is to document your problems. When I'm setting up a failure analysis process for a new product at work, I issue notebooks to all the technicians and instruct them in how to fill them out. This notebook is to be kept forever, as it is a database of problems and what was done to find them and fix them. This is true even if the failures are recorded on a computer database.

In cases where the application starts up and then fails, keeping notes detailing what has happened is critical. For example, if you had an RS-232 interface that worked for a bit and then locked up, you would have to figure out if the problem was based on time, data value, or data volume (the number of incoming characters). This can be found be creating small experiments to help characterize the problem. Once the determination is made on how the locking up was precipitated, this is recorded as one of the characteristics of the problem.

The information that should be recorded includes:

- Date
- Board serial number
- Technician
- Test failure
  - Input state
  - Display output
  - Output state
- Logic output
- V_{cc} level
- Clock speeds
- Reset state
- Individual pin voltages
- Individual pin input waveforms
- Individual pin output waveforms
- Chip temperature
- Result of tapping chip
- Visual anomalies
- And so on

Over time, you will learn the important parameters to track, those that work best for your style of thinking to clearly identify what the problem consists of.

In trying to find a problem, don't be afraid of creating a small application to try to replicate the problem. If you can come up with another way of demonstrating the problem, then you have correctly characterized it, and the additional data can be used to help hypothesize what the problem is.

Hypothesizing and Testing Your Hypothesis

Once you have your problem characterized, you should be able to make a hypothesis on what the cause of the problem is. Chances are the characterization data you collected should point you toward the root cause of the problem.

For example, if the problem in the application appeared to be an input pin that should have been pulled down in a TTL chip but wasn't, the characterization data should point you to the output where the circuit appears to be in error. From this point, you can work your way back to the inputs and look through the characterization data for the actual problem.

Before going on and trying to fix the application using the hypothesis, you should try to test it out, either as a "thought experiment" or on actual hardware to see if you can reproduce the problem and have it behave exactly as on another board.

A major part of making your hypothesis is deciding what the problem actually is. I am saying this because it is not unusual to have multiple problems with an application. In complex circuits I have designed, it is not unusual to have five or more problems. Often these problems will interact, and I will have to try and separate the characterization data for each of them in order to be able to hypothesize about what are the different problems.

When making your hypothesis, you should first list all the observations that seem to be pertinent to the problem at hand. As part of the hypothesis, you may have to make additional observations. For example, if you had a wire that was broken internally (and the break could not be observed by visually inspecting the wire), you may have to go back and check the input and output levels of the net as well as whether or not there is continuity in the wire.

When you are listing the observations that seem relevant to the "current" problem, you will probably be leaving other problems active. Don't worry, you will approach these problems after the first one is fixed.

Before going on to make a repair based on your hypothesis, you should check to see whether your hypothesis encompasses all the deviations from expected readings and behaviors you found in the characterization step. If your hypothesis misses any irregularities, chances are it is wrong or there is another problem you will have to work through. Before going on, you should improve your hypothesis to include the irregularities or you should make a basic hypothesis about the second problem.

There is one point that I would like to make about modern electronics and TTL chips in particular. When you are hypothesizing as to the cause of the problem, the last thing you should probably assume is that a chip in your circuit is bad. Modern electronics are unbelievably reliable and over the course of a 30-year career it is entirely possible that you will never see a defective TTL chip.

The first TTL chips along with the manufacturing processes to build them became available in the 1960s. Since then, they have been continually redesigned to achieve better yields in manufacturing and be more robust in the field. While it is not impossible to find a failing chip, it is sufficiently rare that you should consider a problem with silicon to be one of the last avenues that you will check. I have mentioned this point because one hypothesis that will often fit the problems noted in the characterization step is that a chip is bad. This is the easy way out of the problem and it will probably be wrong.

If you go to the effort of replacing a chip simply because the input value at a "downstream" chip is invalid, then you will probably find that you will have not fixed the problem and you are just where you started.

Fixing the Problem and Verifying the Result

With a hypothesis of what is the problem with the application, you now have to plan the best way to fix the problem. In my mind, I always look for the easiest way to fix the problem. Often this will be the best way to fix the problem, but in the case of damaged components (and component sockets) the proper way will involve a lot of work. Chances are you will get the same performance and life as if you just soldered in a new component and went on from there instead of removing the component and replacing it.

When you go through the process I outline above, you will find that identifying the fix for virtually all problems is very simple and startlingly obvious when you try to decide how to fix it. This is not to say that it is the case for all problems, but if you go through this process of characterizing and hypothesizing about the problem, you can now talk to an expert about the problem without feeling like you are potentially wasting the expert's time. I know of very few people who will not spend a few minutes explaining what a problem is if the upfront work is done to characterize what is actually happening and some hypotheses about the problem have been put forward, tried, and not found to fix the problem. With this background, an expert can often tell you what the problem is or make suggestions on other things to look at to help you find the root cause of the problem.

Before making the correction to the application, you should try to figure out whether the fix will eliminate the problem and whether it will cause additional problems in the application. The second point is really the more important one—you never want to have to implement a fix if it means that some other aspect of the application will stop working.

To test the fix, I will usually just solder on test wires, bend pins, and generally do whatever it takes to reproduce the problem and make sure the fix doesn't affect any other aspects of the application. Generally this test is not very pretty, but it will give me confidence that what I am doing is correct—especially when I have finished the repair and find that a new problem has appeared or, worse, that the original problem was not fixed.

If your test fix reveals a new problem with the board, you have one of two choices to make. The first is to make the repair as planned and go back and work through the new problem using the characterization, hypothesis, and repair steps I have outlined in this chapter. The second choice is to go back and make a new hypothesis about the problem using this data as additional characterization information. Of the two choices, the second is much more preferable.

The first action, to repair the initial problem and then fix the new problem, is suboptimal because it makes the assumption that the two problems are not related. What you will probably find when you look at the second problem is that it was a result of the repair to the first problem. In going through the characterization, hypothesis, and repair of the new problem, you will probably find that it will cause a *third* problem and so on until you find that you will have ended up redesigning a significant part (or all) of the application. Another possible outcome will be that the first repair did not completely fix the problem, and you will find that the second problem is actually a subset of the first one. In either case, you will find that you will have to carry out additional repair actions and that if you had gone back and looked at the first problem in a different way, you would have been able to come up with a repair action that satisfied the first problem and avoided subsequent problems altogether.

With the proper repair specified and an understanding of the part of the application where the problem is manifested, you should spend some time understanding what is the root cause of the problem. The problem will probably be something like a broken wire, a bad solder joint, or a component that wasn't properly seated in its socket. You should work at understanding what the actual problem was and think about taking steps so that it will never happen again.

For example, if you have a wire-wrapped circuit and the problem was caused by a broken wire, you should want to understand why the wire got broken, including checking your tools to see if a sharp edge is causing the wire to get nicked or cut. If the problem was caused by a bad solder joint, then you should probably practice your soldering until you feel that you can solder reliably without any more problems like this one. The whole goal of this failure analysis process is to ensure you understand what was the actual cause of the failure is and take measures to prevent it from ever happening again.

There is one class of problem that I haven't mentioned and one that you will have to spend a lot of time working through, and that is what happens if you discover that the problem is design related. In this case, you will have the information necessary to understand what the problem is with the design and where it fails. It ultimately will result in you going "back to the drawing board," but at least you will have an understanding of what is wrong with the original circuit and what needs to be done for it to properly operate.

The debugging (or *failure analysis*) process outlined in this chapter is not one that you will master the first time you attempt to work through it. But, as you become proficient at it, you will have gained the following skills:

- Developing a process of characterizing applications so that you can record data easily and work through it efficiently
- Developing hypotheses based on the characterization data and effectively questioning whether the hypothesis takes into account all the irregularities presented in the characterization data
- Specifying a repair action that does not cause any new problems and that you can test before the repair is made permanent

Acquiring these skills takes time, but they will make you a lot better at solving not only electronics problems, but will give you an advantage when you are trying to fix problems in all areas of your life.

Appendix

Common Digital Electronics Parts

In this book, I have used a fairly large number of different chips. I have tried to use only parts that are commonly available. In the table below and the diagrams that follow, I have specified the various chips that are presented in this book along with their pin-outs and CMOS part number equivalencies. The actual list is in no way complete; there are literally hundreds of other standard parts that you can choose from, but I wanted to make sure I included the pin-out diagrams for the parts that I used in this book.

The following table cross-references bipolar logic parts to CMOS and provides pointers to the pin-out drawings (if available). In this table, I have listed the different chips by bipolar part number. I have indicated whether there is a CMOS (40xx series) part that performs the same function and what it is. Included in the table is a brief description of each part and any differentiating points relative to other parts. Note that I have not specified whether C, HC, HCT, ALS, AS, L, LS, etc. parts are available. Chances are, there will be a manufacturer that builds the part you are looking for in the technology you need for your application.

Each one of these chips should have a decoupling capacitor as close to the V_{cc} pin as possible. These chips can be used with sockets that have built-in decoupling caps.

The chip operation is very straightforward, although in a number of cases, I have added instructions and comments on specific aspects and features of the chips. The operation of these chips with the PICmicro can be seen in the projects. More complete chip datasheets can be found at National Semiconductor's Web site, http://www.national.com or Philips' Web site, http://www.semiconductors.philips.com/.

The abbreviations used in the table are:

G/P: general purpose

O/C: open collector

Part numbers			
Bipolar	CMOS	Pin-out	Comments
324			Quad op-amp
555	555	See Fig. A.1.	8-pin G/P timer
556			14-pin dual G/P timer
741			G/P op-amp
	4060	See Fig. A.2.	CMOS oscillator and ripple counter
7400	4011	See Fig. A.3.	Quad two-input NAND
7402	4001	See Fig. A.4.	Quad two-input NOR
7404	4049	See Fig. A.5.	Hex inverter
7407	4050		Hex O/C buffer
7408	4081	See Fig. A.6.	Quad two-input AND
7432	4071	See Fig. A.7.	Quad two-input OR
7447	4055		Seven-segment display driver
7474	4013	See Fig. A.8.	Dual D flip-flops
7476	4027		Dual J-K flip-flops
7485	4063		4-bit magnitude comparators
7486	4030	See Fig. A.9.	Quad two-input XOR
74123	4098		Dual single-shots
74125	404502	See Fig. A.10.	Quad tristate driver
74138		See Fig. A.11.	3-to-8 decoder
74139		See Fig. A.12.	Dual 2-to-4 decoder
74151	4097		1-of-8 multiplexer
74153	4052		Dual 1-of-4 multiplexer
74166	4014		4-bit parallel/serial converter
74174		See Fig. A.13.	6-bit D flip-flop
74192	40192		4-bit BCD counter
74193	40193		4-bit hex counter
74244		See Fig. A.14.	8-bit tristate driver
74245		See Fig. A.15.	8-bit bidirectional tristate driver
74283		See Fig. A.16.	4-bit full adder
74373		See Fig. A.17.	Octal latch with data pass-through
74374		See Fig. A.18.	Octal latch
74573		See Fig. A.19.	Octal latch with data pass-through
74574		See Fig. A.20.	Octal latch

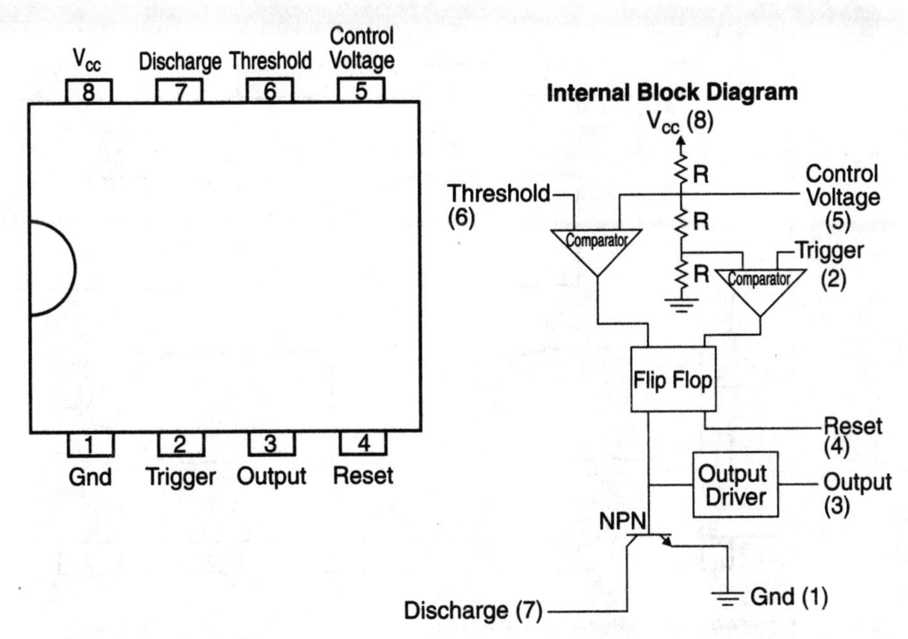

Figure A.1 555 timer chip.

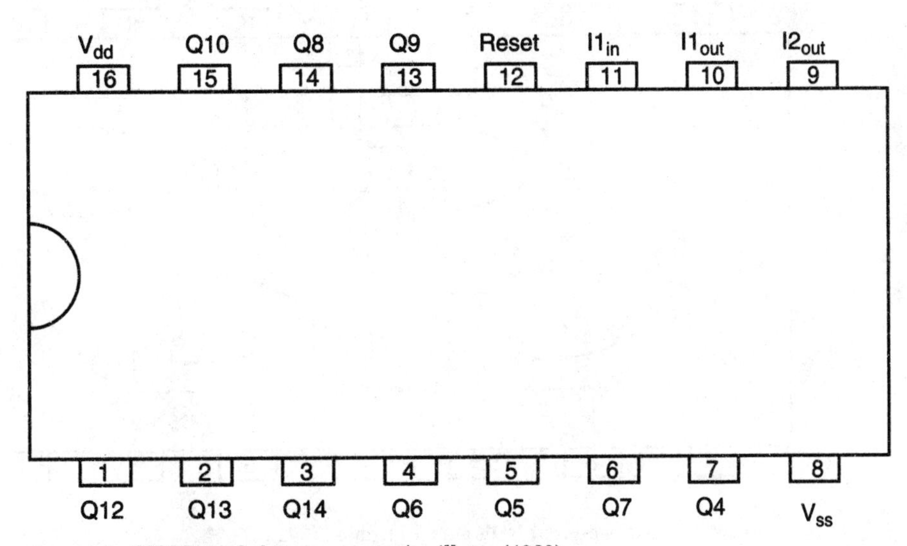

Figure A.2 CMOS ripple binary counter/oscillator (4060).

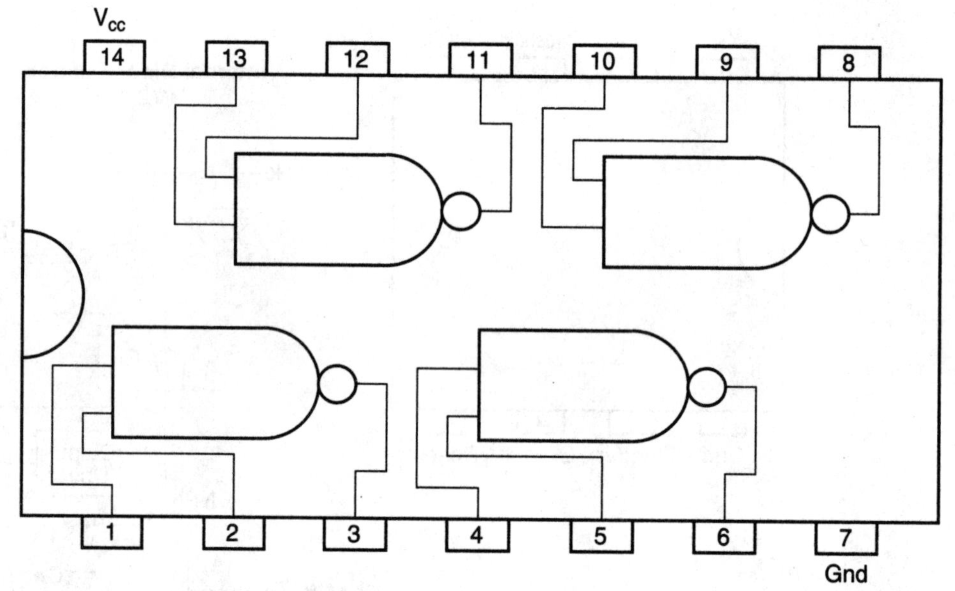

Figure A.3 Quad two-input NAND gate TTL chip (7400).

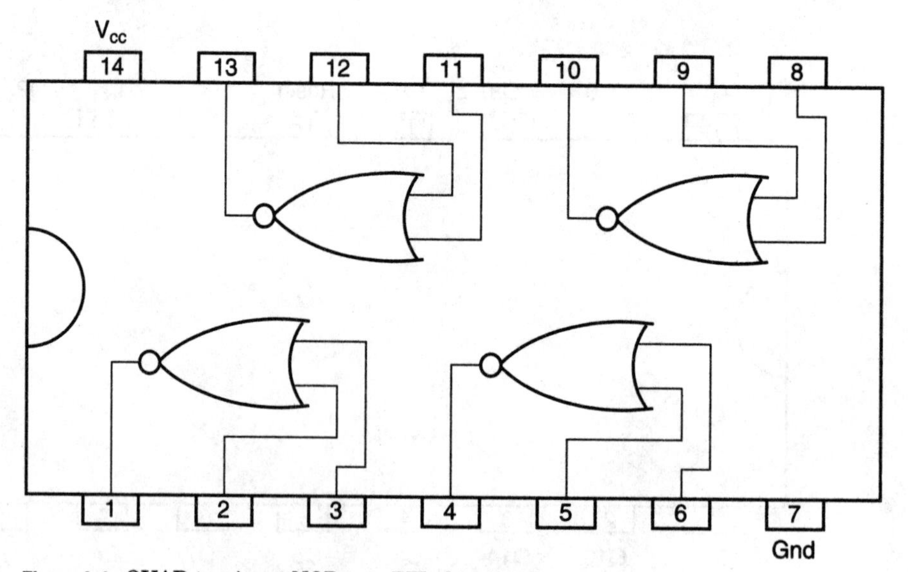

Figure A.4 QUAD two-input NOR gate TTL chip (7402).

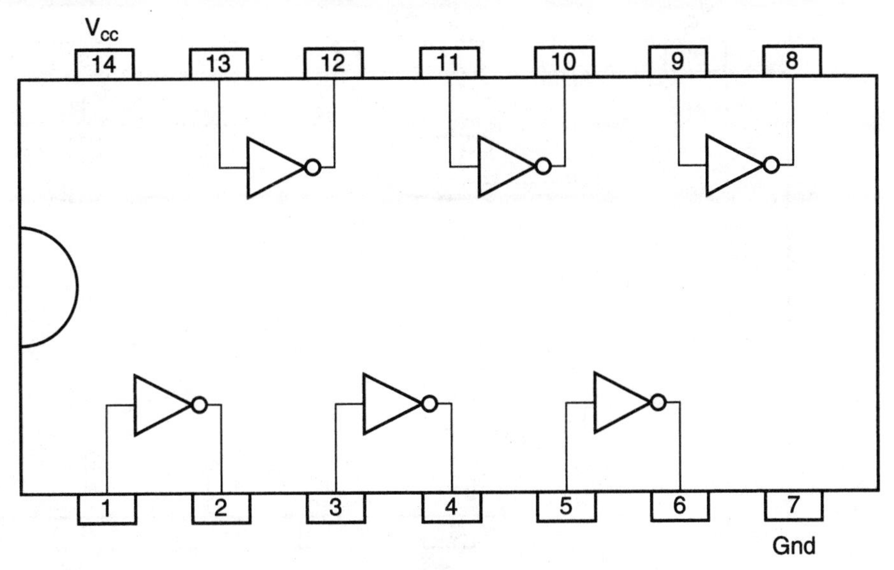

Figure A.5 Hex inverters with totem pole outputs (7404).

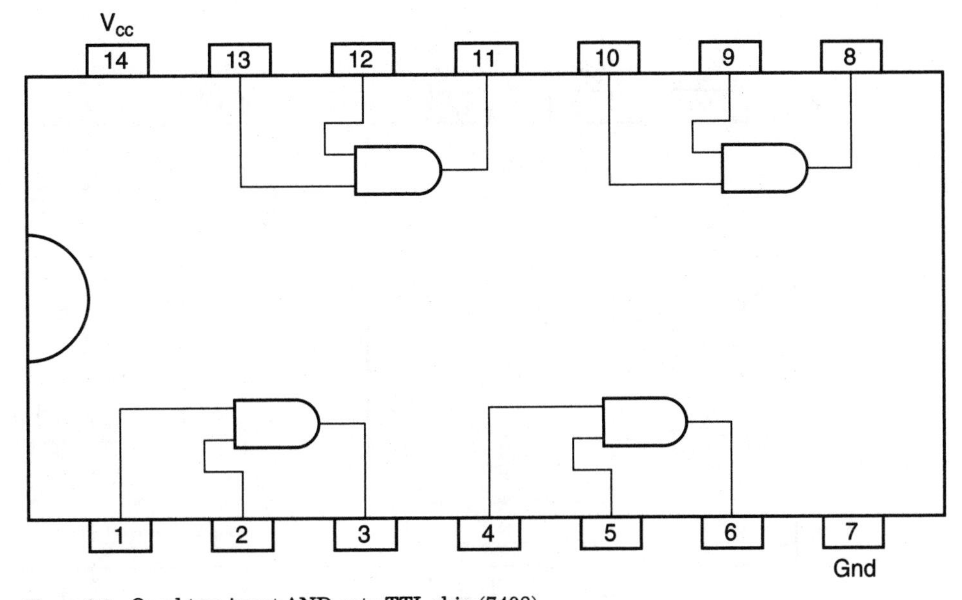

Figure A.6 Quad two-input AND gate TTL chip (7408).

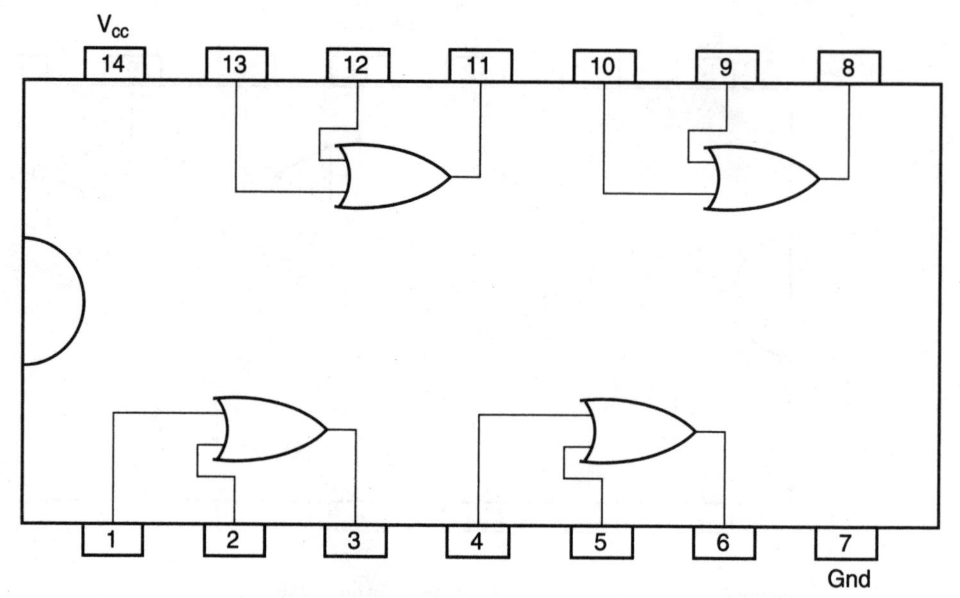

Figure A.7 Quad two-input OR gate TTL chip (7432).

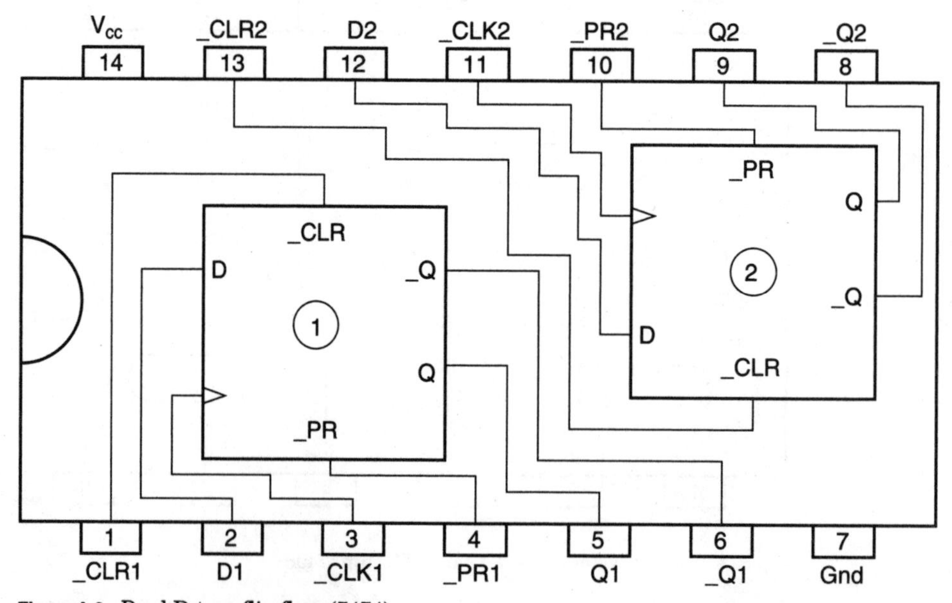

Figure A.8 Dual D-type flip-flops (7474).

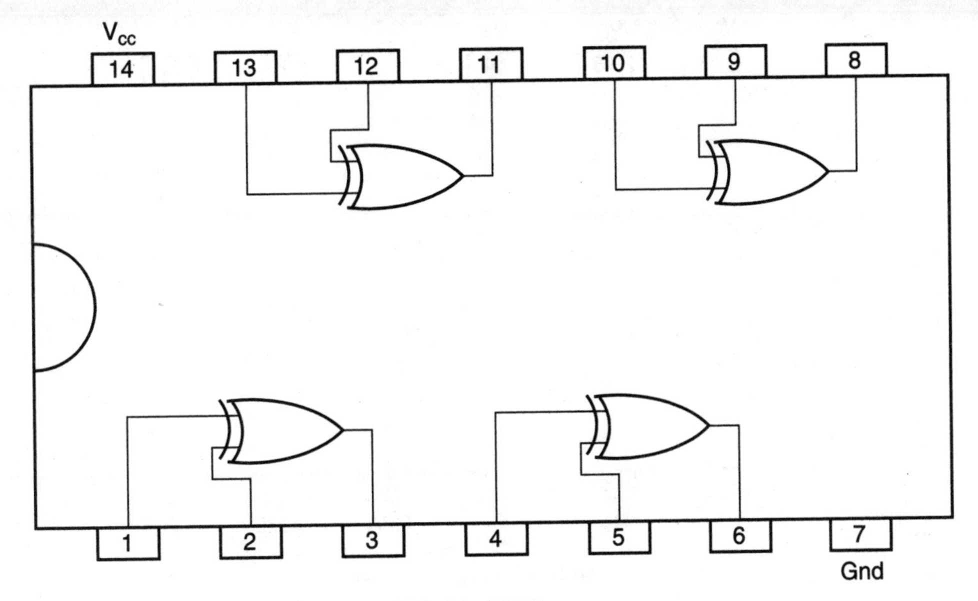

Figure A.9 Quad two-input XOR gate TTL chip (7486).

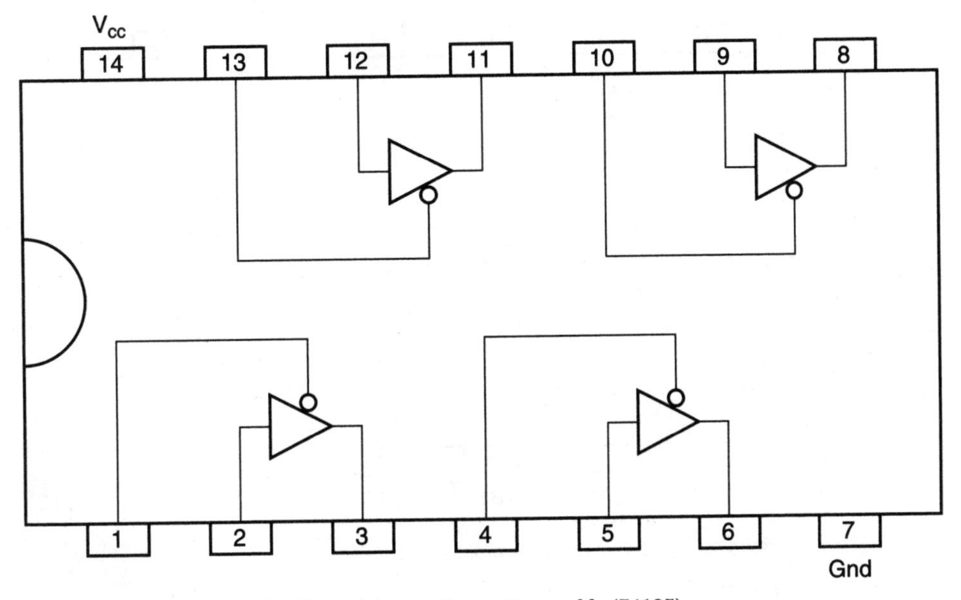

Figure A.10 Quad tristate buffers with negative active enable (74125).

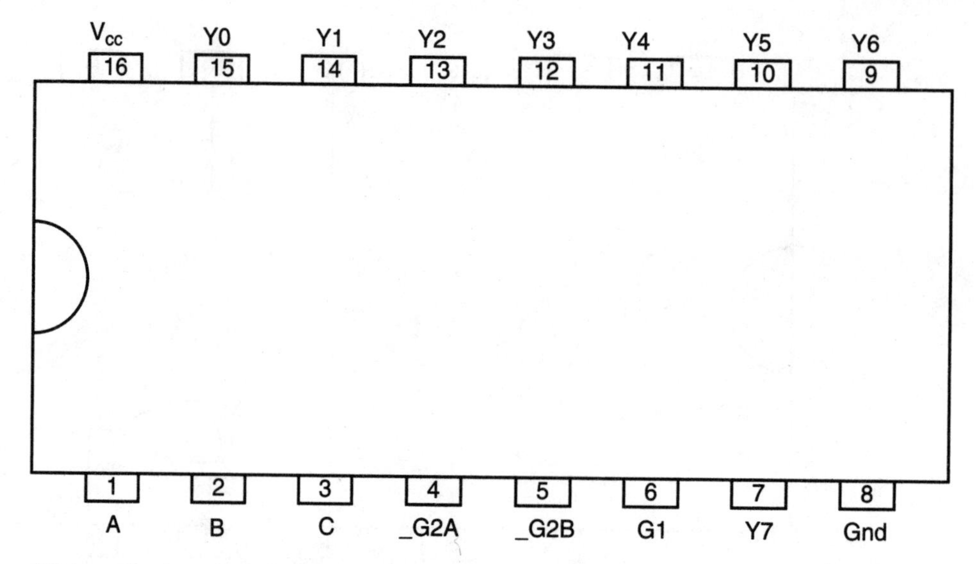

Y# is active low when G1/G2 are set correctly.

= (C x 4) + (B x 2) + (A x 1)

Figure A.11 Three-to-eight decoder (74138).

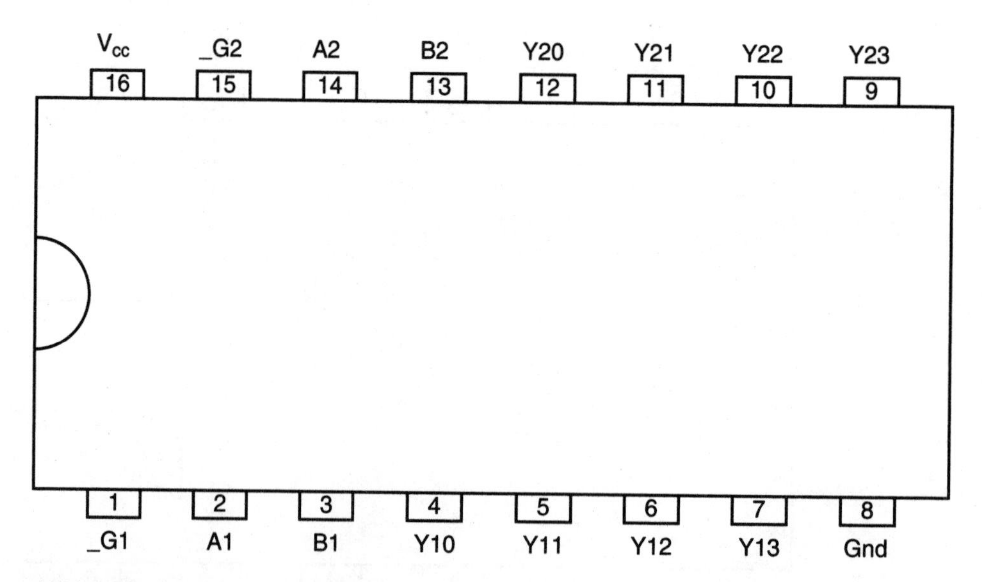

Y# is active low when _G# pulled low.

= (B x 2) + (A x 1)

Figure A.12 Dual two-to-four decoder (74139).

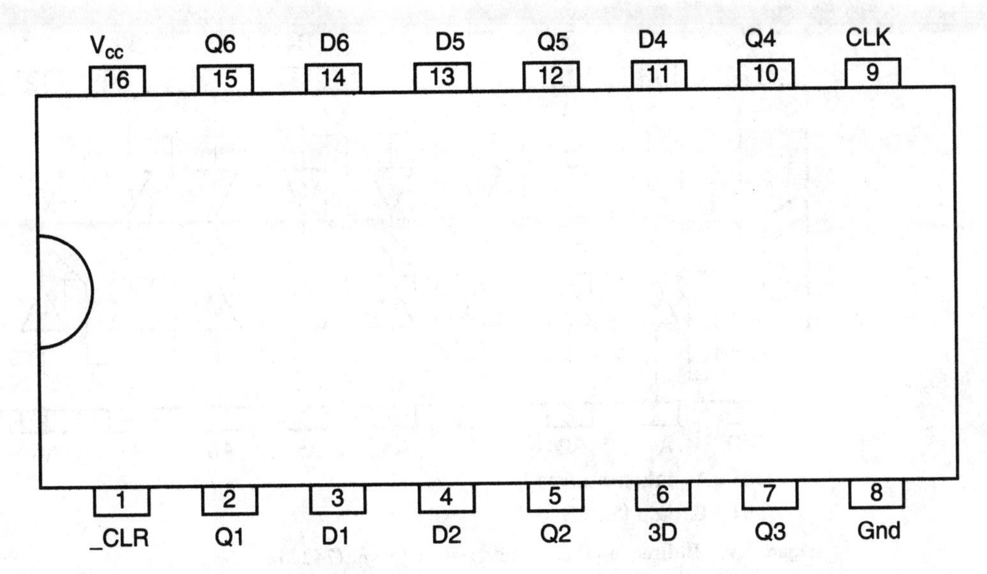

Figure A.13 Hex D-type flip-flops (74174).

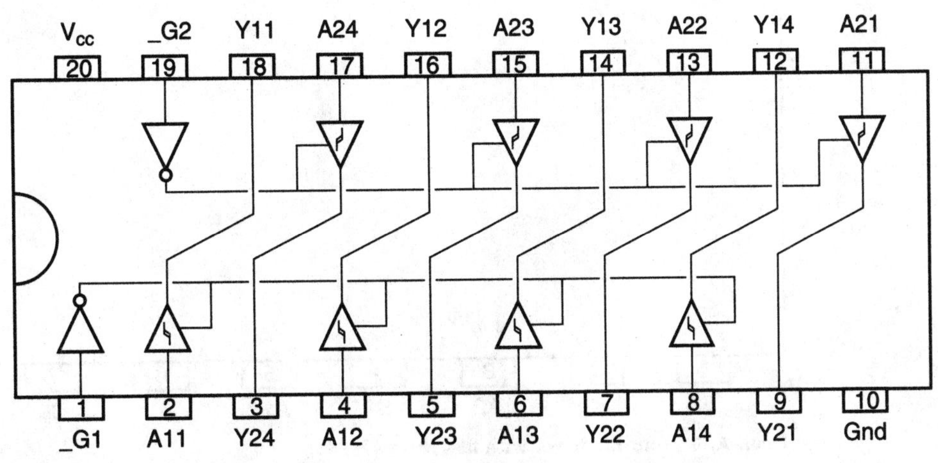

Figure A.14 Eight-bit tristate driver (74244).

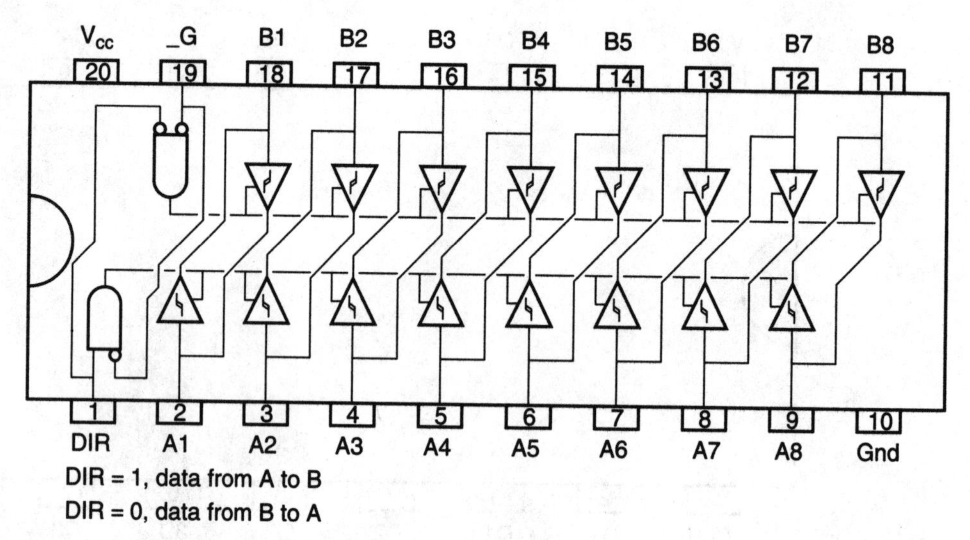

Figure A.15 Bidirectional eight-bit tristate driver (74245).

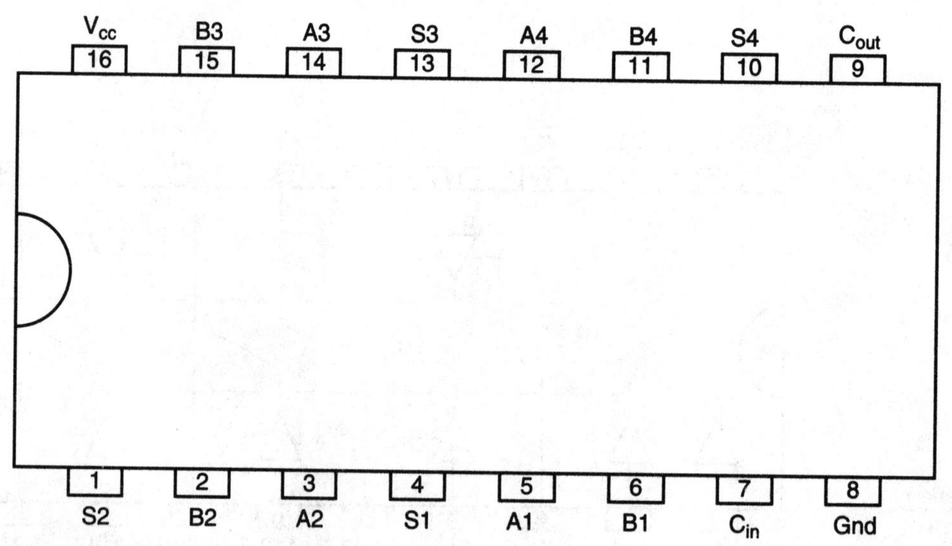

Figure A.16 Four-bit driver with fast carry (74283).

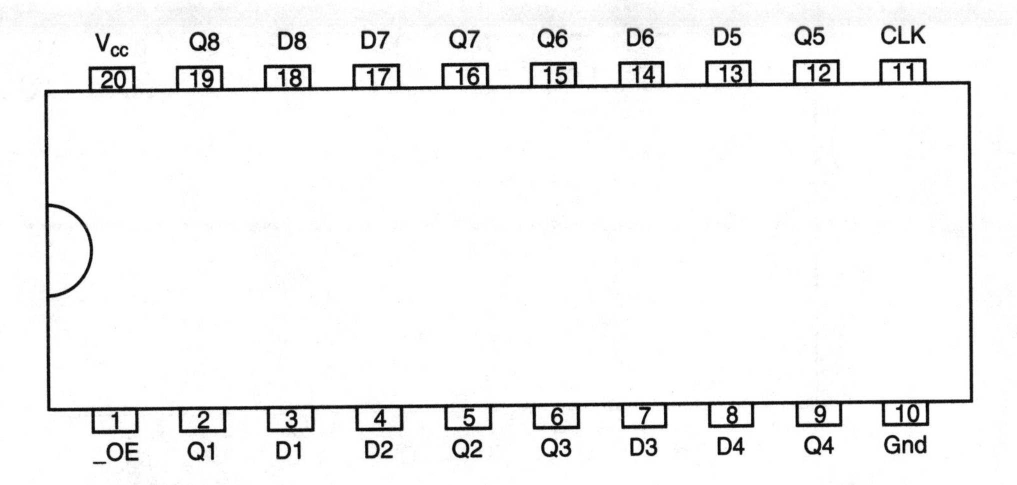

Figure A.17 Eight-bit latch with tristate output driver (74373).

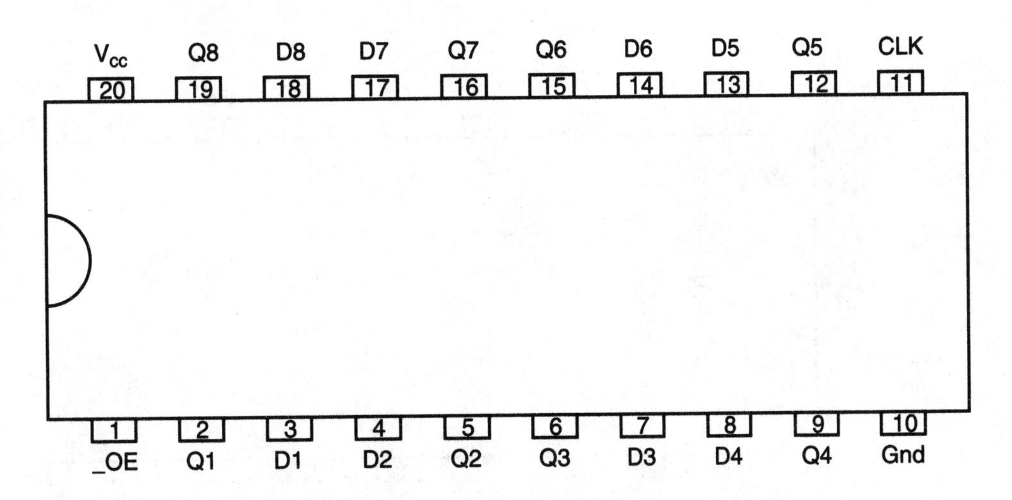

Figure A.18 Eight-bit latch with tristate output driver (74374).

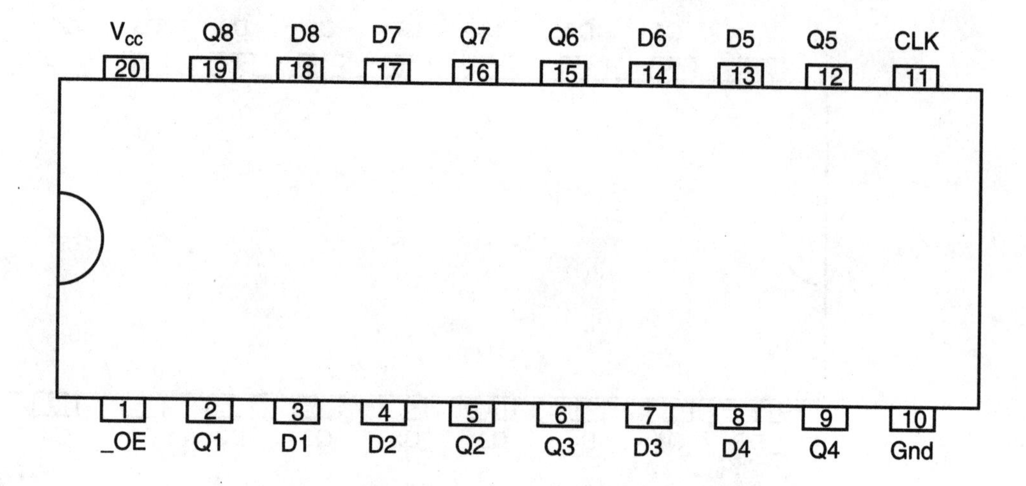

Latches are transparent when CLK is high.
Data latched in on negative edge of CLK.

Figure A.19 Eight-bit latch with tristate output driver (74573).

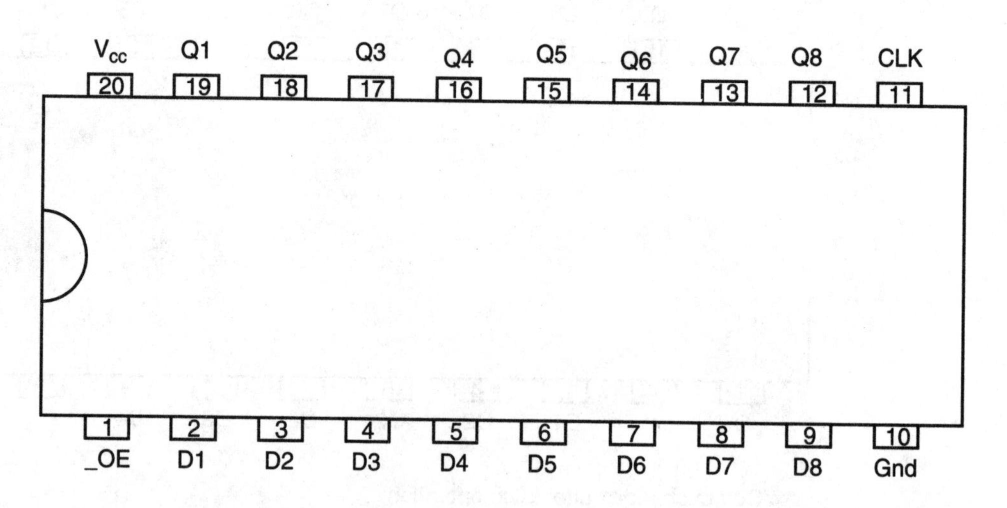

Output changes after latch operation.
Data latched in on positive edge of CLK.

Figure A.20 Eight-bit latch with tristate output driver (74574).

Appendix

B

Useful Tables and Data

Physical Constants

Symbol	Value	Description
AU	149.59787×10^6 km, 92,955,628 mi	Astronomical unit (distance from the sun to the earth)
c	2.99792458×10^8 m/s, 186,282 mi/s	Speed of light in a vacuum
e	2.7182818285	Natural logarithm constant
ε_0	$8.854187817 \times 10^{-12}$ F/m	Permittivity of free space
ev	$1.60217733 \times 10^{-19}$ J	Electron-volt value
g	32.174 ft/s^2, 9.807 m/s^2	Acceleration due to gravity
G	6.67259×10^{-11} m^3/(kg · s^2)	
h	6.626×10^{-34} J · s	Planck's constant
k	1.380658×10^{-23} J/K	Boltzmann entropy constant
m_e	$9.1093897 \times 10^{-31}$ kg	Electron rest mass
m_n	1.67493×10^{-27} kg	Neutron rest mass
m_p	1.67263×10^{-27} kg	Proton rest mass
p_c	2.06246×10^5 AU	Parsec
pi, π	3.1415926535898	Ratio of circumference to diameter of a circle
R	8.314510 J/(K · mol)	Gas constant
sigma, σ	5.67051×10^{-8} W/(m^2 · K^4)	Stefan-Boltzmann constant
mu, μ	1.66054×10^{-27} gram	Atomic mass unit
mu-0, μ_0	1.25664×10^{-7} N/A^2	Permeability of vacuum
Mach 1	331.45 m/s, 1087.4 ft/s	Speed of sound at sea level in dry air at 20°C
Speed of sound	1480 m/s, 4856 ft/s	In Water at 20°C

Physical Formulas

Circumference of a circle = $2r\pi$

Area of a circle = πr^2

Volume of a sphere = $\frac{4}{3} \pi r^3$

Surface area of a sphere = $4 \pi r^2$

Area of a triangle = $\frac{1}{2}hw$

Right angle triangle law: sum of the squares of the sides = square of the hypotenuse

Cosine (angle) = l/hypotenuse

Sine (angle) = h/hypotenuse

Tangent (angle) = h/l

Position = $\frac{1}{2} at^2 + vt + P_0$

Velocity = $at + v$

Rotational acceleration = v^2/r

Acceleration due to gravity = Gm_1m_2/r^2

Frequency = speed/wavelength

For electromagnetic waves: frequency = c/wavelength

Perfect gas law: $PV = nRT$

Audio Notes

Notes around middle C (note that octave above is twice the note frequency and octave below is one-half note frequency).

Note	Frequency, Hz
G	392
G#	415.3
A	440
A#	466.2
B	493.9
C	523.3
C#	554.4
D	587.3
D#	622.3
E	659.3
F	698.5
F#	740.0
G	784.0
G#	830.6
A	880.0
A#	932.3
B	987.8

Electrical Engineering Formulas

V = voltage

I = current

R = resistance

C = capacitance

L = inductance

DC Electronics Formulas

Ohm's law: $V = IR$

Power: $P = VI$

Series resistance: $R_t = \text{R1} + \text{R2} + \cdots$

Parallel resistance: $R_t = \dfrac{1}{1/\text{R1} + 1/\text{R2} + \cdots}$

Two resistors in parallel: $R_t = \dfrac{\text{R1} * \text{R2}}{\text{R1} + \text{R2}}$

Series capacitance: $C_t = \dfrac{1}{1/\text{C1} + 1/\text{C2} + \cdots}$

Parallel capacitance: $C_t = \text{C1} + \text{C2} + \cdots$

Wheatstone bridge (Fig. B.1):

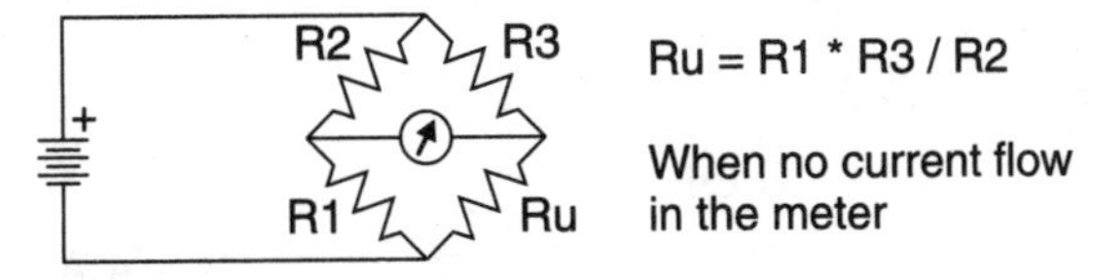

Figure B.1 Wheatstone bridge operation.

AC Electronics Formulas

Resonance: frequency $= \dfrac{1}{2\pi\sqrt{LC}}$

RC time constant: $\tau = RC$

RL time constant: $\tau = L/R$

RC charging: $V(t) = V_f(1 - e^{-t/\tau})$

$i(t) = i_f(1 - e^{-t/\tau})$

RC discharging: $V(t) = V_i e^{-t/\tau}$

$i(t) = i_i e^{-t/\tau}$

Transformer current/voltage: Turns ratio = number of turns on primary (p) side/number of turns on secondary (s) side

Turns ratio $= V_s/V_p = I_p/I_s$

Transmission line characteristic impedance: $z_0 = \sqrt{L/C}$

Mathematical Formulas

Frequency = speed/wavelength

For electromagnetic waves: Frequency = c/wavelength

Perfect gas law: $PV = nRT$

Boolean Arithmetic

Identity functions: A AND 1 = A

A OR 0 = A

Output set/reset: A AND 0 = 0

A OR 1 = 1

Identity law: $A = A$

Double negation law: NOT (NOT (A)) = A

Complementary law: A AND NOT (A) = 0

A OR NOT (A) = 1

Idempotent law: A AND A = A

A OR A = A

Commutative law: A AND B = B AND A

A OR B = B OR A

Associative law: (A AND B) AND C = A AND (B AND C) = A AND B AND C

(A OR B) OR C = A OR (B OR C) = A OR B OR C

Distributive law: A AND (B OR C) = (A AND B) OR (A AND C)

A OR (B AND C) = (A OR B) AND (A OR C)

De Morgan's theorem: NOT (A OR B) = NOT (A) AND NOT (B)

NOT (A AND B) = NOT (A) OR NOT (B)

Note:

AND is often represented as multiplication, nothing between the terms, or the "." or "*" characters between them.

OR is not represented as addition with "+" between terms.

NOT is indicated with a "–" or "!" character before the term. "~" is usually used to indicate a multibit, bitwise inversion.

Gray Code Tables

The Gray codes presented below are useful in developing truth tables for your boolean logic functions:

2-Bit Gray Code

B_0 B_1	Hex
0 0	00
0 1	01
1 1	03
1 0	02

3-Bit Gray Code

B_0 B_1 B_2	Hex
0 0 0	00
0 0 1	01
0 1 1	03
0 1 0	02
1 1 0	06
1 1 1	07
1 0 1	05
1 0 0	04

4-Bit Gray Code

B_0 B_1 B_2 B_3	Hex
0 0 0 0	00
0 0 0 1	01
0 0 1 1	03
0 0 1 0	02
0 1 1 0	06
0 1 1 1	07
0 1 0 1	05
0 1 0 0	04
1 1 0 0	0C
1 1 0 1	0D
1 1 1 1	0F
1 1 1 0	0E
1 0 1 0	0A
1 0 1 1	0B
1 0 0 1	09
1 0 0 0	08

5-Bit Gray Code

B_0	B_1	B_2	B_3	B_4	Hex
0	0	0	0	0	00
0	0	0	0	1	01
0	0	0	1	1	03
0	0	0	1	0	02
0	0	1	1	0	06
0	0	1	1	1	07
0	0	1	0	1	05
0	0	1	0	0	04
0	1	1	0	0	0C
0	1	1	0	1	0D
0	1	1	1	1	0F
0	1	1	1	0	0E
0	1	0	1	0	0A
0	1	0	1	1	0B
0	1	0	0	1	09
0	1	0	0	0	08
1	1	0	0	0	18
1	1	0	1	0	1A
1	1	0	1	1	1B
1	1	0	0	1	19
1	1	1	0	1	1D
1	1	1	1	1	1F
1	1	1	1	0	1E
1	1	1	0	0	1C
1	0	1	0	0	14
1	0	1	0	1	15
1	0	1	1	1	17
1	0	1	1	0	16
1	0	0	1	0	12
1	0	0	1	1	13
1	0	0	0	1	11
1	0	0	0	0	10

6-Bit Gray Code

B_0	B_1	B_2	B_3	B_4	B_5	Hex
0	0	0	0	0	0	00
0	0	0	0	0	1	01
0	0	0	0	1	1	03
0	0	0	0	1	0	02
0	0	0	1	1	0	06
0	0	0	1	1	1	07
0	0	0	1	0	1	05
0	0	0	1	0	0	04
0	0	1	1	0	0	0C
0	0	1	1	0	1	0D
0	0	1	1	1	1	0F
0	0	1	1	1	0	0E
0	0	1	0	1	0	0A
0	0	1	0	1	1	0B
0	0	1	0	0	1	09
0	0	1	0	0	0	08
0	1	1	0	0	0	18
0	1	1	0	1	0	1A
0	1	1	0	1	1	1B
0	1	1	0	0	1	19
0	1	1	1	0	1	1D
0	1	1	1	1	1	1F
0	1	1	1	1	0	1E
0	1	1	1	0	0	1C
0	1	0	1	0	0	14
0	1	0	1	0	1	15
0	1	0	1	1	1	17
0	1	0	1	1	0	16
0	1	0	0	1	0	12
0	1	0	0	1	1	13
0	1	0	0	0	1	11
0	1	0	0	0	0	10

6-Bit Gray Code *(Continued)*

B_0	B_1	B_2	B_3	B_4	B_5	Hex
1	1	0	0	0	0	30
1	1	0	0	0	1	31
1	1	0	0	1	1	33
1	1	0	0	1	0	32
1	1	0	1	1	0	36
1	1	0	1	1	1	37
1	1	0	1	0	1	35
1	1	0	1	0	0	34
1	1	1	1	0	0	3C
1	1	1	1	1	0	3E
1	1	1	1	1	1	3F
1	1	1	1	0	1	3D
1	1	1	0	0	1	39
1	1	1	0	1	1	3B
1	1	1	0	1	0	3A
1	1	1	0	0	0	38
1	0	1	0	0	0	28
1	0	1	1	0	0	2C
1	0	1	1	0	1	2D
1	0	1	0	0	1	29
1	0	1	0	1	1	2B
1	0	1	1	1	1	2F
1	0	1	1	1	0	2E
1	0	1	0	1	0	2A
1	0	0	0	1	0	22
1	0	0	0	1	1	23
1	0	0	0	0	1	21
1	0	0	1	0	1	25
1	0	0	1	1	1	27
1	0	0	1	1	0	26
1	0	0	1	0	0	24
1	0	0	0	0	0	20

Mathematical conversions

1 inch = 2.54 centimeters

1 mile = 1.609 kilometers

1 ounce = 29.57 grams

1 U.S. gallon = 3.78 liters

1 atmosphere = 29.9213 inches of mercury

= 14.6960 pounds per square inch

= 101.325 kilopascals

10,000,000,000 angstroms = 1 meter

1,000,000 micrometers = 1 meter

Tera = 1000 giga

Giga = 1000 mega

Mega = 1000 kilo

Kilo = 1000 units

Unit = 100 centi

Unit = 1000 milli

1 hour = 3600 seconds

1 year = 8760 hours

ASCII

The ASCII Definition uses the 7 bits of Each ASCII Character.

3–0	6–4 →	000	001	010	011	100	101	110	111
V	Control			Characters					
0000		NUL	DLE	Space	0	@	P	`	p
0001		SOH	DC1	!	1	A	Q	a	q
0010		STX	DC2	”	2	B	R	b	r
0011		ETX	DC3	#	3	C	S	c	s
0100		EOT	DC4	$	4	D	T	d	t
0101		ENQ	NAK	%	5	E	U	e	u
0110		ACK	SYN	&	6	F	V	f	v
0111		BEL	ETB	`	7	G	W	g	w
1000		BS	CAN	(	8	H	X	h	x
1001		HT	EM	)	9	I	Y	i	y
1010		LF	SUB	*	:	J	Z	j	z
1011		VT	ESC	+	;	K	[	k	{
1100		FF	FS	‘	<	L	\	l	|
1101		CR	GS	-	=	M	]	m	}
1110		SO	RS	.	>	N	^	n	~
1111		SI	US	/	?	O	_	o	DEL

ASCII Control Characters

The ASCII control characters were specified as a means of allowing one computer to communicate with and control another. These characters are actually commands, and if the BIOS or MS-DOS display or communications APIs are used with them, they will revert back to their original purpose. As I note below, when I present the IBM extended ASCII characters, writing these values (all less than 0x020) to the display will display graphics characters.

Normally, only carriage return/line feed are used to indicate the start of a line. Null is used to indicate the end of an ASCIIZ string. Backspace will move the cursor back one column to the start of the line. The bell character, when sent to an ASCII terminal, will cause the speaker to beep. Horizontal tab is used to move the cursor to the start of the next column, which is evenly distributed by eight. Form feed is used to clear the screen.

Hex	Mnemonic	Definition
00	NUL	Null—used to indicate the end of a string
01	SOH	Message: start of header
02	STX	Message: start of text
03	ETX	message: end of text
04	EOT	End of transmission
05	ENQ	Enquire for identification or information
06	ACK	Acknowledge the previous transmission
07	BEL	Ring the bell
08	BS	Backspace—move the cursor on column to the left
09	HT	Horizontal tab—move the cursor to the right to the next tab stop (normally a column evenly divisible by 8)
0A	LF	Line feed—move the cursor down one line
0B	VT	Vertical tab—move the cursor down to the next tab line
0C	FF	Form feed up to the start of the new page; for CRT displays, this is often used to clear the screen
0D	CR	Carriage return—move the cursor to the leftmost column
0E	SO	Next group of characters does not follow ASCII control conventions so they are shifted out
0F	SI	The following characters do follow the ASCII control conventions and are shifted in
10	DLE	Data link escape—ASCII control character start of an escape sequence; in most modern applications escape (0x01B) is used for this function
11	DC1	Not defined—normally application specific
12	DC2	Not defined—normally application specific
13	DC3	Not defined—normally application specific
14	DC4	Not defined—normally application specific

Hex	Mnemonic	Definition
15	NAK	Negative acknowledge—the previous transmission was not properly received
16	SYN	Synchronous idle—if the serial transmission uses a synchronous protocol, this character is sent to ensure the transmitter and receiver remain synched
17	ETB	End of transmission block
18	CAN	Cancel and disregard the previous transmission
19	EM	End of medium—indicates end of a file; for MS-DOS files, 0x01A is often used instead
1A	SUB	Substitute the following character with an incorrect one
1B	ESC	Escape—used to temporarily halt execution or put an application into a mode to receive information
1C	FS	Marker for file separation of data being sent
1D	GS	Marker for group separation of data being sent
1E	RS	Marker for record separation of data being sent
1F	US	Marker for unit separation of data being sent

Appendix

C

Project Assembly Techniques and Prototyping

While I have focused on using the breadboard prototyping system for the projects presented in this book, there are a number of projects that should be built in a more permanent fashion or by other techniques because they have advantages over the breadboard. Deciding on what is the appropriate method of building the circuits will have some pretty profound implications on the time required to build the application, its cost, and whether or not the circuit will work.

Most of the projects presented in this book are fairly low speed (frequency) circuits to maximize the reliability of the circuits. As you get into higher speeds, different techniques will alter how signals actually propagate in the circuit. When I've drawn logic waveforms, I have tended to show the signals as idealized waveforms. For example, for a clock signal, I use a diagram like the one in Fig. C.1.

If you were to look at an actual signal on a line, chances are you would see something like the oscilloscope picture in Fig. C.2. You can see that the actual edges are actually rounded and there is "ringing" in the middle of the signal. This rounding and ringing are due to the in-line resistance, capacitance, and inductance of the actual wiring between components, along with the

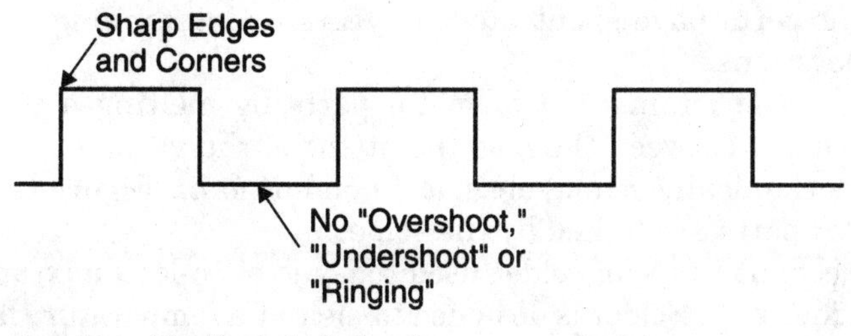

Figure C.1 Idealized clock signal.

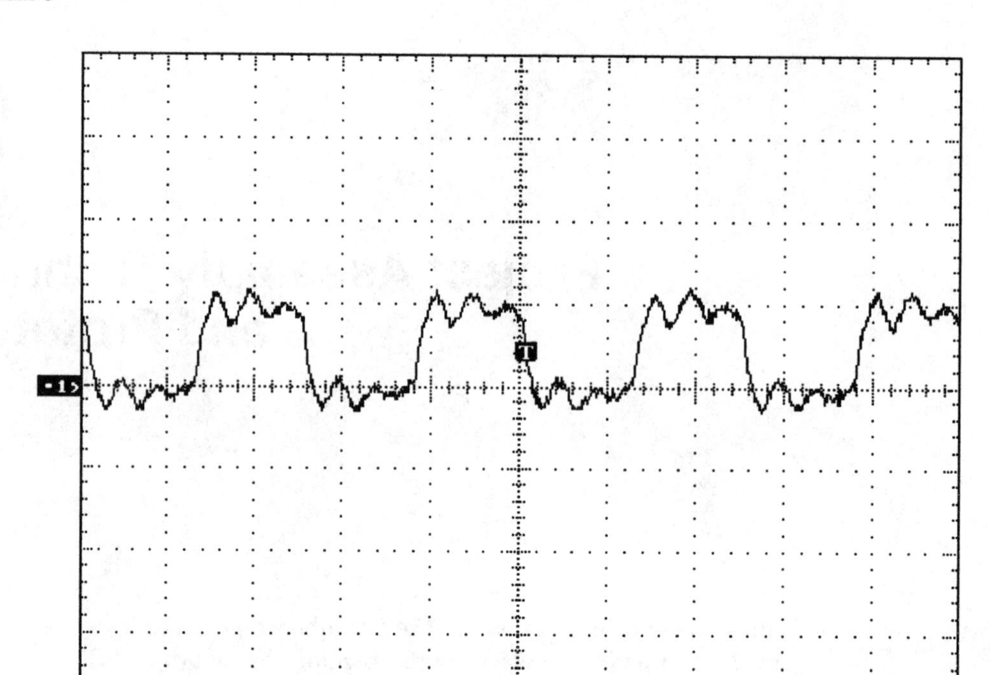

Figure C.2 Actual clock signal.

components' characteristics. To minimize these effects, the devices and wiring should all be *impedance matched*. For this book, I have slowed down the circuit speeds as much as possible so that the edge rounding and level rise time become insignificant and improve the chances that, when you build one of the projects presented here, it will work.

As I review the different methods of building circuits, I'll discuss these issues along with the out-of-pocket cost, reliability, and time spent with each method.

Soldering

Arguably, the most basic skill you will require for working with electronics is the ability to solder components, wires, and PCBs together. At work, the solder process experts have spent literally years understanding heat flows and chemical operations.

Soldeirng is the joining of two metal parts by melting a metal material (known as *solder*) between them so the atoms of the various metals combine, forming a mechanically strong electrical bond or *joint*. Figure C.3 shows how the two metal parts are linked by the solder.

The most common type of solder used today is a tin-lead mixture with silver optionally added to it. Solder is designed to melt at a temperature far below that of the parts (normally copper) that it connects. The lower melting temperature

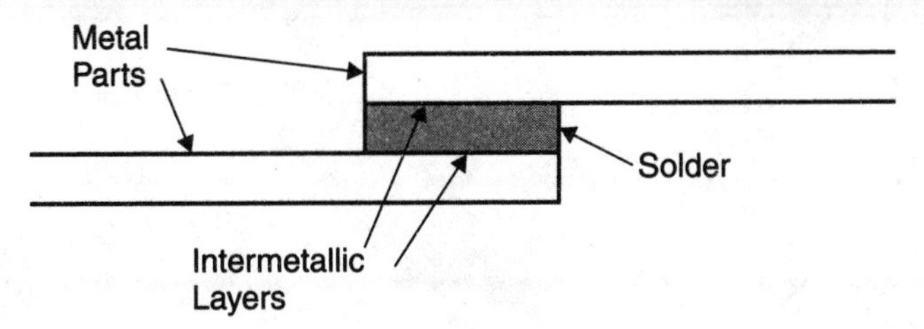

Figure C.3 Solder joint features.

of solder is to ensure that the metal parts to be joined are not deformed and any silicon devices connected to them are not damaged in the process.

When solder is applied to a metal, an *intermetallic layer* is formed, which is an alloy of the solder and the metal part. Ideally, this intermetallic layer should be as thin as possible to allow easy remelting of the solder for rework and for the best electrical characteristics.

With this in mind, I recommend that you buy a temperature-controlled soldering station, which can be purchased for as little as $50. Having a temperature-controlled station will give you the control to solder high-quality joints in a way that works best for you.

Along with the soldering station, you should have as small a tip as possible, a solder sponge, and a rosin-core solder. You may also want to buy a soldering-iron cleaner, which is a small metal tin with a gray sandy substance in it that will clean and re-tin your soldering iron tip. With these tools, you will be able to solder the circuits shown in this book as well as most other electronic circuits.

Plumbing soldering irons, while performing the same function, are not appropriate, as they provide too much heat. It may sound ridiculous, but I have been approached by several people asking for help when they've burned a hole through a PCB with an iron that worked fine on a 3-in copper drain pipe the week before.

The tip should be kept clean and shiny by wiping it on the damp solder sponge and using the soldering iron cleaner above. Placing the solder against the hot tip should cause it to melt, give off some smoke, and flow over the tip. If there are parts of the tip that solder doesn't flow to, then replace the tip. Never file down the tip. If it has lost its shininess, then replace it. The tip is specially coated, and when this shininess (the tinning) is gone, the tip is useless.

To solder two pieces of metal together, hold the hot tip against both pieces of metal to heat them both up, wait a couple of seconds, and then touch the solder to one of the metal surfaces. This operation looks like Fig. C.4. The solder should flow between the two pieces of metal, and there should be a curl of smoke. The smoke is the *flux* built into the solder. Flux is a weak heat-activated acid that cleans away any oxides on the metal surfaces before the solder melts on it.

Plumbing solder, like plumbing soldering irons, is not appropriate to use. The flux built into plumbing solder is much more aggressive and will literally eat away at electronic circuits. In case you haven't guessed, *no* plumbing materials should be used with electronics.

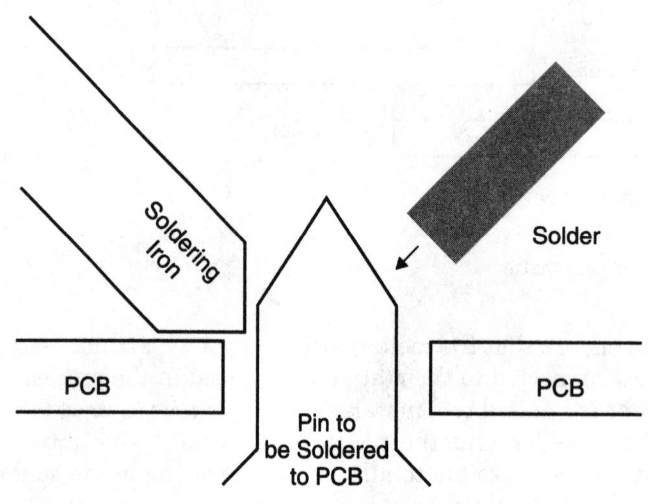

Figure C.4 Creating a solder joint.

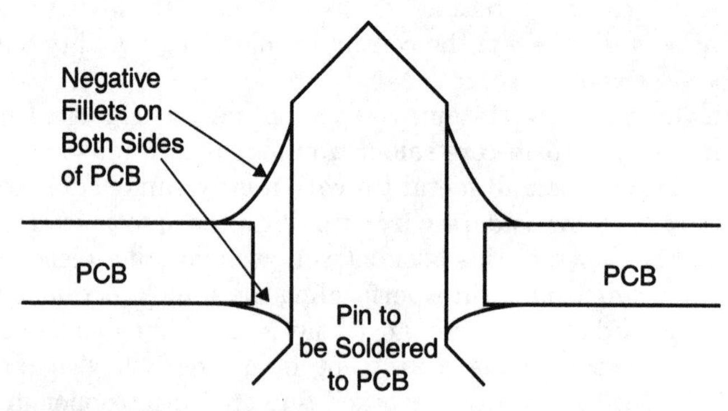

Figure C.5 Good solder joints.

After the solder has flowed, remove the solder and the iron. After a few moments the solder should harden into a smooth, shiny surface with a negative *fillet*. If it doesn't, reapply the soldering iron and try again. Good solder joints are shown in Fig. C.5.

Bad soldering joints will have a dull, crinkly finish and are known as *cold solder joints*. These joints can break over time, and they may not be connecting the two pieces of metal in the first place.

When you first start soldering, expect to make a mess and potentially damage (burn) a few joints and boards. It may be a good idea to buy a few cheap boards at a surplus store to learn the techniques of soldering before going on and building the circuit that uses the PCB provided in this book.

You should find that with a bit of experimentation with the heat control of your soldering iron, you will be soldering like a pro in no time.

Note that, while soldering is not a terribly dangerous operation, there are a few things to watch out for. The first is that the iron is hot. Test it by melting solder on it or making the damp solder sponge sizzle. Do not, as a former employee of mine did, hold it up to your lips to see if it is hot.

Soldering is a manufacturing process that involves some moderately dangerous chemicals. Work in a well-ventilated room and don't smoke while soldering (hot cigarette smoke and lead fumes generate cyanide). Keep the solder away from children and wash your hands after soldering.

Embedded PCB Boards

It probably seems unusual to say that an embedded PCB could be used as a prototyping circuit method, but it can be a very cost- and time-effective way of building your initial circuits. As I will show in this section, there are a few tricks you can use to create PCBs for your projects that take a similar amount of time and cost about the same as more traditional methods, but result in circuits that are much easier to assemble and debug.

As I work through this section, I will show how I might design a prototype card and what sorts of features I would design into it to make it suitable first for prototyping and then as a production part. The schematic capture and layout tools that I will be using are Ultimate's ULTICap and Protel's EasyTrax, respectively. EasyTrax is available free of charge for download off the Internet. I want to create a relatively complex circuit in which I can connect two CMOS SRAM memory chips to a Microchip PIC17C44 with an RS-232 interface. Along with this, I want to have the ability to try out PIC17Cxx In Circuit Serial Programming (ICSP), so I have added a reset control circuit on the card as well. The goal of the exercise is to come up with a prototype PCB for this application that will allow me to experiment with external SRAMs or other peripheral devices.

The circuit that I started with is shown in Fig. C.6. The reason why I say "started" with is because as I laid the circuit out onto a PCB, I made a few changes to the original circuit to simplify the wiring that I had to do. As I work through the PCB development, I will discuss what these changes are.

In Fig. C.6, I have directly copied the power supply (for +5 V and +13.4 V) and RS-232 interface from other circuits that I had already designed. I did this because they work, and they eliminated the need for me to come up with unique copies of circuits that I have already designed and laid out. As will be discussed, I also copied in this area when I laid out the PCB to avoid the need to add it to the circuit as well.

With the circuit designed and peripheral functions passed to a prototyping connector, I was ready to start laying out the PCB. I decided on a 4- by 4-in card size to avoid having to "squish" the circuit onto the card (although, as you will see, there will be some portions of the board that will become quite congested with circuitry).

To design the PCB, I used Protel's EasyTrax, which is a layout tool, and a Gerber file generator (which is used by the PCB manufacturer to build the PCB) that can be downloaded free of charge from the Internet. Figure C.7 is an EasyTrax screen shot of the start of the prototype PCB layout with the RS-232

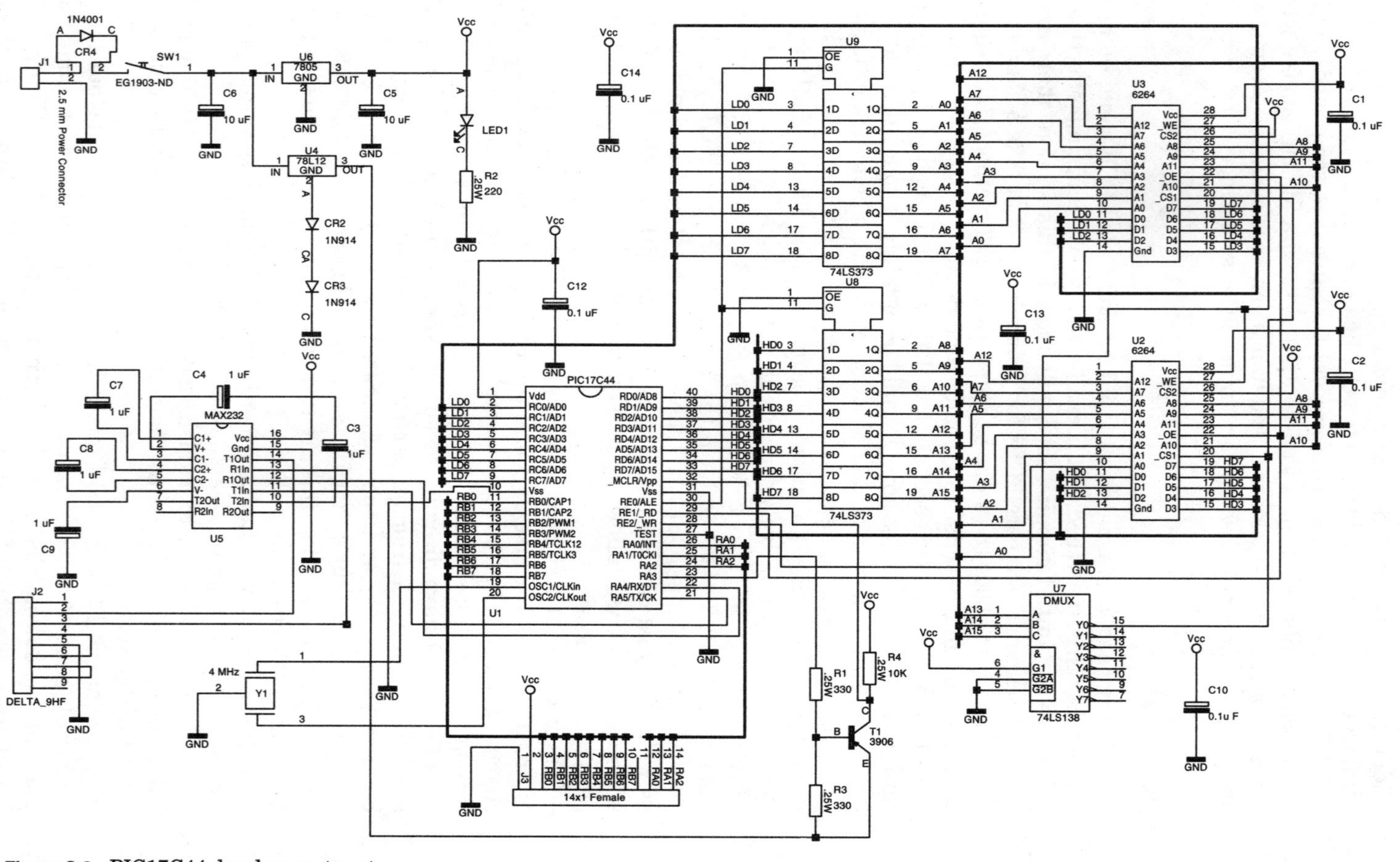

Figure C.6 PIC17C44 development system.

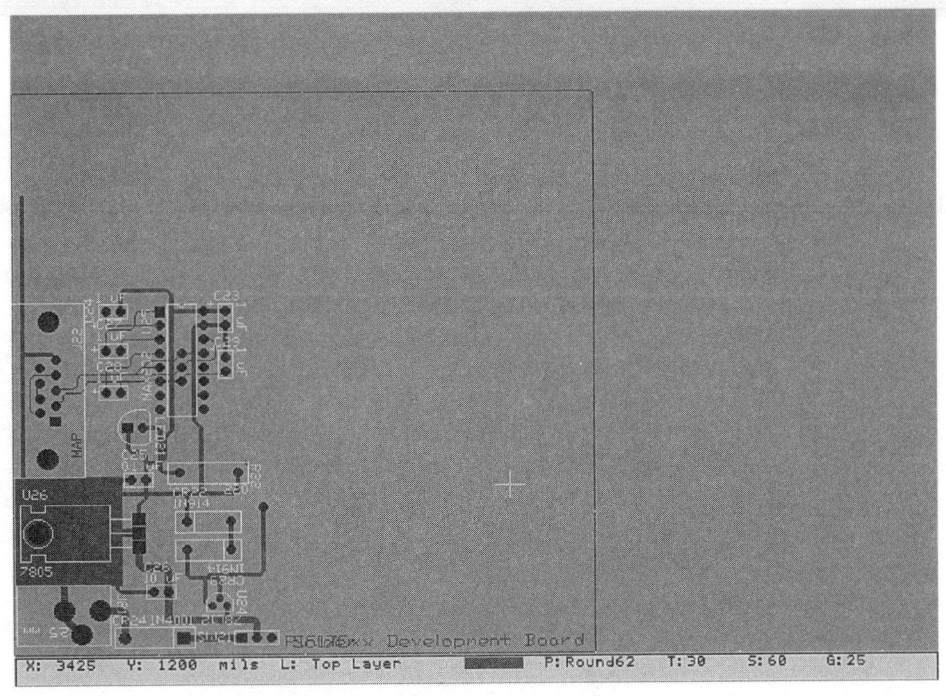

Figure C.7 Start of prototyping—power and RS-232 interface copied from previous PCB design.

interface and power supply copied from another application that had the same requirement as this circuit.

Next, I put down the major parts that I want to use in the circuit. Going back to the schematic, these parts are the PIC17C44 microcontroller, the 74LS373 buffers, the two 6264 SRAMs, and the 74LS138 decoder. When I first lay out a circuit, I first try placing the functional blocks together as is shown in Fig. C.8. Also when I place the circuits, I make sure that all the components are oriented in the same direction and the programmable part (the PIC17C44) can be easily removed from the circuit for reprogramming.

In Fig. C.8, the parts are laid out as far away from each other as possible to make wiring easier. In addition, I have arranged the 6264s to be in line to see if this makes the bus (address and data) wiring easier. When I initially lay out the components, I work through what I consider to be a small "thought experiment" and visualize how each component will interconnect with others. As a result of this thought experiment, I realized that U2 and U8 as well as U3 and U9 should be paired together to take advantage of the common multiplexed address and data buses that will be wired to each chip. The resulting design is shown in Fig. C.9.

I should point out that each time I change the PCB design significantly, I also save the PCB design as a different file name (usually I add a 2- or 3-digit number to the end of the filename for this purpose). This way, if I make any

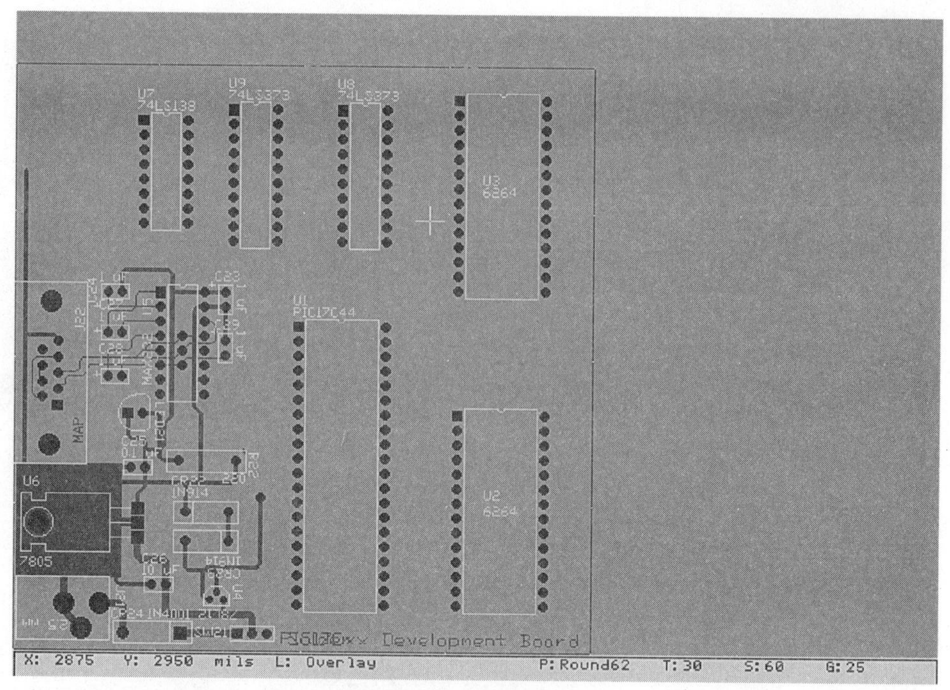

Figure C.8 Prototyping—adding DIP components to circuit.

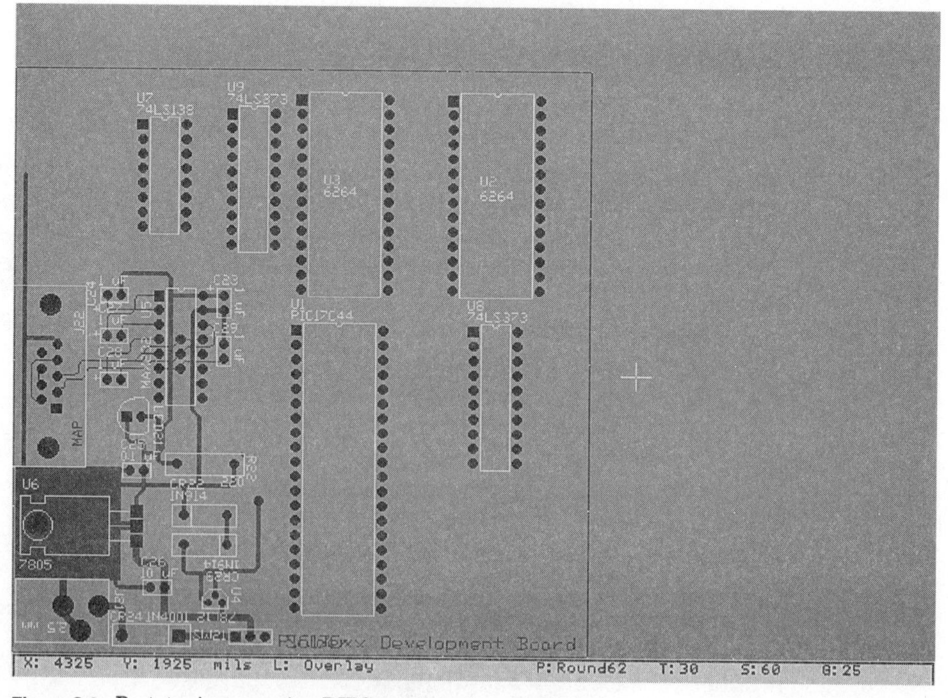

Figure C.9 Prototyping—moving DIPS to take advantage of buses.

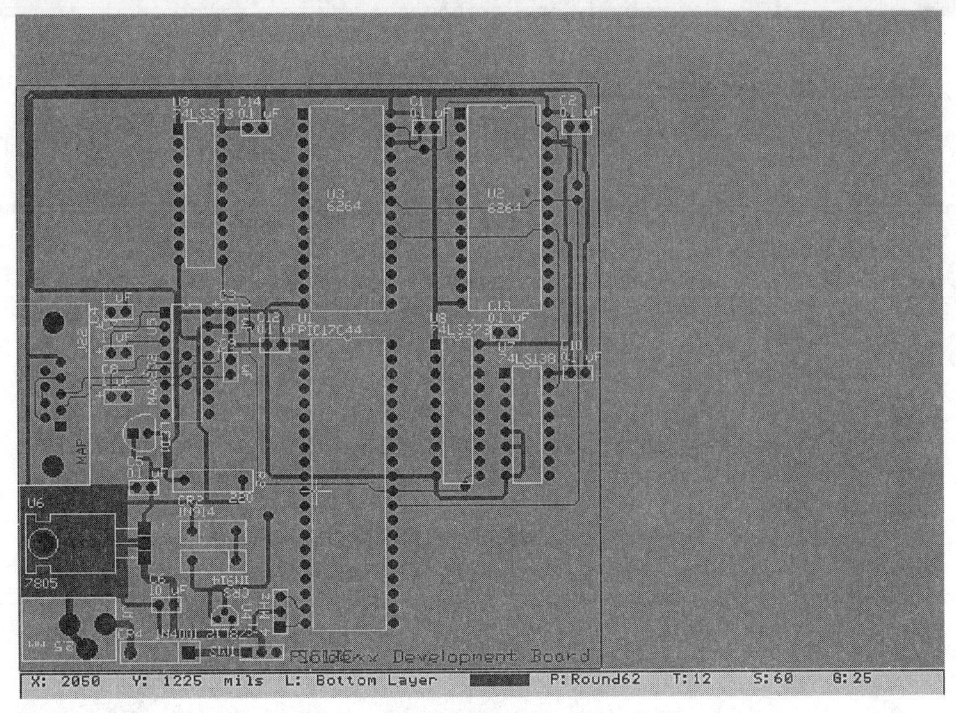

Figure C.10 Prototyping—putting power control signals onto the PCB.

mistakes, I can go back and find out where the problem was first put in and go through the subsequent circuits with this change in place.

Now, I can make the basic connections of power, I/O reset, and clocking. These traces and any extra components needed to support them are easily laid down with as many traces as possible put on the bottom side of the board. The power traces are given a 0.030-in width (the signal traces are 0.012 in wide) and the V_{cc} traces are put on the top side of the card to make sure they stay physically separate from the bottom side. The updated card is shown in Fig. C.10.

If you are familiar with the PIC17C44, you should be able to find three errors. This is common, especially with large chips like the PIC17C44, which has a lot of pins on one side. These problems will be fixed up as I add more wires to the PCB and discover things where they shouldn't be. You will also notice that I moved the 74LS138 to be closer to U8 (this was something I missed when I did the original thought experiment).

Depending on what you want to do with the PCB, you may consider yourself finished. If wire-wrap sockets are going to be used to make the interconnections on the board between the chips, you could hand-wire the circuit very easily. Many people stop at this point to avoid incurring the costs of unneeded PCB masks, but personally, I don't believe that this saves you a lot in the long run.

If wire wrapping at this point is going to be used, the total number of wires required is about 100. Using the calculations presented later in this appendix about wire wrapping, it would take about 2 hours to wire-wrap it. In actuality, it would probably be significantly less because all the wire wraps are easy ones, requiring no special connections to power or discrete components added to the board. Since it took me about a half-hour to get to this point, the total time to build the prototype circuit will be about 3 hours.

Since I am this far and I don't like wire wrapping or point-to-point wiring more than 25 wires or so on a PCB, I will continue and finish the board with all the traces connected properly. Initially, I would want to do the multiplexed traces (from the PIC17C44 to the 74LS373s and 6264s). These are the basic connections between the PIC17C44 and the 16-bit bus. I tried to put as much on the back side of the PCB to make rework (i.e., fixing the board) as simple as possible. Figure C.11 shows the 16-bit multiplexed address and data bus connected on the board.

When I was making the connections in Fig. C.11, I was not able to follow the schematic exactly. The differences are in the two 74LS373s (U8 and U9). To simplify the wiring and minimize the need for vias, I rerouted some of the signals going to the 373s. When I made these changes, I documented them and then updated the schematic.

Another change I could have made in this circuit that would have made the layout easier would be to use 74LS573s, which do not have staggered input and output pins and have instead all the input and output pins on one side of the chip.

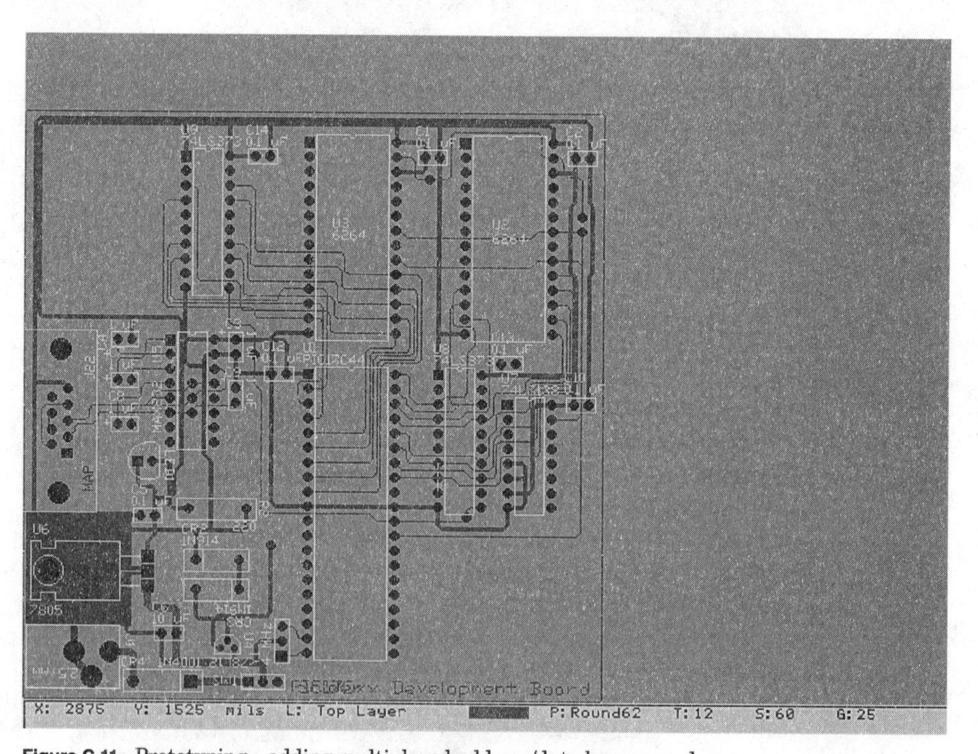

Figure C.11 Prototyping—adding multiplexed address/data bus on card.

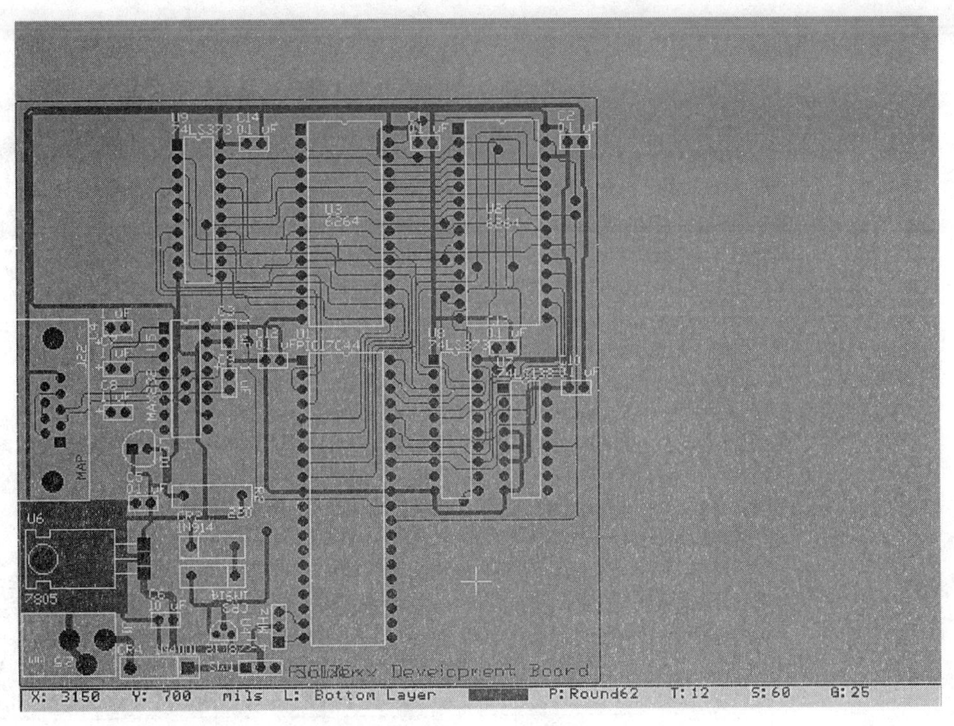

Figure C.12 Prototyping—adding buffered address to SRAM on card.

With the bused connections made, I then proceeded to make the connections to the address lines from the 74LS373 buffers and the 74LS138 address selector chip. In Fig. C.12, you can see the result. Note that some of the traces that were put down earlier in the design had to be changed to allow "throughways" for the common address signals between the two 6264 SRAMs.

With the 6264 SRAM's address data and control lines connected, the majority of the work to wire the board is finished. To finish it off, I added three parts to the PCB design. The reset circuit is designed to allow code to write to the PIC17C44's internal EEPROM and was taken out of the Microchip "In-Circuit Serial Programming Guide" (Microchip document DS91015B). In addition, I added a 14-pin prototype connector to give me easier access to the PORTB and unused PORTA I/O pins and a "prototyping" area that uses up the unused space in the top left corner of the board. This is shown in Fig. C.13.

Putting down the reset circuit was very simple (as would be expected with a single transistor and three resistors). A bit more difficult was putting down the multiple vias, and I had to relocate some of the power traces in the area. The final result is a 13- by 8-in area where sockets and components can be inserted into the board to try out new circuits. Adding a prototyping area like this is something that I always try to do when developing first-off or time-sensitive boards. Having an area in which I can add new circuits to an application has saved my reputation on more than one occasion at work, when application

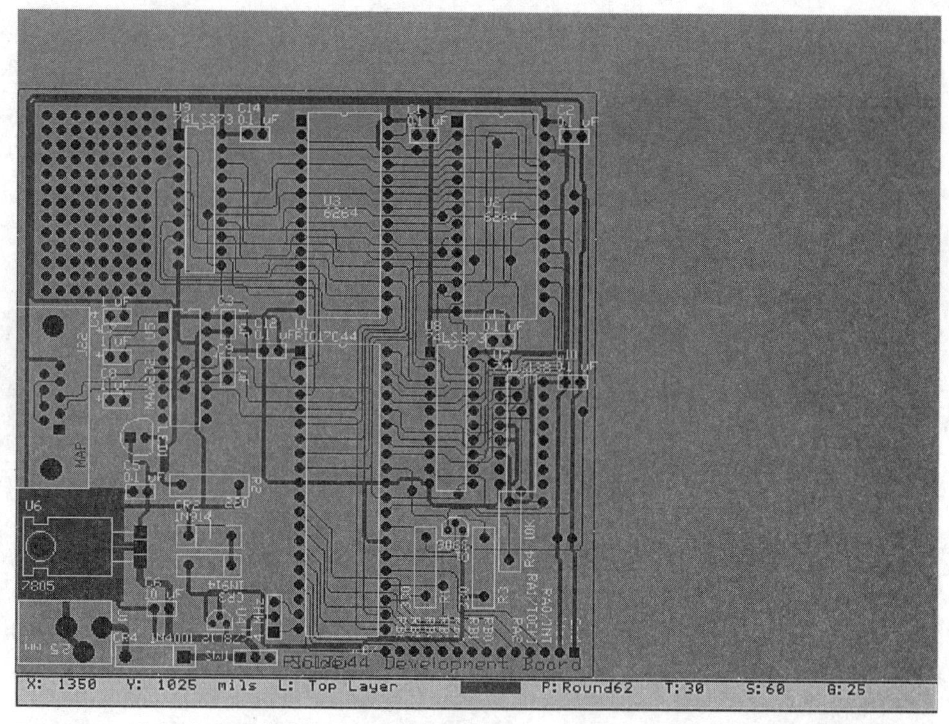

Figure C.13 Finishing off the board with prototyping areas and a PORTA/PORTB connector.

requirements have changed (of course it's never *my* fault). Adding these vias to the board generally doesn't add to the cost of the final PCB.

When I wired the PORTA and PORTB connector into the bottom right-hand corner of the board, I discovered that it would be much easier if I reversed the PORTA connections with the PORTB. This change was made and the prototype PCB design completed. This should not be regarded as a significant change to the board because the connector was put on to make access to the advanced functions of the board simpler.

With all the changes made to the circuit documented, I went back and updated the schematic. The final schematic is shown in Fig. C.14 and just took a few moments to update.

The total time to design this PCB was about 2 hours. Building the circuit requires about a half-hour. Surprisingly enough, there was only one mistake in this PCB design, which took less than a minute to correct with point-to-point wiring (two nets on the back side of the board crossed each other). While this is only about two-thirds of the time required to wire-wrap a prototype, by developing a PCB prototype board, you can update, fix, or even replicate the board as often as you need to. The investment in developing the board is not lost and will pay for itself more and more as you get additional copies made.

In terms of costs, the difference between a wire-wrapped prototype and a PCB board is not as significant as you would probably expect. If you were to

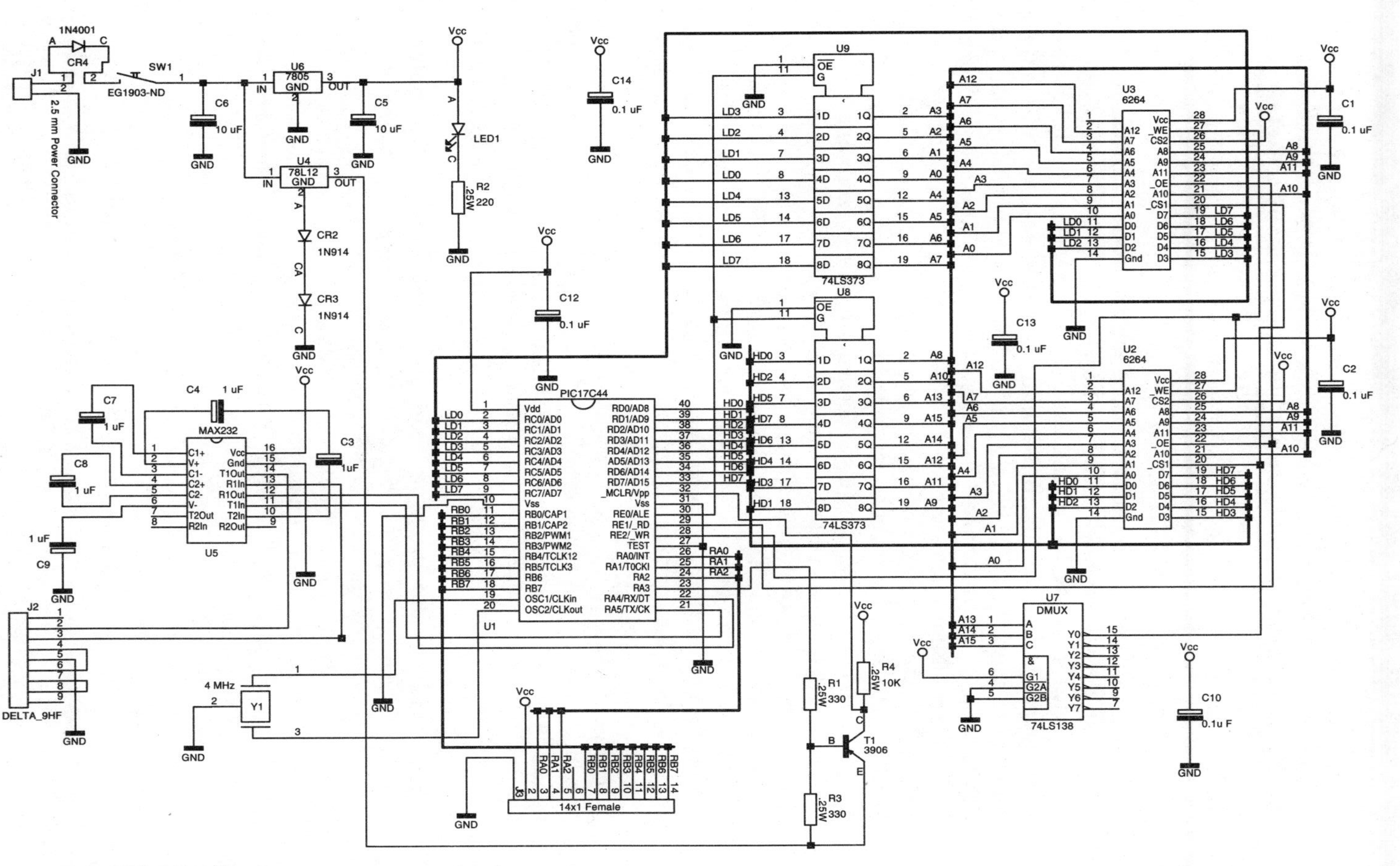

Figure C.14 PIC17C44 development system—updated.

assume that a "quick-turn" prototype PCB house (like AP Circuits) produces boards at a cost of $4.50 per square inch (for two-sided bards with plated through-holes and no solder mask or silk screening), the total board cost would be about $72. When I build circuits like this, I usually socket only the expensive parts like the Microchip PICmicro MCU and the SRAMs. Everything is soldered to the board to avoid socket costs (which can be as much as the "glue" logic chips). Total cost for the three sockets will probably be about $5 for a total of about $77 for the PCB.

For wire wrapping, the socket prices are about $2 each for 16- and 20-pin wire-wrap sockets and $4 each for the 28- and 40-pin wire-wrap sockets. If sockets with internal decoupling capacitors are used, the socket costs will at least double. A good-quality prototyping card will cost a minimum of $5 (and could be as much as $20). Wire-wrap wire will be about $5. Finally, there will be the issue of how to connect discrete components and the RS-232 connector. Using a Scotchflex connector for this function will run you about $10 in small quantities. The total cost for the wire-wrapped solution will range anywhere from $40 to $75, take you a lot longer to assemble than the PCB, and result in a maximum of only one copy of the board for your investment.

In this calculation, I haven't included PCB issues like shipping and handling, taxes, and other incidentals (the price above does include mask development, though), so there may be some differences in the actual price when you finally get the boards. For wire wrapping, I have not included the time required to build the prototype. If you consider that professional wire wrappers get about $25 per hour, you can see that from an out-of-pocket perspective, an embedded PCB is the better deal.

Looking at the $4.50 cost per square inch PCB quick-turn assembly costs (production costs are considerably less), you might feel that you could minimize the prototype costs even more by building the PCBs yourself. Kits are available that provide everything that you need for making your own prototype PCBs for less than $20—which will put the cost of a PCB below that of wire wrapping.

I would like to discourage you from building your own PCBs unless you are going to do it for a large number of boards and are willing to make the investment in the proper tools to do the job. The kits are difficult to work with and often require that you draw out your design with a pen on the back side of the board. This can be simplified by buying component masks, but this adds to the cost. Another issue is that these kits provide only single-sided boards, and without dual-sided boards, there is no way to add plated-through vias. PCB drill bits are quite costly and easily broken.

A new option is a copper sheet that can be printed in a laser printer with the design of the circuit and then ironed onto a piece of phenolic or fiberglass PCB board material. Once the copper is ironed onto the board, it is then etched as if it were a regular copper board. If the PCB is already predrilled with holes at 0.100-in centers, this can be quite an easy way to build your own boards. This method avoids having to draw the traces onto the copper of the PCB board or expose the PCB board with a mask and develop it before etching.

The only issue with this method is that, while two sheets can be laid on top of each other, they cannot be connected by plated-through vias or holes. When laying out the board with these sheets of printed copper, you should plan for vias to be placed in between the components and small pieces of wire soldered between the sides to bridge them. There should not be any top-side copper pads at component interfaces, as these will be difficult to solder reliably.

Last, the chemical etchant that is provided with the kits is a toxic substance and must be disposed of as hazardous waste in many jurisdictions. You should *never* pour etchant down a drain; not only is it illegal, but the chemicals will dissolve your home's pipes.

This is not to say that it isn't interesting or fun to make your own PCB boards—I have done it a number of times over the years and you do end up with a good sense of satisfaction when you are finished. I just want to say that there is a lot more to it than buying a $10 kit at Radio Shack, and your time, money, and energy would be better spent designing, debugging, and building circuits.

Circuit Layout Rules

One of the most neglected aspects of application design is how a prototype or even a product's circuits are laid out. Like many aspects of application design, the circuit design and layout is a result of developer experiences and preferences. The success of a circuit design, like the software and overall concept, is a result of the time spent planning how the circuit is built before laying it out onto a PCB. In this section, I want to review some points and tips I have discovered that make a circuit board layout successful.

These rules are similar for prototypes that are built into a breadboard, hand-wired together, or built into an embedded circuit card. Some of these points will be more applicable to specific methods than others, but they are all good points to keep in mind.

1. Make sure all the parts can fit on the board. This may seem basic, but I have seen (and unfortunately designed myself) a few boards where the carrier isn't large enough for all the parts. The result is chips glued "dead bug" on the backs of others or cards that have "daughter cards" that are held together by solder and wire.
2. Put the components on the board in the following order:
   a. Connectors
   b. Programmable parts/chips
   c. Decoupling capacitors
   d. Oscillators
   e. Discrete components
   f. Remaining parts

   When you follow this order, you will be putting the connectors and programming parts where they can be easily accessed. For programmable parts, placing them on the edges of the board will allow them to be easily removed for reprogramming. Decoupling caps have to be as close to chips as possible. Once these parts are done, the positions of the remaining ports can be chosen.

3. When doing a PCB board, lay out the power tracers first, using 0.030- to 0.050-in-width traces (normal signal traces are 0.010 to 0.01 in wide). Once these are done, you can do the signal traces. Note that I always put positive voltage traces on the top side with negative (ground) voltage traces on the bottom side of the board. Also, work to only have one path for voltage and ground. Make sure you avoid loops in the traces that can pick up electrical noise.
4. When planning your wiring, do power first and make sure nothing is blocked. Never lay a power trace over another that will prevent signal traces from following through. It is important to remember to always stagger them so traces can pass between them, as is shown in Fig. C.15.
5. Another good idea for embedded boards (as well as hand-wired boards) is to put as many signal traces on the back side as possible to allow tracing of signals and reworks with components in place.
6. When wiring a board, keep decoupling capacitors as close to the chips' V_{cc} (V_{dd} for CMOS devices) as possible. The distance to end ground (V_{ss}) is less important. When I'm hand wiring boards, I'll solder the decoupling cap to the back side of the chip socket. You can also buy sockets with decoupling caps built in, which eliminates the need for this wiring.
7. Don't plan on getting everything right the first time. For example, the breadboard power supply PCB that comes with this book underwent two circuit designs and three PCB board designs until I was happy enough to include it in this book.

Continuing this thread; if you wire a circuit incorrectly, remember that, while problems are maddening, they are never the end of the world. There is

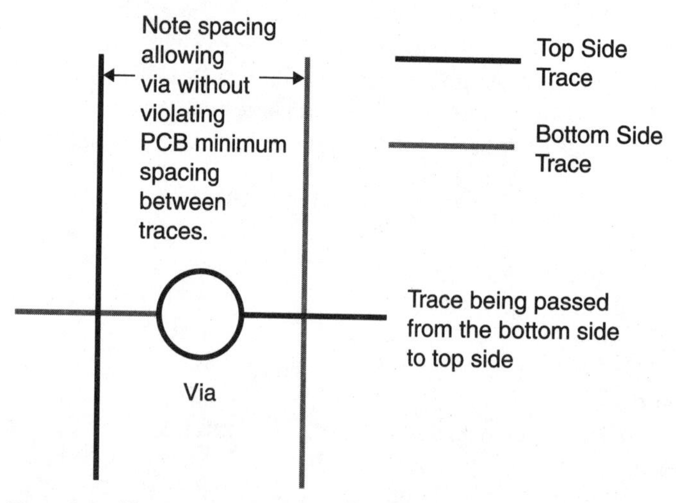

Figure C.15 Vias to pass a trace between PCB layers.

nothing that can't be fixed or reworked, and you can fix the problem in the next pass of the board. Yes, you don't want to do any extra work, but chances are you will have learned something that will make it easier to prevent making the mistake again in the future.

Wire Wrapping and Point-to-Point Wiring

When I'm talking to people about things I've had to replace over the years, I usually focus on how many PCs I've owned. What I should probably discuss is the number of wire-wrap tools I've owned. For some reason, I have gone through 50 or more wire-wrap tools in the last 20 years; this is in contrast to my still having the same pliers and clippers that I bought when I was a teenager. I tend to wear out or jam wire-wrap tools at a rate of two or three a year, probably because I use them so often when building prototype circuits.

For most of my prototype circuits, I primarily wire-wrap them or, if they are very simple, use point-to-point wiring. I find that using these methods produces a reasonably robust circuit that can be fairly easily modified and enhanced.

The basic tools needed for wire wrapping and point-to-point wiring are a wire-wrap tool, a wire stripper (which is often integrated into the wire-wrap tool), 32- to 28-gauge wire, and sockets. Reasonable quality wire-wrap tools and wire can be purchased from Radio Shack for just a few dollars.

Motor-driven wire-wrap tools and precut and stripped wire lengths are available from many vendors. While this increases your efficiency, it also dramatically increases your costs. If you are starting out, I recommend that you buy as cheap tools as possible and try to determine whether or not wire wrapping is something you want to do a lot of.

As the name implies, *wire wrapping* consists of wrapping a wire around a post to make an electrical connection. In Fig. C.16, I show a sample wire-wrap joint.

To make the joint, ¾ in to 1 in of wire is stripped and placed in a small off-center hole in the wire-wrap tool. The tool itself has a second, larger hole in the middle of the post that is placed over the pin to be connected and turned until all the bare wire has been wrapped around the pin's post. Figure C.17 shows the wire wrapping operation in more detail.

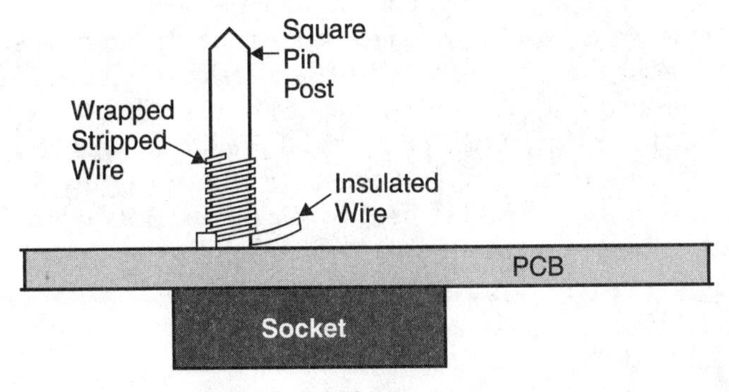

Figure C.16 Wire-wrap electrical joint.

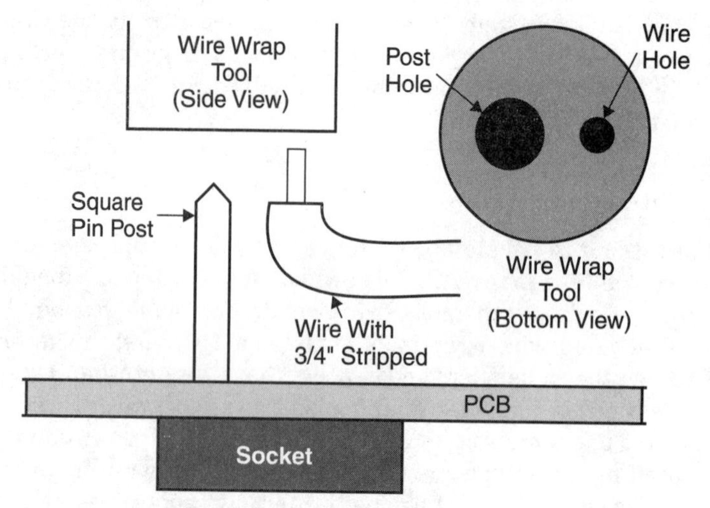

Figure C.17 Tool for making wire-wrap joints.

The wire-wrap tool's wire hole is normally designed for placing a short length of insulated wire, so there is no chance of adjacent pins being shorted together. The wire wrap socket has very long pins leading from them that the wire can be soldered to. The post should have a square cross section, as this makes a better connection with the wrapped wire.

Note: Some unscrupulous dealers will try to sell you sockets with round posts. These sockets will not work as well as square ones, and chances are the wire will slide right off them. When the wire is wrapped onto the pin, it forms a gastight bond that is extremely reliable and is not likely to corrode or loosen. Gold-plated, machined, square cross-section wire-wrap sockets are the most expensive sockets—but they are the most reliable. If you are expecting to keep your circuit for an indefinite period of time, you should spend the extra money for the reliability the best sockets will give you.

Another option to be aware of when buying sockets is that there are some that have decoupling capacitors built in. These capacitors are wired to the top right and bottom left corners; this wiring arrangement is perfect for TTL chips and makes wire wrapping the application much easier.

When circuits are to be wire wrapped, some time should be spent determining where the sockets are to go. With a little foresight, you can make a big difference in how much work is required to wire the board.

For creating the circuit, a *net list* must be produced. This can be produced manually or, if you are using an automated schematic capture tool, it will produce the net list for you. The net list is typically an ASCII text file that you can read and follow to wire your circuit.

For example, a section of a net list could be:

```
Bit0  U1.6
      U2.11
      U5.7
```

The three pins on chips U1, U2, and U5 are wired together in one continuous string to form the Bit0 net. They should not be wired in a kind of triangular shape where each pin is wired to the other two. Just one path should be used at all times to avoid inductive loops that pick up noise.

There are two other points you should be aware of. The first is to wire the pins in the order that is given in the net list. In many cases, this will be extremely inefficient in terms of wiring (which is why I suggested that you think about how the chips are arranged). The reason for following the order of the net list is to make sure that all terminations are correctly wired. This is not an issue for any of the circuits in this book, but it can be an issue for high-speed circuits.

The second point is how the multiple pins are wired in a "daisy chain." In Fig. C.18, I show two methods of wiring daisy chains. The first ("Wrong") is probably the most obvious way of doing it, but can be a problem when you have to unwrap a wire. The second ("Right"), stacked-wiring, method is recommended to ease reworking the circuit. In the stacked-wiring method, if a net is to be rewired, then a maximum of three wires are changed for each pin to avoid leaving a wrapped pin or having to unwrap *all* the pins, which can happen in a daisy chained circuit. In any case, there should not be more than two wires wrapped onto each post.

In the diagrams I have shown fairly neat wraps with straight wires. With a bit of practice, you can be doing this as well. The problem that trips up most new wire wrappers is keeping track of the pins; it is important to remember that you are working with the mirror image of the pins, and they are essentially reversed from how they look on the top side. You can buy plastic templates that fit over the pins before wrapping to help you keep the wires straight.

When you go to an electronics store, you will find that, along with simple wire-wrap tools (which will probably look like a pen or a block of aluminum with black anodized steel posts pressed into it), there will be electrically powered wire-wrap hand tools available. These tools will have pistol grips and cost $200 or more. These tools will make the work quite a bit easier, but before investing in them you should decide how much you want to wire wrap. Even if an electric tool makes it 100 times easier, wire wrapping is still an unreasonable amount of work.

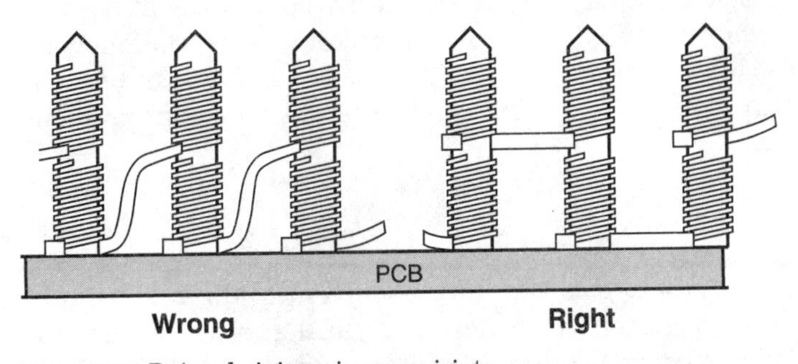

Figure C.18 Daisy chaining wire-wrap joints.

Wire wrapping individual pins is quite easy, but it is easy to get bogged down in large projects. For example, if your circuit has about 150 nets and if you assume that each net will require 1.5 wires and each wire requires 1½ to 2 minutes to wrap, the total time required is somewhere between 5½ to 7½ hours.

This probably seems like an exaggeration, but it is actually a pretty good estimate. Professional wire wrappers quote a half-minute per wire, but they tend to have literally years of experience. Moreover, discrete parts like resistors, capacitors, and transistors do take a lot more time than DIP pins to place on a board and wire.

In contrast, getting a quick-turn PCB designed and built will be much cheaper in terms of time and possibly out-of-pocket expenses for sockets, special connectors, and carriers for discrete parts. Having a quick-turn PCB could be less expensive and, on the plus side, the design can be easily replicated or modified.

Point-to-point wiring is a particularly onerous method of wiring prototype circuits. Instead of wrapping pins with very thin wire, the pins are soldered to the wire directly. While I seem to be the only person dumb enough to use this method of circuit prototyping, there are useful skills that are developed when you have to modify or debug a circuit.

I personally find point-to-point wiring to be about as fast as wire wrapping, but it avoids the need for keeping wire-wrap sockets on hand and avoids the problem of trying to figure out how to handle discrete components.

Point-to-point wiring is a useful skill when reworking—changing a circuit or fixing a PCB board. By baring a cut trace, you can add a wire to another net. Reworking a trace is most easily accomplished by first scraping off any "solder mask" that is on the board. Next, the trace is cut in two places by using an X-Acto blade and then rubbing a hot soldering iron to remove the trace, as shown in Fig. C.19. The cuts are best made close to a pin or a via on the PCB, but not so close that there is no trace left to solder a wire to. After the cuts are made, the wire is soldered to the trace near the cut (Fig. C.20).

Note that I cut both sides of at race. This is done to avoid long stubs, which can cause problems with the operation of the circuit because of reflections from the open line. Doing the second cut usually just takes seconds and avoids any potential problems on the net.

Being able to reliably rework boards by point-to-point soldering techniques comes in very useful with quick-turn prototypes. If you follow the instructions given above, you will find that the reworked net will have virtually the same impedance of the original net and no stubs to cause problems with signal reflections.

Cooling

The subject of cooling can be very important to digital electronics and power supply designs, although nowhere as critical as the cooling requirements for a PC, in which modern titanium processors can dissipate over 100 W of heat. For

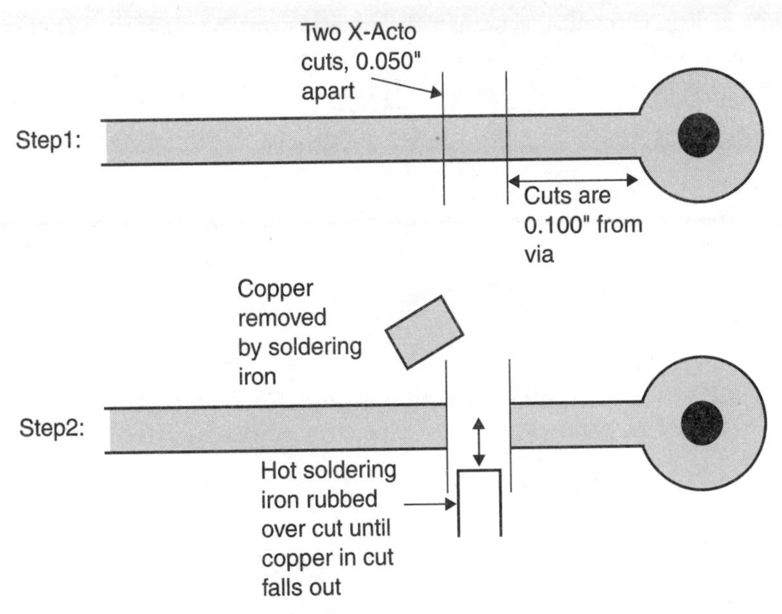

Figure C.19 Cutting a PCB trace.

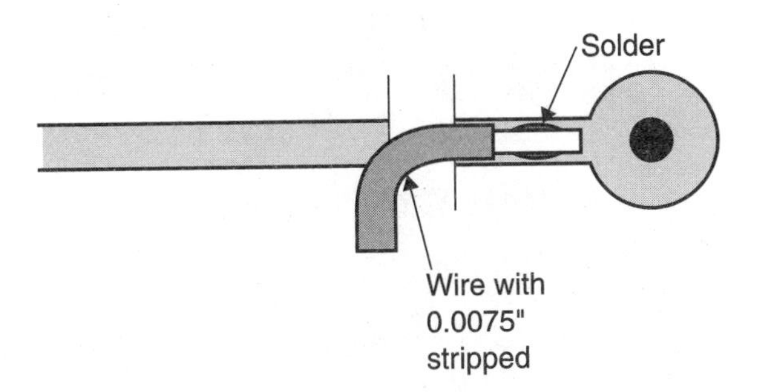

Figure C.20 Adding a wire to a cut trace.

some parts in your applications, you will have to make sure you understand whether you have to provide features and options that will help cool them. The parts that you will have to most actively cool are voltage regulators and motor and relay drivers. These devices can dissipate several watts of power; enough to cause the parts to become hot enough to burn you, cause thermal shutdown, or even unsolder themselves from the board.

In this section, I'm going to talk about *passive cooling,* which does not involve any parts that aid cooling and have to be powered themselves. I'm going to discuss the need for cooling along with two methods of passively cooling parts.

As I have discussed, DC power is defined by the equation

$$P = VI$$

where V is the voltage across the device and I is the current through it. In a linear regulator or bipolar motor driver, this equation can be used directly. For example, in a 7805, which has 9-V input and is sourcing 1 amp, the power dissipated is defined as:

$$\begin{aligned} P &= VI \\ &= (9\text{ V in} - 5\text{ V out}) \times 1\text{ amp} \\ &= 4\text{ V} \times 1\text{ amp} \\ &= 4\text{ W} \end{aligned}$$

For MOSFET motor drivers, there is no defined voltage drop; instead, there is an internal resistance. In this case, the power equation replaces I with the Ohm's law equivalent:

$$\begin{aligned} P &= VI \\ &= V\,(V/R) \\ &= V^2/R \end{aligned}$$

So, for an N-channel MOSFET driver that has an on resistance of 0.5 Ω with a ¼ V drop in a motor circuit, the power dissipated is:

$$\begin{aligned} P &= V^2/R \\ &= (\tfrac{1}{4}\text{ V})^2/0.5\ \Omega \\ &= \tfrac{1}{8}\text{ W} \end{aligned}$$

This example shows two things: First, MOSFETs can be much more power-efficient in providing motor control than bipolar devices which have on resistances on the order of 10 Ω. Second, Ohm's law and the other basic DC laws are something you should always keep in the back of your mind.

With the power dissipated by each component in a circuit calculated (and this is a step you should take for every circuit), you should determine whether any components require a heat sink. My rule of thumb is that if more than ½ W is not going to be dissipated by the part itself, then it should have a heat sink connected to it.

This statement is a bit hard to understand, as it takes into account the ability of the component's package to act as a heat sink itself. A heat sink is any device that is used to pass heat from a component into the surrounding environment (normally the air naturally circulating around the component). You have seen heat sinks in a variety of different applications; a good example of one is the finned cylinders of an air-cooled motorcycle engine. These fins pass heat from the device (whether it is a voltage register or a motorcycle engine) and transfer it to the surrounding air.

For devices that operate at room temperature (20 to 25°C or 68 to 77°F), I use the rule of thumb that 1 square centimeter of heat sink surface area is required for each watt of heat to dissipate. The important part of this statement is the term *surface area.* Surface area refers to the area of the heat sink that is exposed to the outside environment. For the situation where you have a voltage regulator in a TO-220 package that is bolted to a PCB, the surface area available to act as the heat sink is the exposed sides of the part to the surrounding air, as is shown in Fig. C.21.

According to my calculations, the surface area of the TO-220 package is 3.452 cm^2, which means the package itself can dissipate 3.45 W of power. Going back to the voltage regulator example above (4 W output), the amount of power that is not accounted for is 4 − 3.45 W, or 0.55 W, and will have to be dissipated by a heat sink.

The term *surface area* is also an important concept to remember because it affects how a heat sink is designed. In Fig. C.22, I show two different types of heat sinks, both taking up the same amount of space, but the one with the cutouts has a lot more surface area. By adding cutouts to a heat sink's irregular surface, you can double or triple the surface area of a heat sink quite easily.

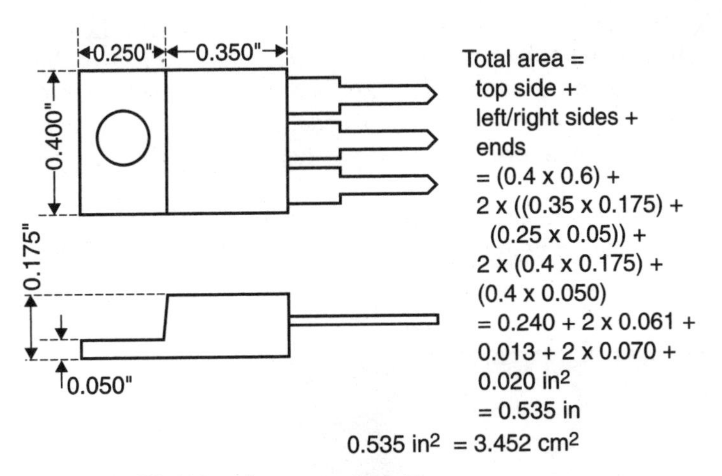

Figure C.21 TO-220 surface area calculation.

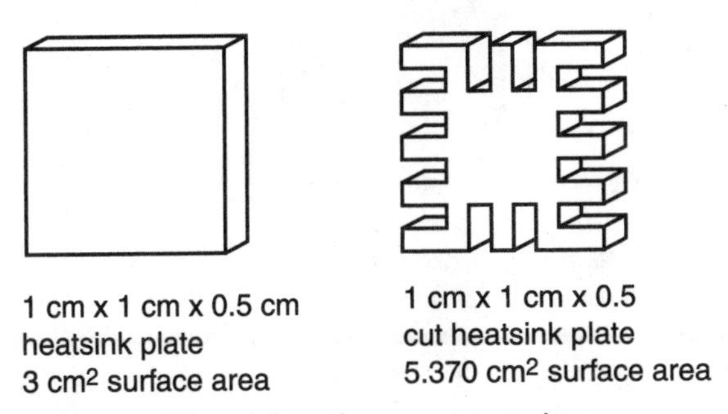

Figure C.22 Heat sink surface area comparison.

As you learned in grade school, black is the best color for heat transfer. Commercial heat sinks use black-anodized aluminum with a rough surface, which provides more surface area than a smooth surface, and the black color allows better heat transfer.

Most heat sinks are designed to operate in an open-air environment or in a cabinet that has holes to allow heat to leave via convection. If an opening is not provided, then heat can build up within the product and cause the electronics to fail or age prematurely.

If an air path is not available, then the components can be bonded to large heat sinks. When using large heat sinks, there are two aspects you should be aware of: heat conductance and heat capacity. Heat capacity is the amount of heat a sink can absorb and is directly related to the mass of the heat sink.

Heat conductance is the most important parameter, as it relates to how heat is transferred through the sink. As Fig. C.23 shows, a good heat conductor will have the most even heat distribution through the heat sink whereas a poor heat conductor will be hot close to the component and cool far away from it. A terrible heat conductor is a Space Shuttle thermal tile, which can be 5000°F on the outside, but less than 100°F on the inside. A perfect heat conductor would have even temperature throughout.

Ideally, in a solid heat sink, the heat sink passes the heat to air or water, which will transfer the heat out of the heat sink. Otherwise, the heat sink will have to be designed so that it does not become saturated during operation and between high-load operations; it will cool sufficiently to prevent any kind of heat buildup from happening so that the part does not go beyond specified temperature.

Physical heat sinks add cost to a project, in addition to often increasing the height of the components. If you have an application that has modest (less

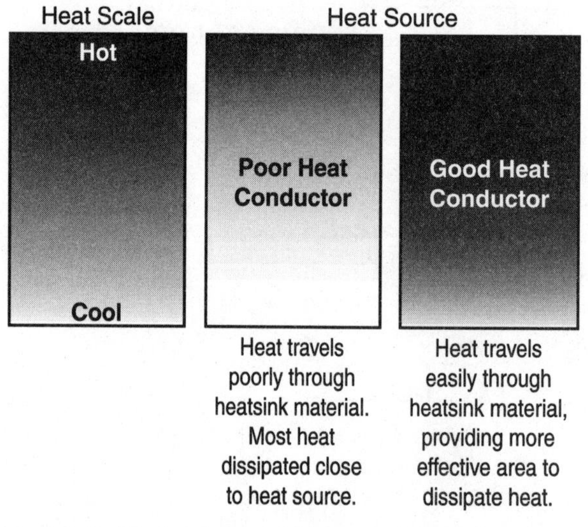

Figure C.23 Heat sink material heat conduction.

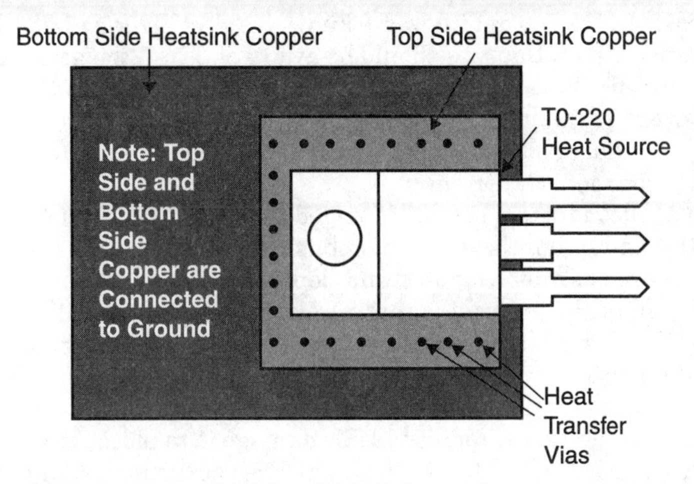

Figure C.24 Using a PCB for a TO-220's heat sink.

than 5 W) dissipation needs, the PCB that the application is mounted on can be used as the heat sink. For a TO-220 package, the heat sink can be put on the PCB back side as shown in Fig. C.24. This type of heat sink is provided on the PCB that comes with the book.

With this type of heat sink, an outside PCB layer must be used with a direct metal (i.e., no solder mask) interface between the top-side heat sink interface and the part. When a back-side heat sink is provided, then additional vias should be built into the card to help transfer heat from the top side to the bottom.

For this type of heat sink, instead of using the rule of thumb of 1 cm^2/W, I use 0.5 cm^2/W for planning the amount of area required in the PCB. This derating is used to reflect the lower airflow that can be expected across the heat sink. This may seem like a significant derater, but remember that each square inch has 6¼ cm^2; dissipating 5 W by this method requires less than 2 in^2.

ESD

If you are familiar with modern electronics at all, you will know that static electricity can be a major concern and could potentially affect the operation of your applications and even damage the semiconductor circuits used in them. In this section, I would like to introduce the concept of *electrostatic discharge* (ESD) and discuss the damage it can do and the measures that can be taken to avoid it.

Static electricity should not be a new concept to anyone; I'm sure everyone has shuffled across a synthetic fiber carpet and touched a door handle and received a shock at one time or another. This series of actions consists of building up the static electrical charge (shuffling across the carpet) and discharging the energy (the blue spark and the snap you hear). The energy discharge is caused by electrons at one potential jumping to an object at a different potential. After the discharge, both objects are at the same electrical potential and the opportunity for the charge to move between the objects is greatly diminished.

The spark and the current flow are known as *electrostatic discharge* and have some characteristics that you should be aware of. First, the actual current flow between the objects is miniscule, often on the order of millionths of a coulomb of charge, but the voltage potentials are very large. For you to feel the ESD shock, a potential of at least 2500 V has been produced—a potential of 100 V is enough to damage a CMOS input.

Silicon semiconductors can be damaged by electrostatic discharges of 100 V of potential and billionths of a coulomb of charge. The surge from the electrostatic discharge can pierce gates and doping junctions (Fig. C.25), degrading the operation of the transistor that received the charge. ESD causes "microholes" in the silicon, which can allow invalid current flows, changing the operation of the transistor. As the transistor is hit with repeated ESD events, the surface area of the holes increases, degrading the transistor's operation.

One popular method of reducing ESD damage is to place clamping diodes on I/O pins as I show in Fig. C.26. When the ESD surge is passed to the I/O pin, it is shunted through the diodes the same way a kickback surge is shunted through clamping diodes. These diodes greatly reduce the opportunity for ESD damage.

I have found most modern silicon devices to be very immune to damage by ESD, but this does not mean they cannot be damaged, and execution upsets are possible. To avoid any problems at all due to ESD, there are a number of

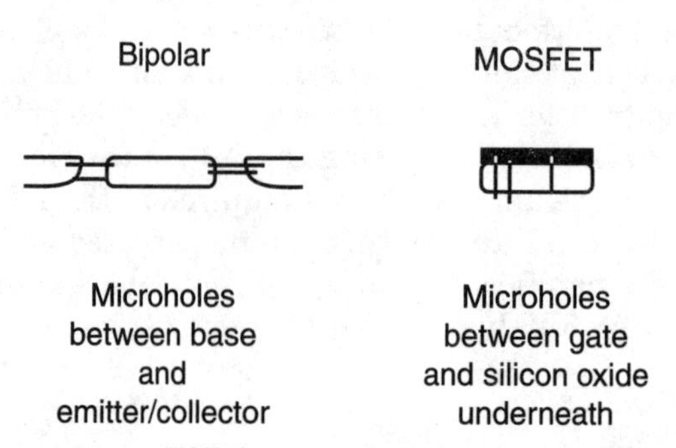

Figure C.25 ESD damage to transistors.

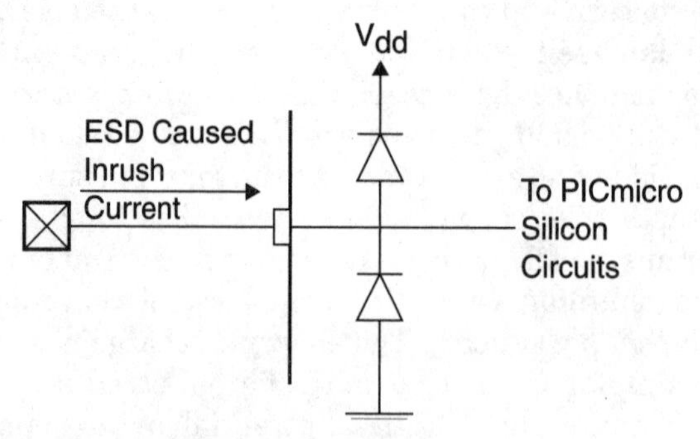

Figure C.26 ESD clamping diodes.

things you can do to minimize the opportunity for ESD shocks reaching the application circuits.

The most basic thing you can do is to buy an antistatic wrist strap and use it any time you are handling electronic components. By connecting the strap to a wall outlet's earth ground, you will protect your circuits from being damaged by any static charges that build up on you. Wrist straps (and the coiled cables that come with them) cost just a few dollars and are available from most electronic stores; they are often labeled as *field technician's ESD kits.*

The earth ground can normally be found as the center screw in a wall outlet. Before connecting the wrist strap to the earth ground screw, use a three-pin socket tester to make sure that the wall socket is wired correctly. The socket tester can usually be bought at a hardware store for $5 or less. Please check the operation of the socket before wiring your ESD straps to it. While there is generally a 1-MΩ resistor built into the wrist strap and ESD cable, it is not a good idea to wire your wrist to 110 V AC.

A better solution is to have an antistatic mat that is connected to earth ground and attach the wrist strap to it. This will cost a few dollars more, but it provides you with a lot more protection.

When you are not working with electronic projects, place them in ESD static-safe bags or containers. This will minimize the opportunity for casual handling of parts causing problems and damaging parts. Large numbers of these bags can be bought for just a few dollars—or if you're cheap like me you can save the ESD bags that come with chips that I have ordered and reuse them for completed projects.

Something that is fairly easy to do is to make sure the air is continually humidified. For the North American Southwest, a "swamp cooler" is definitely a necessity in the summer and in northern climates, a humidifier should be added to your furnace. The moister the air is, the less the opportunity for a static charge to build. Ideally, the relative humidity should be 60 to 70 percent to prevent both static buildup and condensation.

Once your project is finished, it should be put into an enclosed box. Not only will this prevent static-filled fingers from causing ESD damage to the circuit, but it will also protect those fingers from shocks from the high-voltage circuits of the application.

These precautions are pretty rudimentary; if you are going to be creating a professional lab, then you may want to invest in ESD smocks, shoes, tiling, and other specialized apparatus to prevent the product you are working with from being damaged.

Appendix

D

Resources

This book, or any one text, should not be considered to be the ultimate resource on digital electronics. In this appendix, I have listed a number of different sources for you to get more information on digital electronics. When you are working on your own applications and you are running into problems or think that there are better ways of doing something, I suggest that you look through these resources to see what else you can find out.

I can be contacted by sending an e-mail to myke@passport.ca, or visit my Web site at http://www.myke.com. If you do send me an e-mail, please try to be as concise as possible and list everything that you have tried and the failure analysis that you have already completed. This helps me to understand exactly what the problem is and not give you suggestions that you have already worked through.

Useful Books

Here is a collection of books that I have found useful over the years for developing electronic applications. Some of these are hard to find, but definitely worth it when you do find them in a used bookstore.

The Art of Electronics, 1989. The definitive book on electronics. It's a complete engineering course wrapped up in 1125 pages. Some people may find it to be a bit too complex, but just about any analog electronics question you could have will be answered in this book. I find the digital information in this book to be less complete. ISBN: 0-521-37095-7.

Bebop to the Boolean Boogie, 1995. Somewhat deeper in digital electronics (and less serious) than *The Art of Electronics,* Clive Maxwell's introduction to electronics stands out with clear and insightful explanations of how things work and why things are done the way they are. I bought my copy when it first became available in 1995 and still use it as a reference when I'm trying to explain how something works. It distinguishes itself from other books by explaining printed wiring assembly technology (PCB Boards, components, and soldering). This book complements *The Art of Electronics* very nicely. ISBN: 1-878707-22-1.

Teach Yourself Electricity and Electronics, 1997. An excellent resource for learning the basics of electronics—it saved my oldest son when he was taught basic electrical theory in science. The focus is on theory with not a lot of practical information and circuits. This book is laid out in course fashion with questions that you can test yourself on. ISBN: 0-07-024579-7.

The Encyclopedia of Electronic Circuits, Volumes 1 to 6. Rudolf Graf's Encyclopedia Series of Electronic circuits is an excellent resource for circuits and ideas, which have been cataloged according to circuit type. Each book contains thousands of circuits and can really make your life easier when you are trying to figure out how to do something. Each volume contains an index listing circuits for the current volume and the previous ones.

Volume 1 ISBN: 0-8306-1938-0

Volume 2 ISBN: 0-8306-3138-0

Volume 3 ISBN: 0-8306-3348-0

Volume 4 ISBN: 0-8306-3895-4

Volume 5 ISBN: 0-07-011077-8

Volume 6 ISBN: 0-07-011276-2

CMOS Cookbook, revised 1988. In *CMOS Cookbook,* Don Lancaster introduces the reader to basic digital electronic theory. He also explains the operation of CMOS gates, provides hints on soldering and prototyping, lists common CMOS parts (along with TTL pin-out equivalents), and provides a number of example circuits (including a good basic definition of how NTSC video works). The update by Howard Berlin has made sure the chips presented in the book are still available. In the 1970s, Don Lancaster also wrote the *TTL Cookbook* (which was also updated in 1998) and the *555 Cookbook,* but I find the *CMOS Cookbook* to be the most complete and useful for modern applications. ISBN: 0-7506-9943-4

The TTL Data Book for Design Engineers, Texas Instruments. I have a couple of 1981 printed copies of the second edition of this book, and they are all falling apart from overuse. The Texas Instruments TTL data books have been used for years by hundreds of thousands of engineers to develop their circuits. Each datasheet is complete with pin-outs, operating characteristics, and internal circuit diagrams. While the data books are not complete for the latest HC parts, they will give you just about everything you want to know about the operation of small-scale digital logic. The latest edition I have references for was put out in 1988 and is no longer in print, but you can pick it up in used book stores for a relatively modest price.

Periodicals

Here are a number of electronics magazines for the hobbyist and professional that will have discussions and sample applications for you to look at. At the very least, I suggest that you make a stop at your local bookshop every month to see if there are any applications that interest you.

Circuit Cellar Ink
Subscriptions
P.O. Box 698
Holmes, PA 19043-9613
1 (800) 269-6301
Web site: http://www.circellar.com/
BBS: (860) 871-1988

Poptronics
Subscriptions
Subscription Department
P.O. Box 459
Mt. Morris, IL 61054-7629
1 (800) 999-7139
Web site: http://www.gernsback.com

Microcontroller Journal
Web site: http://www.mcjournal.com/
This is published on the Web.

Nuts & Volts
Subscriptions
430 Princeland Court
Corona, CA 91719
1 (800) 783-4624
Web site: http://www.nutsvolts.com

Everyday Practical Electronics
Subscriptions
EPE Subscriptions Dept.
Allen House, East Borough,
Wimborne, Dorset, BH21 1PF
United Kingdom
+44 (0) 1202 881749
Web site: http://www.epemag.wimborne.co.uk

Suppliers

The following companies supplied components that are used in this book. I am listing them because they all provide excellent customer service and are able to ship parts anywhere you need them.

Digi-Key. Digi-Key is an excellent source for a wide range of electronic parts. They are reasonably priced and most orders will be delivered the next day. The company is a real lifesaver when you're on a deadline.

Digi-Key Corporation
701 Brooks Avenue South
P.O. Box 677
Thief River Falls, MN 56701-0677
Phone: 1 (800) 344-4539 [1 (800) DIGI-KEY]
Fax: (218) 681-3380
http://www.digi-key.com/

AP Circuits. AP Circuits will build prototype bare boards from your Gerber files. Boards are available within 3 days. I have been a customer for several years, and the company has always produced excellent quality and been helpful in providing direction for learning how to develop my own bare boards. Their Web site contains the EasyTrax and GCPrevue MS-DOS tools necessary to develop your own Gerber files.

Alberta Printed Circuits Ltd.
#3, 1112 40th Avenue NE
Calgary, Alberta T2E 5T8
Phone: (403) 250-3406
BBS: (403) 291-9342
Email: staff@apcircuits.com
http://www.apcircuits.com/

Tower Hobbies. Excellent source for servos and R/C parts useful in home-built robots.

Tower Hobbies
P.O. Box 9078
Champaign, IL 61826-9078
Toll-free ordering in the United States and Canada: 1 (800) 637-4989
Toll-free fax in the United States and Canada: 1 (800) 637-7303
Toll-free support in the United States and Canada: 1 (800) 637-6050
Phone: (217) 398-3636
Fax: (217) 356-6608
Email: orders@towerhobbies.com
http://www.towerhobbies.com/

Jameco. Components, PC parts/accessories, and hard-to-find connectors.

Jameco
1355 Shoreway Road
Belmont, CA 94002-4100
Toll-free in the United States and Canada: 1 (800) 831-4242
http://www.jameco.com/

JDR. Components, PC Parts/Accessories, and hard-to-find connectors.

JDR Microdevices
1850 South 10th Street
San Jose, CA 95112-4108

Toll-free in the United States and Canada: 1 (800) 538-5004
Toll-free fax in the United States and Canada: 1 (800) 538-5005
Phone: (408) 494-1400
Email: techsupport@jdr.com
(408) 494-1430
Compuserve: 70007,1561
http://www.jdr.com/JDR

Newark. Components, including the Dallas line of semiconductors.

Toll-free in the United States and Canada: 1 (800) 463-9275 [1 (800) 4-NEWARK]
http://www.newark.com/

Marshall Industries. Marshall is a full-service distributor of Philips microcontrollers as well as other parts.

Marshall Industries
9320 Telstar Avenue
El Monte, CA 91731
1 (800) 833-9910
http://www.marshall.com

Mouser Electronics. Mouser tends to have parts that are different from the run-of-the-mill supplier's.

Mouser Electronics, Inc.
958 North Main Street
Mansfield, TX 76063
Sales: 1 (800) 346-6873
Sales: (817) 483-6888
Fax: (817) 483-6899
Email: sales@mouser.com
http://www.mouser.com

Mondo-tronics Robotics Store. Self-proclaimed as "the world's biggest collection of miniature robots and supplies," and I have to agree. This is a great source for servos, tracked vehicles, and robot arms.

Order Desk
Mondo-tronics Inc.
524 San Anselmo Avenue #107-13
San Anselmo, CA 94960
Toll-free in the United States and Canada: 1 (800) 374-5764
Fax: (415) 455-9333
http://www.robotstore.com/

Radio Shack. Over the past few years, Radio Shack has done a lot to improve its image as a full service supplier to the electronics hobbyist. Virtually all the parts that are required for this book can be bought either from your local store or from the Web page.

http://www.radioshack.com

The Internet

I am amazed at the number of electrical engineers I know who do not use e-mail and are uncomfortable "surfing" the Web. It would probably surprise you to find out that many of these individuals are quite young (in their twenties) and seem to be uncomfortable with words and diagrams on a screen rather than on a piece of paper. The ultimate irony for me is, many of these people who refuse to access the Internet are designing servers and routers—the backbone of the Internet. In an effort to try and keep you from becoming like these people, I would like to note that these engineers are viewed with suspicion by their coworkers and are rarely considered for promotion.

Over the last 10 years, the Internet, with e-mail and the World Wide Web, has become an important tool for all engineers to get information on the designs they are working on as well as to interface with other engineers. In this section, I would like to discuss some aspects of interfacing with other engineers, and, in the next section, I will introduce you to a number of Web sites that I have found useful for working with digital electronic circuits.

In my previous books, there have generally been a number of list servers that I recommended that readers subscribe to. A *list server* is an Internet server that takes mail on a specific topic and redirects it to a list of e-mail ids. List servers are outstanding tools for asking questions and receiving help from more knowledgeable application developers.

Unfortunately, there are no list servers for digital electronics. Instead, I recommend that you take a look at the different newsgroups for help with your queries. The primary newsgroup for digital electronics design is sci.electronics. There are a number of different newsgroups that are based on this one, and each one focuses on a different area of electronics.

Before looking at the newsgroups, I recommend that you first take a look at http://www.repairfaq.org/. This site has an explanation about the different sci.electronics newsgroups. The ones that I recommend that you look at first are:

sci.electronics.basics—Elementary questions about electronics.

sci.electronics.cad—Schematic drafting, printed circuit layout, simulation.

sci.electronics.components—Integrated circuits, resistors, capacitors.

sci.electronics.design—Electronic circuit design.

sci.electronics.equipment—Test, lab, and industrial electronic products.

sci.electronics.misc—General discussions of the field of electronics.

sci.electronics.repair—Fixing electronic equipment. This newsgroup has a lot of useful debugging information and has an excellent FAQ on a number of different electronic devices.

There are a few things that I would like to mention to you about contacting other people or experts and asking questions about your projects or basic information. Please remember that nobody is being paid to help you and try to avoid the situation where you could be regarded as a pest or someone who doesn't take answers well. The following suggestions will make the process of requesting help much more efficient and positive for everyone involved.

Outline in as much detail as possible what your problem is. If possible, make some graphics of your circuit schematic and what you are observing. Questions like "I'm trying to design digital dice and nothing works, can you help me?" will probably not get very many responses and the ones you do get will not be very positive.

When you are replying to a message and quoting previous text to set up a new question, please keep the quoted material to a minimum. Try to quote just the information that you do not understand or are having problems implementing. Long replies with large amounts of quoted text are not helpful and are actually very annoying to wade through to find specific information.

Last, be sensitive to the person on the other end of the message and anybody who can see what you are writing. This is especially true for news groups and list servers where arguments, off-color jokes, and racist comments *will* cause problems for you. Unless you know exactly whom you are talking to and there is no opportunity for anybody else to see the message, be as PC (politically correct) as possible.

You may have heard a great joke in a bar the night before and it may even be appropriate to the discussion at hand, but if you think that somebody might be offended, do not repeat it. A comment sent out without sufficient thought could get you banned from a list server or from posting to a newsgroup.

If someone posts something that is offensive to you, take a few minutes to calm down and analyze your feelings. Was the comment upsetting because *you* had a bad day? Is English the first language of the sender? If not, could there be a translation problem or a cultural difference? The term *Web* is short for *World Wide Web,* and you will have to be sensitive to others. Send a personal e-mail to the person who made the original comment and explain to them calmly why their words upset you.

Please don't be too upset if the other person does not understand why you would find the comment offensive—if the other person had been sensitive to others in the first place, the comment wouldn't have been made.

Arguments often come up when two people have different ideas about how something should be done. If this happens between you and another person, take the conversation off line (i.e., do not argue publicly in a news group or list server) and try to find common ground. If you cannot reach a compromise, then just drop the conversation.

One of my favorite quotes is "Don't argue with an idiot; others will have trouble telling the difference between the two of you."

Web Sites of Interest. Here are a number of Internet Web sites that I have found useful in designing my own circuits. Please note that these sites were up and active when this book was written—I cannot guarantee that these sites will still be available when you look for them or if the information will be the same as what I saw.

You might want to check my web page, http://www.myke.com, for the latest links and comments on different sites.

I was surprised at the difficulty I had finding good digital electronics sites—it seems that popular technology has moved beyond working with simple TTL and CMOS logic chips. This is disappointing because, as you can see in this book, there are a *lot* of useful circuits that can be built with simple TTL logic, without the need for microcontrollers or device programmers. Despite this difficulty, there are still quite a few excellent sites that you can access to find out more about digital electronics.

http://www.play-hookey.com/digital/ A very attractive introductory site with excellent JavaScript examples of how digital logic works. I found JavaScript to be an excellent tool to illustrate how the different logic gates work.

http://cs.ru.ac.za/func/boolJava/indexn.htm Boolean arithmetic tutorial with JavaScript tests that help illustrate what is happening. This site complements the site I previously listed, as boolean arithmetic is the focus of the information, rather than the actual logical gates.

http://www.ti.com/sc/docs/products/logic/index.htm This is a Texas Instruments Web page that you can use to access the datasheets for different TTL parts.

http://www.ee.ic.ac.uk/hp/staff/dmb/courses/dig2/dig2.htm Course notes on digital electronics. An excellent resource with information on different logic technologies, state machines, and analog-to-digital conversions.

ftp://ftp.ee.ualberta.ca/pub/cookbook/index.html The University of Alberta's "circuit cookbook." One of the Web's classic reference sites.

http://members.tripod.com/~schematics/electro.htm Diana's electronics reference page. Quite good with lots of basic information and pointers to useful circuits.

http://www.iserv.net/~alexx/lib/tutorial/.htm#TOP Alex's Library of Electronics Information. Lots of good pointers to tutorials and reference guides.

http://www.mc.edu/courses/csc/110/ Craig Lowrey's digital information page. Course notes explain in detail many different aspects of binary mathematics.

http://www.webelectricmagazine.com/01/1/we.htm *Webelectric Magazine* is one of the more useful "Webzines" available and has some pretty interesting applications that don't necessarily use microcontrollers/microprocessors. It has the type of projects I would like to see in the mainstream electronics magazines.

http://ee.ntu.edu.au/staff/saeid/teaching/tbe254/lectures/cldch01/sld001.htm Course notes explain how digital logic circuits are wired.

http://www.epemag.wimborne.co.uk/resources.htm Articles available from *Everyday Practical Electronics*. A good soldering tutorial can be found on this page.

http://www.optimagic.com/faq.html Nice FAQ for programmable logic devices.

http://eem.com/ *Electrical Engineer's Master* on-line catalog of parts and information.

http://www.hut.fi/Misc/Electronics/circuits/index.html Tom Engdahl's page is primarily concerned with designing interface devices for the PC, but along with this, there are a number of projects that you will find useful and can help you in designing your own applications.

http://www.bithose.com/serfaq/REPAIR/F_samschem.html Sam Goldwasser's sci.electronics.repair FAQ source and list of useful circuits. Very nice collection of circuits with explanations and warnings when high voltages/high power is used or produced. This site is continually updated with new information/circuits.

http://www.hkstar.com/~hkiedsci/ Introductory tutorial to digital electronics. The information on this page is really at an introductory level.

http://www.cs.may.ie/~dvernon/cs1103/sld135.html Slide presentation on implementing/using Karnaugh maps.

http://www.bithose.com/serfaq/REPAIR/F_Pinouts.html List of connector specifications. This is a good site to keep bookmarked within your Web browser to understand how to wire between devices.

http://www.uoguelph.ca/~antoon/circ/circuits.htm "Electronic Circuits for the Hobbyist" is a good collection of circuits for basic needs, and many can be built into more complex circuits. What I particularly like about this page is the amount of space dedicated to the 555 timer, including an outstanding tutorial about the part, how it works, and some example circuits for it.

http://www.eelab.usyd.edu.au/digital_tutorial/ A rather animation-intensive site with a tutorial on basic digital electronics concepts and the circuits used to implement them. I found this site to be somewhat annoying because of all the flipping and flashing icons. You may want to access this site in a text-only mode.

http://www.educatorscorner.com/index.shtml This Agilant (formerly Hewlett Packard) reference page was originally written for educators using Agilant's products. Lots of reference and background information including explanations of how things work.

http://www.outsim.com/ On-line JavaScript logic simulator. This is a quick tool to see how different gates (and resulting subsystems) work.

http://www.mecanique.co.uk/digital-works/ Windows (Win/32) based digital logic simulator that has a demo version for download for 30 days. The product is reasonably priced and very accurate for simulating complex digital circuits.

http://www.bseng.com/debug.htm Charles Braun's "Mastering the Art of Test and Debug." Great words to live by when you have a problem or you want to test something.

http://www2.eng.cam.ac.uk/~dmh/e1/spice.htm This is the list of free SPICE implementations on the Internet.

(1) *http://www.ee.washington.edu/circuit_archive/models* (2) *http://www.intusoft.com/slinks.htm* (3) *http://www.orcad.co.uk/europe/techserv/spicemod.htm* To go along with the list of free SPICE implementations, here are component vendors and sources for SPICE models that you can download from the Internet.

http://socrates.berkeley.edu/~phylabs/bsc/Pspice/SpiceFiles.html Example SPICE files and hints on implementing applications.

http://pcb.cadence.com/product/analog/eval.asp Download a demonstration copy of the Cadence PSpice product.

http://www.cise.ufl.edu/~fishwick/dig/DigSim.html Interesting Java applet that demonstrates the operation of digital logic. This is what I would call a *simple* simulator. At time of writing, this tool does not allow saving of circuits and could be considered somewhat flaky.

Glossary

If this book is your first experience with digital electronics, I'm sure there are a lot of terms I've used that are unfamiliar to you. Here, I've given an extensive list of acronyms, terms, and expressions that are frequently used. Acronyms are expanded before they are described.

Accumulator Register used for temporary data storage in a computer.

Active Components Generally integrated circuits and transistors. Active components require external power to operate. See Passive Components.

ADC Analog-to-digital converter. Hardware devoted to converting the value of a DC voltage into a digital representation. See *DAC*.

Address The location of a bit, register, RAM byte, or instruction word within a bus memory space.

ALU Acronym for arithmetic logic unit. Circuit within a processor that carries out mathematical operations.

Amps Measure of current. One amp (or ampere) is the movement of one coulomb of electrons in one second.

Analog A quantity at a fractional value rather than a binary 1 or 0. Analog voltages are the quantity most often measured.

AND Logic gate that outputs a 1 when all inputs are a 1.

ASCII American Standard Character Interchange Interface. Bit-to-character representation standard most used in computer systems.

ASCIIZ A string of ASCII characters ended by a null (0×000) byte.

Assembler A computer program that converts assembly language source code to object code.

Assembly Language A set of word symbols used to represent the instructions of a processor. Along with a primary instruction, there are parameters that are used to specify values, registers, and addresses.

Asynchronous Serial Data sent serially to a receiver without clocking information. Instead, data synching information for the receiver is available inside the data packet or as part of each bit.

Bare Board See *raw card*.

BCD Binary coded decimal. Using 4 bits to represent a decimal number (0 to 9).

Binary Numbers Numbers represented as powers of 2. Each digit is 2 raised to a specific power. For example, 37 decimal is $32 + 4 + 1 = 2^4 + 2^2 + 2^0 =$ 00010101 binary. Binary can be represented in the forms: 0b0nnnn, B'nnnn', or %nnnn where *nnnn* is a multidigit binary number comprising 1s and 0s.

Bipolar Logic Logic circuits made from bipolar transistors (either discrete devices or integrated onto a chip).

Bit Banging Simulating Interface functions with code.

Bit Mask A bit pattern that is ANDed with a value to turn off specific bits.

Bounce Spurious signals in a changing line. Most often found in mechanical switch closings.

Bus An electrical connection between multiple devices, each using the connection for passing data.

Capacitor Device used for storing electrical charge. Often used in microcontroller circuits for filtering signals and input power by reducing transient voltages. The different types include *ceramic disk, polyester, tantalum,* and *electrolytic.* Tantalum and electrolytic capacitors are polarized.

Ceramic Resonator A device used to provide timing signals. More robust than a crystal but with poorer frequency accuracy.

Character Series of bits used to represent an alphabetic, numeric, control, or other symbol or representation. See *ASCII.*

Chip Package The enclosure by which a chip is protected from the environment (usually made of either ceramic or plastic), with wire interconnects to external circuitry (see *PTH* and *SMT*).

CISC Complex instruction set computer. A type of computer architecture that uses a large number of very complete instructions rather than a few short instructions. See *RISC.*

Clock A repeating signal used to run a circuit's operation sequence.

Clock cycle The operation of an oscillator going from a low voltage to a high voltage and back again. This is normally referenced as the speed the device runs at.

CMOS Logic Logic circuits made from N-channel and P-channel MOSFET (metal oxide semiconductor field-effect transistor) devices (either discrete devices or integrated onto a chip).

Comparator A device that compares two voltages and returns a logic 1 or 0 depending on the relative values.

Compiler A program that takes a high-level language source file and converts it to either assembly language code or object code for a computer processor.

Concatenate Joining two pieces of data together to form a single *contiguous* piece of data.

Contiguous When a set amount of data is placed altogether in memory and can be accessed sequentially using an index pointer, it is said to be *contiguous.* Data is noncontiguous if it is placed in different locations that cannot be accessed sequentially.

Control Store See *program memory.*

Constant Numeric value used as a parameter for an operation or instruction. This differs from a variable value that is stored in a RAM or register memory location.

Coulomb Charge of 1.6×10^{19} electrons.

CPU Central processing unit; often referred to as *processor.*

Cross Assembler A program written to take assembly language code for one processor and convert it to object code while working on an unrelated processor and operating system. See *Assembler.*

Crystal Device used for precisely timing the operation of a circuit.

Current The measurement of the number of electrons that pass by a point each second. The units are *Amps,* which are coulombs per second.

DAC Digital-to-analog converter. Hardware designed to convert a digital representation of an analog DC voltage into that analog voltage. See *ADC.*

DCE Data communications equipment, the RS-232 standard by which modems are usually wired. See *DTE.*

Debouncing Removing spurious signals in a noisy input.

Decimal Numbers Base 10 numbers used for constants. These values are normally converted into hex or binary numbers for a microcontroller.

Decouping Capacitor Capacitor placed across V_{cc} and ground of a chip to reduce the effects of increased/decreased current draws from the chip.

Demultiplexer Circuit for passing data from one source to an output specified from a number of possibilities. See *multiplexer.*

Demux Abbreviation for *demultiplexer.*

Digital A term used to describe a variety of logic families where values are either high (1) or low (0). For most logic families, the voltage levels are approximately either 0 V or 5 V with a switching level somewhere between 1.4 and 2.5 V.

Driver Any device that can force a signal onto a *net.* See *receiver.*

DTE Data terminal equipment. The RS-232 Standard your PC's serial port is wired to. See *DCE.*

D-Shell Connectors A style of connector often used for RS-232 serial connections as well as other interfaces. The connector is D shaped to provide a method of polarizing the pins and ensuring that they are connected the correct way.

Duty Cycle In a pulse-width-modulated digital signal, the duty cycle is the fraction of time the signal is high over the total time of the repeating signal.

Edge Triggered Logic that changes on the basis of a change in a digital logic level. See *level sensitive.*

EEPROM Electrically erasable programmable read-only memory (also known as *flash memory*). Nonvolatile memory that can be erased and reprogrammed electrically (i.e., it doesn't require the UV light of *EPROM*).

EPROM Erasable programmable read-only memory. Nonvolatile memory that can be electrically programmed and erased by ultraviolet light.

FIFO First in, first out. Memory that will retrieve data in the order in which it was stored.

Flash A type of *EEPROM.* Memory that can be electrically erased in blocks, instead of as individual memory locations. True flash is very unusual in microcontrollers; many manufacturers describe their devices as having flash, when in actuality they use EEPROM.

Flip-flop A basic memory cell that can be loaded with a specific logic state and read back. The logic state will be stored as long as power is applied to the cell.

Floating The term used to describe a pin that has been left unconnected and is "floating" relative to ground.

Flux Weak acid added to solder to clean metal surfaces to be joined.

Frequency The number of repetitions of a signal that can take place in a given period of time (typically 1 second). See *period* and *hertz.*

FTP File transfer protocol. A method of transferring files to/from the Internet.

Ground Negative voltage to a microcontroller/circuit. Also referred to as V_{ss} and has the abbreviation Gnd.

Hertz A unit of measurement of frequency. One hertz (or *Hz*) means that an incoming signal is oscillating once per second.

Hex Numbers A value from 0 to 15 that is represented by 4 bits or the characters 0 through 9 and A through F.

Horizontal Synch A pulse used to indicate the start of a scan line in a video monitor or TV set.

Hysteresis Characteristic response to input that causes the output response to change based on the input. Typically used in digital circuits for *debouncing* input signals.

Hz See *hertz.*

Inductor Wire wrapped around some kind of form (metal or plastic) to provide a magnetic method of storing energy. Inductors are often used in oscillator and filtering circuits.

Infrared A wavelength of light (760 nanometers or longer) that is invisible to the human eye. Often used for short-distance communications.

Instruction A series of bits that are executed by a computer processor to perform a basic function.

Instruction Cycle The minimum amount of time needed to execute a basic function in a computer processor. One instruction cycle typically takes several clock cycles. See *clock cycle.*

I2C "Inter-intercomputer" communication. A synchronous serial network protocol allowing microcontrollers to communicate with peripheral devices and each other. Only common lines are required for the network.

kHz Kilohertz. Abbreviation for frequency in thousands of cycles per second.

Latency The time or cycles required for hardware to respond to a change in input.

LCD Liquid crystal display. A device used for outputting information from a microcontroller. Typically controlled by a Hitachi 44780 controller, although some microcontrollers contain circuitry for interfacing to an LCD directly without an intermediate controller circuit.

LED Light-emitting diode. Diode (rectifier) device that will emit light of a specific frequency when current is passed through it. When used with microcontrollers, LEDs are usually wired with the anode (positive pin) connected to V_{cc} and the microcontroller I/O pin sinking current (using a series 200- to 270-Ω resistor) to allow the LED to turn on. In typical LEDs in hemispherical plastic packages, the flat side (which has the shorter lead) is the cathode.

Level Conversion The process of converting logic signals from one family to another.

Level Sensitive Logic that changes according to the state of a digital logic signal. See *Edge triggered.*

LIFO Last in, first out. Type of memory in which the most recently stored data will be the first retrieved.

Logic Analyzer A tool that graphically shows the relationship of the waveforms of a number of different pins.

Logic Gate A circuit that outputs a logic signal on the basis of input logic conditions.

Logic Probe A simple device used to test a line for being high, low, transitioning, or in a high impedance state.

Manchester Encoding A method for serially sending data that does not require a common (or particularly accurate) clock.

Mask-Programmable ROM A method of programming a memory that takes place at final assembly of a microcontroller. When the aluminum traces of a chip are laid down, a special photographic mask is made to create wiring that will result in a specific program being read from a microcontroller's control store.

Matrix Keyboard A set of push button switches wired in an *x-y* pattern to allow button states to be read easily.

MCU Abbreviation for microcontroller.

Memory Circuit designed to store instructions or data.

Memory Array A collection of memory devices (such as *flip-flops*) arranged in a matrix format that allows consistent addressing.

MHz Megahertz. Abbreviation for frequency in millions of cycles per second.

MPU Acronym/abbreviation for microprocessor.

msec Millisecond. One thousandth of a second (0.001 second). See *nsec* and *μsec.*

Multiplexer Device for selecting and outputting a single stream of data from a number of incoming data sources. See *Demultiplexer, demux,* and *mux.*

Mux Abbreviation for *Multiplexer.*

Negative Active Logic A type of logic where the digital signal is said to be asserted if it is at a low (0) value. See *positive active logic.*

Net A technical term for the connection of device pins in a circuit. Each net consists of all the connections to one device pin in a circuit.

NiCad Abbreviation for nickel-cadmium battery. These batteries are rechargeable, although they typically provide only 1.2 V per cell output compared to 1.5 to 2.0 V for standard "dry" or alkaline radio batteries.

NMOS Logic Digital logic in which only N-channel MOSFETs are used.

Noise High-frequency variances in a signal line that are caused by switch *bouncing* or electrical signals picked up from other sources.

NOT Logic gate that inverts the state of the input signal (1 NOT is 0).

nsec Nanosecond. One billionth of a second (0.000000001 second). See *μsec* and *msec.*

Octal Numbers Numbers represented as the digits from 0 to 7. This method of representing numbers is not widely used, although some high-level languages, such as C, have made it available to programmers.

One's Complement The result of XORing a byte with 0 × 0FF, which will invert each bit of a number. See *two's complement.*

Op Codes The hex values that make up the processor instructions in an application.

Open Collector/Drain Output An output circuit consisting of a single transistor that can pull the net it is connected to the ground.

OR Basic logic gate in which, when any input is set to a 1, a 1 is output.

ORT Acronym for ongoing reliability testing. A continuing set of tests that are run on a manufactured product during its life to ensure that it will be as reliable as originally specified.

Oscillator A circuit used to provide a constant-frequency repeating signal. This circuit can consist of a crystal, ceramic resonator, or resistor-capacitor network for providing the delay between edge transitions.

Oscilloscope An instrument used to observe the waveform of an electrical signal. The two primary types of oscilloscopes in use today are the analog oscilloscope, which writes the current signal onto the phosphors of a CRT, and the digital storage oscilloscope, which saves the analog values of an incoming signal in RAM for replaying on either a built-in CRT or a computer connected to the device.

OTP One-time programmable. This term generally refers to a device with EPROM encased in a plastic package that does not allow the chip to be exposed to UV light. Note that EEPROM devices in a plastic package may also be described as OTP when they can be electrically erased and reprogrammed.

Parallel Passing data between devices with all the data bits being sent at the same time on multiple lines. This is typically much faster than sending *serial* data, but more difficult to wire.

Parameter A user-specified value for a subroutine or macro. A parameter can be a numeric value, a string, or a pointer, depending on the application.

Passive Components Generally resistors, capacitors, and inductors. Components that do not require a separate power source to operate.

PCA Printed circuit assembly. A *bare board* with components (both active and passive) soldered onto it. See Active Components.

PCB Printed circuit board. See *raw card.*

PDF Files Files suitable for viewing with Adobe Postscript.

Period The length of time that a repeating signal takes to go through one full cycle. The reciprocal of *frequency.*

Poll A programming technique in which a bit (or byte) is repeatedly checked until a specific value is found.

Positive Active Logic Logic that becomes active when a signal becomes high (1). See *negative active logic.*

PPM Measurement of something in parts per million. An easy way of calculating the PPM of a value is to divide the value by the total number of samples or opportunities and multiplying by 1,000,000. One percent is equal to 10,000 PPM; 10 percent is equal to 100,000 PPM.

Princeton Architecture Computer processor architecture that uses one memory subsystem for instructions (*control store*), variable memory, and I/O registers. See *Harvard architecture* and *Von Neumann.*

Program Counter A counter within a computer processor that keeps track of the current program execution location. This counter can be updated by the counter and have its contents saved/restored on a stack.

Program Memory Memory (usually nonvolatile) devoted to saving the application program for when the computer processor is powered down. Also known as *control store* and program storage.

PROM Programmable read-only memory. Originally an array of fuses that were "blown" to load in a program. Now PROM can refer to *EPROM* in an *OTP* package.

PTH Pin through-hole. Technology in which the pins of a chip are inserted into holes drilled into an FR4 printed circuit card before soldering.

Pull-Down A resistor (typically 100 to 500 Ω) that is wired between a digital input pin and ground. See *pull-up.*

Pull-Up A resistor (typically 1 k to 10 k) that is wired between a digital input pin and V_{cc}. A switch pulling the signal at the input pin may be used to provide user input. See *pull-down.*

PWB Printed wiring board. See *raw card.*

PWM Pulse-width modulation. A digital output technique where a single line is used to output analog information by varying the length of time a pulse is active on the line.

Raw Card Fiberglass board with copper "traces" attached to it that allows components to be interconnected. Also known as *PCB, PWB,* and *bare board.*

RC Resistor-capacitor network used to provide a specific delay for a built-in oscillator or reset circuit.

Receiver A device that senses the logic level in a circuit. A receiver cannot drive a signal.

Register A memory address devoted to saving a value (like "RAM") or providing a hardware interface for the computer processor.

Resistor A device used to limit current in a circuit.

Resistor Ladder A circuit consisting of a number of resistors that can be selected to provide various voltage divider circuits and analog voltage outputs.

Reset To place a digital electronics circuit in a known state before allowing it to execute.

RISC Reduced instruction set computer. This is a philosophy in which the operation of a computer is sped up by reducing the operations performed by the processor to the absolute minimum for application execution and making all resources accessible by a consistent interface. The advantages of RISC include faster execution time and a smaller instruction set. See *CISC*.

ROM Read-only memory. This type of memory is typically used for a control store because it cannot be changed by a processor during the execution of an application. *Mask-programmable ROM* is made by the chip manufacturer with specific software as part of the device and cannot be programmed in the field.

Rotate A method of moving bits within a single or multiple registers. No matter how many times a rotate operation or instruction is carried out, the data in the registers will not be lost. See *shift*.

RS-232 An asynchronous serial communications standard. Normal logic level for a 1 is −12 V and for a 0, +12 V.

RS-485 A differential pair, TTL voltage level communications system.

Scan The act of reading through a row of matrix information for data rather than interpreting the data as a complete unit.

Serial Passing multiple bits one at a time by a serial line. See *parallel*.

Shift A method of moving bits within a single or multiple registers. After a shift operation, bits are lost. See *rotate*.

SMT Acronym for surface mount technology [also known as surface mount devices (SMD)]. Technology in which the pins of a chip are soldered to the surface of a printed circuit card.

Solder A tin/lead/flux, and occasionally silver, combination that is used to connect two pieces of metal together.

Soldering The process of connecting two pieces of metal together by using *solder* and applying heat.

SPI A synchronous serial communications protocol.

Splat Asterisk (*). Easier to say and spell and funnier than asterisk.

SRAM Static random-access memory. A memory array that will not lose its contents while power is applied.

Stack *LIFO* memory used in a computer processor to store program counter and other context register information.

Stack Pointer An index register available within a processor; used for storing data and updating itself to allow the next operation to be carried out with the index pointing to a new location.

State Analyzer A tool used to store and display state data on several lines. This is an option often available in a logic analyzer that makes a separate instrument unnecessary.

State Machine A programming technique that uses external conditions and state variables for determining how a program is to execute.

String Series of *ASCII* characters saved sequentially in memory. When ended with 0×000 to note the end of the string, known as an *ASCIIZ* string.

Subroutines A small application program devoted to carrying out one task or operation. Usually called repeatedly by other subroutines or the application mainline.

Synchronous Serial Refers to data transmitted serially along with a clocking signal, which is used by the receiver to indicate when the incoming data is valid.

Timer A counter incremented by either an internal or an external source. Often used to time events, instead of counting instruction cycles.

Traces Electrical signal paths etched in copper in a printed circuit card.

Transistor An electronic device by which current flow can be controlled.

Two's Complement A method for representing positive and negative numbers in a digital system. To convert a number to a two's complement negative, it is complemented (converted to *one's complement*) and incremented.

UART Universal asynchronous receiver/transmitter. Peripheral hardware inside a microcontroller used to asynchronously communicate with external devices. See *USART* and *asynchronous serial.*

USART Universal synchronous/asynchronous receiver/transmitter. Peripheral hardware inside a microcontroller used to communicate synchronously (using a clock signal either produced by the microcontroller or provided externally) or asynchronously with external devices. See *UART* and *synchronous serial.*

μsec Microsecond. One-millionth of a second (0.000001 seconds). See *nsec* and *msec.*

UV Light Ultraviolet light. Light at shorter wavelengths than the human eye can see. UV light sources are often used with windowed EPROM chips for erasing the chips' contents.

Variable A label used in an application program to represent an address that contains the actual value to be used by the operation or instruction in a computer processor. Variables are normally located in *RAM* and can be read from or written to by a program.

V_{cc} Positive power voltage applied to a TTL digital electronics circuit. Generally 2.0 to 6.0 V, depending on the application. Also known as V_{dd} for CMOS logic.

V_{dd} See V_{cc}.

Vertical Synch A signal used by a monitor or TV set to determine when to start displaying a new screen (field) of data.

Vias Holes in a printed circuit card.

Volatile RAM is considered to be volatile if, when power is removed, the contents are lost. EPROM, EEPROM, and PROM are considered to be nonvolatile because the values stored in the memory are saved, even if power is removed.

Voltage The amount of electrical force placed on a charge.

Voltage Regulator A circuit used to convert a supply voltage into a level useful for a circuit or microcontroller.

Volts Unit of voltage.

V_{ss} Ground for CMOS logic.

Wattage Measure of power consumed. If a device requires 1 amp of current with a 1-volt drop, 1 watt (W) of power is being consumed.

Word The basic data size used by a computer processor.

XOR A logic gate that outputs a 1 when the inputs are at different logic levels.

ZIF Zero insertion force. ZIF sockets will allow the plugging/unplugging of devices without placing stress on the devices' pins.

Index

Note: boldface numbers indicate illustrations; italic *t* indicates a table.

ABOUT THE AUTHOR

MYKE PREDKO is the author of *The Microcontroller Handbook: Programming & Customizing the 8051 Microcontroller* and *Programming & Customizing the PICmicro™ Microcontrollers*, also from McGraw-Hill. As a New Product Test Engineer at Celestica in Toronto, Ontario, Canada, Myke works with innovative electronic product designs. He has also been a test engineer, product engineer, and manufacturing manager for some of the world's largest computer manufacturers. Mr. Predko has a patent pending on an automated test for PC motherboards as well as patents pending on microcontroller architecture design. He has a degree in electrical engineering from the University of Waterloo.